# SECOND EDITION

# PRINCIPLES OF HUMAN EVOLUTION

**Roger Lewin and
Robert A. Foley**

**Blackwell**
Publishing

© 2004 by Blackwell Science Ltd
a Blackwell Publishing company

350 Main Street, Malden, MA 02148-5020, USA
108 Cowley Road, Oxford OX4 1JF, UK
550 Swanston Street, Carlton, Victoria 3053, Australia

The right of Roger Lewin and Robert Foley to be identified as the
Authors of this Work has been asserted in accordance with the UK
Copyright, Designs, and Patents Act 1988.

First published 1998 by Blackwell Science, Inc.
2004 by Blackwell Science Ltd

*Library of Congress Cataloging-in-Publication Data*

Lewin, Roger.
 Principles of human evolution / Roger Lewin and Robert Foley. – 2nd ed.
  p. cm.
Includes bibliographical references and index.
  ISBN 0-632-04704-6 (pbk. : alk. paper)
 1. Human evolution.  I. Foley, Robert.  II. Title.

 GN281.L489 2004
 599.93'8–dc21                                    2003004461

A catalogue record for this title is available from the British Library.

Set in 10/12$\frac{1}{2}$pt Meridien
by Graphicraft Limited, Hong Kong
Printed and bound in the United Kingdom
by The Bath Press

For further information on
Blackwell Publishing, visit our website:
http://www.blackwellpublishing.com

# CONTENTS

# PREFACE FOR STUDENTS: A GUIDE TO STUDYING HUMAN EVOLUTION

The science of human evolution is a unique one for many reasons. One is that it is a genuine science, which requires a systematic approach, hard facts, and general theories, and yet it also deals with humans, who are notoriously difficult to study objectively. The history of the discipline is littered with examples where the balance between these two aspects has been hard to keep. This means that this book must tread a fine line between the recognition that some of what we study is little more than good conjecture, and the need to seek out solid facts and good logic to determine what really did happen. In studying human evolution it is possible to either wallow in the uncertainty of it all – all theories are equal and everyone's opinion is equally valid – or to rigidly limit yourself to the sturdy facts of the fossil record – what I see or can measure is what I know. In what follows in this book we try to show that there are real facts out there, that there is an actual history to be unraveled, but that it requires thinking and a knowledge of the wider fields of evolution.

Another unique feature of the study of human evolution is that it is a historical science. Those of you from a "hard science" background will miss the reassurance of the experiment. As evolution happened in the past, it is not possible to run it again, under different conditions, to test hypotheses (although computer simulations are bringing that closer). We are dependent upon observations and statistical analyses. This does not mean that it is not a hard science – in the phrase of the great evolutionary biologist John Maynard Smith, it is certainly a difficult science. Penetrating the past means that we need to be able to extend the information as far as possible, and that requires using models and inferences, and applying general theories. In a similar science – cosmology – this has been done with extraordinary success. Again, in this book, unlike many textbooks, we therefore encourage an approach where the particular problems are approached in terms of general evolutionary models.

Linked to this need to place human evolution in a broader evolutionary framework is a further demand. At first sight it might seem that the way to study human evolution is to find a suitable site, hope for some luck, and dig up some fossils, which can then be studied using some basic anatomical principles, adding more to the edifice of human evolution. Alas, it is more complicated than that. Choosing an area to work depends upon a detailed geological knowledge; the excavation requires expertise in stratigraphy, paleoenvironments, plant and animal fossils. Many dating techniques, all using complex physical and chemical methods, are needed. Nowadays the fossils can be subjected to scanning electron microscopy, histology, imaging and CAT (computerized axial tomography) or MRI (magnetic resonance imaging) scanning, isotope analysis. Placing the fossils in the right context requires statistical analysis, the application of computerized phylogenetic techniques. Interpreting their functional anatomy and behavior requires a broad knowledge of primates and mammals more generally, ranging from biophysics to socioecology. If tools are associated with the fossil, then a whole other battery of approaches and methods is required, drawn from archeology. To add yet another dimension, perhaps ancient DNA can be extracted from the fossils, requiring a knowledge of biochemistry, molecular biology, and genetic analysis, not to mention the whole framework of human genetic diversity.

It follows from this that a textbook on human evolution should be the sum of textbooks in archeology, anatomy, inorganic chemistry, biochemistry, genetics, molecular biology, sedimentology, geophysics, zoology, and ecology. Such a task is certainly beyond the skills of the authors, and probably beyond the endurance of the student. For this reason we have attempted here not to provide a comprehensive introduction to all aspects, but to provide an introduction to the *principles* of human evolution. These principles should provide a basis for further investigation.

In summary, this is the approach we encourage in this book:

♦ Be prepared to tackle problems in terms of both theory and empirical data; each is essential to "doing paleoanthropology."

♦ Treat the information here as a platform on which to build; there is a vast literature out there on the subject, and this book provides a way into it, in terms of both the sketch we provide and the references.

♦ Take a broad-minded approach to what constitutes "human evolution" – the methods, approaches, and questions are constantly shifting, and it is an inherently multidisciplinary subject.

♦ Remember that even though there is much that is disputed, and much that has been wrong in the past, and no doubt will be again, nonetheless this is a science, and you should adopt a scientific approach to it.

## NEW TO THE SECOND EDITION

In this second edition of *Principles of Human Evolution* we have made the text more accessible and easy to use. The book has been completely restructured into nineteen chapters, organized by issue rather than chronology, integrating behavior, adaptation, and anatomy. Figure numbers and references have been added to the art program in this edition for easier classroom use.

This edition also brings a new emphasis on ecological and behavioral evolution to the discussion of evolutionary theory. The book is thoroughly updated with scores of new examples covering both the most recent archeological finds and the latest evolutionary theories, integrating genetic and paleoanthropological approaches.

The second edition also features the following pedagogical features:

♦ **"Beyond the Facts"** boxes, at the end of each chapter, explore the ideas behind key scientific debates in greater depth

♦ **Margin questions** highlight key points in each section

♦ **Key questions** in the margins review and test students' knowledge of central chapter concepts

♦ **Dedicated website** – www.blackwellpublishing.com/lewin – provides study resources and artwork downloadable to Powerpoint files

# ACKNOWLEDGMENTS

We would like to thank the following people who helped in the development of this book. Many people read all or sections of it, and helped to improve it: Debbie Guatelli-Steinberg, Henry Harpending, Katerina Harvati, Jim Kidder, John Relethford, Karen Rosenberg, Karen Schmidt, Liza Shapiro, Robin Smith, Shelley Smith, Christian Tryon, Christina Turner, Russell Tuttle, and Carol Ward. William McGrew was especially helpful in correcting errors and discussing the larger issues of human evolution with great enthusiasm. Nathan Brown of Blackwell Publishing should be especially thanked for his persistence and patience, both of which in their different ways were appreciated, and for allowing us to interpret his word "urgent" in our own idiosyncratic way. Fiona Sewell also provided much assistance in the production of the text and figures. R. F. would especially like to thank Marta Mirazón Lahr, for her tolerance of this distraction away from other things that were equally pressing, as well as, as always, endless expert advice and guidance.

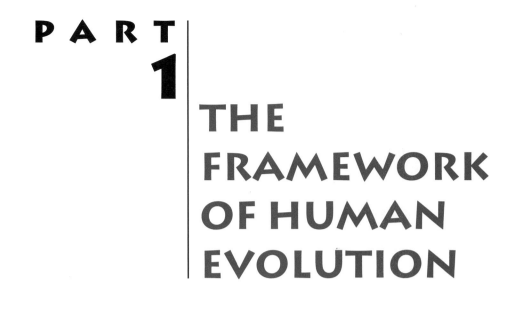

# PART 1

# THE FRAMEWORK OF HUMAN EVOLUTION

# CHAPTER 1

# THE GROWTH OF THE EVOLUTIONARY PERSPECTIVE

## OUR PLACE IN NATURE

As the train doors open at many stations on the London Underground, a disembodied voice can be heard saying "Mind the gap" to warn passengers that there is a larger than usual step between the train and the platform. This helpful announcement can act as a rather surprising motto for anyone about to embark on a course on human evolution.

> **KEY QUESTION** How should evolutionary biology approach the problem of human uniqueness versus the continuities inherent in the evolutionary process?

The reason for this is very simple. If one asks the average person to come up with terms they associate with evolution, then after "survival of the fittest," "progress" (of which more later), and "missing link," another one that is highly likely to figure is "continuity." Evolution is a continuous process, and so provides a link between all organisms in such a way as to place them all on a continuum, from the simplest single-cell organism to the most complex social mammal. Through evolution, plants and animals slide endlessly from one form to another. Continuity is therefore a major part of nature. However, the same average person, if asked whether there is a continuity between humans and other animals, is likely to answer no. Certainly there are many things that humans and other animals share, from their basic genetic code to the broad body plan of the vertebrate skeleton, but the gulf between humans and, say, chimpanzees, no matter how smart the latter appear to be, remains large, and to some unbridgeable. Humans are the species of Shakespeare and Dante, of Galileo and Einstein, of Wallace and Darwin, of Michelangelo and Picasso, of Beethoven and Bach; or alternatively, the species of Hitler and Stalin. No chimpanzee can come close to these sorts of achievements.

In the contrast between the continuity of evolution and the uniqueness of humans lies the challenge and interest of human evolution. It is the paradox that lies at the heart of the discipline – how is it possible to simultaneously "mind the gap" that exists between humans and other species and be true to the continuous nature of the evolutionary process? It is that challenge that has fueled much of the research into human evolution.

The problem of the gap has long been recognized. In 1859 Charles Darwin published his epoch-making book, *The Origin of Species*, in which he provided an account of how evolution worked, and how science might explain the patterns of life without recourse to supernatural beings and processes. While Darwin made little or no reference to humans, but confined himself to ordinary plants and animals, the implications were plain to see. Within four years his friend and supporter, Thomas Henry Huxley, published one of the first books on human evolution, *Evidences as to Man's Place in Nature*. The book was based on evidence from comparative anatomy among apes and humans, embryology, and fossils of early

**FIGURE 1.1 Ptolemy's universe:** Before the Copernican revolution in the sixteenth century, scholars' views of the universe were based on the ideas of Aristotle as elaborated by Ptolemy. The Earth was seen as the center of the universe, with the Sun, Moon, stars, and planets fixed in concentric crystalline spheres circling it.

humans (few were available at the time). Huxley's conclusion – that humans have a close evolutionary relationship with the great apes, particularly the African apes – was a key element in a revolution in the history of Western philosophy: humans were to be seen as being a part of nature, no longer as apart from nature – hence the title of Huxley's book. What both Darwin and Huxley, as well as many other scientists of that time, were keen to demonstrate was the continuity between humans and the rest of the biological world, and that all were the product of the same evolutionary processes. In other words, evolution underpinned the continuity of nature, including humans.[1]

Although Huxley was committed to the idea of the evolution of *Homo sapiens* from some type of ancestral ape, he nevertheless recognized that humans were a very special kind of animal. In his book he wrote:

No one is more strongly convinced than I am of the vastness of the gulf between . . . man and the brutes, for, he alone possesses the marvelous endowment of intelligible and rational speech [and] . . . stands raised upon it as on a mountain top, far above the level of his humble fellows, and transfigured from his grosser nature by reflecting, here and there, a ray from the infinite source of truth.[2]

## Continuity and discontinuity

It is worth noting that the problem of continuity versus breaks in the chain of life is one that both continues through to the present day and existed in pre-evolutionary scientific thought. The reason for this goes back to the intellectual upheavals of the seventeenth and eighteenth centuries. The revolution wrought by Darwin's work was, in fact, the second of two such intellectual upheavals within the history of Western philosophy.[3] The first revolution occurred three centuries earlier, when Nicolaus Copernicus replaced the geocentric model of the universe with a heliocentric model (Fig. 1.1). Although the Copernican revolution deposed humans from being the very center of all of God's creation and transformed them into the occupants of a small planet orbiting in a vast universe, they nevertheless remained the pinnacle of God's works. From the sixteenth through the mid-nineteenth centuries, those who studied humans and nature as a whole were coming close to the wonder of those works (Fig. 1.2).

This pursuit – known as natural philosophy – positioned science and religion in close harmony. What linked them was the notion of design.[4,5] The living world could be seen to be admirably efficient and well organized, with each organism playing a role for which it was well suited. This was taken as evidence for a remarkable design, and consequently as evidence for a designer – in other words, the hand of God.

In addition to design, a second feature of God's created world was a virtual continuum of form, from the lowest to the highest, with humans being

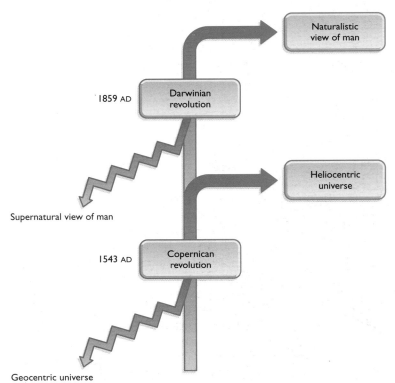

FIGURE 1.2 **Two great intellectual revolutions:** In the mid-sixteenth century the Polish mathematician Nicolaus Copernicus proposed a heliocentric rather than a geocentric view of the universe. "The Earth is not the center of all things celestial," he said, "but instead is one of several planets circling a sun, which is one of many suns in the universe." Three centuries later, in 1859, Charles Darwin further changed people's view of themselves, arguing that humans were a part of nature, not apart from nature.

Naturalistic view of man

1859 AD  Darwinian revolution

Heliocentric universe

Supernatural view of man

1543 AD  Copernican revolution

Geocentric universe

MAN · IS · BVT · A · WORM ·

**FIGURE 1.3 The Chain of Being:** Both pre-evolutionary and early Darwinian perceptions of the relationship among living creatures involved the idea of a Chain of Being, or *scala naturae*, which implied a progressive development of greater complexity and change in the direction of humanity.

near the very top, just a little lower than the angels. This continuum – known as the Great Chain of Being – was not a statement of dynamic relationships between organisms, reflecting historical connections and evolutionary derivations (Fig. 1.3). This focus on continuity, echoed in evolutionary ideas, was in fact one of the platforms on which Darwin built his theory. The difference between the pre-evolutionary idea of the Great Chain of Being and the later concept of an evolving lineage, though, is that the former was fixed. According to Stephen Jay Gould, the essence of the Chain of Being is the fixed positions of biological organisms in an ascending hierarchy.[6]

However, powerful though the Chain of Being theory was, it faced problems – as it happens, exactly the same problems as Huxley recognized, namely the gaps that occurred between certain parts of this Great Chain. One such discontinuity appeared between the world of plants and the world of animals. Another separated humans and apes.

Knowing that the gap between apes and humans should be filled, eighteenth- and early nineteenth-century scientists tended to exaggerate the humanness of the apes while overstating the simianness of some of the "lower" races. For instance, some apes were "known" to walk upright, to carry off humans for slaves, and even to produce offspring after mating with humans. By the same token, some humans were "known" to be brutal savages, equipped with neither culture nor language. Basically, the natural philosophers were using the tales of sailors and the myths of the ancient world to fill in the gaps in their model scheme of creation.

This perception of the natural world inevitably became encompassed within the formal classification system, which was developed by Carolus Linnaeus in the mid-eighteenth century. In his *Systema Naturae*, published first in 1736 and finally in 1758, Linnaeus included not only *Homo sapiens* – the species to which we all belong – but also the little-known *Homo troglodytes*, which was said to be active only at night and to speak in hisses, and the even rarer *Homo caudatus*, which was known to possess a tail[7] (Fig. 1.4).

## Evolution and progress

In one sense the theory of evolution provided a solution to the difficulties faced by the natural philosophers, namely a better understanding of the dynamic nature of the links between the entities on the Great Chain of Being. The dynamism of evolution did not really remove the Great Chain, but added a new dimension to it

– that of progress, which in turn provided an explanation for the hierarchy that many saw in both the natural world and humanity. For instance, humans were still regarded as being "above" other animals and endowed with special qualities – those of intelligence, spirituality, and moral judgment. And the gradation from "lower" races to "higher" races that had been part of the Great Chain of Being was now explained by the process of evolution.

"The progress of the different races was unequal," noted Roy Chapman Andrews, a researcher at the American Museum of Natural History in the 1920s and 1930s. "Some developed into masters of the world at an incredible speed. But the Tasmanians . . . and the existing Australian aborigines lagged far behind, not much advanced beyond the stages of Neanderthal man."[8] Such overtly racist comments were echoed frequently in literature of the time and were reflected in the evolutionary trees published then (Fig. 1.5).

In other words, inequality of races – with blacks at the bottom and whites at the top – was explained away as the natural order of things: before 1859 as the product of God's creation, and after 1859 as the product of evolution.

In the same vein, early discussions of human evolution incorporated the notion of progress, and specifically the inevitability of *Homo sapiens* as the ultimate aim of evolutionary trends. In the words of the prominent British anthropologist Sir Arthur Keith, written in 1927, "Progress – or what is the same thing, Evolution – is [Nature's] religion," [9,10] or, as Robert Broom put it in 1933, "Much of evolution looks as if it had been planned to result in man, and in other animals and plants to make the world a suitable place for him to dwell in."[11] (Broom was responsible for some of the more important early human fossil finds in south Africa during the 1930s and 1940s.)

In this brief historical sketch we can see the main themes of human evolution and its controversies, and what is perhaps striking is the extent to which the issues that form the primary subject matter of this book were

**FIGURE 1.4 The anthropomorpha of Linnaeus:** In the mid-eighteenth century, when Linnaeus compiled his *Systema Naturae*, Western scientific knowledge about the apes of Asia and Africa was sketchy at best. Based on the tales of sea captains and other transient visitors, fanciful images of these creatures were created. Here, produced from a dissertation of Linnaeus' student Hoppius, are four supposed "manlike apes," some of which became species of *Homo* in Linnaeus' *Systema Naturae*. From left to right: *Troglodyta bontii*, or *Homo troglodytes*, in Linnaeus; *Lucifer aldrovandii*, or *Homo caudatus*; *Satyrus tulpii*, a chimpanzee; and *Pygmaeus edwardi*, an orangutan.

Humans, race, and progress

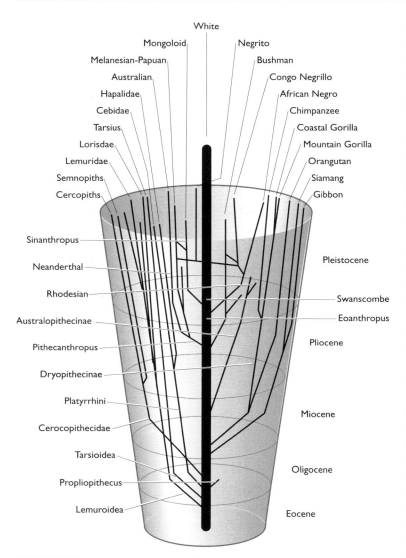

White
Mongoloid
Melanesian-Papuan
Australian
Hapalidae
Cebidae
Tarsius
Lorisdae
Lemuridae
Semnopiths
Cercopiths

Negrito
Bushman
Congo Negrillo
African Negro
Chimpanzee
Coastal Gorilla
Mountain Gorilla
Orangutan
Siamang
Gibbon

Sinanthropus
Neanderthal
Rhodesian
Australopithecinae
Pithecanthropus
Dryopithecinae
Platyrrhini
Cerocopithecidae
Tarsioidea
Propliopithecus
Lemuroidea

Pleistocene
Swanscombe
Eoanthropus
Pliocene
Miocene
Oligocene
Eocene

**FIGURE 1.5 Racism in anthropology:** In the early decades of the twentieth century, racism was an implicit part of anthropology, with "white" races considered to be superior to "black" races, through greater effort and struggle in the evolutionary race. Here, the supposed ascendancy of the "white" races is shown explicitly, in Earnest Hooton's *Up from the Ape*, second edition, 1946.

present not only among the founding fathers of evolutionary biology, but even prior to that. In both the Great Chain of Being and evolutionary trees we have the strong idea that nature can be seen as a continuity of form, on which humans can be placed. Among both natural philosophers and evolutionary biologists there is the problem of how to find a place within these schemes for humans that can reflect both their unique abilities and their evolutionary heritage. And finally, there is the idea of change, or progress to some, whereby something that was not present at one stage of evolution, or history, does emerge, and comes to thrive.

Modern paleoanthropology, the study of human evolution, has amassed a huge amount of evidence to help solve these problems (Fig. 1.6), and has available to it methods entirely undreamt of by the Victorians, but nonetheless it is worth remembering that it was within four years of the publication of *The Origin of Species* that Thomas Huxley had put his finger on the central problem of human evolution – namely our place in nature, or how we can both "mind the gap" and still remain faithful to evolutionary biology. As we shall see, archeology, fossils, and genetics have all provided ways of filling the gap between humans and the apes, which Huxley had thought unbridgeable. The science of paleoanthropology has emerged to fill the gap. In particular, it can set out to answer two major questions: first, whether the differences between humans and other animals are ones of degree or ones of kind, and second, the extent to which humans are not only unique in the sense that they are different as any species might be,

but also uniquely different in the way they have acquired their basic characteristics.[12]

That these questions remain to be answered can be illustrated with reference to two biologists' thoughts on the subject of the gap. At one extreme is Julian Huxley, grandson of Thomas Henry, who suggested that humankind's special intellectual and social qualities were such that they should be recognized formally by assigning *Homo sapiens* to a new grade, the Psychozoan. "The new grade is of very large extent, at least equal in magnitude to all the rest of the animal Kingdom, though I prefer to regard it as covering an entirely new sector of the evolutionary process, the psychosocial, as against the entire non-human biological sector."[13] At the other end lies Jared Diamond, who argued on the basis of genetic evidence that humans should actually be placed in the same category as chimpanzees, and that we are in fact nothing more than the "third chimpanzee."[14] The power of the study of human evolution is that the answers to these questions can now be treated empirically as well as philosophically.

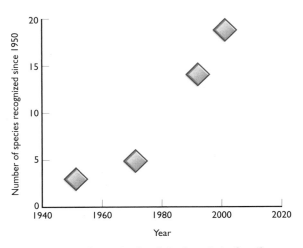

**FIGURE 1.6 The growth of the hominin fossil record:** While changes in approach and perception have been important in the development of our understanding of human evolution, one of the most important factors has been the growth of the fossil record. When Darwin wrote *The Origin of Species* there were virtually no fossils known; now they number in their thousands. This graph shows changes in the number of hominin species recognized. (Courtesy of Robert Foley.)

## ESTABLISHING THE LINK BETWEEN HUMANS AND APES: HISTORICAL VIEWS

Debate over human origins has advanced substantially in recent years, particularly in broadening the scientific basis of the discussions. Nevertheless, many of the issues addressed in current research have deep historical roots. A brief sketch of the subject's progress during the past 100 years or so will put modern debates into historical context.

Two principal themes have been recurrent in the last century of paleoanthropology (Fig. 1.7). First is the relationship between humans and apes: how close, how distant? Second is the "humanness" of our direct ancestors, the early hominins – do the characteristics of humanity go back to the very earliest hominins and beyond, or are they a recent acquisition? ("Hominin" is the term now generally used to describe species in the human family, or clade; until recently, the term "hominid" was used, as discussed in chapter 8.)

**KEY QUESTION** How has the way scientists have perceived the relationship between humans and other animals changed over a century and a half of research and discovery?

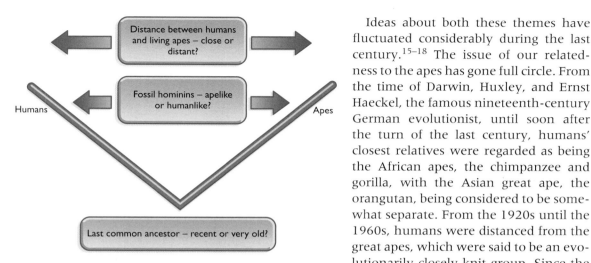

**FIGURE 1.7 Two themes in human evolutionary research:** Two questions have dominated research in human evolution, the first being how close or distant humans and apes are, and the second the extent to which fossil hominins are more humanlike or more apelike.

Ideas about both these themes have fluctuated considerably during the last century.[15–18] The issue of our relatedness to the apes has gone full circle. From the time of Darwin, Huxley, and Ernst Haeckel, the famous nineteenth-century German evolutionist, until soon after the turn of the last century, humans' closest relatives were regarded as being the African apes, the chimpanzee and gorilla, with the Asian great ape, the orangutan, being considered to be somewhat separate. From the 1920s until the 1960s, humans were distanced from the great apes, which were said to be an evolutionarily closely knit group. Since the 1960s, however, conventional wisdom has returned to its Darwinian cast.

This shift of opinions has, incidentally, been paralleled by a related shift in ideas on the location of the "cradle of mankind." Darwin plumped for Africa, Asia became popular in the early decades of the twentieth century, and Africa has once again emerged as the focus.

## The loss of humanity in the fossil hominins

While this human/African-ape wheel has gone through one complete revolution, the question of the humanness of the hominin lineage has been changing as well – albeit in a single direction. Specifically, hominins, with the exception of *Homo sapiens* itself, have been gradually perceived as less humanlike in the eyes of paleoanthropologists, particularly since the 1980s. The different views on the origin of modern humans are taken from different perspectives on this issue. However, the two themes are in practice deeply intertwined. Determining which ape or monkey humans are most closely related to is dependent upon what traits are considered to be important to "being human," and so the extent to which they can be traced back to a common ancestor.

In his *Descent of Man*,[19] Darwin identified those characteristics that apparently make humans special – intelligence, manual dexterity, technology, and uprightness – and argued that an ape endowed with minor amounts of each of these qualities would surely possess an advantage over other apes. Once the earliest human forebear became established upon this evolutionary trajectory, the eventual emergence of *Homo sapiens* appeared almost inevitable because of the continued power of natural selection. In other words, hominin origins became explicable in terms of human qualities, and hominin origin therefore equated with human origin. It was a seductive formula, and one that persisted until quite recently.

In the early decades of the twentieth century two
opposing views of human origins were current:

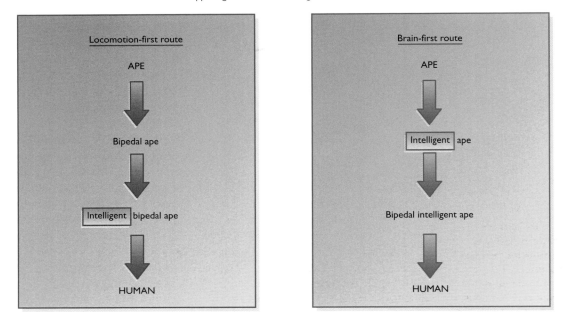

**FIGURE 1.8 Two models of the pattern of human evolution:** Human evolution has, in terms of anatomy, been characterized as involving two major changes – the evolution of upright walking, and the evolution of larger brains. One general set of models has viewed brains as leading the way, while another reverses the sequence. Some models have also suggested that there is simultaneous evolution. Earlier work tended to support brain-led models, whereas more recent work has tended toward bipedalism coming earlier, and no strong linkage between the two.

Darwin's human traits set the agenda for the intellectual debate that occurred at the beginning of the last century concerning the order in which the major anatomical changes occurred in the human lineage (Fig. 1.8). One notion was that the first step on the road to humanity was the adoption of upright locomotion. A second idea held that the brain led the way, producing an intelligent but still arboreal creature. It was into this intellectual climate that the perpetrator of the famous Piltdown hoax – a chimera of fragments from a modern human cranium and an orangutan's jaw, both doctored to make them look like ancient fossils – made his play in 1912 (Fig. 1.9).[20] (In mid-1996 the first material clues as to the identity of the Piltdown forger came to light, pointing to Martin Hinton, the British anthropologist and Arthur Smith Woodward's colleague at the Natural History Museum, London.)

The Piltdown "fossils" appeared to confirm not only that the brain did indeed lead the way (in other words, it was the first important human trait to evolve, and others, such as bipedalism or upright walking, were consequences of having a larger brain), but also that something close to the modern *sapiens* form was extremely ancient in human history. The

Was human evolution driven by intelligence?

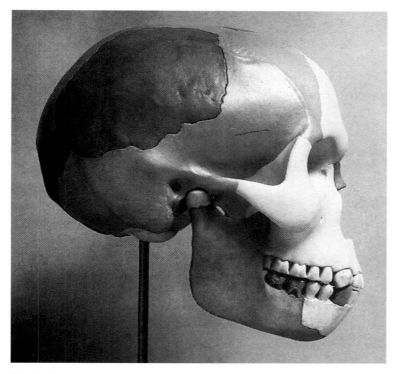

**FIGURE 1.9 Piltdown man: a fossil hybrid and fake:** A cast of the Piltdown reconstruction, based on lower jaw, canine tooth, and skull fragments (shaded dark). The ready acceptance of the Piltdown forgery – a chimera of a modern human cranium and the jaw of an orangutan – derived from an adherence to the brain-first route. (Courtesy of the American Museum of Natural History.)

apparent confirmation of this latter fact – extreme human antiquity – was important to both Sir Arthur Keith and Henry Fairfield Osborn (director of the American Museum of Natural History in the early decades of the twentieth century), because their theories demanded it. One consequence of Piltdown was that Neanderthal – one of the few genuine fossils of the time – was disqualified from direct ancestry to *Homo sapiens*, because it apparently came later in time than Piltdown and yet was more primitive.

For Osborn, Piltdown represented strong support for his Dawn Man theory, which stated that humankind originated on the high plateaus of Central Asia, not in the jungles of Africa. During the 1920s and 1930s, Osborn was locked in constant but gentlemanly debate with his colleague, William King Gregory, who carried the increasingly unpopular Darwin-and-Huxley torch for a close relationship between humans and African apes – the Ape Man theory.

Although Osborn was never very clear about what the earliest human progenitors might have looked like, his ally Frederic Wood Jones espoused firmer ideas. Wood Jones, a British anatomist, interpreted key features of ape and monkey anatomy as specializations that were completely absent in human anatomy. In 1919, he proposed his "tarsioid hypothesis," which sought human antecedents very low down in the primate tree, with a creature like the modern tarsier. In today's terms, this proposal would place human origins in the region of 50 million to 60 million years ago, close to the origin of the primate radiation (see chapter 6), while Keith's notion of some kind of early ape would date this development to approximately 30 million years ago (Fig. 1.10).

## Apes, great apes, and African apes

During the 1930s and 1940s, the anti-ape arguments of Osborn and Wood Jones were lost, but Gregory's position did not immediately prevail. Gregory had argued for a close link between humans and the African apes

on the basis of shared anatomical features. Others, including Adolph Schultz and D. J. Morton (two scientists who laid the foundations of primate evolution and anatomy in the first part of the twentieth century), claimed that, although humans probably derived from apelike stock, the similarities between humans and modern African apes were the result of parallel evolution (Fig. 1.11). This position remained dominant through the 1960s, firmly supported by Sir Wilfrid Le Gros Clark, Britain's most prominent primate anatomist of the time. Humans, it was argued, came from the base of the ape stock, not later in evolution with the specializations developed by the African apes.[21]

During the 1950s and 1960s, the growing body of fossil evidence related to early apes appeared to show that these creatures were not simply early versions of modern apes, as had been tacitly assumed. This idea meant that those authorities who accepted an evolutionary link between humans and apes, but rejected a close human/African ape link, did not have to retreat back in the history of the group to "avoid" the specialization of the modern species. At the same time, those who insisted that the similarities between African apes and humans reflected a common heritage, not parallel evolution, were forced to argue for a very recent origin of the human line. Prominent among proponents

(a)

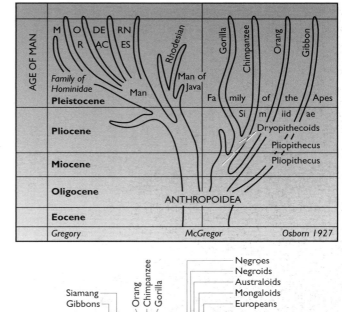

(b)

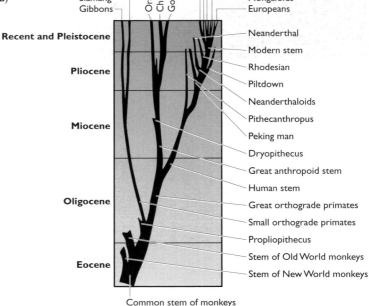

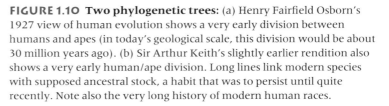

**FIGURE 1.10  Two phylogenetic trees:** (a) Henry Fairfield Osborn's 1927 view of human evolution shows a very early division between humans and apes (in today's geological scale, this division would be about 30 million years ago). (b) Sir Arthur Keith's slightly earlier rendition also shows a very early human/ape division. Long lines link modern species with supposed ancestral stock, a habit that was to persist until quite recently. Note also the very long history of modern human races.

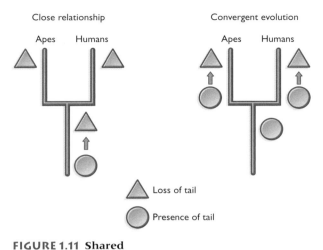

FIGURE 1.11 **Shared descent and parallel evolution:** The shared characteristics of apes and humans were thought by some, such as Gregory and Sir Wilfrid Le Gros Clark, to be the result of the two lineages having a close relationship and thus many traits having evolved in a common ancestor. Others, including Adolph Schultz and D. J. Morton, viewed these traits as having evolved in parallel independently in different lineages, and did not see the traits as evidence of a close relationship.

Do humans and African apes have a special relationship?

of this latter argument was Sherwood Washburn, of the University of California, Berkeley (Fig. 1.12).[22]

One of the fossil discoveries of the 1960s – in fact, a rediscovery – that appeared to confirm the notion of parallel evolution to explain human/African ape similarities was made by Elwyn Simons, then of Yale University. The fossil specimen was *Ramapithecus*, an apelike creature that lived in Eurasia approximately 15 million years ago and appeared to share many anatomical features (of the teeth and jaws) with hominins (Fig. 1.13). Simons, later supported closely by David Pilbeam (now of Harvard University), proposed *Ramapithecus* as the beginning of the hominin line, thus excluding a human/African ape connection.[23]

Arguments about the relatedness of humans and African apes were mirrored by a reconsideration of relatedness among the apes themselves. In 1927, G. E. Pilgrim, a geologist who discovered the important *Ramapithecus* fossils in the 1920s, had suggested that the great apes be treated as a natural group (that is, evolutionarily closely related), with humans viewed as more distant. This idea eventually became popular and remained the accepted wisdom until molecular biological evidence undermined it in 1963, via the work of Morris Goodman at Wayne State University. Goodman's molecular biology data on blood proteins indicated that humans and the African apes formed a natural group, with the orangutan more distant.[24]

As a result, the Darwin/Huxley/Haeckel position returned to prominence, with first Gregory and then Washburn emerging as its champion. Subsequent molecular biological – and fossil – evidence appeared to confirm Washburn's original suggestion that the origin of the human line is quite recent, close to 5 million years ago. *Ramapithecus* was no longer regarded as the first hominin, but simply one of many early apes. (The nomenclature and evolutionary assignment of *Ramapithecus* subsequently were modified, too.)

Meanwhile, discoveries of fossil hominins, and the stone tools they apparently made, had been accumulating at a rapid pace from the 1940s through 1970s, first in south Africa and then in east Africa. Culture – specifically, stone-tool making and tool use in butchering animals – became a dominant theme, so much so that "hominin" was considered to imply a hunter-gatherer life-style. The most extreme expression of culture's importance as the hominin characteristic consisted of the single-species hypothesis, promulgated during the 1960s principally by C. Loring Brace and Milford Wolpoff, both of the University of Michigan.

## The single-species hypothesis

According to this hypothesis, only one species of hominin existed at any one time; human history was viewed as progressing by steady improvement up a single evolutionary ladder (Fig. 1.14). The rationale relied upon a supposed rule of ecology: the principle of competitive exclusion, which states that two species with very similar adaptations cannot coexist. In this case, culture was viewed as such a novel and powerful behavioral adaptation that two cultural species simply could not thrive side by side. Thus, because all hominins are cultural by definition, only one hominin species could exist at any one time. This was a powerful idea developed in the middle of the last century by Theodosius Dobzhansky and Ernst Mayr, two of the great evolutionary biologists of the twentieth century, who developed modern evolutionary theory as it is used today, and the idea came to be very influential in human evolutionary studies.[25]

The single-species hypothesis collapsed in the mid-1970s, after fossil discoveries from Kenya undisputedly demonstrated the coexistence of two very different species of hominin: *Homo erectus* (or *Homo ergaster*), a large-brained species that apparently was ancestral to *Homo sapiens*, and *Australopithecus boisei*, a small-brained species that eventually became extinct. Subsequent discoveries and analyses implied that several species of hominin coexisted in Africa some 2 million or so years ago, suggesting that several different ecological niches were being successfully exploited.

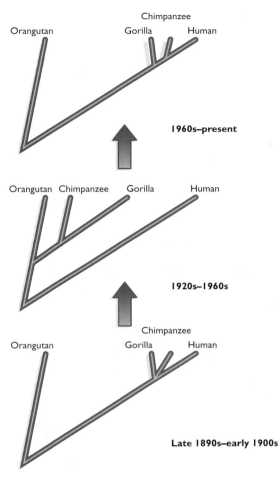

**FIGURE 1.12 Shifting patterns:** Between the beginning of the twentieth century and today, ideas about the relationships among apes and humans have moved full circle.

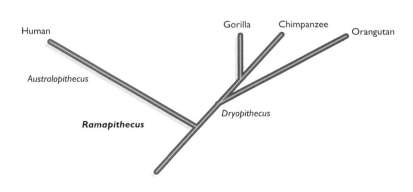

**FIGURE 1.13** *Ramapithecus*: Now considered to be related to the orangutan, *Ramapithecus* was at one time considered to be the first hominin and evidence for an early origin and long time scale for human evolution.

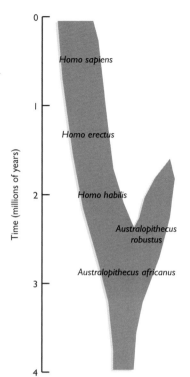

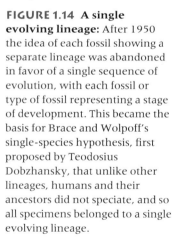

**FIGURE 1.14 A single evolving lineage:** After 1950 the idea of each fossil showing a separate lineage was abandoned in favor of a single sequence of evolution, with each fossil or type of fossil representing a stage of development. This became the basis for Brace and Wolpoff's single-species hypothesis, first proposed by Teodosius Dobzhansky, that unlike other lineages, humans and their ancestors did not speciate, and so all specimens belonged to a single evolving lineage.

**KEY QUESTION** How have paleoanthropologists attempted to explain human evolution?

These findings implied that to be hominin did not necessarily mean being cultural. Thus, no longer could hominin origins be equated with human origins (Fig. 1.15).

From the 1980s onward, not only has an appreciation of a spectrum of hominin adaptations – including the simple notion of a bipedal ape – emerged, but the lineage that eventually led to *Homo sapiens* has also come to be perceived as much less human. Gone is the notion of a scaled-down version of a modern hunter-gatherer way of life. In its place has appeared a rather unusual African ape adopting some novel, un-apelike modes of subsistence.

The two themes identified as being central to debates about human evolution – the specific relationship to the apes, and the antiquity of human characteristics – remain central to current research. The close relationship between humans and African apes seems to have been firmly established, and older ideas can clearly be rejected. The main inference that has now been drawn is that the basal, primate stock from which humans are derived were part of an already well-developed African ape form. However, it is also the case that these earliest hominins were still far from being human, and that the actual traits that so categorically split humanity from the rest of the primate world are not present in the early part of the evolutionary story. The result is that there are now two problems in human evolution where before there had been one – the origin of the hominin lineage, and the origin of humanness. These questions are in turn strongly interlinked to ideas about both the time-depth of human evolution and the diversity of species involved during its course (Fig. 1.16). Questions about the beginning of the hominin lineage are now firmly within the territory of behavioral ecology and do not draw upon those qualities that we might perceive as separating us from the rest of animate nature. Questions of hominin origins must now be posed within the context of primate biology.

# HUMAN EVOLUTION AS NARRATIVE AND AS EXPLANATION

"One of the species specific characteristics of *Homo sapiens* is a love of stories," noted Glynn Isaac, "so that narrative reports of human evolution are demanded by society and even tend toward a common form." Isaac was referring to the work of Boston University anthropologist Misia Landau,[26] who has analyzed the narrative component of professional – not just popular – accounts of human origins.

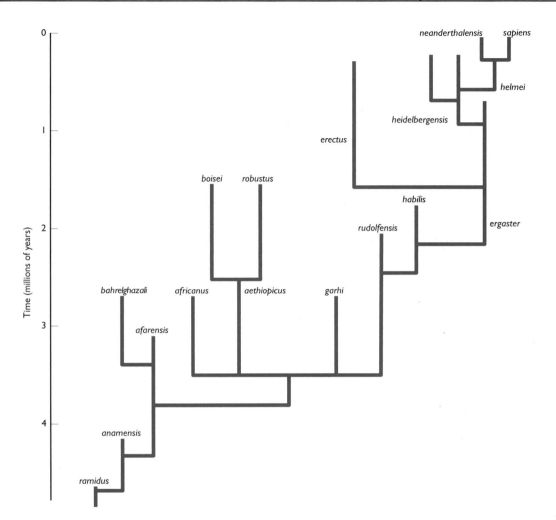

**FIGURE 1.15 The return of multiple species (an extreme radiation view):** With the spectacular growth of the fossil record during the second half of the twentieth century, more than one lineage of hominin was increasingly recognized, particularly the divergence of *Homo* and some australopithecines (types of early hominin).

"Scientists are generally aware of the influence of theory on observation," concludes Landau. "Seldom do they recognize, however, that many scientific theories are essentially narratives." This is true not just of paleoanthropology, but even of the hardest sciences. One way, for example, is to look at how DNA, the most basic biological molecule, acts to make a protein, the building blocks of life. In one sense it is a chemical journey, from genetic code to complex bodily forms. The sequence from DNA to amino acid to protein and the composite individual can be told as a journey, with a starting point (conception) and an end point (adult development). There is therefore nothing "unscientific" about the fact that narrative is common to both fact and fiction; rather, it expresses that in

| DATE | CONSENSUS MODEL FOR HUMAN ORIGINS | KEY ANTHROPOLOGISTS |
|---|---|---|
| 1860–1920 | **Unilineal progressive evolution** <br> All evolution seen as a progressive trend, driven by orthogenetic processes, leading to modern humans. No geography and no point of origin. Key issues related to whether there were any intermediate steps (Neanderthals, etc.) and how they should be ordered. Living human diversity seen as ladders along a progressive continuum. | Manrouvier <br> Cunningham <br> Schwalbe <br> Haeckl |
| 1900–50 | **Typological trees** <br> Most fossils seen as side branches away from the main line, generally becoming extinct. Lack of continuity and an emphasis on parallel evolution. Key controversies related to which if any fossils (e.g. Piltdown) did lie on the true line of descent. Living human diversity seen as part of the tree of hominid variation. The origin of modern humans located with particular fossils. | Keith <br> Boule <br> Breuil <br> Coon |
| 1940–90 | **Anagenetic polytypic evolution** <br> The development of the Modern Synthesis led to a recognition that variation within populations and species could occur, and that populations would be transformed gradually by selection. Emphasis was placed on continuous variation, gene flow, and progressive adaptive change. Living human diversity was part of a spectrum of variation. This model was gradually transformed into the mulitregional model as a means of accounting for spatio-temporal patterns within a population genetics framework. No point of origins, and gene flow is the principal mechanism of evolutionary change. | Weidenreich <br> Mayr <br> Brace <br> Wolpoff/Thorne |
| 1980– | **Divergence and replacement** <br> Emphasis on geographical variation, mechanisms of speciation, and the role of isolation in evolution led to renewed interest in more taxonomically diverse models of human evolution. Emphasis was increasingly placed on localized events such as range fragmentation and genetic bottlenecks underlying evolutionary processes, with a correlated interest in dispersals, replacements, and extinctions. This model was very much driven by a greater use of genetics and a more precise chronology. Living human diversity is insignificant except as a marker of recent historical patterns. Human origins located at a particular point in time and space (Africa, Late Pleistocene). | Stringer <br> Wilson, etc. <br> Howell |

**FIGURE 1.16 Four phases in the history of studies of human origins:** These overlap considerably in time, and to some extent represent an ongoing conflict between unilineal/progressive/polytypic models stressing gradual transformation, and adaptive radiation models emphasizing divergence, isolation, and extinction of populations and species. The current conflict between multiregional and single-origin models reflects the latest manifestation of the debate, and the positions adopted are influenced by both the evidence and interpretations of evolutionary process. (Courtesy of Marta Lahr and Robert Foley.)

What drives changes in perspective on human evolution?

science we are looking for causal links and these occur in temporally related events – in other words, in a narrative. However, Landau identifies several elements in paleoanthropology that make it particularly suscept- ible to being cast in narrative form, both by those who tell the stories and by those who listen to them. These elements arise primarily from the fact that as human evolution, like any evolutionary history, is a sequence of events occurring over time, it falls naturally into narrative form.

Placing events in the correct order is an essential part of any historical science, and should not be underrated, even if it can sometimes be unglamorous. However, if human evolution were simply a case of ordering the events, then it would indeed be the case that human evolution is akin to story telling (although in history it is as well to remember that some stories are true and others are not). There is, though, an essential second element, which is the attempt to explain causally the links between the events, to try to understand why they occurred in one particular order rather than another. This is an essential part of science – seeking causal explanations for key events.

## The key events in human evolution

Traditionally, paleoanthropologists have recognized four key events in human evolution: the origin of terrestriality (coming to the ground from the trees), bipedality (upright walking), encephalization (brain expansion in relation to body size), and culture (or civilization) (Fig. 1.17). While these four events have usually featured in accounts of human origins, paleoanthropologists have disagreed about the order in which they were thought to have occurred and their importance in "causing" human evolution.[15,16]

For instance, Henry Fairfield Osborn considered the order to be that given above, which, incidentally, coincides closely with Darwin's view. Sir Arthur Keith, building on earlier ideas of D. J. Morton, considered bipedalism to have been the first event, with terrestriality following. In other words, Keith's ancestral ape began walking on two legs while it was still a tree dweller; only

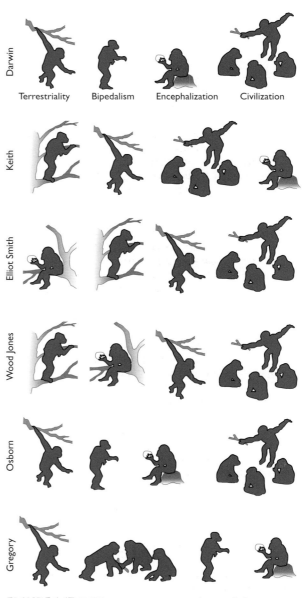

**FIGURE 1.17 Different reconstructions of the sequence of events in human evolution:** Even though anthropologists saw the human journey as involving the same fundamental events – terrestriality, bipedalism, encephalization, and civilization – different authorities sometimes placed these steps in slightly different orders. For instance, although Charles Darwin envisaged an ancient ape first coming to the ground and then developing bipedalism, Sir Arthur Keith believed that the ape became bipedal before leaving the trees. (Courtesy of Misia Landau/*American Scientist*.)

**FIGURE 1.18 Sir Grafton Elliot Smith:** A leading anatomist and anthropologist in early twentieth-century England, Elliot Smith often wrote in florid prose about human evolution. (Courtesy of University College, London.)

subsequently did it descend to the ground. For Sir Grafton Elliot Smith, a contemporary of Keith, encephalization led the way (Fig. 1.18). His student, Frederic Wood Jones, agreed with Smith that encephalization and bipedalism developed while our ancestor lived in trees, but thought that bipedalism preceded rather than followed brain expansion. William King Gregory, like his colleague Osborn, argued for terrestriality first, but suggested that the adoption of culture (tool use) preceded significant brain expansion. And so on.

Thus, we see these four common elements linked together in different ways, with each narrative scheme purporting to tell the story of human origins. And "story" is the operative word here. "If you analyze the way in which Osborn, Keith and others explained the relation of these four events, you see clearly a narrative structure," says Landau, "but they are more than just stories. They conform to the structure of the hero folk tale." In her analysis of paleoanthropological literature, Landau drew upon a system devised in 1925 by the Russian literary scholar Vladimir Propp. This system, published in Propp's *Morphology of the Folk Tale*, included a series of 31 stages that encompassed the basic elements of the hero myth. Landau reduced the number of stages to nine, but kept the same overall structure: hero enters; hero is challenged; hero triumphs.

In the case of human origins, the hero is the ape in the forest, who is "destined" to become us. The climate changes, the forests shrink, and the hero is cast out on the savannah where he faces new and terrible dangers. He struggles to overcome them, by developing intelligence, learning to use tools, and so on, and eventually emerges triumphant, recognizably you and me.

"When you read the literature you immediately notice not only the structure of the hero myth, but also the language," explains Landau. For instance, Elliot Smith writes about "the wonderful story of Man's journeyings towards his ultimate goal" and "Man's ceaseless struggle to achieve his destiny." Roy Chapman Andrews, Osborn's colleague at the American

Museum, writes of the pioneer spirit of our hero: "Hurry has always been the tempo of human evolution. Hurry to get out of the primordial ape stage, to change body, brains, hands and feet faster than it had ever been done in the history of creation. Hurry on to the time when man could conquer the land and the sea and the air; when he could stand as Lord of all the Earth."

Osborn wrote in similar tone: "Why, then, has evolutionary fate treated ape and man so differently? The one has been left in the obscurity of its native jungle, while the other has been given a glorious exodus leading to the domination of earth, sea, and sky." Indeed, many of Osborn's writings explicitly embodied the notion of drama: "The great drama of the prehistory of man," he wrote, and "the prologue and opening acts of the human drama," and so on.[26]

Of course, it is possible to tell stories with similar gusto about non-human animals, such as the "triumph of the reptiles in conquering the land" or "the triumph of birds in conquering the air." Such stirring tales are readily found in accounts of evolutionary history – look no further than every child's hero, the dinosaur. The fact that the hero of the paleo-anthropology tale is *Homo sapiens* – ourselves – makes a significant difference, however. Although dinosaurs may be lauded as lords of the land in their time, only humans have been regarded as the inevitable product of evolution – indeed, the ultimate purpose of evolution, as we saw in the previous chapter. Not everyone was as explicit about this as Broom was, but most authorities betrayed the sentiment in the hero-worship of their prose.

These stories were not just accounts of the ultimate triumph of our hero; they carried a moral tale, too – namely, triumph demands effort. "The struggle for existence was severe and evoked all the inventive and resourceful faculties and encouraged [Dawn Man] to the fashioning and first use of wooden and then stone weapons for the chase," wrote Osborn. "It compelled Dawn Man . . . to develop strength of limb to make long journeys on foot, strength of lungs for running, and quick vision and stealth for the chase."

According to Elliot Smith, our ancestors "were impelled to issue forth from their forests, and seek new sources of food and new surroundings on hill and plain, where they could obtain the sustenance they needed." The penalty for indolence and lack of effort was plain for all to see, because the apes had fallen into this trap: "While man was evolved amidst the strife with adverse conditions, the ancestors of the Gorilla and Chimpanzee gave up the struggle for mental supremacy because they were satisfied with their circumstances."

In the literature of Elliot Smith's time,[27] the apes were usually viewed as evolutionary failures, left behind in the evolutionary race. This sentiment prevailed for several decades, but eventually became transformed. Instead of evolutionary failures, the apes came to be viewed as evolutionarily primitive, or relatively unchanged from the common ancestor they

How is it possible to reconstruct an evolutionary narrative?

shared with humans. In contrast, humans were regarded as much more advanced. Today, anthropologists recognize that both humans and apes display advanced evolutionary features, and differ equally (but in separate ways) from their common ancestor.

Although modern accounts of human origins usually avoid purple prose and implicit moralizing, one aspect of the narrative structure lingers in current literature. Some paleoanthropologists still tend to describe the events in the "transformation of ape into human" as if each event were a preparation for the next. "Our ancestors became bipedal in order to make and use tools and weapons . . . tool-use enabled brain expansion and the evolution of language . . . thus endowed, sophisticated societal interactions were finally made possible." Crudely put, to be sure, but this kind of reasoning was common in Osborn's day and persists in some current narratives.

Why does it happen? "Telling a story does not consist simply in adding episodes to one another," explains Landau. "It consists in creating relations between events." Consider, for instance, our ancestor's supposed "coming to the ground" – the first and crucial advance on the long road toward becoming human. It is easy to imagine how such an event might be perceived as a courageous first step on the long journey to civilization: the defenseless ape faces the unknown predatory hazards of the savannah. "There is nothing inherently transitional about the descent to the ground, however momentous the occasion," says Landau. "It only acquires such value in relation to our overall conception of the course of human evolution."

If evolution were steadily progressive, forming a program of constant improvement, the transformation of ape to human could be viewed as a series of novel adaptations, each one naturally preparing for and leading to the next. Such a scenario would involve continual progress through time, going in a particular direction. From our vantage point, where we can view the end product, it is tempting to view the process in that way because we can see that all those steps did actually take place. This slant, however, ignores the fact that evolution tends to work in a less directional way; the adaptations of one period do not necessarily lead directly on to those of another. There is often a period of rapid change, followed by a radiation of forms, and then these may all disappear or reduce in abundance, to be succeeded by something that does not continue these trends.

Furthermore, the fossil and archeological evidence belies any inevitability to the process of evolutionary change. For instance, one cannot say that the first bipedal ape would inevitably become a stone-tool maker. In fact, if the current archeological record serves as any guide, those two events – bipedality and the advent of stone-tool making – were separated by approximately 2.5 million years. The brain expanded at this time as well. In addition, a more humanlike body structure emerged abruptly. The origin of anatomically modern humans after another 2 million or so years

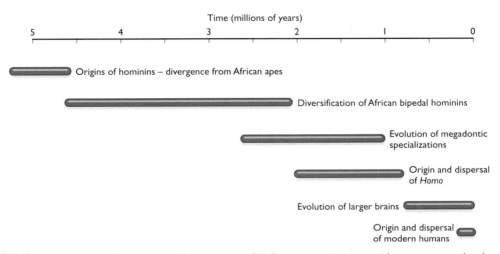

**FIGURE 1.19  The pattern of human evolution as multiple events:** Current evidence suggests that human evolution occurred over more than 5 million years, and that different events were spread across this time, so that there cannot be a single explanation for all of human evolution.

was also probably a relatively rapid event. Thus, although many writers proclaim that our ancestors were propelled inexorably along an evolutionary trajectory that ended with *Homo sapiens*, that scenario simply describes what did happen; it ignores the many other possibilities that did not transpire (Fig. 1.19). As Landau remarks: "There is a tendency in theories of hominid evolution to define origins in terms of endings."

Landau's exposition of the similarities between narrative structure and scientific explanations have many lessons for the discipline of evolution. One simple one is that the language in which we describe evolution is important, and that it is all too easy to elide the language of everyday speech – "early humans were struggling against the drying environment," for example – with the technical explanations of evolutionary biology. This can conjure up images of deserts and thirst, while actually representing a summary of minor change in temperature over hundreds of thousands of years. We need both the force of powerful description and the cold technical language of science. Another is that what the similarity between narrative in folklore and narrative in science reflects is not necessarily the fact that science is "just another form of fiction," but that it must use the tools of everyday thought. In particular, narratives are in a sense another way of describing the creation of models of the past, and it is these models that are tested scientifically. Narratives may be a constant motif of paleoanthropology, but the actual structure of them reflects the accumulating data of the discipline. Finally, what binds together the narrative of the folk tale and the narrative of paleoanthropology is that both are events that occur through time, and so attract the form of "explanation" where one event is causally related to another.

*Is evolutionary change a response to a challenge?*

# Explanation in modern paleoanthropology

We can see that there is much in the history of paleoanthropology that shows that the researchers have been greatly influenced by their own insights and prejudices about the nature of humans and their relationship to the rest of the biological world. This is unavoidable. To some this may mean that science is inevitably subjective and biased, but another approach is to attempt to minimize the effects of subjectivity.[28] In this way, as evolutionary biology has developed, explanations for events in the past are now more formally phrased – an end to the purple passages of triumphalism of Broom and his contemporaries – and placed in the quantitiative framework of hard evidence. Thus the "narrative sequence" of human evolution can actually be tested by measuring the changes in fossil human brain sizes, dating fossil sites, measuring the complexity of stone tools, and so on. Modern paleoanthropology is moving toward a much more quantitative phase than in the past.

This extends to the realm of explanation as well. For example, a common element in earlier writings was the effect of a harsher and drier environment on human evolution. Human traits were forged by the challenges of this environment. While it may be easy to dismiss this as "the hero folk tale," it can equally be formulated in terms of testable hypotheses and examined in relation to climatic data, the rate of evolutionary change, etc. Explicit links between particular types of environment and adaptations can be postulated – for example, between higher temperatures and selective pressure for traits that keep the body cool (thermoregulation), and then these can be sought in either fossil or comparative evidence. This change, from "stories" to "models", in effect, has also coincided with a shift away from an emphasis on the intrinsic and inevitable character of human ancestry, toward one on the environment and the context-dependent nature of evolutionary explanations[29] (Fig. 1.20).

*What makes a good evolutionary explanation?*

At the start of this chapter the main point was made that it is the special nature of humans that makes human evolution both interesting and extremely challenging. This can produce more overlap between the sciences and the humanities than is usual in most disciplines. While this may lead to levels of subjectivity and involvement that can be problematic, modern paleoanthropology attempts to find scientific hypotheses and approaches to test ideas. In doing so the aim is both to build a better historical narrative (the basic data of human evolution) and to seek explanations for why such an important event (human evolution) occurred when it did, and in the way it did.

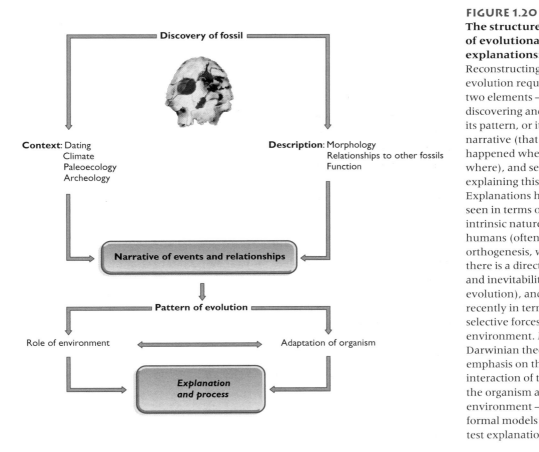

**FIGURE 1.20**
**The structure of evolutionary explanations:** Reconstructing human evolution requires two elements – first, discovering and revealing its pattern, or its narrative (that is, what happened when and where), and second, explaining this pattern. Explanations have been seen in terms of the intrinsic nature of humans (often ideas of orthogenesis, where there is a direction and inevitability to evolution), and more recently in terms of the selective forces of the environment. Modern Darwinian theory places emphasis on the interaction of the two – the organism and the environment – and uses formal models to try to test explanations.

## BEYOND THE FACTS
## SCIENCE AND SENSIBILITY

*The issue: the critique of Landau and others of the scientific objectivity of work in human evolution is part of a wider critique of science as a way of looking at the world. Is science "just another story" or is it a powerful and unique tool to understand the world?*

In 1950 an Englishman called C. P. Snow wrote an essay called "The Two Cultures" in which he contrasted the arts and the sciences as two worlds that had little contact. He was concerned that there were two ways of looking at the world, one based on factual observation, the other based on human sentiment and experience. His essay aroused great controversy, especially among those in the humanities, who saw it as an attack on the arts, and there arose a great concern that the utilitarianism of science would swallow up the less certain world of the humanities. It is certainly the case that the half century since that essay has seen the sciences grow in ways that are extraordinary. The scale of technological achievement is unquestioned.

However, C. P. Snow would perhaps be surprised to see the extent to which this has gone hand in hand with a questioning of the special status of science. Philosophers of science, such as Thomas Kuhn, and social theorists have all played a part in showing that science is not detached from human experience, but very much a part of it, and that science progresses as much through the interplay of political and social factors as it does through experiment and dry observation. The history of the study of human evolution would seem to fit into this. Landau's suggestion that models of human evolution owe as much to folklore as to fossil discoveries is a case in point. John Reader, in his book *Missing Links*,[16] showed that interpretations of fossils reflected the initial prejudices of the anthropologists as well as the actual evidence – hence the idea of "brain-led" evolution and the successful perpetuation of the Piltdown fraud.

All this might easily lead to the notion that current interpretations of human evolution are merely reflections of the current society in which we live, and will pass too as intellectual fashion accessories. At best they are socially sanctioned speculation, at worst, just-so stories to justify the status quo. Such ideas are of course very attractive, not least because they actually give equal authority to all in putting forward theories about human evolution. Why should one theory of human evolution be judged better than another when all they reflect are the prevailing biases of the cultures that produce the authors of those theories?

However, the question to be asked is not whether social and intellectual fashions influence ideas about human evolution – they clearly do – but how significant that influence is compared to others.

The Afrocentric ideas that dominate much of this book might be a test case for asking this question. To what extent has the motor of changing ideas about human evolution been new discoveries – the fossils from Africa over the last quarter of a century, or the growth of molecular genetics, for example – and to what extent has it been the prejudices of the practitioners? Has the change in focus from Asian origins during the early part of the twentieth century been the product of changing geopolitics and social theories, or a consequence of the growing numbers of fossils with earlier dates in Africa than elsewhere? Has the demise of brain-led theories of evolution been driven by a loss of confidence in the power of the human mind to solve problems and so be the driving force in human evolution, or is it due to the empirical evidence showing that the shift to bipedalism preceded any enlargement in brain size? Finding an answer to this question is important if paleoanthropology is to be taken seriously as a science, rather than treated as an oblique form of social history.

# CHAPTER
## 2

# THE PRINCIPLES OF EVOLUTIONARY THEORY

## THE FUNDAMENTALS OF EVOLUTIONARY THEORY

The difficulties faced by earlier anthropologists in tackling the problems of human evolution have to a large extent derived from the interaction of two very contrasting elements. On the one hand there is the uniqueness of humans, which has led to ideas about special creations and a great antiquity to the human lineage. Human evolution is a "one-off" event, and so requires special explanations. That has been one of the major themes in earlier research. On the other hand, though, the reason that humans are studied from an evolutionary perspective is that they are the outcome of general mechanisms that apply equally to all organisms. The key question is therefore how such general processes can produce such a special organism as a human being.[30] This obviously depends upon the nature of those evolutionary processes.

> **KEY QUESTION** What conditions and mechanisms are necessary for evolution to take place?

Evolutionary theory can be traced back to Charles Darwin and Alfred Wallace, its co-discoverer. However, it is necessary to distinguish exactly what they contributed to evolutionary theory. It is important to establish that they were by no means the first to propose evolution.[3] Evolution simply means change through time – in this case applied to biological systems. Evolutionary ideas had been widely discussed throughout the nineteenth century, and indeed before. Many scientists of the time were interested in whether life forms had changed, and if so, how. Although there were many who held to the immutability of life, others had proposed theories of what was generally termed "transformation." Darwin's grandfather, Erasmus Darwin, had been one of those, and had developed his own theories of evolutionary change. The most widely discussed theory of the time was that of Jean-Baptiste Lamarck, a French biologist. He was convinced that there was ample evidence that evolution had occurred – on the basis of comparative anatomy and the fossil record – and he suggested that the mechanism was the "need" of organisms to adapt to changing environments. His theory, generally known as evolution through "the inheritance of acquired characteristics," was that, faced with new and challenging circumstances, those individuals who were able to acquire a better ability to survive did so, and that their offspring inherited these traits. The classically quoted example is that of the giraffe, which, threatened by a drought, would have to eat leaves higher and higher in a tree. Those individuals best able to stretch their necks would do better, and so survive, so that the outcome of that stretching process would be passed on to their offspring.

Lamarck's theory failed, largely because he was unable to show how such characteristics could be passed from one generation to another.[5,31] However, Lamarck's importance lies in the fact that he represented the general intellectual context in which nineteenth-century biologists were working. They were interested in the transformation of life forms, and it was this problem that Darwin and Wallace addressed.

If they did not come up with the idea of evolution, what was their contribution? The answer is that they supplied a coherent theory of how evolution could, in practice, work. In Darwin's case, he also supplied exhaustive documentation to back up his theory.

Darwinism is therefore a theory of how evolutionary change occurs. The theory is in itself very simple – indeed, that has probably been one of the reasons why it has been considered so often to be inadequate, but frequently the most powerful scientific theories are characterized by an elegant simplicity.[6,32,33,34,35]

## Natural selection

Darwinian evolutionary mechanisms are based on four conditions, each of which builds successively on the other (Fig. 2.1). Natural selection – the differential survival of genes – is the outcome of certain conditions that have to be fulfilled. The first of these is that organisms *reproduce*. If there is no reproduction, then the game of life would have to start afresh after each deceased generation. In the early days of life forms perhaps this happened, when the replicating molecules were less efficient, although natural selection would still be operating, selecting for the most efficient replicating molecule.

The second condition is that there should be some mode of *inheritance* – that is, offspring

**FIGURE 2.1 The structure of the Darwinian theory of evolution:** Natural selection is the core mechanism of change (the differential survival of offspring). Its operation depends upon four conditions, each of which is known to occur in nature. Adaptation and evolutionary change are the consequences or outcomes of these four conditions.

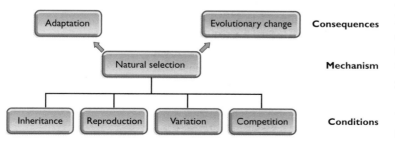

Consequences

Mechanism

Conditions

should resemble their parents more than they do the population as a whole. This is the field of genetics, but when Darwin wrote *The Origin of Species* it was this aspect that was least well understood and caused him the most critical problems. If information that determines the characteristics of the parent can be transmitted to offspring, then those features which enhance the survival and reproductive potential of the parents will occur more frequently in each subsequent generation, dependent upon the number of offspring. If there is no mode of inheritance, then the advantageous features of a parent would simply be lost in each generation. There could be no evolutionary change (Fig. 2.2). It was the absence of this condition that made Lamarckism an unworkable theory of evolution.

There must, third, be *variation* within the population. Even if the first two conditions are fulfilled, if each individual in the population is phenotypically and genetically identical, then natural selection cannot operate. Differential survival will have no effect because all individuals are the same, and so each generation will be identical. It is for this reason that Darwin was himself so concerned with the problem of variation, and why he devoted the first two chapters of *Origin of Species* to "Variation under Domestication" and "Variation under Nature." This was in fact also one of the principal lines of evidence employed against the theory of a special creation of immutable types – if God had created a number of types of plant and animal, then unless he or she was incompetent there was no reason why they should vary at all. Darwin went to extraordinary lengths to show that even the humblest type of creature displayed variation.

Finally – and this condition is perhaps the one most closely associated with Darwin's own ideas – there is *competition*. Imagine a world in which all the conditions outlined above were fulfilled. However, if the resources needed to support all the populations were infinite, then there would be no differential reproduction. An individual could have all the offspring possible, and so there would be no change from one generation to another, just a constant and everlasting expansion. Clearly, though, such a world does not exist. Indeed it is theoretically impossible, as time itself is a resource (time to have offspring, etc.) and so as long as there is time, there will be at least some limitation. In practice of course all resources are limited – energy, water, shelter, potential mates, and so on. It was the eighteenth-century demographer Thomas Malthus who first pointed out the imbalance between the potential of resources to expand and the potential of populations to reproduce. Darwin harnessed this notion as the central condition necessary for natural selection to operate. If resources are limited, then not all individuals will survive and reproduce, or they

**FIGURE 2.2 Natural selection:** Evolution is basically the change in gene frequencies from one generation to the next, and it is the differential survival of offspring through natural selection that is the main mechanism for this in evolutionary theory.

What is necessary for the operation of natural selection?

will reproduce at different rates of success. Given the conditions of reproduction, variation, and inheritance, then those individuals that are better adapted to acquiring the resources necessary to survive and reproduce will leave more offspring, and those offspring will carry the feature of their parents that gave them this competitive edge.

These then are the conditions under which natural selection must occur. Each of them is independently derived and each of them is easily and empirically tested. Organisms can be observed reproducing, the mechanisms of inheritance have been worked out, the occurrence of variation can be and largely is extremely well established, and the finite nature of the resources of the world is virtually a truism. Looking at the theory of natural selection in this way shows that far from being untestable, it is in fact a logical necessity deriving from a number of simple observations.

## Evolution

If these conditions occur, then evolution must be a consequence. It is useful to distinguish in this way between evolution and natural selection. Natural selection is the mechanism of change, dependent upon certain conditions. For example, if natural selection favors the existing forms (known as stabilizing selection) then there will be stasis; if, on the other hand, it favors a new mutant form, then there will be evolutionary change (directional selection) (Fig. 2.3). Evolution is the outcome of those

**FIGURE 2.3**
**Stabilizing and directional selection:** Selection acts to structure gene frequencies in populations. When selection maintains the status quo, it is referred to as stabilizing selection; when it changes over time, it is known as directional selection. Sometimes selection may favor two extremes (divergent selection), which can lead to the separation of a population into two.

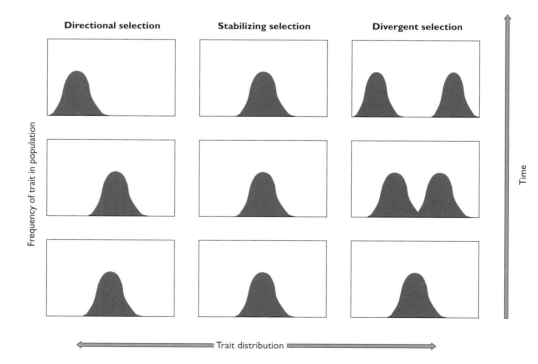

conditions, and evolutionary patterns will vary if those conditions vary. The fact that they do accounts for the enormous diversity of forms of life and the pattern of evolution itself. Some of the conditions – particularly competition – vary more than others. The genetic system, for example, is common to all organisms, and there is a limited amount of variation in how it operates largely dependent upon the presence or absence of sexual systems of reproduction. Genetic systems in fact are not only conditions necessary for evolution, but also major constraints. The particular pattern of inheritance found in all living systems – the Mendelian system – means that inheritance is confined to parents and their offspring. It is not possible to pass on genetic information or characteristics to any individuals other than offspring. If such a thing were possible, then evolution might be expected to occur in a radically different way and perhaps much more chaotically. Indeed, it can be argued that this is the case with cultural change, where items of cultural information can be transmitted from any individual to another.[36,37,38] As a result, what might be referred to as cultural evolution is much more complicated than biological evolution.

## Adaptation

Apart from evolution – change through time – there is another consequence of natural selection. This is the concept of adaptation.[39,35] In its simplest meaning this refers to the goodness of fit between an organism and its environment. The better fitted an individual is to its environment, then the better adapted it is. Adaptation is a consequence of natural selection because it is those individuals that are better adapted to their environment that will leave more offspring, and given the other conditions, then over time a population will come to be adapted to its environment. Without the underlying principle of natural selection, though, there would be no reason to expect an adapted suite of plants and animals.

Adaptation is a key concept in evolutionary theory. One of the implications of the principle of natural selection is that the traits – both anatomical and behavioral – of an organism are adaptive. An adaptive trait is one that helps that individual to survive. For example, the trunk of an elephant is an adaptation in that it enhances the elephant's ability to forage in trees, to eat grass, to communicate, etc. Bipedalism is an adaptation among hominins for moving across open, terrestrial landscapes – the examples are endless. Indeed identifying and explaining adaptations is one of the main themes of evolutionary research in general. However, in its ubiquitous nature, the concept of adaptation has also led to many problems. One such is that an adaptation is always relative – a trunk, for example, is an adaptation relative to some other trait, and specifying this comparative element is often difficult – partly because such alternative adaptations no longer exist. Another problem is that it is all too easy to assume that a trait is adaptive just because it exists – it is necessary to show

What is an adaptation?

how reproduction and survival are enhanced, and this is often difficult, especially when there is only fossil evidence. Despite these difficulties, however, adaptation remains at the core of evolutionary biology, and is used extensively here to explore the pattern of human evolution.[40,41,42]

## Evolutionary explanations

We can see, therefore, that the principle of natural selection – often referred to as differential reproductive success – is both powerful and inevitable given certain conditions, and that evolution and adaptation are outcomes of this mechanism. Natural selection is thus one of the means by which we can seek to explain evolutionary change and the adaptive features that we find in the biological world, and more specifically among humans. While in one sense natural selection is the explanation that can best be used to explain human evolution, it is worth considering briefly exactly what is meant by causality in evolution, and the context in which such explanations are sought.

A comparison with animal behavior will make this clear. In the past there was much conflict within ethology about the various competing explanations for a particular behavior. In a classic paper the Dutch ethologist Nikko Tinbergen showed that in fact much of the conflict arose from the fact that people were answering different questions.[43,44] He proposed that there are four types of evolutionary question, each demanding a different type of answer. First there were questions of immediate causation, which demand a mechanical answer in terms of (usually mechanistic) biological processes; second there were questions on how a particular state of being has come about, which are essentially questions about development; third there are questions about function, which refer to the adaptive value of a trait; and fourth, there were questions about evolutionary history, or what has evolved from what.

We can consider an example in relation to human evolution. In the course of our evolution we have evolved language, taken here to mean actual speech. There are many "evolutionary explanations" of speech, and Tinbergen's questions allow us to distinguish several categories of such explanations (Fig. 2.4). The first type of explanation would address how speech is actually produced, which would require a functional and mechanistic understanding of the anatomy of the oral cavity and the vocal chords, and of the diaphragm, as well as one of the neurobiology of sound production. This would be an explanation in terms of Tinbergen's mechanistic questions. The second type addresses how an individual actually acquires language, which is to do with development, again, both anatomical and neurobiological, in practice occurring in the first three or four years of life. The third type of explanation is directly addressed to the issue of selection and adaptive value – why language evolved, in the sense of what advantage (reproductive and survival) the possession of speech gave

Why are there different types of evolutionary explanation?

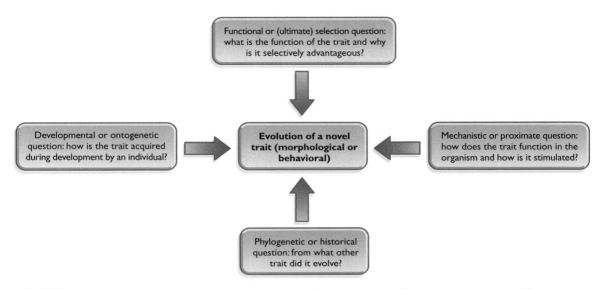

**FIGURE 2.4 Tinbergen's four evolutionary questions:** Evolutionary explanations may address different issues. With respect to speech and language, explanations may be in terms of: understanding the anatomical mechanics of speech production; phylogenetic history; development; and adaptive success. (Courtesy of Robert Foley.)

relative to those that did not have this capacity, or in whom it was less effective. And finally there are explanations in terms of what might be the evolutionary antecedents of speech – whether it is derived from primate vocalizations, or adapted from gestural language, and so on.

As can be seen, each explanation is equally valid, and they do not necessarily compete one with another, but rather address different questions. While each is equally important in some respects, because the functional/ selection explanation is the one that most directly addresses the question of "why" something should have evolved (rather than "how" or "from what"), it is the one that has captured the most interest, and will be at the forefront of much of this book.

## The four Cs of evolution

Tinbergen's four questions are a useful discipline in ensuring that the right evolutionary questions are being asked. For evolutionary theory more generally it is possible to identify four types of factor involved in any particular event (Fig. 2.5). These are the four Cs of evolutionary explanation: conditions, causes, constraints, and consequences.[29] Conditions represent the context in which natural selection occurs, or more particularly the context in which competition is occurring. These may be environmental or social, behavioral or anatomical, and incorporate what many people have referred to as phylogenetic heritage or the pre-existing characteristics of any particular lineage. Causes then refer to the actual selective

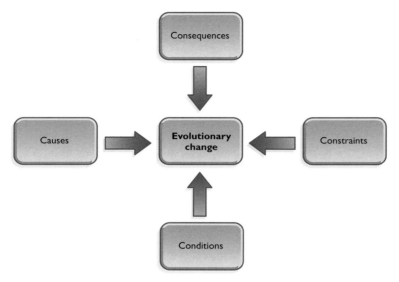

**FIGURE 2.5**
**Conditions, causes, constraints, and consequences in evolution:** In explaining any evolutionary event, it is useful to consider what might be the underlying conditions (such as the environmental context), the causes (the actual selective pressures), the constraints (which may be developmental or from the phylogenetic heritage), and the consequences.

pressures operating to favor particular characteristics of the individual organism, especially those that promote reproductive success. The way in which these selective pressures lead to specific evolutionary changes depends in turn upon the presence of particular constraints. These constraints are the mechanical context in which selection must operate. They may be biochemical, physiological, anatomical, or ecological. Finally, the consequences of the interaction of these three factors is evolutionary change, which itself is a consequence that will feed back into the interplay of the other factors.

It should be stressed that this categorization is heuristic rather than biological. Biological attributes may move from one category to the other depending upon the particular situation. Nothing is inherently one or the other. The point here is that any evolutionary explanation must incorporate all these components, and must recognize the difference between them.

The fundamentals of evolutionary theory provide the major framework for studying humans in a biological perspective. These basic principles have proved to be remarkably robust over the last century and a half, and developments such as population and molecular genetics, ecology, and behavior have, on the whole, strengthened the discipline. However, although these broad principles would be accepted by virtually all biologists, there is nonetheless considerable debate within the subject as to exactly how evolution operates, and the nature of the biological mechanisms involved.

## MODERN EVOLUTIONARY THEORY: THE DEVELOPMENT OF NEO-DARWINISM AND THE POWER OF NATURAL SELECTION

**KEY QUESTION** How have developments in genetics and molecular biology changed evolutionary theory from the original concepts of Darwin, and created the "Modern Synthesis"?

In his *Origin of Species*, Darwin explained the purpose of the book as follows: "I had two distinct objects in view; firstly to show that species had not been separately created, and secondly, that natural selection had been the chief agent of change." Natural selection, Darwin believed, explained how species became adapted to their environments.

The notion that species do, in fact, change through time was already widely debated. Consequently, Darwin readily succeeded with his first goal, given the volume of evidence he presented in *Origin* in support of the reality of evolution. The second goal, showing that natural selection was an important engine of evolutionary change, remained elusive until the 1930s, when it became the central pillar of newly established evolutionary thinking, known as Neo-Darwinism (Fig. 2.6) or the Modern Synthesis.[45]

As described in the previous section, natural selection depends on four conditions. First, members of a species differ from one another (variation), and second, this variation is heritable (heritability). Third, all organisms produce more offspring than can survive (reproduction). (Although some organisms, most notably large-bodied species and those that bestow a lot of parental care, produce few offspring while others may produce thousands or even millions, the same rule applies.) Fourth, given that not all offspring survive, those that do are, on average, likely to have an anatomy, physiology, or behavior that best prepares them for the demands of the prevailing environment (competition). Because of this notion of survivorship in a challenging environment, the principle of natural selection came to be known (inaccurately) as survival of the fittest, even though Darwin did not use that term.

Natural selection, then, is differential reproductive success, with heritable favorable traits bestowing a survival advantage on those individuals that possess them. Generation by generation, favorable traits will become ever more common in the population, causing a microevolutionary shift in the species. Such traits will remain favored, however, only if prevailing conditions remain the same. A species' environment usually does not remain constant in nature. A shift in climate may modify available vegetation, for instance. A new species may enter the ecosystem, or an existing species may disappear. An existing species may itself evolve, becoming a more efficient predator or a harder-to-harvest prey. Each of these changes, and others that can be imagined, may alter a population's adaptive landscape, perhaps rendering a previously advantageous trait less beneficial or making a less advantageous trait more favorable. As a result, the heritable traits that become important in a species' population depend on the prevailing environment: natural selection, or an individual's "struggle for existence," as Darwin put it, is a local process consisting of a generation-by-generation adjustment to local conditions.

The power of natural selection can be seen in the phenomenon of convergent (or parallel) evolution. Typically, when two species share characteristics, it is thought that this reflects shared ancestry – they both inherited the trait from their common ancestor. The shared traits are thus homologies, in that they have a common origin. The basic bones of the

**The hierarchy of evolutionary concepts**

**FIGURE 2.6**
**The history of evolutionary theory:** It is useful to distinguish between *evolution* (which is the actual change in a biological system) and explanations for that change. Darwinism is the basis for modern evolutionary theory (descent with modification + natural selection); Neo-Darwinism (also known as the Modern Synthesis) is the current form of the theory, developed in the middle of the twentieth century (Darwinism + Mendelian genetics).

How did the Darwinian theory of evolution develop?

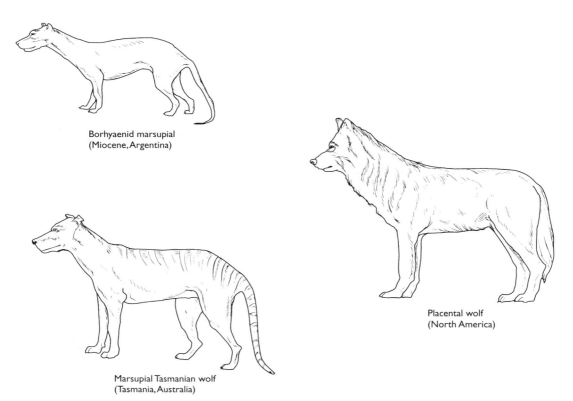

Borhyaenid marsupial
(Miocene, Argentina)

Placental wolf
(North America)

Marsupial Tasmanian wolf
(Tasmania, Australia)

**FIGURE 2.7 Convergent evolution:** The power of natural selection is seen in its ability to produce similar morphologies in widely different species. Here we see a Miocene hyena from South America (a marsupial mammal), the Tasmanian wolf (a marsupial mammal), and the North American wolf (a placental mammal). Although marsupial and placental mammals diverged more than 100 million years ago, their morphologies have become very similar through similar adaptations as large, terrestrial carnivores. The Tasmanian wolf is closer evolutionarily to the kangaroo than it is to the North American wolf.

Why is convergent evolution important to evolutionary theory?

vertebrate body plan are excellent examples of homologies. Convergent evolution, on the other hand, is where distantly related species come to resemble one another very closely by adapting to similar ecological niches. Indeed, Simon Conway Morris of Cambridge University has argued that the extensive nature of convergence in the biological world is in fact the strongest evidence for the power of natural selection. If selection was not a major force, then the same solutions to survival problems would not be found so frequently. The anatomical similarity of the North American wolf and the Tasmanian wolf is a good example (Fig. 2.7). The former is a placental mammal and the latter is a marsupial, making the two species extremely distant genetically, having been evolutionarily separate for at least 100 million years. Likewise, many bird species of Australia were assumed to be close relatives of anatomically similar species in Europe and Asia until recently gathered genetic evidence revealed that they were quite separate. Once again, the anatomical similarities reflected

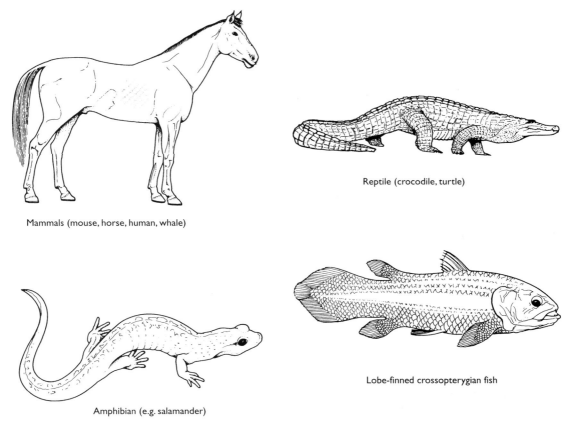

Mammals (mouse, horse, human, whale)

Reptile (crocodile, turtle)

Lobe-finned crossopterygian fish

Amphibian (e.g. salamander)

**FIGURE 2.8 The principle of historical constraint:** Evolution is in many ways a conservative process. The preservation of a four-limbed body over vast tracts of time and through very different environmental circumstances illustrates the power of historical constraint. For example, the horse has four legs not just because it is a very efficient way of moving about on dry land but because its fish ancestors also had four appendages.

convergent evolution, or analogy – not shared ancestry. Anatomical similarities that result from shared ancestry are examples of homology. Homologous structures are especially important in the reconstruction of evolutionary history on the basis of morphological characters.

Natural selection should not be envisioned as being able to shape any anatomical form according to the demands of the environment, however. There are no animals with wheels, for instance. In addition to physical constraints, evolution is limited by historical constraints (Fig. 2.8). As the French biologist François Jacob put it, evolution works like a tinkerer – it works with what is available – not like an engineer, who would start from scratch with each new design.[46] Moreover, changing morphology in one part of the body usually produces correlated changes in another part that are unrelated to adaptation. Fundamentally, these constraints are associated with the limited number of pathways by which embryonic development can operate, both in absolute terms and at any point in the history of a particular lineage.[47]

## Establishment of population genetics

Darwin was well aware that members of a species vary, and that these variations are heritable: his observations of natural populations and his experiments with domestic breeding were proof of that ability. He was not familiar with the basis of inheritance, however. Although the rules of inheritance were discovered by the Austrian monk Gregor Mendel in the early 1860s, the results of his work remained generally unknown until the turn of the century, two decades after Darwin's death.[45]

From observations on the progeny from experimental crossing of pea plants, Mendel discovered that physical traits are determined by stable inheritance factors (what we now call genes). He also found that each plant has two genes for each trait, one from the female parent and one from the male. The variants of each gene, known as alleles, may be identical (in which case the individual is homozygous) or different (the individual is heterozygous). When the two alleles differ, one form may be dominant and the other recessive; this means that the allele that will actually determine the phenotype (what the trait looks like in an individual) is the dominant one, while the recessive one will only affect the phenotype when it is in a homozygous condition (that is, both alleles, from each parent, are recessive; this is the case with many rare, genetically determined diseases). Mendel also found that, as alleles are passed from generation to generation, they remain intact and do not blend their effects. Gametes, or sex cells, receive one or the other of the two alleles with equal probability.

Mendel's experiments were very simple from a genetic standpoint, with just one or two genes affecting one trait. This choice was fortunate, because otherwise the outcome of the crosses would have been far too complex to be understood at the time. Thanks to the simplicity of the system, Mendel's insights were able to lay the foundation for the new science of genetics. Before long it became apparent that most traits are influenced by many genes, not just one or two. Nevertheless, the system was amenable to mathematical analysis, and the selection of favored traits (the phenotype) could be studied in terms of the selection of genes that underlay them (the genotype). This approach formed the foundation of modern population genetics.

## Microevolution and the Modern Synthesis

The change in frequency of particular alleles within a population as a result of natural selection on them provides the basis of microevolution, which in turn is the central dogma of the Modern Synthesis (see Fig. 2.6 above).

If the range of alleles available within a species' population remains constant, the scope of evolution will necessarily remain limited. From time to time, however, the sequence of chemicals on the DNA – the molecule on which genetic information is coded – that represents the information encoded in a gene becomes changed, often when a mistake occurs as the gene is copied within the germline. Such a mutation introduces the potential for further genetic variation within the population.

No simple relationship exists between a mutation and the degree of phenotypic change it might produce. For instance, a single base mutation in the gene of a serum albumin might marginally modify the physical chemistry of the blood, perhaps with some impact on adaptation or perhaps not. On the other hand, a similar mutation in a gene that affects the timing of the program of embryological development might have dramatic consequences for the mature organism. The slowing of embryological development and subsequent prolongation of the growth period, a phenomenon known as neoteny, was apparently important in the evolution of humans from apes (Fig. 2.9). This "uncoupling" of magnitude of mutation from magnitude of phenotypic change produced has clear consequences for the ability to infer genetic distance from degree of morphological difference.

The fate of mutations, and therefore their importance in future evolution, was the topic of intense debate in the early years of population genetics. (In this discipline, it is important to distinguish between the mutation rate of a gene, which may be quite common, and the retention, or fixation, of those mutations in the species' populations, which is much less common and dependent upon factors such as genetic drift and selection.) Initially, the process of mutation itself was suspected to fuel evolutionary change, with the role of natural selection envisioned as simply the removal of deleterious alleles. Consequently, evolutionary change was viewed as the outcome of mutation pressure, an internal force that drove evolution constantly forward. This concept was not what Darwin had described. In Darwinian evolution, natural selection acted to retain beneficial traits (alleles) and was therefore a creative process, not just a cleaning-up process that eliminated disadvantageous traits (alleles).

Until the mid-1940s, evolutionary theory remained distinctly at odds with strict Darwinism, and many different views were put forth to explain how the pattern of life was shaped. Then, following the creative melding of natural history, population genetics, and paleontology, a consensus of sorts appeared, known as the Modern Synthesis. This theory encompassed three principal tenets. First, evolution proceeds in a gradual manner, with the accumulation of small changes over long periods of time. Second, this change results from natural selection, with the differential reproductive success founded on favorable traits, as described earlier. Third, these processes explain not only changes within species but also higher-level processes, such as the origin of new species, producing the great diversity of life, extant and extinct.

What is the role of mutation in evolution?

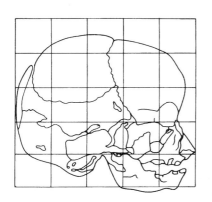

Chimp fetus

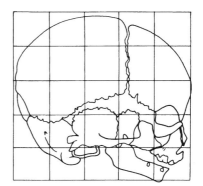

Human fetus

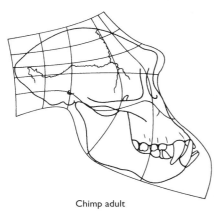

Chimp adult

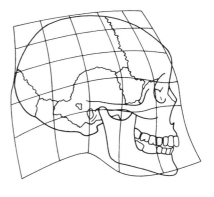

Human adult

**FIGURE 2.9 Neoteny in human evolution:** Although the shape of the cranium in human and chimpanzee fetuses is very similar, a slowdown in development through human evolution has produced adult crania of very different forms, varying principally in the shape of the face and the size of the brain case. The changes in grid shapes indicate the orientation of growth.

## The extent and source of variation

Nevertheless, strong differences of opinion remained among evolutionary theorists, particularly over the extent and source of variation at the genetic level. One school argued for little variation, the other for a lot. The former group believed that most mutations were deleterious and were therefore removed; the inimical genetic variation within the population was viewed as genetic load. The latter group argued that many mutations offered advantages for survival in different environmental circumstances and would therefore be retained among different populations of the species. Although variation could be discerned between individuals at the phenotypic level – that is, in their anatomy – its relationship to differences in the genes and their initial products, proteins, remained unclear.

Not until the mid-1960s could this issue be addressed directly, when techniques (such as gel electrophoresis) were developed to measure genetic variation at the level of certain differences in protein structure. These investigations revealed that genetic variation is extensive. In most natural populations, at least 40% of genes are polymorphic (that is, they exist as several alleles); any individual is heterozygous for approximately 14% of its gene loci. This insight prompted questions about the origin of that variation, and the ensuing discussion evolved in the 1970s and beyond into what came to be known as the neutralist–selectionist debate.

Specifically, the selectionist school argues that genetic variation is the product of natural selection, which selects favorable new variants. By contrast, proponents of the neutral theory of molecular evolution contend that the great majority of variants are selectively neutral.[48] Such variants are therefore invisible to natural selection and will be neither removed nor favored, instead accumulating passively. Under the neutral theory, the accumulation of genetic variants is driven principally by the rate of mutation itself, and their eventual frequency in a population is determined by chance. (Although this idea may sound similar to the notion of mutation pressure mentioned earlier, it differs in two important ways: most variants are viewed as selectively neutral, not deleterious, and the postulated process leads to high levels of genetic variation in populations, not low levels.)

*Why is variation essential to evolutionary process?*

Today, most evolutionary theorists agree that a significant proportion of variation at the genetic level represents selectively neutral alleles. Thus, certain traits in a population may become common, or even dominant, by chance processes, known as genetic drift. Chance sorting of neutral alleles over many generations may lead them to become common, just as the toss of a coin may occasionally produce "heads" five times out of six. Remember that this scenario assumes that many alleles (and traits) are selectively neutral. If, through a change of environmental conditions, such traits lose that neutrality, they will then be either removed or favored by natural selection, depending on whether they are advantageous or disadvantageous. Genetic drift – random changes in gene frequencies – therefore must be added to natural selection as a source of phenotypic and genetic variation within populations.

A few species exist as a single population or a few populations in close contact with one another. In these cases, the frequency of alleles within the population or populations effectively represents the profile of genetic variation of the species as a whole. In many instances, however, species exist as many populations distributed over numerous separate geographical areas that vary ecologically. As a result, each population has the opportunity to develop its own distinct genetic profile, through adaptation to local conditions and by genetic drift. This distinctiveness leads to the regional variants of the species. The greater the effective separation (by geographical barriers, such as river systems or mountains, for instance, or a lack of mobility of the species' members), the greater the genetic difference that will develop. Gene flow among populations, through the migration of individuals between the populations and interbreeding within them, will reduce genetic differences among populations as the alleles from the separate populations become mixed. Species whose geographical populations display distinct genetic profiles are said to be polytypic.

## Other aspects of selection

In his second major work on evolution, *The Descent of Man and Selection in Relation to Sex*, which was published in 1871,[19] Darwin described a variant of natural selection that has an important impact on the anatomy and behavior of many species. Where natural selection adapts an organism for survival in its environment, sexual selection adapts it to the needs of obtaining a mate (Fig. 2.10). In many species, the females select their mates, and the male's role is to be as attractive as possible. In birds, for instance, this consideration often leads to males having brightly colored plumage, which they show off in elaborate displays. In other species, such as red deer and impala, males fight one another for the possession of a herd of females. (Even in these cases, females often exercise some choice of whether to stay with the herd "owner.") Male–male competition often leads to a larger body size in males (sexual dimorphism of body size) and enhanced fighting equipment, such as large antlers. In primate societies, this type of competition (plus social skills) is common and has usually resulted in the evolution of increased body size in males, as well as enlarged canine teeth. This pattern is also observed among hominins, although canine size is not exaggerated in males, probably for reasons of diet.

FIGURE 2.10 **Sexual selection:** The struggle for access to females often involves aggressive competition among males. This competition is reflected physically in male anatomy – for example, large antlers in deer, large canines in primates, and large body size in many groups, including primates. Here, the great difference in body size observed between male and female deer provides an example of sexual selection.

Natural selection works at the level of the individual, in its struggle for existence in competition with other members of its own species (Fig. 2.11). This essential selfishness of the processes therefore raises questions about apparently altruistic behaviors in social species. For instance, worker bees forgo their own opportunity for reproduction in order to help the queen, their mother, raise more sisters. In some bird species (such as the Florida scrub jay), young males often help provision the young of other individuals, rather than raise their own. Vervet monkeys give alarm calls when specific predators are present.[49] If natural

selection demands selfish behavior so that an individual can maximize its reproductive success, why are behaviors that compromise this Darwinian imperative so commonly observed in social species?

One answer has been that selection may also act at the level of the social group, via a process known as group selection. According to this argument, if a group's members act altruistically toward one another, then they may be more successful than other groups whose members react hostilely toward one another. The notion of group selection enjoyed considerable popularity until the mid-1960s, pro-

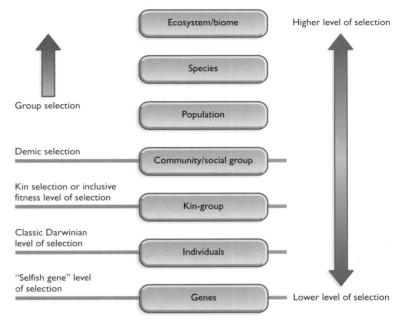

**FIGURE 2.11 Levels of selection:** In theory, natural selection can occur at any level of organization. What determines whether it does depends upon whether the conditions outlined in the text (variation, inheritance, reproduction, and competition) occur at that level. In evolutionary theory there has been considerable debate as to which levels operate through selection. The higher the level, the greater the controversy.

moted by V. C. Wynne-Edwards in his book *Animal Dispersal in Relation to Social Behavior*.[50] Soon, however, explanations of apparently altruistic behavior began to be formulated in terms of individual interests, not group interests. Principal among these concepts were kin selection, proposed by Oxford University evolutionary biologist W. D. Hamilton, and reciprocal altruism, proposed by Robert Trivers, then at Harvard University.[51,52]

Hamilton pointed out that an individual's reproductive success is measured in terms of passing one's genes to the next generation. Because an individual shares half its genes with its siblings (and a smaller proportion with increasingly distant relatives), acts of altruism toward siblings and other relatives are, in effect, promoting one's own genes. An individual's fitness should therefore be measured in terms of the genes it passes directly to the next generation and those of its relatives. This idea is known as inclusive fitness. On closer examination, many acts of apparent altruism came to be viewed as kin helping relatives, which did not require a group selection interpretation.

Trivers extended this notion to non-kin, however. In long-lived, highly social creatures, individuals are able to help non-kin, with the expectation that the favor will be returned. In this case, alliances may be formed between non-kin, based on the principle that "I'll scratch your back now because you scratched mine a while ago, and I expect you to scratch mine again soon." This kind of behavior is commonly observed among large primate species.

Because kin selection and reciprocal altruism successfully explained apparently altruistic behaviors in terms of individual – not group –

What do sexual selection and kin selection add to evolutionary theory?

interests, the theory of group selection fell out of favor. Just recently, however, the theory has been revived in a more critical formulation than that of the earlier, naive explanations and has been applied to modern human behavior. This latter association is particularly problematical because of the heavy overlay of culture in modern human behavior.[53]

This section has shown how Darwin's basic idea of adaptation through selection has been extended and modified through the development of population genetics and the growth of more sophisticated ideas as to how and at what levels adaptation might occur. Throughout this whole period, though, there has been a consistent acceptance of the importance of adaptation and selection as primary mechanisms of evolutionary change. However, as the brief discussion of the selectionist–neutralist debate indicated, this has not been universally held, and these debates have influenced the way in which human evolution has been investigated (Fig. 2.12). Over this period there have been a number of challenges to Neo-Darwinism. These are not the challenges mounted by creationists against the whole

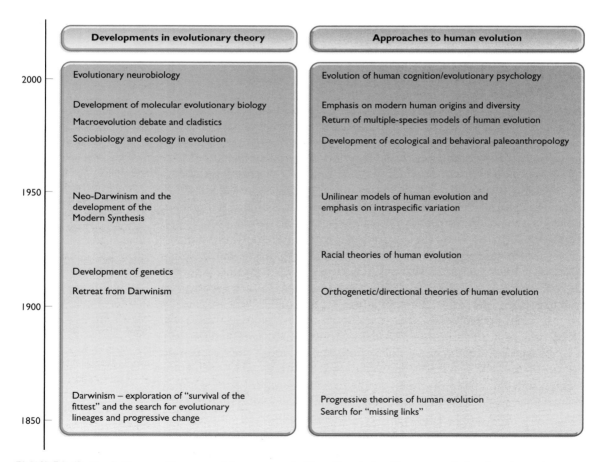

FIGURE 2.12 **Evolutionary theory and human evolution:** The study of human evolution has always been influenced by what has been happening in the broader field of evolutionary theory.

theory of evolution, but technical challenges on the question of whether selection is the only or the major process underlying evolutionary change. It is this question that will be discussed in the following chapter.

## BEYOND THE FACTS
# IS LAMARCKISM DEAD?

*The issue: we know that Darwinism triumphed where Lamarckism did not, as discussed in this chapter. However, Lamarck raised a number of important questions about evolutionary mechanisms, not all of which have been resolved, and which may be particularly important in human evolution and in the light of molecular mechanisms.*

Whenever the history of evolutionary ideas is recounted, one central element of it is the way in which the ideas of Wallace and Darwin triumphed where those of the great French biologist Lamarck had failed. Natural selection rules and the evolution of acquired characteristics, the ideas most associated with Lamarck, are treated with a gently patronizing air – nice try, but not good enough. In fact there are two parts to Lamarck's ideas. One was that the driving force of evolution came from within the organism, rather than from external factors. The challenges of the environment may have set up the condition in which evolution could occur, but the response came from inside the organism – its inner need, as it were, to survive and adapt. From this came the second component of Lamarckism, the inheritance of acquired characteristics. Through this inner need individuals acquired – absorbed into their biology – the adaptive traits that would help them to survive, and these would be passed on to their offspring.

Lamarckism collapsed in the face of the clear evidence that the biological information that an individual acquired during its life could not be passed back to the genes, and so could not be transmitted to the next generation. There is little to challenge this in the main corpus of biology, and so the Darwinian dogma remains to this day. In light of this, there may be little reason to look back at Lamarck except as a figure of historical curiosity. However, with any historical idea it is important to focus on the general principles, rather than the details. The scientists who failed to come up with the right idea may have lost out in some ways, but that does not mean that the underlying principles were entirely wrong. In Lamarck's case there are two general principles that are important.

The first is the question of whether evolution is driven by internal characteristics of the organism, or by the environment. Darwinism of course showed that the latter was of major importance, and Neo-Darwinism emphasized that because the source of variation internal to the organism – mutation – was random, therefore this was not an important element. While modern biology has largely supported this view, it is also clear that in the emerging field of developmental molecular genetics, the processes occurring internally are more complex than the term "random" suggests, and that an understanding of these mechanisms, some of which may occur across generations, needs to be taken into account.

The second is the question of acquired inheritance. Certainly for anything related to classical genetics, this is not a sustainable idea. However, the theory of evolution is a broad one, and applies not just to genetic change. Effectively, Mendelian genetics constrains biological evolution to a Darwinian rather than a Lamarckian mode. However, where Mendelian genetics does not operate – for example, in cultural inheritance, in certain aspects of immunology, in aspects to do with maternal condition – it appears there may be a chink in the Darwinian armor through which some element of Lamarck may enter evolutionary biology.

# CHAPTER 3

# PATTERN AND PROCESS IN EVOLUTION

FROM MICRO- TO MACROEVOLUTION:
DEBATES IN MODERN
EVOLUTIONARY THEORY

**KEY QUESTION** To what extent is the unitary and microevolutionary theory of the Modern Synthesis sufficient to explain all evolutionary patterns?

The Darwinian theory of evolution is often described as a unitary theory. This term reflects the fact that one mechanism, albeit operating under variable conditions and constraints, accounts for all aspects of evolutionary change. Another way of putting this is that evolutionary mechanisms, as we have seen, are microevolutionary – that is, they operate at the level of genes and individuals, in populations, over short-term generational periods. Long-term evolution, or macroevolution, is nothing more than the repeated outcome of microevolutionary processes. This unitary nature of the theory has been seen as both a great strength and a great weakness of the subject. The central question is whether natural selection can account for all aspects of evolutionary change, or whether there are other mechanisms in play.

## Species and speciation

Criticisms of the unitary theory of Neo-Darwinism have come from two major directions. The first of these is that of molecular biology, and from genetics more generally. As we have seen in the previous chapter, these relate to whether variation is under the influence of natural selection, or whether it is neutral, not expressed in the phenotype, and more subject to the chance effects of drift in small populations than to the deterministic

| SPECIES CONCEPT | CHARACTERISTICS OF SPECIES |
|---|---|
| Biological species concept (Mayr) | Groups of actually or potentially interbreeding natural populations which are reproductively isolated from other such groups |
| Specific mate recognition concept (Paterson) | Members sharing a specific mate recognition system to ensure effective syngamy within a population of organisms |
| Cohesion species concept (Templeton) | The most inclusive population of individuals having the potential for phenotypic cohesion through intrinsic cohesion mechanisms |
| Evolutionary species concept (Simpson) | A lineage evolving separately from others and with its own unitary evolutionary role and tendencies (and historical fate) |
| Ecological species concept (Van Valen) | A lineage which occupies an adaptive zone minimally different from that of any other lineage in its range and which evolves separately from all lineages outside that range |
| Phylogenetic species concept (Cracraft) | The smallest diagnosable cluster of individual organisms within which there is a parental pattern of ancestry and descent |

FIGURE 3.1 **Species concepts:** Although the idea that species are defined as reproductively isolated groups (the biological species concept) is widely accepted and used, nonetheless there are both problems with this concept and various alternatives in existence.

effects of selection. The other direction in which the power of selection has been questioned is from paleontology. The essence of the argument here is that the pattern of major changes, the formation of new species and trends among related species, etc., are not the product of natural selection. The key argument here is that while natural selection is a powerful mechanism below the level of the species, it does not account for what may be occurring above that level – that is, at the macroevolutionary level.[54]

It is useful at this point to define a species, an issue that has long taxed biologists' ingenuity (Fig. 3.1). The most commonly used definition is known as the biological species concept, which was developed in 1942 by the Harvard biologist Ernst Mayr:[55] "Species are groups of actually or potentially interbreeding natural populations, which are reproductively isolated from other such groups." An alternative definition, proposed in 1983 by Joel Cracraft of the University of Chicago, is the phylogenetic species concept:[56] "A species is the smallest diagnosable cluster of individual organisms within which there is a parental pattern of ancestry and descent." The first concept focuses on behavioral relationship, the second on evolutionary relationship. Other definitions of species also exist, each of which focuses on a different element as being the most significant. George Gaylord Simpson, another of the architects of the modern synthesis, emphasized the idea of a species as an evolving lineage, recognizing the importance of continuity through time.[57] Colin Paterson, then a biologist at the Natural History Museum in London, presented the idea of "species

Why are species difficult to define?

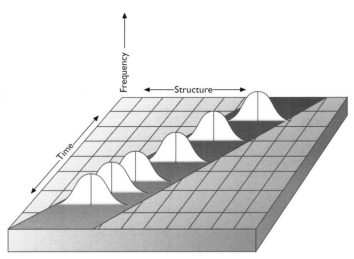

**FIGURE 3.2 Phyletic gradualism:** Each of the bell curves represents the frequency distribution of genes for a certain structure within populations at different times in history. As time passes, the distribution shifts, changing the morphology of those structures. At some point, the nature of the structures in descendant populations is sufficiently different from that in earlier populations to deserve recognition as a new species. Thus, species 1 is gradually transformed into species 2. This pattern of phyletic gradualism is often called a "march of means." Once considered the principal mode of evolution, phyletic gradualism is now considered to be of less importance than cladogenesis.

mate recognition" – that is, that species should be defined on the basis of the traits that actually signal reproductive compatibility.[58]

The reason there are so many definitions of a species is that it is an extremely difficult unit to define theoretically in a way that allows both continuity through time – an essential component of evolution – and clear-cut demarcation – an equally essential component when dealing with issues of reproductive isolation. There are also practical difficulties. For biologists dealing with living species, even the apparently solid definition of the biological species concept is not always as straightforward as might be imagined, as the question of producing fertile offspring is often a gradual rather than instantaneous one. For paleontologists dealing with extinct species, the difficulties are obvious: you simply cannot know whether one fossil specimen was able to breed with another specimen that it resembled physically. Evolutionary relationships among fossil species must be based on a logical system of resemblances.

The definition of "species" is important if we are going to ask how they arise. New species may arise in two ways. First, an existing species may be transformed by gradual change through time, so that the descendant individuals are sufficiently differentiated from their ancestors to be recognized as a separate species (Fig. 3.2). This mode is known as anagenesis (Fig. 3.3), and it results in one species evolving into another over time. Second, a population of an existing species may become reproductively isolated from the parent species, producing a second, distinct species. This mode, known as cladogenesis, comprises a splitting event, yielding two species where previously only one existed, thus increasing the diversity of existing species. This process has obviously been important in the history of life, because the fossil record shows that biodiversity has increased steadily (with fluctuations and occasional mass extinctions, as discussed later in this chapter) since complex forms of life evolved, a little more than half a billion years ago. (Cladogenesis is also called speciation, which is a somewhat confusing term because anagenesis also produces a new species; speciation applies only in cases involving evolutionary branching, not simply transformation.)

On a shorter time scale, cladogenesis plays an important role in adaptive radiation. Adaptive radiation is a characteristic pattern of evolution following the origin of an evolutionary novelty, such as feathered flight (for birds), placental gestation (for eutherian mammals), or bipedal locomotion (in hominins). The original species bearing the evolutionary novelty very quickly yields descendant species, each representing a variant on the new adaptation. Under certain circumstances, more new species arise than existing species become extinct. It is adaptive radiations that give the characteristic "bushy" shape to evolutionary trees. The result, drawn graphically, is an evolutionary bush, with an increasing number of coexisting species through time that have all descended from the same ancestor. The sum total of descendants of that common ancestor is known as a clade – hence the term "cladogenesis."

According to Ernst Mayr, cladogenesis is most likely to occur when a small, peripheral population of a species is separated from the parental population.[55] Small populations contain less genetic variation and are less stable genetically than large populations, and they can change much faster as new variants arise. Such small populations may become established in one of several ways, such as through the origin of new physical barriers, the colonization of islands, or the rapid crash of a subpopulation to small numbers. When a small population becomes established in one of these ways and then expands, it exhibits what is termed as a founder effect. A founder population that gives rise to a new species in separation from other populations of the same species produces allopatric speciation ("allopatric" means "in another place") (Fig. 3.4). Allopatric speciation is the most common means by which new vertebrate species arise. Under the rare condition in which a new species arises from a subpopulation that is not separated from the main population, the process is termed sympatric speciation ("sympatric" means "in the same place").[32]

The origin of species is therefore normally a process in which geographical isolation leads to the formation of two or more clades (populations), and over time these become distinctive morphologically and behaviorally, and genetically distant to the point where they can no longer interbreed. On this there is general agreement. However, in terms of the mechanism that produces this outcome there is considerable debate.

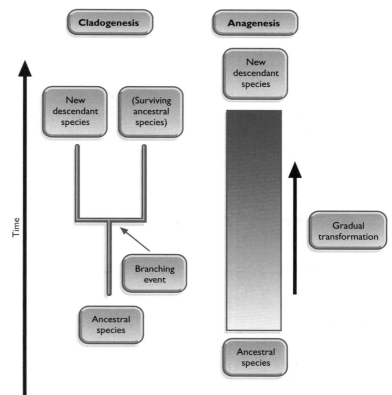

**FIGURE 3.3**
**Anagenesis and cladogenesis:** When a new species evolves by gradual change within a lineage, it is referred to as anagenesis; when branching occurs in the formation of new species, it is referred to as cladogenesis. Most theoreticians consider cladogenesis to be the major form of new species evolution.

Is geographical isolation essential to speciation?

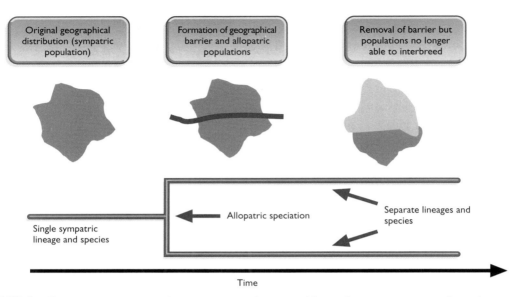

**FIGURE 3.4 Allopatric speciation:** The most commonly accepted form of speciation occurs when a barrier of some form comes between members of a population/species, such that interbreeding is no longer possible and divergence occurs. Where speciation occurs with no such barrier it is referred to as sympatric speciation.

## Micro- and macroevolution

On the one hand there is the perspective that all macroevolutionary change is a product of microevolutionary change – the Neo-Darwinian view. Speciation would be the product of natural selection operating on these small populations, producing minor (or perhaps major) differences in adaptation. The Modern Synthesis proposes that macroevolution is simply an extrapolation of microevolutionary processes; that is, gradual evolutionary change of a population is driven by natural selection, resulting eventually in a new species. Because this process is driven by the gradual process of natural selection, it creates new adaptations that, when sufficiently different from those in the ancestral species, lead to a new species that is characterized by those adaptations. Within this framework either an entire species may be transformed into something new (anagenesis, although usually only through a small demographic bottleneck), or else it can occur through the splitting of populations and the formation of new species as well as the persistence of the existing one (cladogenesis).

In principle, this transformation should be evident in the fossil record, whether anagenesis or cladogenesis is the end result. However, such occurrences are relatively rare, particularly for very ancient parts of the fossil record. Often the new species appears abruptly, either replacing the parental species or appearing concurrently with it, with no transitional forms present.

It is the lack of transitional forms that has led to the macroevolutionary debate. The lack of transitional forms in the fossil record worried Darwin, and he explained the absence as the result of the incompleteness of the record. (The fossil record does not capture year-by-year change but, because of gaps, gives incomplete snapshots of the passage of long-separated periods of time. It is analogous to taking a photograph of a football game every 10 minutes, and then trying to piece together the flow of action of the entire 90 minutes from 10 frames.) There is little doubt that the fossil record does contain many gaps, and this remains an important empirical problem, but in the early 1970s, Niles Eldredge, of the American Museum of Natural History, and Stephen Jay Gould, of Harvard University, challenged this interpretation. On the basis of their work related to various fossil species, they argued that, incomplete though the fossil record may be, it presents an accurate view of the tempo of evolutionary change. Instead of undergoing continual, gradual change, species remain relatively static for long periods of time; when change comes, it occurs rapidly ("rapidly" means a few thousand years). Apart from rare occasions in unusual geological circumstances, the bursts of change go unrecorded in the fossil record because they are rare in geological terms and occur over short time spans. Eldredge and Gould gave this tempo of evolution – that is, long periods of stasis interspersed with brief intervals of rapid change – the name of punctuated equilibrium[59,60] (Fig. 3.5).

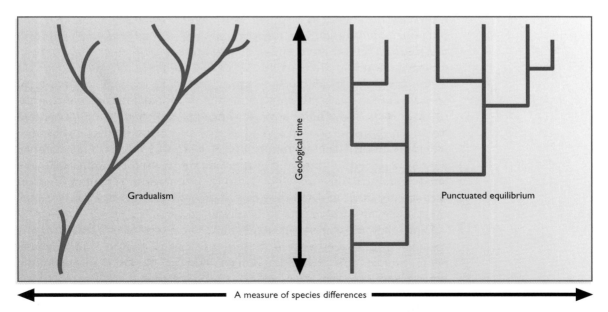

**FIGURE 3.5 Two modes of evolution:** Gradualism and punctuated equilibrium. Gradualism views evolution as proceeding by the steady accumulation of small changes over long periods of time. In contrast, punctuated equilibrium sees morphological change as being concentrated in "brief" bursts of change, usually associated with the origin of a new species. Evolutionary history reflects the outcome of a combination of these two modes of change, although considerable debate has arisen as to which mode is the more important.

An important difference between punctuated equilibrium and the traditional explanation of species formation relates to the nature of change that occurs at that time. According to the Modern Synthesis, the gradual accumulation of new adaptations brings about the genetic separation of a daughter species from the ancestral species. Under punctuated equilibrium, this scenario is not necessarily the case, although it can occur. Like Mayr, Gould and Eldredge argued that speciation is most likely to arise in allopatric populations. Such populations may become unable to interbreed (that is, become reproductively isolated) through genetic changes due to drift in the smaller, isolated population. Once populations are reproductively isolated in this manner, distinctive adaptations may then accumulate. Thus, the Modern Synthesis saw adaptation as the cause of speciation, whereas punctuated equilibrium sees it as a potential consequence.

Much of the heat of the punctuated equilibrium debate has been dissipated in recent years. On the one hand, it can be seen that the architects of the Modern Synthesis, especially G. G. Simpson, fully recognized the importance of variation in the tempo of evolution as part of the evolutionary process (indeed Simpson wrote a book called *The Tempo and Mode of Evolution*[57] in which he defined different rates and discussed the implications). Empirically it is clear that there is evidence for both modes of change, and that the problem lies in the way in which the term "gradual" became synonymous with the term "constant." Evolutionary rates are clearly variable, under both selection and other forces. On the other hand, it is clear that the focus on small, peripheral populations emphasized by the punctuated equilibrium theorists does imply that the potential for drift is significant, and therefore that adaptation may not be the lead factor in differentiation.[6]

**Is evolutionary change gradual or sudden?**

The first issue in macroevolution is that of speciation, and we have seen that while this is not simply a matter of gradual and constant change, neither is it something entirely unconnected with microevolution. However, there is a second and equally interesting issue that is central to macroevolution – that of trends within groups of species. For example, over long periods of time, the horse clade shows an increase in body size and a decrease in the number of toes (Fig. 3.6). A second example involves the increase in brain size during human evolution, at least once the genus *Homo* had evolved, some 2.5 million years ago.

How might these trends be explained? The classic Neo-Darwinian perspective would see the trend in terms of cumulative selection, which, because it is cumulative, leads to longer-term directional change.[61] For the horse clade, larger bodies and fewer toes were sequentially adaptive – more efficient ways of being a horse. Similarly, the brain-size increase evident with the first species of *Homo* is often described as the beginning of brain enlargement, as if it were a progressive change that was nurtured steadily by natural selection. Through the lens of the modern synthesis, while there is no inherent reason why change should be directional, and indeed it is known to reverse, on occasions it will be so and will produce

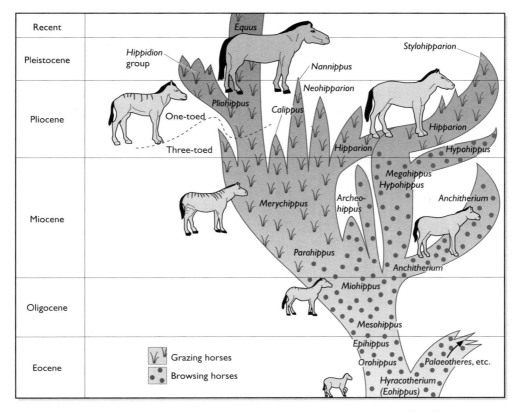

**FIGURE 3.6 Evolutionary trends:** The evolutionary history of horses was once considered as a series of evolutionary trends (to larger body size, more complex teeth, and fewer toes) that marked steady, directional progression. In fact, the evolution of horses is more like a bush than a directional ladder. The differential survival rates of certain species with certain characters gives the impression of steady progression, but does not represent reality.

progressive evolution. Punctuated equilibrium, however, provides a different explanation.

If, as noted earlier, species persist unchanged for most of their duration, then evolution is not directional in this sense. Trends within groups are viewed as the outcome of the nature of the species themselves and their differential tendency to go extinct. Many factors can influence species' tendencies for extinction (and speciation) because the two trends are linked, but biologists are only beginning to scratch the surface of this phenomenon.

One such factor is the nature of a species' adaptation. The fossil record shows that species with highly specialized environmental and subsistence requirements are more likely to speciate and become extinct than those with much broader adaptations. The reason is that any change in prevailing environment is likely to push specialists beyond the limits of their tolerances, promoting both speciation and extinction. Clearly, generalists can

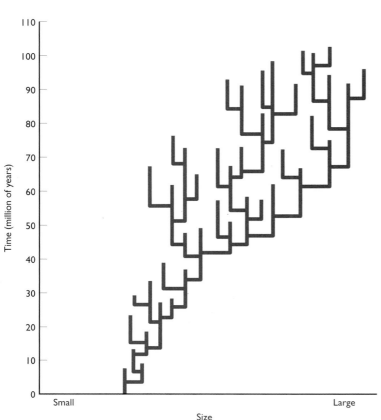

**FIGURE 3.7 Species-selection hypothesis of macroevolutionary trends:** In this hypothetical evolutionary history of a clade, the greater longevity of larger body size results in a directional trend of body size within the clade through differential survival. Note that no trend toward larger body size exists within individual species.

accommodate much broader shifts in conditions, making speciation and extinction rarer for them. The principal point is that species often become extinct for reasons unrelated to how well they are adapted to a particular habitat.[62,63]

Suppose, for example, that horse species with large body size survive longer, for some reason. The differential survival rates of species along these lines would produce a trend toward larger horses, not because it made better horses in the sense of adaptation but as a consequence of the properties of species. Similarly for hominin species and large brain size, no persuasive evidence indicates increasing encephalization within species, merely a trend toward larger brain size within the clade. If large brain size endowed species with greater longevity, a history of encephalization within the group would result.

This pattern, which has been called species selection (Fig. 3.7), is analogous to Darwinian natural selection within species. The differential survival rates in this case relate to species, not individuals within a species, and the evolutionary change that takes place occurs within a group of species, not within a single species.

More than the issue of rates of evolution, the question of whether there are levels of selection operating on species as groups is a major challenge to classical models. As was stated at the beginning of this chapter, the modern synthesis is unitary in mechanism, whereas this model derived from punctuated equilibrium can be described as hierarchical or pluralistic – different mechanisms operating at different levels. Again, the intensity of this debate has reduced over the last few years as it has become recognized that most of the species-selection models can be reformulated more simply in terms of natural selection (for example, horse species with large body size survive longer because the individuals concerned can maintain better differential reproductive rates) and with the growing emphasis on developmental genetic mechanisms which operate on individuals leading to major change.[47] However, the debate was important in focusing evolutionary biologists' attention on how long-term patterns could be integrated with short-term processes. What is important to recognize is that macroevolutionary patterns do exist and require explanation – whether there are macroevolutionary processes remains controversial.

Debates about evolutionary process and pattern are important for paleoanthropolgy,[30,64,65] for they are the context in which human evolution must be assessed and explained. As discussed in earlier chapters, the impact of evolutionary thinking has been to challenge whether humans are unique, and this applies as much to the process by which they evolved as the outcome itself. For example, in thinking about the shape of human evolution, the issues of population size and breadth of adaptation are important factors in assessing how many hominin species might have existed at any one time. The typical shape of evolution through time for any group is bushy, with multiple species existing at any point, rather than linear, with just one species existing at any one time. Hominins and horses are unusual in nature in that each group is represented in today's world by a single genus. The fossil record of horses has shown, however, that this group was once a luxuriant evolutionary bush. The fossil record of hominins, once notable for its ladder-like simplicity, is becoming increasingly bushy, and in that sense is looking more and more like any other mammalian group. However, the fact that there is currently only one species of hominin becomes a matter of major theoretical interest.

Are there mechanisms in evolution beyond natural selection?

## THE PHYSICAL CONTEXT OF EVOLUTION

Regardless of the precise mechanism involved, an important question is what drives evolutionary change in general. Three factors are recognized as influencing the evolution of new species and the extinction of existing species. First is the inherited properties of a lineage (the historical constraint discussed in earlier chapters). Second is the biotic context – that is, the interactions between members of a species and between different species, principally in the form of competition and associated natural selection. This context includes competition for resources such as food and space (both intraspecies and interspecies), predation, and mutualist behaviors. Third is the physical context, such as geography and climate, which determines the types of species that can thrive in particular regions of the world, according to their adaptations.

**KEY QUESTION** To what extent is evolution driven or influenced by changes in the climate and environment?

Biologists have long debated the relative contributions of the latter two factors in driving evolutionary change. Not surprisingly, Darwin emphasized the power of biotic interaction because it lies at the core of natural selection. As he wrote in *Origin of Species*: "The theory of natural selection is grounded in the belief that . . . each new species is produced by having some advantage over those with which it comes into competition; and the consequent extinction of less-favoured forms almost inevitably results." Darwin did not ignore the effects of the physical environment, but saw them as merely tightening the screws of competition: "As climate chiefly acts in reducing food, it brings about the most severe struggle between the individuals." To put this another way, the effects of natural selection could produce evolutionary change in the absence of any actual change in the climate or the abiotic (physical or non-biological) environment; competition produced its own internal evolutionary dynamic. Thus for Darwin and many later biologists, natural selection was not only necessary but also sufficient for driving major evolutionary change.

This viewpoint was central to the Modern Synthesis. Even in the absence of change in the physical environment, it was assumed, evolution would continue, driven by the constant struggle for existence. When one individual (or species) gained a slight adaptive advantage over others, the Darwinian imperative to catch up would fuel the evolutionary engine. Predators and prey, for instance, were viewed as being engaged in a constant battle, or arms race. According to this concept, if a prey species evolves ways of avoiding its predator more often, for example, the predator species is likely to evolve countermeasures that boost its chances of catching the prey; this development, in turn, puts pressure on the prey to evolve other predator-avoidance behaviors; and so on. In the early 1970s, the Chicago University biologist Leigh Van Valen termed this idea the Red Queen hypothesis;[66] the name is derived from the character in *Alice Through the Looking Glass*, who tells Alice that it is necessary to run faster and faster in order to stay in the same place. The same evolutionary dynamic would apply to the effect of competition among species for resources.

*Does evolution require environmental change?*

The Red Queen theory has been applied and tested in many contexts,[67] and has been particularly useful for looking at background rates of speciation and extinction. However, there has also been a growing recognition that the Earth's climate, and indeed its geographical configuration, have not been stable over geological time, or indeed over quite short periods of time, and so the issue of the role of global and local environmental perturbation in driving evolution has come to the fore.

In recent years, however, interest has grown in the physical context of life and its possible role in evolution at all levels, from promoting change within species to being a forcing agent in speciation, and even shaping the entire flow of life. This shift in perspective comes from two sources. The first is greater precision in putting dates to climatic and environmental change, and hence the possibility of linking such change to the appearance and disappearance of species – in other words, a better insight into major evolutionary mechanisms. The second source is the possibility that mass extinction is more than simply an interruption in the flow of life, and instead is a creative influence on that flow.

# The influence of plate tectonics

All evolutionary theorists agree on the importance of the isolation of small populations in the processes leading to speciation, extinction, and diversity. Thus, if new species preferentially arise in small, isolated populations (allopatric speciation) rather than in large, continuous populations (sympatric speciation), then processes that promote the establishment of such populations can be regarded as a potential engine of evolution. The physical environment provides two means by which this process of fragmentation and isolation might occur. (A third factor would be the effect of extra-terrestrial events, such as asteroid impacts.) First, topography on local and global scales may change, principally through the mechanism of plate tectonics. Second, global climate change may be driven by many factors, including some of the effects of plate tectonics (Fig. 3.8).

The Earth's crust is a patchwork of a dozen or so major plates whose constant state of creation and destruction keeps them in continual motion relative to one another. The rock that constitutes continental landmasses is less dense than that of the plates, and so the continents ride atop them. As a result, they are also in a constant state of (extremely slow) motion, shuffling around the globe like a mobile jigsaw puzzle. Continents occasionally come together or separate, sometimes producing smaller fragments. As a result, communities of plants and animals that were once united have been divided, and previously independent biotas have been brought together (Fig. 3.9).

For instance, Old World and New World monkeys derive from a common stock, but followed independent paths of evolution as South America and Africa drifted apart some 50 million years ago. Australia's menage of marsupial mammals evolved in isolation from placental mammals, as the island continent lost contact with Old World landmasses more than 60 million years ago, before placental mammals were introduced by humans. By contrast, when the Americas joined some 3 million years ago via the Panamanian Isthmus, an exchange mingled biotas that had evolved separately for tens of millions of years (Fig. 3.10). Indian and Asian species migrated into one another's lands when the continents united approximately 45 million years ago. India's continued northward movement eventually caused the uplift of the massive Himalayan range, producing further geographic

**FIGURE 3.8 Scales of change:** Habitats can be altered by external changes at several levels. Global climate change, which may result from many agencies, including global tectonic action, can cause habitat change globally. Local geographic change can come about through local effects of plate tectonics, which may modify a region's topography. Finally, asteroid or comet impact may drive catastrophic global responses.

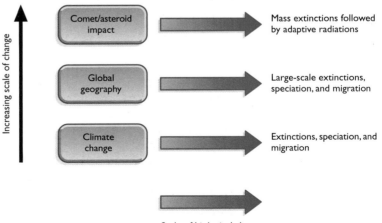

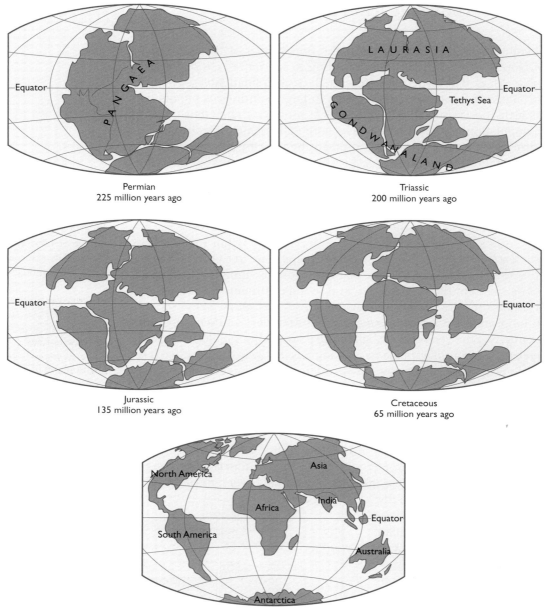

**FIGURE 3.9  Plate tectonics:** Over time the continents have fragmented and moved through the process of plate tectonics. The geography of the world has therefore varied considerably.

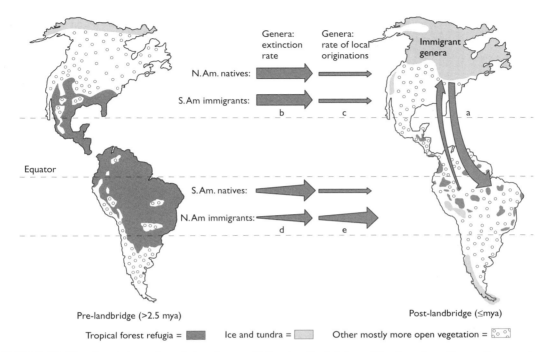

**FIGURE 3.10 The Great American Interchange:** Following the joining of the Americas, the northern and southern biota had different fates. More northern species entered South America than the reverse; the speciation rate of northern immigrants to the south exceeded that of the native species; the extinction rate of native species in the south exceeded that of northern immigrants. In North America, rates of speciation and extinction changed little for both native species and immigrants. These differences have traditionally been explained in terms of an assumed adaptive superiority of northern species. An alternative explanation, based on biogeography and habitat theory, is also plausible.

and climatic modification on a grand scale. Africa and Eurasia exchanged species when the landmasses made contact approximately 18 million years ago; in the process, many groups of animals expanded either from Africa to Asia or vice versa.

Whenever landmasses become isolated as a result of plate tectonics, the environment – and therefore the evolutionary fate – of the indigenous species is influenced by the simple fact of isolation. The isolation of ancestral mammalian species some 100 million years ago, when landmasses were particularly fragmented, has recently been suggested as having prompted the development of the modern mammal orders. Based as it is on a comparison of gene sequences in a handful of modern mammals, this conclusion is at odds with previously accepted views of mammalian evolution.[68] This theory posited the origination of modern orders of mammals as a result of ecological niches having been opened up following the extinction of the dinosaurs 65 million years ago.

When previously isolated landmasses unite, a complex evolutionary dynamic ensues, with some species becoming extinct. This fate befell

*How has global geography changed?*

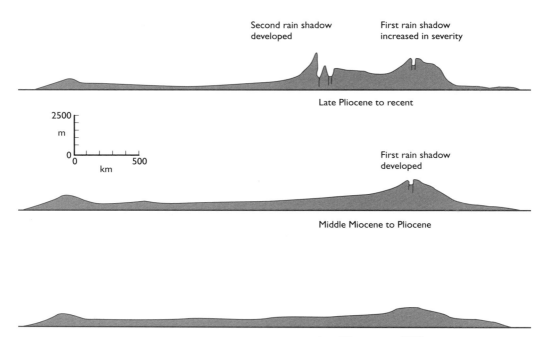

**FIGURE 3.11 Topographic section of Africa along the equator:** During the past 20 million years, tectonic activity beneath east Africa caused uplift and subsequent faulting, forming the modern Great Rift Valley. The effect was twofold. First, it threw the continent east of the uplifted highlands into a rain shadow, causing once-continuous forest cover to shrink and fragment. Second, it produced great topographic diversity, which generated a mosaic of fragmented habitats. These effects are thought to have been influential in the evolution of the hominins, among other evolutionary changes. (Courtesy of T. Partridge et al.)

many South American mammals during the Great American Interchange. Other species may enjoy a burst of speciation during this process, as did many of the North American mammals when they populated South America, the apes as they spread into Eurasia, and the antelopes as they thrived in Africa. In addition to influencing evolution by shuffling land-masses, plate tectonics can modify the environment within individual continents. A prime example of this phenomenon occurred in Africa, where it may have affected the evolution of the hominin clade. Broadly speaking, 20 million years ago, the African continent was topographically level and carpeted west to east with tropical forest; tectonic activity greatly modified this pattern.[69]

A minor tectonic plate margin runs south-to-north under east Africa. Beginning 15 million years ago, it produced localized uplift that yielded tremendous lava-riven highlands that reached 2000 meters and were centered near Nairobi in Kenya and Addis Ababa in Ethiopia (Fig. 3.11). These highlands were the Kenyan and Ethiopian domes. Weakened by the separating plates, the continental rock then collapsed in a long, vertical fault, snaking several thousand kilometers from Mozambique in the south to Ethiopia in the north, and on to the Red Sea. The immediate effect of

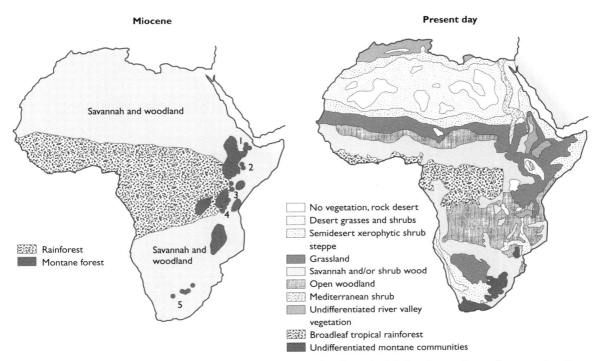

FIGURE 3.12 **African environments – early Miocene and today:** The left-hand map shows the reconstruction of habitats in Africa during the Miocene compared to today (right). In the Miocene forest, distribution was more extensive than today. (Courtesy of Robert Foley.)

the newly elevated highlands was to throw the eastern part of the continent into a rain shadow, dramatically altering the vegetation there. Continuous forest was replaced by a patchwork of open woodlands and, eventually, grassland savannah. Such a habitat fragmentation and transformation would have fragmented the range of forest-adapted animal species living there, encouraging allopatric speciation. More important, the once topographically even terrain became extremely diverse, ranging from hot, arid lowland desert to cool, moist highlands, and a range of different types of habitat in between (Fig. 3.12).

All species can tolerate only a limited range of environmental conditions, as defined by temperature, availability of water, and type of terrain. For animal species, the kinds of plant species that are available influence their ability to occupy a biome or a particular kind of environment. Some species' range of tolerance is greater than that of others; the former will, therefore, be able to live across several biomes. Overall, however, a topographically diverse terrain will also be biologically diverse.

Why are barriers important?

In addition, topographical diversity creates barriers to population movement. For instance, a species that is adapted to the conditions of high elevation may be prevented from migrating from one highland to another because the intervening terrain is inhospitable to it. As a result, a region that is topographically diverse harbors small, isolated populations and therefore represents a potential factory of the evolution of new species. The tectonic uplift and vertical faulting that formed the Great Rift Valley in east Africa produced such a topography, and may well have created conditions conducive to the evolution of hominins from an apelike ancestor.

## Climate change during hominin evolution

A considerable body of data has been amassed during the past decade relating to the Earth's climate during the Cenozoic era, 65 million years ago (mya) to the present, and particularly for the time period most relevant to human evolution, the last 5 million years.[70] The climatic picture is one of continual and sometimes dramatic change (Fig. 3.13). Based

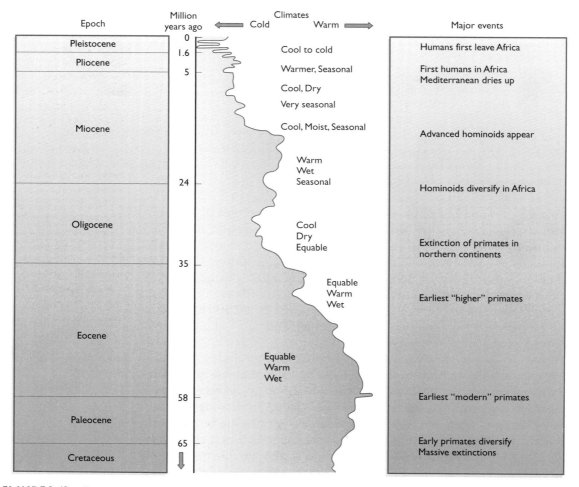

**FIGURE 3.13  Climate patterns since the end-Cretaceous:** An overall cooling trend with local fluctuations marks the Cenozoic era, which culminates in the Pleistocene ice age. Major events of primate evolution are shown in the right-hand column. (Courtesy of I. Tattersall.)

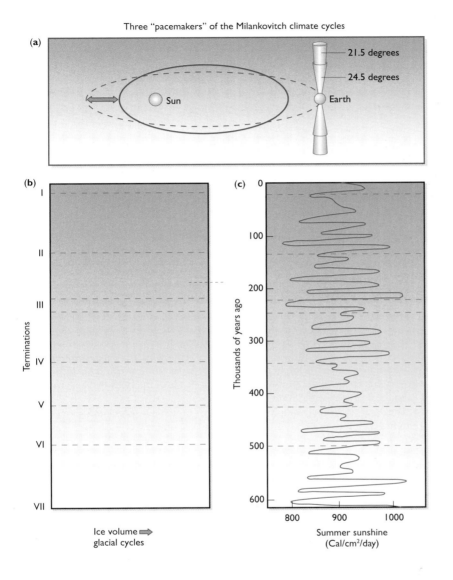

Three "pacemakers" of the Milankovitch climate cycles

**FIGURE 3.14**
**Milankovitch climate cycles of the past 600,000 years:**
Superimposed on long-term global climate change are regular cycles driven by three "pacemakers":
(a) changes in orbital eccentricity, and tilt and orientation of the Earth's spin axis, which results in a 100,000-year cycle;
(b) changes in the volume of the Earth's ice sheets, giving a 41,000-year cycle; and (c) the effect of the intensity of summer sunshine at northern latitudes, yielding a 23,000-year cycle.

principally on the ratio of oxygen isotopes in calcium salts taken from deep-sea Foraminifera, which indicates prevailing temperature, large fluctuations appeared to occur during a net cooling trend. Superimposed on this pattern are global cooling and warming cycles, the so-called Milankovitch cycles, with periodicities of approximately 100,000, 41,000, and 23,000 years (Fig. 3.14). Each of these cycles dominates climate fluctuation at different times in Earth history. For example, prior to 2.8 million years ago, the shortest cycle was dominant; between 2.8 and 1 million years ago, the 41,000-year cycle prevailed; from 1 million years onward, the dominant cycle has been 100,000 years.[71]

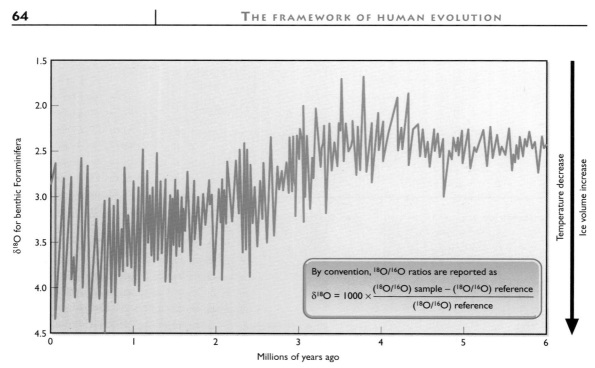

**FIGURE 3.15 Temperature change of the past 6 million years:** A significant decrease in the global temperature of the oceans occurred between 2 million and 3 million years ago. Following the drop, the amplitude of the Milankovitch cycle increased substantially, giving larger differences between minimum and maximum temperatures in the cycles. Some researchers believe that this increase in amplitude of fluctuation was more important in driving evolutionary change than any general climate trend.

Temperature fluctuations are measured vicariously, through the differences in two isotopes of oxygen, O-18 and O-16, which become incorporated in the skeletons of Foraminifera. Cold water contains less O-16 than warm water and thus is incorporated into the skeletons to a lesser extent. Long cores of ocean floor effectively contain a temperature record of the past, locked up in the oxygen isotope ratios of foraminifer skeletons.

What is the pattern of climatic change?

The context in which the hominin clade emerged and evolved – that is, the last 5 million years – is one of major climatic change and variability, and so ideas concerning the role of physical change in evolution are particularly critical. The period of human evolution is generally associated with "ice ages" – one of several such phases in the history of the Earth when global temperature declined very markedly (Fig. 3.15). However, this period is not one prolonged glaciation, but rather consists of a whole series of climatic oscillations, varying from periods warmer than today to periods considerably colder. More than 20 such oscillations have been identified. Some, however, were more marked than others, and show up strongly against the background of the frequent Milankovitch cycles; the existence of these episodes has been inferred from oxygen isotope data and more recently from measures of wind-blown dust in the oceans around Africa. The first event, appearing at 5 million years, involved significant cooling. The second, between 3.5 and 2.5 million years ago, was associated with the first major buildup of Arctic ice and substantial expansion of Antarctic ice. The modern Sahara's roots lie at this point, too. This beginning of the modern ice age may have been initiated by a change in circulation patterns in the atmosphere and oceans as a result of the rise of the Panamanian Isthmus, which joined North and South America some 3.5 million years ago. The third event occurred nearly 1.7 million years ago. The fourth, arising approximately 0.9 million years ago, was possibly caused by uplift in western North America and of the Himalayas and the Tibetan Plateau (Fig. 3.16). The overall pattern of climate change is therefore extremely complicated, driven by several different forcing agents.

If evolution occurs against this background of environmental change occurring at a number of scales, then how does it operate? There is considerable debate on this point. At one extreme are those who see this climatic change as the primary motor of evolution. For them climatic change is a necessary and sufficient cause of evolutionary change. This perspective is associated with two other ideas – one, that because the change is focused on periods of climatic perturbation, then there will be a broad synchronicity of evolutionary events (the turnover-pulse hypothesis); the other that speciation and extinction are focused in these periods, associated with changing distributions of habitat (habitat theory). These ideas are most strongly presented by Elizabeth Vrba of Yale University.

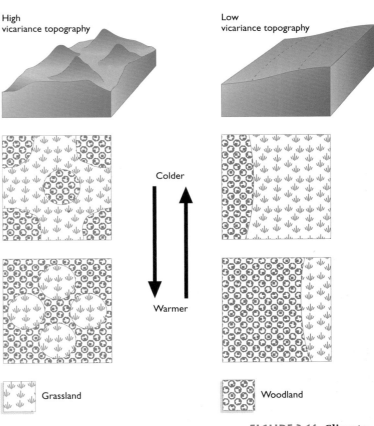

**FIGURE 3.16 Climate change and high topographic diversity:** During times of climate cooling, regions of high topographic diversity will host many vicariant populations (see text), which become isolated through the inability of organisms to track congenial habitats through dispersal. (Courtesy of E. Vrba.)

## Habitat theory

Although it has many components, the habitat hypothesis can be stated simply:[69,72,73] species' responses to climate change represent the principal engine of evolutionary change. The major mechanism of such change is vicariance, or the creation of allopatric populations from once continuous populations, either by the establishment of physical barriers or by the dispersal of populations across such barriers. After such populations become

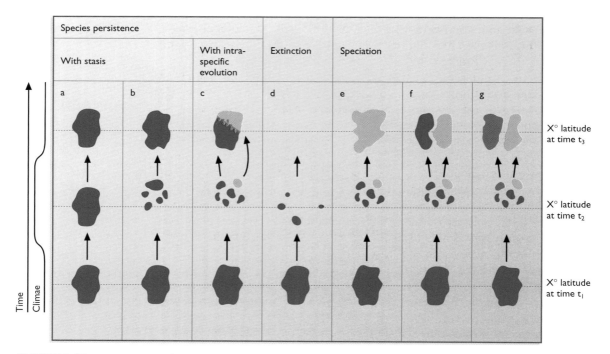

**FIGURE 3.17 Responses to climate change:** Species may persist during, for example, climate cooling, by being able to tolerate the effects of climate change on local habitats. (a) They might migrate, tracking their preferred habitat (in a northerly direction during cooling in the southern hemisphere, and the reverse in the northern hemisphere). (b) Following a period of vicariance, the return to the original latitude with a reversal of the climate change would reconstitute the previous species' distribution. (c) Alternatively, vicariance might promote intraspecies evolution, producing subpopulations with genetic variation between them, but no speciation. Dispersal and reconstitution are the most common response to climate change. Extinction is common as well, especially when the climate shift exceeds a species' threshold of habitat tolerance. (d) Speciation, the least likely outcome, may produce various end results: (e) the replacement of the original species by a single descendant; (f) the persistence of the parental species and the production of a daughter species; or (g) the production of several daughter species, with the parental species becoming extinct.

established, they are vulnerable to extinction and have an opportunity for speciation.

Because of their variable adaptations, different types of species exhibit different vulnerabilities to climate change (Fig. 3.17). Warm-adapted species, such as tropical forests and the animals living there, cluster around the equator and will be extensive in warm times. Temperate forests and grasslands become increasingly dominant at higher latitudes. A fall in global temperature will produce a general equator-ward migration, drastically reducing the area available for tropical forest, which responds by becoming reduced in extent and fragmented. In their equator-ward migration, grasslands may be able to occupy an area similar to that in previous climes, leaving behind patches of vicariant habitat encroached upon by tundra. During such climatic times, therefore, warm-adapted species are likely to undergo higher rates of extinction and speciation than cold-adapted species. The reverse should be true during times of global warming. Because of the general cooling trend of the past 20 million years, the former pattern will have been predominant.

Differences are observed among warm-adapted and cold-adapted species, of course. Some species are habitat specialists, while others are generalists (these terms refer to the availability of required food resources, not just the breadth of diet). Anteaters, for instance, are food specialists; because their food is plentiful in many different ecosystems, however, they can tolerate significant habitat change. Food generalists, such as large carnivores and omnivores, can also tolerate habitat change because of their breadth of diet. Species that can survive in different kinds of habitats, or biomes, are known as eurybiomic; those with narrow biomic tolerance are deemed stenobiomic. Not surprisingly, stenobiomic species are more vulnerable to climate change than are eurybiomes – a pattern that is seen in the evolutionary history of African mammals, for instance. All clades of exclusive grazers and all clades of exclusive browsers consistently show higher speciation and extinction rates than species that can both graze and browse. As a result, biome generalist species are less numerous than biome specialists.

How does habitat theory work?

Vrba argues that the habitat hypothesis offers an alternative explanation of the outcome of the Great American Interchange that followed the joining of the Americas 3.5 million years ago. North American species fared far better than their South American counterparts during this period, with more northern species moving south than the reverse, a higher rate of extinction among native southern species in South America, and higher speciation rates in northern species than in southern species in South America. This pattern has been explained in terms of superior adaptive qualities in the northern species, an idea originally proposed by Darwin.

When landmasses join, the most likely consequence is the migration of biome generalists, because they can find suitable habitats. An important factor in the dynamics of the interchange, suggests Vrba, is that it took place in the context of a dramatic global cooling period. As a result, warm-adapted species migrated toward the equator, which runs through the top of South America. By itself, this trend could produce an imbalance in migration of all types of species. For specialist species, however, it presented an important barrier to migration from South to North America. According to the habitat hypothesis, the higher rates of extinction of South American species are explained by the types of habitat that existed there prior to the formation of the landbridge. Because of the relatively warm climes, the southern continent was extensively forested and was populated with numerous biome specialists. With global cooling, the forests and dense woodland contracted, causing extensive extinction of native species. Because of barriers to movement, only a few North American biome specialists that were adapted to forest habitats migrated to South

America. Many of the northern migrants to the south were woodland–savannah species, and their preferred habitat increased in area in South America during the cooling period.

The habitat hypothesis explains the higher rate of speciation of northern species than of southern species in South America as follows. Native biome generalists would have survived the cooling event preferentially over biome specialists; these generalists have an inherently low rate of speciation. Many of the northern immigrants were biome specialists, having an inherently high speciation rate.

Although the aftermath of the interchange is rather more complicated than presented here, the underlying pattern of the biotic response is at least consistent with many predictions of the habitat hypothesis. This is not to say that local competition does not play a role, for such competition is one of the means by which successful occupation of niches occurs, and forms the microevolutionary basis for macroevolutionary patterns.

## The turnover-pulse hypothesis

One of the most important, and controversial, aspects of the habitat hypothesis is the suggestion that, because extinction and speciation are postulated as being driven by major climatic fluctuations, this pattern should be apparent in all species' groups, including the hominin clade (Fig. 3.18). This suggested synchrony of evolutionary change has been termed the turnover-pulse hypothesis by Vrba. Others have suggested that periods of global cooling initiated major evolutionary changes in hominins and other groups. In contrast, Vrba has promoted the notion of the synchrony of extinction and speciation across taxa.

Several researchers have tested this hypothesis by looking for the predicted synchrony, with mixed results. Vrba has shown it to be valid for African antelopes, for instance, with a major burst of extinction and speciation at the 2.9 to 2.5-million-year event.[74] (An expansion of savannah occurred at this time as well.) This climatic event is also suggested as responsible for a burst of species turnover among hominins, including the origin of the *Homo* lineage. The origin of the hominin clade coincides with the cooling event at 5 million years.[69,75] Data collected by Henry Wesselman on turnover of rodent species in the Omo Valley of Ethiopia are also consistent with climate forcing in the 2.9 to 2.5-million-year event.[76]

Other researchers have been unable to see correlations predicted by the hypothesis.[77] For instance, Andrew Hill, of Yale University, found little to support the hypothesis in the mammalian record of the past 15 million years in the Lake Baringo region of Kenya.[78] Similarly, Tim White, of the University of California, Berkeley, saw no evidence of synchronicity of evolutionary events linked to climate change in the fossil record of pigs in east Africa during the past 4 million years.[79] John Barry, of Harvard University, noted bursts of first appearances and last appearances of

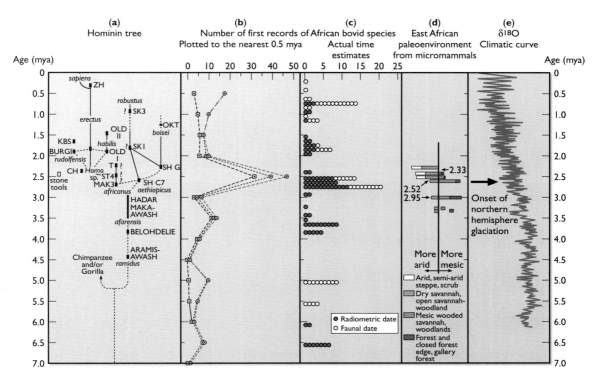

**FIGURE 3.18** **The turnover-pulse hypothesis:** This hypothesis predicts that climate change will drive synchronous origination and extinction in all groups. Such synchrony appears to be present in African antelopes and rodents. Hominins have been said (by Vrba) to fit this pattern, too, but others (for example, Foley) have failed to find the predicted correlation with speciation, although there is such correlation with extinction. (Courtesy of E. Vrba.)

mammalian species in the Siwalik Hills of India, but they are not synchronous (Fig. 3.19).[80] Laura Bishop of John Moores University, Liverpool, has also shown that there is little synchronicity among pigs across the later Pliocene and early Pleistocene.[81]

In an analysis specifically of hominins and baboons, the lineage that shares most in terms of habitat distribution with hominins, Foley showed that evolutionary change, in terms of first and last appearances, was spread across all periods, and could not be said to be confined to periods of excessive climatic change.[82,83,84] This was a statistical analysis, and one of the observations, which does give a clue to evolutionary process, was that the relationship between climatic change and speciation was especially weak – in other words, that novelty did not arise in relation to climatic change per se. However, there was a somewhat stronger relationship between climatic change and extinction. What this implies is that climatic change has the effect of knocking species out of an ecosystem, but that the evolution of novel forms after such events is a much more complex process, and is probably related to local conditions.

Is evolutionary change across lineages synchronized?

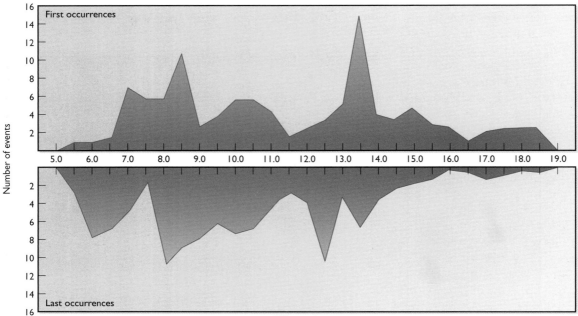

**FIGURE 3.19 A test of the turnover-pulse hypothesis:** The graph records the first and last occurrences of vertebrate species in the Siwalik Hills of India during the past 20 million years. These occurrences may be taken as origination and extinction events. Although some tracking of climate change occurs in conjunction with changes in the biota, the synchrony predicted by the turnover-pulse hypothesis is not strongly indicated. In particular, bursts of extinction follow originations rather than precede them, as the hypothesis argues. (Courtesy of John Barry.)

## Evolutionary geography

Vrba's models belong to the macroevolutionary school of thought outlined in the first part of this chapter. These models downplay the role of competition and natural selection, and see species as relatively fixed entities that shift their distribution in response to environmental change, but otherwise remain in stasis. Evolutionary change is confined to short bursts in small populations during periods of greater than usual change. However, there are other ways of looking at the relationship which draw more strongly on the classical ideas of evolutionary theory, and in effect reiterate the basic element of Van Valen's Red Queen model – that evolution is driven by the interactions of organisms that are themselves subject to evolution.[66]

What the more microevolutionary models share with Vrba's ideas is a focus on geography. Evolution is of course defined as change through time, but in fact its chronological patterns all start as geographical ones. All populations have a spatial distribution; they exist in the areas that they can tolerate, and are terminated by the presence of barriers, or the decline in tolerability (Fig. 3.20). In common with Vrba's ideas, all biogeographers have recognized that populations, and hence species, will expand during periods when tolerable habitats are more widespread, and contract as those habitats become limited. The history of any lineage is therefore a history of changing spatial distributions.

Eitan Tchernov of Tel Aviv University has expressed this most cogently with his model of lineage biogeography over time.[85] The history of any lineage has a recurrent pattern (Fig. 3.21). Its origins lie in a small, limited geographical area (isolation or a bottleneck). For most populations that might also represent their total lifespan. However, for some lineages either a competitive edge or a change in habitat distribution, or a combination of the two, will lead to a breakdown of geographical barriers, and a spatial expansion based on dispersals. Multiple dispersals over a sustained period of time will result in a more widespread distribution. However, conditions can and do change, and so either the expanded populations may become disrupted, so that they again form small isolated units, or else they will suffer local extinctions, and the sum of those population extinctions will be the extinction of the lineage (species) as a whole.

There are a number of key elements in this model that differ from Vrba's. First, the process is very much based on microevolutionary processes at the level of the population, rather than species as a whole, and in that sense it is a micrevolutionary process (Fig. 3.22). Second, where Vrba's species are passive respondents to environmental change, in this model populations are dynamic opportunists which respond to changing conditions and also to shifts in their own adaptive potential. And third, and most important, the key mechanisms are the twin ones of the potential for dispersal – that is, populations actively expand in relation to change, rather than being limited to specific habitats – and proneness

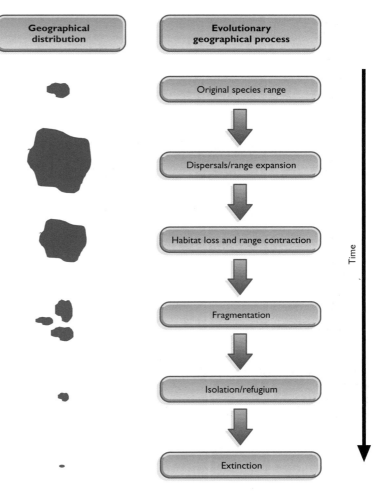

**FIGURE 3.20**
**The geography of evolution:** The evolutionary history of any species can be seen as a change in distribution, from a small local origin, to its full extent following dispersal, to a subsequent contraction and fragmentation, and ultimately extinction. Different geographical distributions will result in different evolutionary conditions and outcomes. (Courtesy of Robert Foley.)

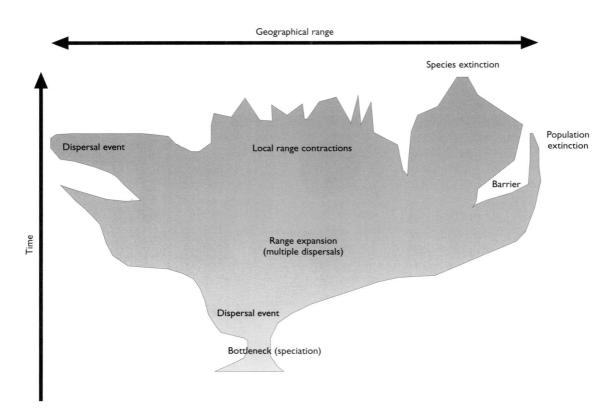

**FIGURE 3.21 The evolutionary geography of a lineage:** Eytan Tchernov's model of the evolutionary history of a lineage emphasizes that it has a geographical component, beginning with a small localized distribution (bottleneck) at the point of speciation, followed by range expansion with multiple dispersals. The ultimate extinction of a species is associated with range contraction, fragmentation into refugia, local extinction, and ultimately final extinction. (Courtesy of M. Lahr and R. Foley.)

Why is evolution a geographical process?

to extinction during times of habitat change that is detrimental. This evolutionary geography model, which has been extensively applied to human evolution, offers a microevolutionary alternative to tacking major evolutionary change.[82,86]

The turnover-pulse hypothesis, when Vrba proposed it over a decade ago, was stated in a strong form; that is, it predicted close synchrony of speciation and extinction, driven by climate change. This strong version of the hypothesis has not stood up to testing. Although no one doubts that the physical context of evolution is important in terms of geography and climate change, the mechanics of biotic response, particularly as related to climate change, are more complex than the original hypothesis suggested. Furthermore, environmental fluctuation, not just direction of environmental change, is increasingly regarded as potentially important in causing evolutionary change. However, the questions posed in models such as the turnover-pulse hypothesis are key to developing general evolutionary models.

# EXTINCTION AND PATTERNS OF EVOLUTION

**L**ife first evolved on Earth almost 4 billion years ago, in the form of simple, single-celled organisms. Not until half a billion years ago did complex, multicellular organisms evolve, in an event biologists call the Cambrian explosion. An estimated 100 phyla (body plans) arose in that geologically brief instant, with few, if any, new phyla arising later. The products of this initial, intensely creative moment in the history of life included all of the 30 or so animal phyla that exist today. The remaining 70 or so phyla disappeared within a few tens of millions of years of their origin.

In the 530 million years since the Cambrian explosion, 30 billion species have evolved. Some represented slight variants on existing themes, while others heralded major adaptive innovations, like the origin of jaws, the amniote egg, and the capacity of flight. Given that an estimated 30 million species exist today, it's clear that 99.9% of species that have ever lived are now extinct. Some extinctions occur at a steady, background rate of approximately one species every four years; others are part of mass extinction events, during which a great proportion of extant species disappear in a geologically brief period, measuring from a few hundred to a few million years (Fig. 3.23). Although extinction – and particularly mass extinction – is an important fact of life, until recently evolutionary biologists have virtually ignored the topic, choosing instead to focus on mechanisms by which new species arise. However, as the previous chapter showed, the loss of species through habitat change is as much a central part of the factors leading to novelty in evolution.

Several reasons explain this neglect. For many years, mass extinctions were perceived as very difficult to study, being singular events of great complexity. Consequently, they were regarded as something of an impenetrable mystery. Two factors have since changed this perception. The first was the suggestion, made in 1979,[87] that the dinosaurs met their end when a huge asteroid collided with the Earth 65 million years ago; the second was the growing realization that we are witnessing modern extinctions on a cataclysmic scale, as a consequence of

> **KEY QUESTION** What role does extinction play in evolution?

**FIGURE 3.22**
**Phylogeny and geography:** Over time a lineage will diversify in response to changing geographical conditions. Sometimes this will lead to new species, and at other times the isolated populations will be reunited. The evolutionary history of any lineage is made up of multiple such events. (Courtesy of M. Lahr and R. Foley.)

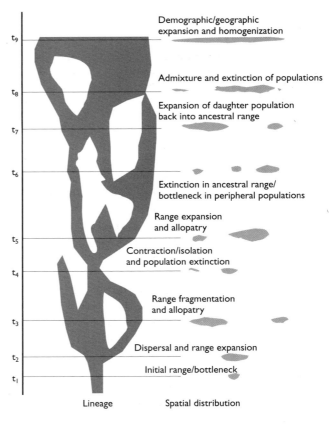

Demographic/geographic expansion and homogenization

Admixture and extinction of populations

Expansion of daughter population back into ancestral range

Extinction in ancestral range/ bottleneck in peripheral populations

Range expansion and allopatry

Contraction/isolation and population extinction

Range fragmentation and allopatry

Dispersal and range expansion

Initial range/bottleneck

Lineage          Spatial distribution

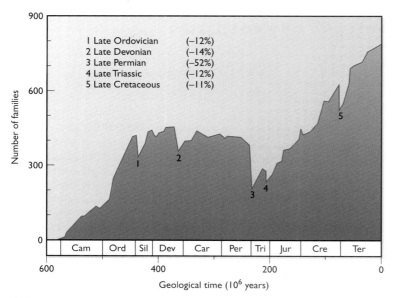

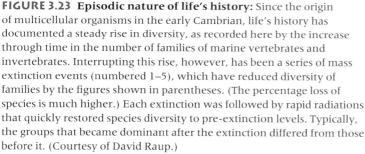

| | | |
|---|---|---|
| 1 Late Ordovician | (−12%) | |
| 2 Late Devonian | (−14%) | |
| 3 Late Permian | (−52%) | |
| 4 Late Triassic | (−12%) | |
| 5 Late Cretaceous | (−11%) | |

**FIGURE 3.23 Episodic nature of life's history:** Since the origin of multicellular organisms in the early Cambrian, life's history has documented a steady rise in diversity, as recorded here by the increase through time in the number of families of marine vertebrates and invertebrates. Interrupting this rise, however, has been a series of mass extinction events (numbered 1–5), which have reduced diversity of families by the figures shown in parentheses. (The percentage loss of species is much higher.) Each extinction was followed by rapid radiations that quickly restored species diversity to pre-extinction levels. Typically, the groups that became dominant after the extinction differed from those before it. (Courtesy of David Raup.)

human-caused destruction and fragmentation of habitat.

As a result of the recent burst of research, biologists' assumptions about mass extinction – about its causes and, more important, its effects – have been overturned. Once mass extinctions had been accepted as a fact of life – as mere interruptions in the slow, steady increase in biological diversity that began after the Cambrian explosion established complex life forms. Now, however, they are recognized as playing a major role in guiding the path of evolution.

## The influence of catstrophism

As with many issues in evolutionary biology, Darwin's influence on modern thought concerning extinction has been profound. In *Origin of Species*, Darwin essentially denied the fact of mass extinction, stating that extinction is a slow, steady process, with no occasional surges in rate. He also argued that species become extinct because they prove adaptively inferior to their competitors. Darwin's equation of extinction with adaptive inferiority clearly derives from his theory of natural selection, and it powerfully shaped biologists' thinking.

*Is extinction a creative evolutionary process?*

The fact of extinction had been demonstrated before Darwin's time, by the French anatomist Baron Georges Cuvier in the late eighteenth century. Cuvier definitively showed that mammoth bones differ from those of the modern elephant. The inescapable conclusion was that the mammoth species no longer existed. Through his extensive study of fossil deposits in the Paris Basin, Cuvier went on to identify what he thought were periods of mass extinctions, or catastrophes, in Earth history when large numbers of species went extinct in very short periods of time.

Cuvier's observations inspired a great volume of geological work in the early part of the nineteenth century. This research identified intervals of apparent major change in the history of life, which formed boundaries

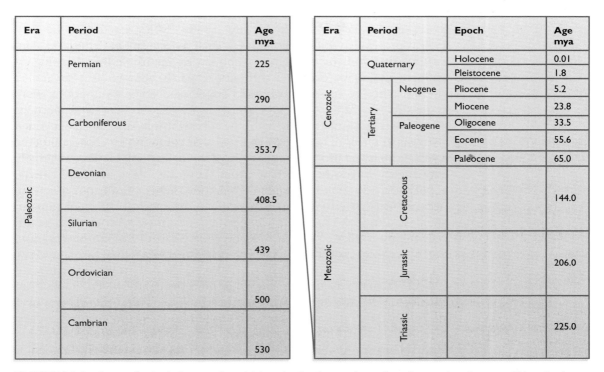

| Era | Period | | Age mya |
|---|---|---|---|
| Paleozoic | Permian | | 225 |
| | | | 290 |
| | Carboniferous | | |
| | | | 353.7 |
| | Devonian | | |
| | | | 408.5 |
| | Silurian | | |
| | | | 439 |
| | Ordovician | | |
| | | | 500 |
| | Cambrian | | |
| | | | 530 |

| Era | Period | | Epoch | Age mya |
|---|---|---|---|---|
| Cenozoic | Quaternary | | Holocene | 0.01 |
| | | | Pleistocene | 1.8 |
| | Tertiary | Neogene | Pliocene | 5.2 |
| | | | Miocene | 23.8 |
| | | Paleogene | Oligocene | 33.5 |
| | | | Eocene | 55.6 |
| | | | Paleocene | 65.0 |
| Mesozoic | Cretaceous | | | 144.0 |
| | Jurassic | | | 206.0 |
| | Triassic | | | 225.0 |

**FIGURE 3.24  The geological time scale:** Divisions in the time scale are based on major changes of biota in the fossil record.

between geological periods that were given the following names: Cambrian, Ordovician, Silurian, Devonian, Carboniferous, Permian, Triassic, Jurassic, Cretaceous, Paleocene, Eocene, Oligocene, Miocene, Pliocene, Pleistocene, Holocene (Fig. 3.24).

Two particularly devastating catastrophes divided the history of multicellular life, known as the Phanerozoic, or visible life, into three eras: the Paleozoic (ancient life), from 530 to 225 million years ago; the Mesozoic (middle life), from 225 to 65 million years ago; and the Cenozoic (modern life), from 65 million years ago to the present. Cuvier lived in pre-evolutionary theory times, of course, and he therefore saw the catastrophes as individual events (some 30 in all) that wiped out all of existing life, setting the stage for new waves of creation. This world view was known as catastrophism.

## The triumph of uniformitarianism

Even before Darwinian theory emerged, catastrophism came under attack, principally by the Scottish geologist Charles Lyell, who was following arguments made earlier by his fellow countryman James Hutton. In his *Principles of Geology*, published in the 1830s, Lyell argued that the geological processes we observe today – such as erosion by wind and rain, earthquakes and volcanoes, and so on – are responsible for all geological changes that have occurred throughout Earth history. He also denied the existence of mass extinctions of species. Lyell's assumptions were impeccable for an empirical scientist, but his inferences were perhaps overdrawn. In combating the extremes of catastrophism, which explained patterns in terms of forces (floods, etc.) that could not be observed today, Lyell emphasized processes that were observable. As these were relatively low-energy forces, he concluded that they must have occurred slowly, repeatedly, over very long periods of time. What he did not take into account was that some forces may have occurred at different rates in the past, and so the pattern of change in the Earth would not be constant. It is highly unlikely that, during the short span of human observation, we have witnessed all possible geological processes that can shape the Earth.

Lyell's scheme came to be known as uniformitarianism. For a while, an intellectual battle pitted it against catastrophism. Uniformitarianism won decisively, and catastrophism was banished from the intellectual arena as a relic of earlier thinking – governed, it was implied, by religious rather than scientific argument. Darwin's intellectual perspective on the evolution of life was very closely allied to Lyell's view of the evolution of the geological context against which life unfolds. Darwin had long struggled to replace a supernatural explanation of life's exquisite adaptation to its circumstances with a naturalistic one, that of natural selection. As a result, he balked at admitting anything that smacked of the supernatural, such as the sudden extirpation of millions of species followed by waves of creation.

**What are the limits of uniformitarianism?**

Catastrophism may have been defeated as an idea, but paleontologists stubbornly continued to find evidence of major events in the fossil record, rather than a pattern of constant change, and extinctions were a major part of this. Earth history evidently is not one of constant change but instead a mixture of continuous change with a litany of sporadic and spasmodic convulsions. Some of these events have moderate impact, with 15 to 40% of marine animal species disappearing, but a few others are of much larger extent, constituting the mass extinctions.

This last group – known as the Big Five – comprises biotic crises in which at least 75% of such species became extinct in a brief geological instant.[88] In one such event, which brought the Permian period and the Paleozoic era to a close, more than 95% of marine animal species are calculated to have vanished. This handful of major events, from oldest to most recent, include the following: the end-Ordovician (440 million years ago), the Late Devonian (365 million years ago), the end-Permian (250 million years ago), the end-Triassic (210 million years ago), and the end-Cretaceous (65 million years ago).

## Causes of mass extinctions

Numerous causative agents of mass extinction events have been suggested over the decades. These putative sources include a drastic fall in sea levels (sea-level regression) (Fig. 3.25), global cooling, predation, and inter-species competition. Of these, sea-level regression and global cooling have traditionally been held as most important.

In 1979, however, Luis Alvarez, a physicist at the University of California, Berkeley, and several colleagues suggested that the end-Cretaceous extinction, which marked the end of the dinosaurs' reign, was the outcome of Earth's collision with a giant asteroid.[87] They based their conclusion on the presence of the element iridium in the layer that marks the Cretaceous/Tertiary boundary. Iridium is rare in crustal and continental rock, but common in asteroids. The impact, striking with the force of a billion nuclear bombs, was postulated to have raised a dust cloud high into the atmosphere, effectively blocking out the sun for at least several months. The ensuing catastrophic results affected plant life first and then the animals that depend on it.

This idea was not well received initially, not least because it sounded too much like a return to catastrophism. In the years since its proposal, a large body of evidence has been gathered in its support, including evidence of an impact crater at the pertinent time, 65 million years ago. While the impact hypothesis for the dinosaurs' demise remained in its infancy, David Raup and Jack Sepkoski, of the University of Chicago, suggested that biological crises occur every 27 million years, with five major extinctions and a dozen lesser ones (Fig. 3.26). Periodic impact by asteroids was the most likely explanation for these events, they claimed. If correct, this theory would mean that fully 60% of all extinctions through the Phanerozoic were caused by impact (5% in the Big Five, and the remainder in smaller events). It would also mean that the history of life on Earth is significantly influenced by external events.[89]

Although skepticism persists about periodic impact and crises, the dinosaur extinction, and several other mass extinctions, are now more widely accepted as resulting from extraterrestrial impacts. Such impacts might not be the sole cause, however; the meteors might have struck a biota that was already fragile for other reasons, or they might have weakened it, making it vulnerable to secondary mechanisms of extinction. Furthermore, they do not explain which taxa (categories of organisms: see chapter 5) tended to become extinct, and which survived. Differential survival at these critical times would have been influenced by micro-evolutionary processes such as adaptation and competition.

## Biotic responses to mass extinctions

The history of life on Earth can therefore be seen as having been punctu-ated by occasional bursts of increased rates of extinction – some moderate,

**Phanerozoic sea-level curve**

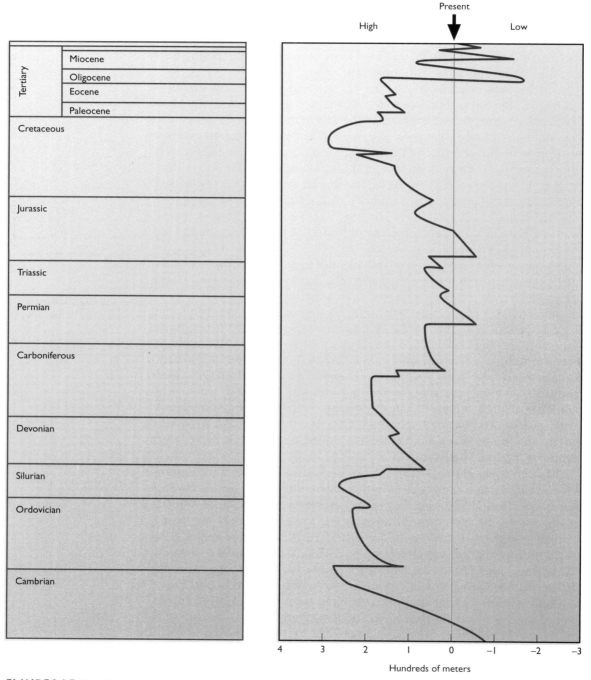

**FIGURE 3.25 Sea-level changes:** Sea-level regression is a probable factor in some extinctions, and is associated with many of them. Many falls do not coincide with extinction, however, indicating that some mass extinctions are complex events, involving the interplay of several agencies.

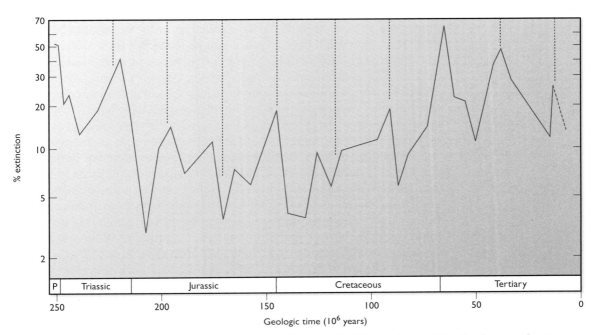

**FIGURE 3.26 Periodic extinction:** Raup and Sepkoski's analysis of the fossil record implies that significant extinctions occur every 27 million years. The only agency that could inflict extinction periodically, they conclude, is extraterrestrial impact. The notion remains controversial. (Courtesy of D. Raup and J. Sepkoski.)

some catastrophic. Whatever the cause of these events, the next question becomes: how do Earth's biota respond? And what determines which species survive through these crises and which do not?

A striking feature of the biota's response is that, following a rapid collapse, species diversity rebounds quickly. Within 5 million to 10 million years of the event, the diversity equals and often exceeds pre-extinction levels. During this brief period, the rate of speciation greatly exceeds the rate of extinction. This aspect of response is qualitative as well as quantitative. Typically, the groups of species that come to dominate the marine and terrestrial ecosystems differ from those that dominated prior to the collapse. Consider, for example, the end-Cretaceous extinction, which saw the disappearance of the dinosaurs as the major terrestrial animal species and their replacement by mammals. Mammals had coexisted with dinosaurs for more than 100 million years, but they were small and probably few in number. Modern orders of mammals may have originated 100 million years ago, but not until after the extinction did larger species evolve and become more numerous; these mammals eventually came to occupy the niches previously occupied by large reptiles.

*Mass extinctions – a significant factor in evolution?*

The shift of major players does not always involve the complete extirpation of an existing group, however. Sometimes a dominant groups survives the extinction event, but in a much reduced fashion. In any event, mass extinctions lead to a change of cast of characters on life's stage.

This concept raises questions about what makes some groups of species vulnerable to extinction, or partial extinction, while others fare better. As Raup has so succinctly put it: was it bad genes or bad luck that consigned the losers to evolutionary oblivion? For a very long time, this question was not asked, because the answer was found in Darwin's *Origin*: survivors are adaptively superior to victims. In other words, bad genes explain the failure to survive. In recent years, however, biologists have begun to question the assumption of the ascendancy of competition, in the ecological dynamics of local population interactions and at the level of evolution and extinction.

Most biologists agree that the prevailing force in times of background extinction is natural selection, in which competition plays an important part. But what about the bursts of higher rates of extinction? Is mass extinction merely background extinction writ large? Do marine regressions, climate cooling, and the effects of asteroid or comet impact merely tighten the screws of competition as times get tough? While the answer to this question may be that competition does intensify, it is also the case that chance may play a significant role. Counterintuitively, random processes can produce patterns. Raup and several colleagues tested the hypothesis that mass extinction events might represent such a pattern.[89] In computer simulations of species communities over long periods of time, in which speciation and extinction were allowed to happen randomly with no external force operating, they found patterns similar in form, but not in magnitude, to the contents of the fossil record. In other words, species numbers fluctuated significantly with no external driving force, but only rarely crashed in a way that could be termed a mass extinction. Thus, bad luck cannot be the sole cause of a species' demise in a mass extinction event. This research also partly inspired the realization that bad genes could not provide the sole explanation of the pattern of life. Instead, some combination of selection and bad luck operated in tandem.

The University of Chicago paleontologist David Jablonski has investigated the nature of that selection by comparing the pattern in background and mass extinction periods.[90] During background extinction, several factors conspire to protect a species from extinction. Species that are geographically widespread resist extinction, for instance. Likewise, marine species that send their larvae far and wide (drifting with the currents) resist extinction, for similar reasons. A group of related species, a clade, resists extinction if it contains many species rather than only a few. Thus, the chance disappearance of a few species is more likely to threaten the survival of a clade that includes only three species, for example, than one that has 20. Of course, it may not be entirely random that a clade has 20 species rather than three, so selection and adaptation raise their head again even in this model.

When Jablonski examined the fate of mollusc species and species' clades across the end-Cretaceous extinction, he saw a very different picture. None of the above rules applied. The only rule he could discern was valid for groups of related species, clades. Once again, geographic distribution

*What is the role of chance in evolution?*

played a part in survival. If a group of species occurred over a wide geographic range, then they fared better in the biotic crisis than those that were geographically restricted, no matter how many species made up the clade. "During mass extinctions, quality of adaptation or fitness values . . . are far less important than membership in the particular communities, provinces, or distributional categories that suffer minimal disturbance during mass extinction events," wrote Jablonski. This finding was a landmark result, because it was the first to clearly indicate that the rules changed between background and mass extinction. Biotic crises are not simply background extinctions writ large.

This idea makes sense because, in the history of life, many successful species or groups of species have met abrupt ends in mass extinctions. The dinosaurs dominated their realms for more than 100 million years and were as diverse as they had ever been when they vanished at the end-Cretaceous extinction. No evidence suggests that the mammals were better adapted in any way than dinosaurs, which mammals subsequently replaced as the major terrestrial tetrapod group.

In another example, reef communities in the oceans have been transformed periodically, as existing dominant organisms were wiped out, coinciding in four cases with the major extinction crises. After each devastation, reefs reappeared, sometimes with calcareous algae dominating, sometimes bryozoans, sometimes rudist molluscs, and sometimes corals. The coral reefs with which we are familiar today are simply the temporary inhabitants of that adaptive zone.

Two interpretations of these patterns are possible, which reflect the same issues as were discussed earlier in this chapter. The first of these rejects a major role for natural selection in these major events. Natural selection operates cogently at the level of the individual, in relation to local conditions, reflecting the impact of competitors and prevailing physical conditions. It is a powerful response to immediate biological experience, but it cannot anticipate future events. And it certainly cannot anticipate rare events. The average longevity of an animal species is 2 million to 4 million years, and extinction bursts occur on average no more than every 27 million years. Consequently, most species never experience such bursts. The mass extinction episodes are rarer still, making them invisible to natural selection. Species cannot adapt to conditions they do not experience. The Darwinian view that the history of life is one of persistent adaptation led by natural selection is therefore incomplete.

The second view does not question that there is a major external factor in these mass extinctions, or that there is an element of chance in which groups survive. It makes a big difference, when an asteroid hits the Earth, where you are in relation to it, and the further away the better is probably a reasonably good rule! However, although these external events do have a major impact, it can be argued that they still operate through the same basic rules. There is a shift not in whether selection operates, but in its intensity and direction. The rapidity of such events is, it should be remembered, relative. Although dinosaurs became extinct over a short period of time relative to their taxonomic lifespan, it was still a process that lasted several million years – hundreds of thousands of generations, and so a long enough period for the event to be played out under the normal rules of evolutionary change.

Whatever their cause, however, mass extinctions, then, restructure the biosphere, with a particular set of survivors finding themselves in a world of greatly reduced biological diversity. With at least 15% and as much as 95% of species wiped out, ecological niches are opened or at least made much less crowded. This time provides an evolutionary opportunity offered to a lucky few. However, whichever are the ones that survive, what is striking is the fact that the same general patterns tend to recur – what is known as convergent evolution. Although dinosaurs and mammals differ markedly in many ways, there are strong parallels in the adaptations of many of them – the teeth of the carnivores, the size of the herbivores, etc. These indicate that although there may be strong chance elements in survivorship, selection will still shape those surviving lineages to the demands of the environment.

## Extinction and human evolution

What is the relevance of this to human evolution? First, it is important to remember that human evolution does not occur in isolation, but in concert with the evolutionary history of all other organisms, especially those with which the hominins were sympatric. The pattern of extinction of these species is important in determining what opportunities and competitive challenges faced the hominins. Second, as we have seen, extinction is an important part of the evolutionary process, and seems to be one which is linked to environmental change, and thus we can expect it to play a role in hominin evolution. Third, and perhaps most important of all, while there is only one extant species of hominin, the fossil record shows us that in the past there were many, and they have all become extinct except our own direct ancestors. An explanation of human evolution must account not just for the successes but also for the failures.

Finally, does the growing recognition of the importance of mass extinctions undermine Lyell's notions of uniformitarianism? The answer is certainly that one should be aware of the limiting assumptions of uniformitarianism, but it is still an important and powerful idea. Uniformitarianism does not mean that only what occurs now could have occurred in the past; what it does mean is that the past must be explained using principles that are compatible with our understanding of the way in which geological and biological systems operate. Old-fashioned catastrophism failed on that account, but the idea of dramatic changes at points in the past remains important in modern evolutionary biology.

## BEYOND THE FACTS
# DEUS EX MACHINA VERSUS THE GHOST IN THE MACHINE

*The issue: in this chapter we have seen that while all biologists work within an evolutionary framework, and broadly within a Darwinian one, there is still considerable disagreement about the precise mechanisms involved. The specific debates, however, often reflect a more general difference in approaches to explanation.*

If one were to boil evolutionary biology down to its most basic problem, it would probably be expressed simply as "how do we explain change?" (This of course takes for granted that there is change that needs explaining, but that is another issue.) The success of the Darwinian revolution lay in the fact that Darwin and Wallace had come up with a plausible and powerful mechanism – natural selection – to explain change. However, at a more general level we can ask where the motor for change comes from. Leaving aside the precise mechanism involved, we can pose three general answers to this question.

First, change can come from something internal to the "lineage" itself – a sort of inner motor. This kind of evolutionary model is usually referred to as orthogenesis, the innate characteristics of the organism determining its evolutionary future. This is largely an unfashionable view, having been the mainstay of pre-modern synthesis biologists such as Teilhard de Chardin and Franz Weidenreich, and such ideas are associated with the notion that there is an end point to evolution (usually humans). However, modern alternatives have been put forward, such as Gabriel Dover's idea of "molecular drive" – the notion that evolutionary change is internally driven by molecular processes.

A second model is that change can only come from something entirely external to the system. Macroevolutionary theories often rely on this type of explanation – change brought about by external perturbation of the system – an asteroid striking the earth and wiping out the dinosaurs, or Vrba's ideas of climatic forcing. Biological change is largely brought about by non-biotic factors.

A third model, and one most closely associated with Darwin, is that change is neither entirely external nor entirely internal, but comes from the interactive effects of organisms and the environment in which they live. Thus the motor of change is external in the sense that it is not an innate tendency of the organism, but is context specific; change is induced by external conditions, but these conditions are themselves biological. This is the essence of Van Valen's Red Queen hypothesis discussed earlier.

These differences represent major theoretical issues, and ones that are not easily solved. Furthermore, they are issues that are not specific to evolution. A brief examination of any historical problem will show the same tension between internal and external factors. For example, was the American Revolution the result of the internal dynamics of an evolving colonial society, or due to the external impact of the French Revolution? And in turn, was the French Revolution the result of the internal dynamics of class conflict in the eighteenth century, or due to declining harvests brought about by a decrease in global temperatures, totally outside the influence of the ancien régime? Virtually any problem associated with the explanation of change can reflect this intellectual conflict, and the future of evolutionary biology depends on resolving it.

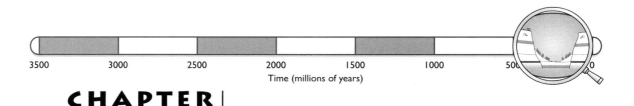

# CHAPTER 4

# THE GEOLOGICAL CONTEXT

The preceding chapters have focused strongly on the principles and problems of evolutionary theory – in other words, they are firmly placed in the biological or life sciences, and the issues involved in understanding living forms. For most of biology this would be a sufficient framework. However, because such a substantial portion of paleoanthropological work is done on fossils, it is also necessary to consider briefly some aspects of the earth sciences. Fossils are the remnants of dead organisms, and technically have undergone transformation from the biological material to a form of rock (transformation from the biosphere to the lithosphere). This means that to study them it is necessary to understand some of the methods by which fossils are preserved and placed into a useful chronological framework. Of particular importance are dating techniques – essential for providing a sound chronological framework – and the biases involved in having only fossils and stone tools as a source of information about evolutionary history.

## DATING METHODS

**KEY QUESTION** How can fossils be accurately dated, and with what limitations?

**A**nthropologists and archeologists are interested in understanding biological relationships among our ancestors, and their behavior. An accurate time scale is a crucial aspect of reconstructing the pattern of evolution of the anatomical and behavioral characteristics of early hominins. The application of reliable methods of dating has the potential to alter radically interpretations of evolutionary relationships. At the same time, uncertain dating may lead to confusing and contentious conclusions. At least half a dozen methods of dating are now available that have the potential to cover events from 100 years ago to many billions of years, albeit with some frustrating gaps. Paleoanthropologists' focus is on the last 10 million years or so, which includes some of those gaps.

Researchers who want to know the age of particular hominin fossils and/or artifacts in principle have two options for dating them: direct methods and indirect methods.

Direct methods apply the dating techniques to the objects themselves, which ultimately is the preferred option. Two types of problem arise with this approach, however. First, for most material of interest, no methods are as yet available for direct dating. Ancient fossils and most stone tools, for example, remain inaccessible to direct dating. Some methods, such as carbon-14 dating and electron spin resonance, may be applied directly to teeth or young fossils, and indeed to the pigments of rock shelter and cave paintings; in addition, thermoluminescence dating may be applied directly to ancient pots, flint, and sand grains. Overall, however, few opportunities are offered by current techniques for direct dating (Fig. 4.1). Second, fossils and artifacts are often too precious to risk destroying any part of them in the dating process. (Archeologists have often tried to develop chronologies for artifacts based on the style of production, for both tools and painted and engraved images. This approach to dating is potentially treacherous, as styles may vary without regard to the passage of time.)

In practice, indirect dating methods represent the typical approach. Here, an age for the fossil or artifact is obtained by dating something that is associated with it. This strategy may involve direct dating on non-human fossil teeth that occur in the same stratigraphic layer, by electron spin resonance, for instance, or by thermoluminescence dating of flints associated with human fossils. Both these approaches have been applied in recent years to fossils relating to the origin of modern humans. Fossils or

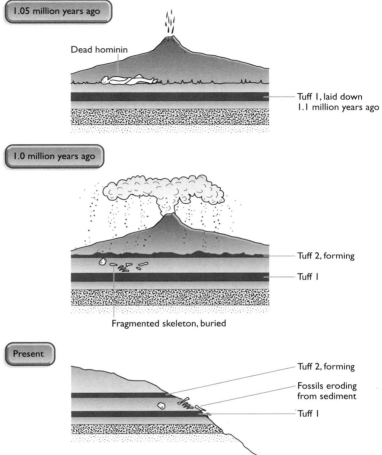

**FIGURE 4.1 The life and date of a fossil:** Fossils can seldom be dated directly. A date may be produced by dating the layers in which the fossil is embedded – in ash layers that lie just below and just above the fossil, formed as shown.

artifacts may be attributed an age through information about the evolutionary stage of non-human fossils associated with them, a technique known as faunal correlation.

The most common indirect approach, where feasible, is to date stratigraphic layers that lie below and above the object in question. Stratigraphic layers accumulate from the bottom up, so that the lower layers are oldest and the upper layers youngest. An object may then be said to be, for example, no older than 1.1 million years and no younger than 1 million years. These two dates, taken from below and above the object, provide brackets that include the date at which the object became incorporated into the stratigraphic system.

This chapter will survey briefly the principal techniques available and identify where they are best applicable. The techniques may be classified into two types: those that provide relative dates and those that provide absolute dates. Relative dating techniques give information about the site in question by referring to what is known at other sites or other sources of information. As the term implies, they provide a date only in relation to the age of something else ("older than" or "younger than"), rather than an actual chronological estimate in years. Absolute dating techniques provide information by some kind of physical measurement of the age of material at the site in question (Fig. 4.2). Radiometric dating techniques, which

*Why is the principle of stratigraphy still important?*

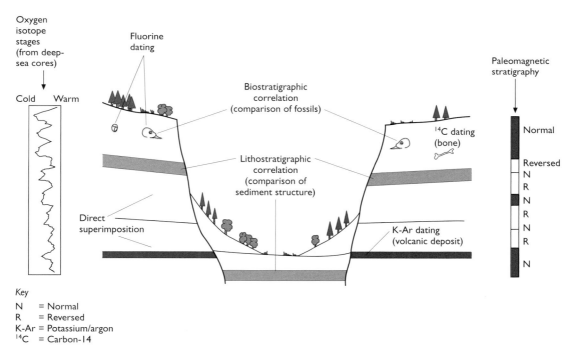

**FIGURE 4.2 Integrating relative and absolute dating techniques:** The principal methods of dating (relative and absolute) are shown here, with the nature of correlation and calibration. (Courtesy of Robert Foley.)

depend upon some aspect of radioisotopic activity, are an important component of absolute dating.[91]

## Relative dating techniques

Relative dating techniques include faunal correlation and paleomagnetism. As stated above, they place events and items into a sequence, rather than giving real ages, although they can be very effectively used with absolute dating techniques (which are often expensive) to extend chronological measurement. Geologists and paleontologists have long used fossils to structure prehistory. For instance, the geological time scale for the history of life on Earth is built upon major changes in fossil populations, such as appearances and disappearances of groups. Because they are interested in a finer-scale approach, archeologists and anthropologists often look for evolutionary changes within groups. For this reason rapidly evolving lineages such as elephants, pigs, and horses have often been used to provide a biostratigraphic framework for earlier hominin evolution (Fig. 4.3). More generally, microfossils often provide important indicators of relative age.

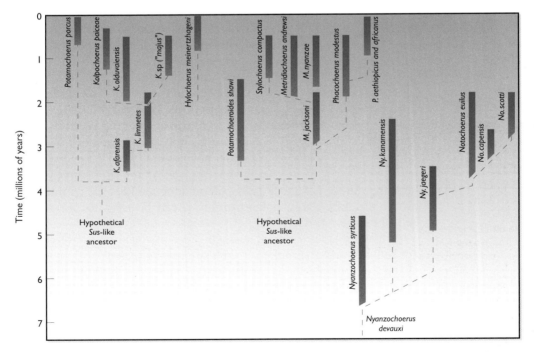

**FIGURE 4.3 Pigs and faunal correlation:** The evolutionary histories of pigs, elephants, and horses have been extremely useful tools for dating by faunal correlation, particularly in Africa. This figure shows the known evolutionary history of pigs. Hominin fossils found in association with a known pig fossil may be dated by reference to the timing of pigs' evolutionary history.

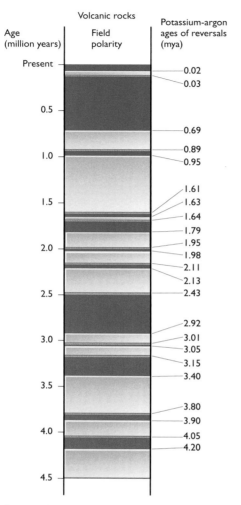

Volcanic rocks

| Age (million years) | Field polarity | Potassium-argon ages of reversals (mya) |

Present — 0.02
0.03

0.5

0.69
1.0
0.89
0.95

1.61
1.5
1.63
1.64
1.79
2.0
1.95
1.98
2.11
2.13
2.5
2.43

2.92
3.0
3.01
3.05
3.15
3.40
3.5

3.80
3.90
4.0
4.05
4.20

■ Normal
□ Reversed

4.5

**FIGURE 4.4**

**Paleomagnetic dating:** Occasional reversals in the direction of the Earth's magnetic field leave an imprint on iron-containing rocks as they form. The stripe pattern seen here represents the main reversals, and reference to it can help date sites.

The fossil records for many appropriate species are now known in some detail, and each displays important speciations and extinctions as well as directional evolutionary trends (such as in tooth size and morphology) in the period that covers human evolution. The principle behind the dating method is simple. If a hominin fossil is found in sedimentary layers that also include fossil pigs that are known to have lived between 2 million and 1.6 million years ago, for instance, then they provide a bracket for the date of the hominin. With a battery of evolutionary information derived from elephant, pig, and horse species, faunal correlation can be quite precise.[92]

The principle behind paleomagnetism is based on the fact that the Earth's magnetic axis reverses periodically. We are currently in what is known as "normal" polarity, where magnetic north coincides with geographical north. During reversals, which occur every few hundred thousand or million years, a magnetic needle would point south. As rocks form, particularly after volcanic eruptions or during deposition of fine-grained material, the direction of the magnetic field is recorded in the orientation of iron-containing particles. Geologists have accumulated much information about past polarities and have constructed a chart showing the dates of reversals (Fig. 4.4). Because there is such a firm paleomagnetic chart, in conjunction with some absolute dates, paleomagnetism can be used to provide virtually absolute age estimates.[93]

In paleomagnetic dating, a single piece of volcanic rock or certain types of sedimentary rock taken from a site can be tested for their polarity. By itself this information is insufficient to date a site, because the knowledge that a particular layer has reversed or normal polarity leaves many options open. A series of layers that reveal a relatively large section of the overall pattern is sometimes sufficient to provide a more secure date. In general, however, paleomagnetic dating is rather imprecise and is used in combination with other methods, particularly radiometric dating.

In principle, both faunal correlation and geomagnetic dating may be applied until at least the beginning of the Cambrian, 543 million years ago, as long as an absolute dating method has provided the calibration. In any case, both techniques are highly suitable for the period covering human evolution.

## Absolute dating techniques: radiopotassium dating

The majority of absolute dating methods are radiometric. All methods share the same two principles. First, some action sets a "clock" to zero, such as the heating that rock experiences during volcanic eruption or burial in the earth. Second, the consequences of some kind of radioactive decay steadily accumulate, thus recording the passage of time.

The most important radiometric technique that has been applied in earlier phases of paleoanthropology is potassium (potassium/argon) dating.[94] This technique is based on the fact that potassium-40, a radioactive isotope of potassium that makes up 0.01% of all naturally occurring potassium, slowly decays to argon-40, an inert gas. Rocks that contain potassium, such as volcanic rocks, slowly accumulate argon-40 in their crystal lattices. The high temperature experienced during eruption drives out the argon (and other gases) from the mineral, and the clock is set to zero – the time of the eruption. As time passes, argon-40 builds up, with the amount in any particular rock depending on the initial potassium concentration and the time since the eruption. The age calculation is based on measurements of the potassium concentration and the accumulated argon-40 in potassium-rich minerals, such as feldspar. A hominin fossil or artifact that is bracketed by layers of volcanic ash, known as tuffs, can therefore be dated.

What are the assumptions of absolute dating techniques?

A problem that lurks constantly with radiopotassium dating is contamination with older rock, which happens all too easily when ash is erupting from a volcano, for instance, or mixing with other minerals as it accumulates on the landscape. Even a few crystals of, for example, Cambrian-age rock in a gram of 2-million-year-old ash can produce an erroneously old date.

The first major application of the potassium/argon technique to paleoanthropology occurred in 1960, in an assessment of ash layers at Olduvai Gorge. In 1959, Mary Leakey found the famous *Zinjanthropus* fossil, the first early hominin discovered in east Africa, at this site. The date produced for the fossil, 1.75 million years, was double the age inferred by indirect means. Both the discovery of the fossil and the application of the dating technique represented major milestones for paleoanthropology.[95]

Since that time two important advances have taken place with radiopotassium-based dating. The first, developed in the 1960s, allows measurements to be taken in one sample rather than in two separate samples (one to measure potassium, the second to measure argon-40). The rock is initially irradiated with neutrons, which transforms the stable potassium-39 into argon-39; when the rock is then heated, the two argon

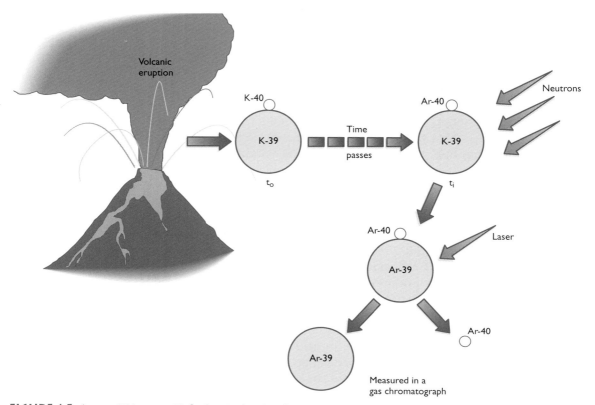

**FIGURE 4.5** **Argon-39/argon-40 dating:** Volcanic ash contains potassium-rich minerals, such as feldspar. A small percentage of the potassium exists as a radioisotope, potassium-40, which has argon-40 as one of its decay products. In the laboratory, crystals of feldspar are irradiated with neutrons, which converts the stable potassium-39 isotope to argon-39. The crystals can then be individually heated by laser beam, and the emitted argon-39 and argon-40 measured in a gas chromatograph. The argon-39 represents a measure of the total amount of potassium that was present in the crystal, and the argon-40 provides a measure of the time since eruption.

isotopes, 39 and 40, are released together and can be measured simultaneously on a gas chromatograph. The potassium-39 level provides a vicarious measure of the potassium originally in the rock, and the argon-40 measures the decay of potassium-40 since the rock was ejected from the volcano. This technique is known as argon-40/argon-39 dating (Fig. 4.5).

The second advance, developed during the 1980s, allows the technique to be applied to single crystals taken from volcanic ash, compared with the several grams required for the conventional technique. The advantages of the new technique, known as single-crystal laser fusion, are several, including avoiding the problem of contamination. Until recently the youngest rocks that could be dated with radiopotassium techniques were approximately 0.5 million years old. Recent work, however, has shown that rocks containing potassium-rich minerals can theoretically be accurately dated with ages as young as 10,000 years – a range that overlaps with the limits of radiocarbon dating. There is no effective upper limit of age estimation. However, the primary limitation of the potassium/argon family of methods is that it can only be employed in rocks of volcanic origin.

## Absolute dating techniques: fission track, radiocarbon, uranium series

The second radiometric technique is fission track dating,[91] which is often used in combination with radiopotassium methods. Naturally occurring glass often contains the isotope uranium-238, which decays through powerful fission. This event effectively burns a tiny track in the glass, which represents the ticking of the clock. Once again, the clock is set to zero during volcanic eruption, which expunges existing tracks. The longer the time after eruption, the more tracks that will accumulate, depending on the concentration of uranium in the glass.

The preparation of glass for the technique is tedious, however, and the counting of tracks not always reliable. In principle, this dating method can be applied to rocks as young as a few thousand years; in practice, the older the material, the more reliable the counting procedure.

Both the radiopotassium and fission track techniques depend principally on the presence of volcanic material. In east Africa, where many early hominin sites occur, volcanic activity has been common throughout the period of human evolution. By now, detailed chronologies of the many volcanic layers have been produced for the important fossil sites, such as the Awash, Hadar, and Omo regions of Ethiopia, Koobi Fora in Kenya, and Olduvai Gorge in Tanzania.

In addition, Frank Brown, of the University of Utah, has developed a sensitive chemical fingerprint method for identifying volcanic tuffs. Some eruptions in Africa extended many hundreds of thousands of square miles. By his technique, Brown has been able to identify which ash layers at one site were produced by the same volcanic eruption as ash layers at another site. Detailed correlation of chronologies is now possible among the sites. In contrast, volcanic episodes have been much less common in Eurasia throughout human occupation, so these techniques have found less application in those locations.

Radiocarbon dating is the best known of all radiometric techniques,[91] but because of its short time-depth can only be applied to the latest phases of human evolution. Most of the carbon dioxide in the atmosphere exists as a stable isotope, carbon-12. Some small percentage consists of carbon-14, a radioactive isotope that decays relatively rapidly. As plants incorporate carbon into their tissues, the ratio of the two isotopes in the tissues mirrors that found in the atmosphere. The same ratio applies for animal tissues, which effectively are built from plant tissues. Once an organism dies, however, the equilibrium between the isotopes in the air and in the tissues begins to change as carbon-14 continues to decay and is not replenished. As time passes, the ratio of carbon-14 to carbon-12 becomes increasingly smaller, a decline that forms the basis of the clock. Researchers can measure the proportions of the two isotopes in the organism's tissues and calculate when it died.

Although straightforward in theory, this technique nevertheless is plagued with variables that reduce its reliability. In principle, any organic material can be dated by the carbon-14 technique; in practice, many tissues decay too quickly to use this approach. The preferred material for dating by this technique is charcoal, as has recently been done on pigments in rock paintings in Europe and the United States. In Australia, rock paintings have recently been dated from blood that formed part of the pigment.

Contamination can represent a serious problem with radiocarbon dating (even a small amount of young material can substantially reduce the apparent age of older material). With the recent application of accelerator mass spectrometry to increase the sensitivity of measuring carbon-14, the useful range of the technique can be from a few hundred years to perhaps 60,000 years or a little more. A further problem with radiocarbon dating is that it assumes that the level of atmospheric carbon-14 is constant over time; it is now known that this is not the case, and it has been necessary to introduce a calibration to take this into account.[96,97] This recalibration problem means that some periods theoretically within the range of radiocarbon dating are notoriously difficult to date. This includes the period when modern humans were first moving into Europe.

Other methods of absolute dating include the uranium series technique, which relies on the decay of the radioisotopes uranium-238, uranium-235, and thorium-232, all of which decay ultimately to stable isotopes of lead. In addition, amino acid racemization has been used to date materials; this method depends on the slow transformation of the conformation of amino acid molecules used in living organisms (left-handed forms) to a non-living mixture (right- and left-handed forms). Neither the uranium series technique nor amino acid racemization is as powerful or as applicable to paleoanthropology as the other absolute dating techniques.

## Absolute dating techniques: thermoluminescence and electron spin resonance

Two other dating techniques depend on the principle that electrons in certain minerals become excited to higher energy levels when irradiated by radioisotopes of uranium, thorium, and potassium, which occur naturally in the ground and in cosmic rays.[98] The radioactive rays knock off the negatively charged electrons from atoms, leaving positively charged "holes." These electrons diffuse through the crystal lattice and usually recombine with other holes, returning to the ground state. But all minerals contain impurities, such as lattice defects and atoms that can "trap" roving electrons, keeping them at an intermediate energy level. Exposure to heat, such as fire (in the case of burned flint or fired pottery) or even sunlight (in the case of sand grains), dislodges trapped electrons; these particles then return to nearby holes, setting the clock to zero as in radiopotassium dating. The number of trapped electrons in a newly unearthed mineral therefore provides a measure of the time that has passed since the mineral was last exposed to heat. These dating techniques, known as thermoluminescence and electron spin resonance, measure these trapped electrons by different means – the former indirectly, and the latter directly.

In the thermoluminescence technique, the artifacts are heated under controlled conditions to release the electrons. As they return to the ground state the electrons release photons (light), which can be detected by sensitive instruments. Electron spin resonance detects the trapped electrons *in situ*, where they act as minute magnets that become oriented when exposed to a strong magnetic field. Microwave energy flips the orientation of the electrons, yielding a characteristic signal. The strength of the signal provides a measure of the number of trapped electrons. The electron spin resonance technique can be applied to tooth enamel, but not, as yet, to bone.

What are the chronological limits of different radiometric dating techniques?

In principle, both thermoluminescence and electron spin resonance techniques can reveal dates between a few thousand and 1 million years ago. This application range is particularly useful in paleoanthropology, because it fills a gap for material that is too old for radiocarbon dating and in practice too young for radiopotassium dating.

Recently the techniques were applied to date Neanderthal and modern human fossils in the Middle East. The results showed that the modern humans were at least 100,000 years old – not 50,000 years as had been inferred by other techniques. Until these results were produced, many scholars believed that the Neanderthals of the region, dated at 60,000 years old, were ancestral to the modern humans. The new dates revealed that this relationship was not possible, as the modern individuals were older than the Neanderthals. Although the situation then became even more complicated, the developments and subsequent rethinking of theories reveal how important accurate dates are for reliable interpretations.[99–103] The techniques are also especially useful for determining the time of the colonization of Australia, which is currently thought to have occurred 55,000 to 60,000 years ago.[104]

The suite of dating techniques available to paleoanthropologists in principle covers the past 5 million years (the period of primary interest) completely. Unfortunately, many important fossil and archeological sites lack material suitable for dating, are embedded in a stratigraphy too complex to unravel, or both. This problem has particularly affected the dating of the hominin sites from south Africa. These provide key evidence for the early stages of human evolution, and although their broad chronological position is known, their detailed dates are not. This is partly because they lack volcanic material suitable for dating, and partly because their stratigraphy is so complex.[105]

Even with the most precise methods of dating, there are potential snares for anthropologists who wish to know the age of a fossil. For

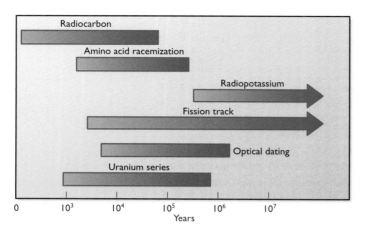

**FIGURE 4.6 Range of absolute dating methods:** The full range of absolute dating methods available to paleoanthropologists begins at a few hundred years and extends to many millions of years. The recent development of thermoluminescence and electron spin resonance dating filled an important gap, between radiocarbon dating and radiopotassium dating in practice.

instance, an animal may die on a land surface, become quickly buried and its bones ossified. Later, various geological forces may uncover the fossil and redeposit it in layers deeper than the original, making the fossil appear older than it really is; or in layers higher up in the section, giving an erroneously young age. Incorrect dating can apply to artifacts, for the same reasons. One development that is helping to overcome these problems is the ever-finer physical and chemical analysis of mineral matrix that may be accreting to the fossil or artifact. With this type of assessment, it is sometimes possible to identify the layer that originally contained the object, if the layer still exists at the site.

The problems that paleoanthropologists still face with obtaining secure dates for important objects from the prehistoric record will become apparent throughout the pages of this book, which feature the frequent litany, "of uncertain age." However, new techniques and refinements of existing techniques are continually developing, so that the chronological framework of human evolution is incomparably more secure than it was half a century ago, or indeed a quarter of a century ago (Fig. 4.6). It is worth bearing in mind that in 1950 there were no absolute dating techniques available (with the exception of some very chronologically limited ones such as varve chronology or dendrochronology, based on counting annual layers in periglacial meltwater deposits and tree trunks respectively). Estimates of the actual time-depth of human evolution were necessarily vague and prone to error, and the construction of the time scale of evolution is one of the major achievements of paleoanthropology's sister discipline of the earth sciences.

Finally, it is important to remember that although a site may be dated, it does not necessarily mean that that date is very precise. It is always important to bear in mind that all dates have errors associated with them, and these errors limit the confidence with which material can be discussed.

# THE SCIENCE OF BURIAL

The fossil and archeological records serve as the principal sources of evidence upon which human prehistory is reconstructed. Unless that evidence can be interpreted with some confidence, the reconstruction – however

**KEY QUESTION** How is information about the past (evolutionary history) affected by the way in which fossils are formed and preserved?

convincing – may not be valid. In recent years, a tremendous emphasis has been placed on understanding the multifarious processes that impinge on bones and stone artifacts that become part of the record. The science of taphonomy, the study of the processes of burial and fossil formation, has revealed that the prehistoric record is littered with snares and traps for the unwary.[106]

A taphonomist in a pessimistic mood has been heard to argue that, because of the countless complicating factors that can plant false clues in the record, the chances of reconstructing the past are virtually nil. More generally, however, a sense of optimism prevails that specific problems in taphonomy are being solved step by step. Through a combination of ever more careful study of material from the prehistoric record and the development of ingenious experiments and observations on modern material, it is becoming possible to scrutinize the material evidence of human history with the required degree of confidence.

Death is a bewildering, dynamic process in the wild. First, many animals meet their end in the jaws of a predator rather than passing away peacefully in their sleep. Once the primary predator has eaten its fill, scavengers, which in modern Africa would include hyenas, jackals, vultures, and the like, move in. The carcass is soon stripped of meat and flesh and the softer parts of the skeleton, such as vertebrae and digits, are crushed between the devourers' powerful jaws. The remaining bones dry rapidly under the sun. Even in this initial phase the skeleton is probably partially disarticulated, with hyenas having torn off limbs and other body parts to be consumed in the crepuscular peace of their dens. Passing herds of grazing animals bring a new phase of disarticulation and disintegration as hundreds of hooves kick and crush the increasingly fragile bones.

*How do fossils form?*

Thus, within a few months of a kill, the remains of a zebra, for example, might be scattered over an area of several hundred square meters, and a large proportion of the skeleton would apparently be missing. Some of the skeleton may indeed be miles away, lying among the cache of bones in a hyena's den. Some bones will have been shattered and disintegrated into minuscule pieces. Others will have been compressed into the ground by the pressure of passing hooves, often being splintered in the process. Only the toughest skeletal parts, such as the lower jaw and the teeth, remain intact.

To this can be added the more human factors that can affect fossil preservation and the formation of archeological sites. Hominin behavior is an important determinant of what survives – for example, whether a community lives in caves or in open sites, whether it stays in one place or moves camps frequently, whether it hunts (and so has bone residues) or whether it gathers plant foods (which stand much less chance of preservation). Indeed some people have suggested that hominins in many parts of Asia during the Pleistocene depended far more on bamboo for technology than on stone, and so there is a much poorer archeological record.[107]

Given these factors and the fate that awaits most animals in the wild, then, it is perhaps unsurprising that the fanfared announcements of ancient hominin discoveries typically mean an interesting jaw, an arm bone, or, rarely, a complete cranium. The most complete specimen found to date is the famous "Nariokotome (Turkana) boy," whose virtually complete skeleton was found in deposits on the west side of Lake Turkana in 1984.[108] Dated at approximately 1.5 million years old, this *Homo erectus* specimen lacks only a few limb bones and most of the bones of the hands and feet. The individual, who was about 11 years old when he died, came to rest in the shallows of a small lagoon. Even this case is marred by evidence of passing animals, in the form of a limb bone that was snapped in two as a hoof stood on it, pressing it into the soft sand.

## The dynamics of burial

To become fossilized, a bone must first be buried, preferably in fine alkaline deposits and soon after death. Rapidity of burial following death is surely the key factor in determining whether a bone will enter the fossil record. Most hominin specimens have been found near ancient lakes and rivers, partly because our ancestors (like most mammals) were highly dependent on water, and partly because these sites provide the depositional environments favoring fossil formation.

As it happens, the forces that can bury a bone – for example, layers of silt from a gently flooding river – can later unearth it as the river "migrates" back and forth across the floodplain through many thousands of years. When this removal occurs, the bones become subject once again to sorting forces. Light bones will be transported some distance by the river, perhaps to be dumped where flow is slowed, while heavier bones are shifted only short distances. Anna K. Behrensmeyer, a leading taphonomist at the Smithsonian Institution (Washington, DC), identifies transport and sorting by moving water as one of the most important taphonomic influences.[109] Abrasions caused when a bone rolls along the bottom of a river or stream provide tell-tale signs of such activity, as do the characteristic size profiles and accumulations in slow-velocity areas of an ancient channel. For hominin remains, this activity often results in accumulation of hundreds of teeth and little else, as the researchers working along the lower Omo River in Ethiopia know only too well.

Large numbers of hominin fossils have been recovered from the rock-hard breccia of a number of important caves in south Africa. At one time, hominins were thought to live in these caves, and the bones of other animals found with them were suspected to represent remains of food brought there to be consumed in safety. In addition, the fractures and holes present in virtually all hominin remains were considered to be the outcome of hominin setting upon hominin with violent intent. In many ways, the south African caves present one of the most severe taphonomic problems possible, but with years of patience a group of workers (in particular, C. K. Brain) has cut through the first impressions and progressed a little closer to the truth.[110]

Is the fossil record biased?

Most of the bone assemblages in the south African caves were almost certainly the remains of carnivore meals accumulated over very long periods of time. The profile of skeletal parts present matches what would be expected after carnivores had eaten the softer parts. In addition, the damage recorded in the hominin crania found at these sites simply reflected the compression of rocks and bones into them as the cave deposits mounted. Exactly how much time is represented in these fascinating accumulations, and when they occurred, are difficult to determine. But the question, as in many taphonomic investigations, is a key one.

One area of investigation in which taphonomic analysis has been particularly crucial in recent years is in the study of ancient assemblies of bones and stones – in other words, putative living sites. Some of the best-known and oldest of these sites occur in the lowest layers of Olduvai Gorge, Tanzania, and are dated to almost 2 million years ago.[111] These concentrations of broken bones and chipped stones have long been assumed to be the product of hunting and gathering activity such as that seen among surviving foraging peoples. The occurrence of such sites appears to increase in frequency through time, giving the impression of an unbroken trail of litter connecting people ancient and modern who shared a lifestyle.

In some cases, however, careful taphonomic analysis of the geological setting and the composition of the bone and stone assembly has shown such "sites" to result from water flow, with the material having been dumped by a stream in an area of low energy – in other words, the assembly is not an archeological site, but a hydrological jumble. Even when a collection of bones and stones can be shown not to be produced by water flow, there remains the task of deciding how the various materials reached the site, and whether they were related. For example, did early hominins use the stones to butcher carcasses?

Taphonomists have determined the stages through which bones go as they lie exposed to the elements. This process, known as weathering, can be calibrated, and by looking at the degree of weathering evident in a fossil bone, it is therefore possible to determine how long the bone lay on the surface before its burial. Applying this technique to the sites at Olduvai reveals that in many cases bones accumulated over periods of 5 to 10 years. This would be unheard of in modern hunter-gatherer sites, which are occupied only briefly.[112]

## Clues from marks on bones

In the late 1970s and early 1980s, several researchers discovered on the surface of a small percentage of the Olduvai bones what appeared to be marks made by stone tools (Fig. 4.7). Thus, although the sites might not have been typical hunter-gatherer home bases, it did appear that a connection existed between the bones and the stones: the hominins almost certainly were eating meat. By looking at the pattern of distribution of

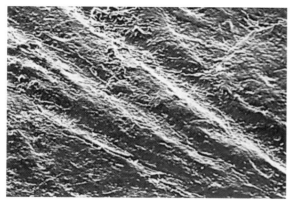

(a)

(b)

(c)

**FIGURE 4.7 Bone surfaces under the electron microscope:** (a) The surface shows the round-bottomed groove made by a hyena gnawing at a modern bone. (b) A sharp stone flake makes a V-shaped groove in a bone surface (modern). (c) This fossil bone from the Olduvai Gorge carries carnivore tooth marks (t) and stone flake grooves (s); the scavenger activity followed the hominin's activity on this occasion. (Courtesy of Pat Shipman and Richard Potts.)

cutmarks over a bone – on the shaft as compared with the articular ends, for example – investigators can obtain some idea of whether the marks were made during the disarticulation of a carcass or during the removal of meat or skin from the bone.

Determining the identity of marks on the surface of fossil bones is an important taphonomic activity: gnawing carnivores and nibbling porcupines can all leave their signatures. Likewise, sand grains can leave behind tell-tale signs. In 1986, Behrensmeyer and two colleagues from the Smithsonian Institution reported that bones trampled in sandy sediment can sustain abrasions that are virtually indistinguishable from genuine stone-tool cutmarks.[113] "Microscopic features of individual marks alone provide insufficient evidence for tool use versus trampling," warn Behrensmeyer and her colleagues. "If such evidence is combined with criteria based on context, pattern of multiple marks and placement on bones, however, it should be possible to distinguish the two processes in at least some cases bearing on early human behavior."

Not all taphonomists agree about the difficulty of distinguishing between the effects of trampling and genuine cutmarks, however.[114] For instance, Sandra Olsen and Pat Shipman have examined the problem experimentally and stated: "Macroscopic and microscopic comparison of experimentally trampled bones and those which have had soft tissue removed with a flint tool demonstrate significant differences between the surface modifications produced by the two processes."

Taphonomic studies can appear to be a pessimistic approach to the science, often undermining previous interpretations. While some aspects of the work do take this form, it is also important to take note of other developments in the way fossils are studied which are greatly increasing the level and accuracy of information obtainable from fossils and archeological remains. Several of these are based on new scientific techniques; for example, trace elements of chemicals in bone have been extracted to determine either the environment in which a hominin lived, or else the diet it had.[115] Another example is the use of high-powered microscopy to look at wear patterns on teeth, again to determine diet.[116] Similar techniques have been used on stone tools, showing what they were used for, and in some cases it has even been possible to retrieve plant or animal residues from the edges of these tools.[117] Greater use of scientific techniques is enhancing the information obtainable from the often sparse fossil remains.

## The transformation of human behavior into the fossil and archeological record

Taphonomy adds yet another caveat to the unwary reading of the prehistoric record, but it is part of a larger process by which prehistorians must think about how the information they use is actually produced. The underlying principle is that observations of life in the past are seldom if ever direct, but are the product of a series of transformations, and it is the role of the prehistorian to work back through those transformations in reverse. Much of the work related to early hominins has focused on what might be termed the geological processes involved in transforming a living animal into a fossil. However, equally important are the behavioral components.[118] An archeological site, for example, is the end product of humans living at a particular point on the landscape for a period of time. This might be a cave or an open site, and different things might happen in each; in a cave the hominins might push all the debris over the edge, so that the actual archeological material recovered would be secondary to where they lived. In an open site things might be left where they are dropped, in the middle of the habitation, and the whole area might then be abandoned when another settlement is started. In this case the material might retain more of its original living structure.

The only way in which a reliable knowledge of the past can be gained is by working at two levels. At one level it is a question of using as many scientific techniques drawn from the earth sciences as possible to understand what has happened to archeological and fossil material once it has entered

How does human behavior influence the archeological record?

the ground. At the other level is a focus on what humans (and other primates) actually do, and how this may then have the potential to appear in the record. This approach, using analogies, can be what is called actualistic (observing and measuring what people do today) or else experimental.[119,120,121] Furthermore, this applies not just to the remote past, but equally to the more recent archeological record of hunter-gatherers and the first farmers.

## BEYOND THE FACTS
# ORDER AND RATE

*The issue: dating and chronology are central to any evolutionary reconstruction, and new scientific techniques are greatly enhancing our ability to know what happened when in the past. However, we need to think about whether it is more important to know exactly when things happened, or the order in which things occurred.*

At first sight it would seem immediately apparent that absolute dating is greatly superior to relative dating. To know that a fossil is 1 million years is surely better than to know that it is older than another one. Naturally, in an ideal world that is the case, and we can hardly overemphasize the impact of a sound chronological framework for human evolution.

It is, though, worth considering for a moment the meaning of the terms absolute and relative. An absolute date provides a number – for example, 1 million years old. However, no matter how good the technique, that number is of course not quite the same as reading from a clock; it is the result of a whole series of calculations and estimates. There are a number of steps from a crystal to a date. With each inferential step along the way there are potential errors, such that in the end a date is actually a statistical estimate. That means that rather than the date being 1 million years exactly, it is actually $1.0 \pm 0.1$, where the $\pm$ term is the standard error on the date. Thus this date really means that there is a 95% probability that the date lies between 0.9 and 1.1 million years ago. In other words, in terms of certainty, the date actually has an error of 200,000 years. Where the 95% confidence limits of two dates overlap, it is a way of saying that there is no way of knowing which is the older, which the younger.

Compare this with a relative date. True, we do not know the actual age, but we do know, if one fossil is stratigraphically superimposed over another, that one is older than the other. In other words, with absolute dating we know something with the potential to be highly precise, but only with an inbuilt error; with relative dating there is less precision, but what is known is known with far greater certainty.

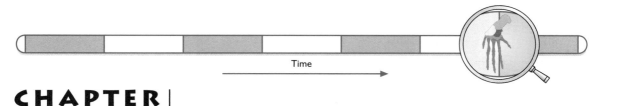

# CHAPTER
# 5

# THE SYSTEMATIC CONTEXT

## SYSTEMATICS

O ne of the major challenges in learning about human evolution is the abundance of names, either in Latin or anglicized from Latin – *Australopithecus*, australopithecines, *Ardipithecus ramidus*, etc. Furthermore, these names are often either ephemeral in the literature, or else change their meaning over time. The genus *Zinjanthropus*, for example, a name coined by Louis Leakey to describe his first major find, has virtually disappeared from the literature, while the term *Homo erectus*, which used to be applied generally across all Middle Pleistocene hominins, is now restricted by many to a small group of Asian fossils. Even worse, the same words can be used by different people to describe different sets of fossils. *Homo sapiens* means to some only the hominins who are anatomically identical to us, while to others it refers to a whole range of larger-brained types, including Neanderthals.[122,123] If it is any consolation to the student, this is not unique to paleoanthropology, but is rife throughout evolutionary biology.

However, the names are important, for they represent evolutionary relationships and history. The names used to describe fossil or living groups are the outcome of analyses of their similarities or differences, and so biologists have devoted considerable energy to determining how to assess biological relationships in evolutionary perspective – otherwise known as systematics.

Systematics is the study of the diversity of life and the relationships among taxa at all levels in the hierarchy of life, from species to genus to family to order, and so on up to kingdom. A taxon (singular of "taxa") is a category of organisms at any level in that hierarchy: a species is a taxon, as

> **KEY QUESTION** How do we determine evolutionary relationships?

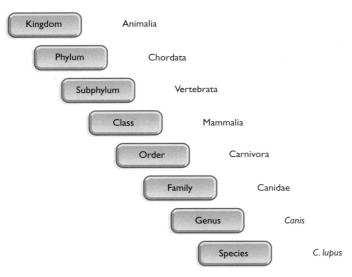

**FIGURE 5.1 Linnaean classification:** This system is hierarchical, with higher groups being inclusive of all those below. Developed in the mid-eighteenth century, the system is still used today.

Why is classification important?

is a genus, a family, an order, and so on. Conventionally, taxa above the level of genus are referred to as higher taxa.

To communicate unambiguously about the diversity of life and relationships within it, biologists require a consistent method of classifying the taxa of interest. The modern system of classification goes back to 1758, when the Swedish scientist Linnaeus published his definitive edition of *Systema Naturae*. This section will first address principles of Linnaean classification, followed by theories and practice of classification. Traditionally, classification has been based on anatomical characters. More recently, molecular data have been used. The advantages and disadvantages of both approaches will be discussed.

Linnaean classification is hierarchical, with every rank in the hierarchy including all those below it. We can illustrate this with the grey wolf, *Canis lupus* (Fig. 5.1). Species are grouped into genera; in this case the grey wolf appears in the same genus as the golden jackal, *Canis aureus*. Genera are grouped into families; here, the wolf and the jackal are in the same family as foxes (genus *Vulpes*), with the family name Canidae. Several families constitute the order Carnivora; and the Carnivora combine with other mammalian orders (including primates) to form the class Mammalia. The class Mammalia joins with other vertebrate classes (such as Reptilia and Aves – reptiles and birds) to form the subphylum Vertebrata. This and other subphyla form the phylum Chordata, which is one of approximately 30 animal phyla that constitute the kingdom Animalia. Between these major taxonomic groups, several other categories exist, such as subfamily and superfamily, for instance, and tribe (intermediate between genus and subfamily), allowing a subtle and flexible classification to emerge. The inclusive hierarchy applies throughout.

The basic unit of Linnaean classification is the species, whose identification includes two parts: the genus name and the specific name. In the *International Code of Zoological Nomenclature*, this identification is termed a binomen. It represents the combination of the genus and specific name that is unique to a species, such as *Australopithecus africanus* (an early hominin). Different species may share the same specific name but are linked to a different genus name, such as *Proconsul africanus* (a fossil ape). The laws governing the naming of species are quite strict under the Code, so that if a species is reclassified (on the basis of new discoveries, for instance), the genus name may be changed but the specific name must remain the same. The swamps of nomenclatural minutiae are not to all tastes, but they are important for consistency.[124]

# Philosophies of classification and systematics: phenetics and cladistics

So much for the mechanics of classification. But how do we arrive at it? For Linnaeus, the criterion was simply anatomical similarity and, naturally, had nothing to do with evolution. After 1859 and the publication of Darwin's *Origin of Species*, however, biologists could approach classification with evolution explicitly in mind, if it was considered appropriate. Darwin argued that because species are related by common descent, genealogy represented the only logical basis for classification. Recent years have witnessed surprisingly heated debate over precisely how classification should be performed. Should it emphasize the results of evolution, in terms of adaptation? Or should it reflect relatedness, or phylogeny, as Darwin argued? This issue is particularly pertinent when classifying the great apes and humans.

The issue can be seen easily by considering what happens in evolution. As a species or lineage evolves, it become different from its "parent" taxon, and it branches from it so that it evolves independently. Differences in methods of classification basically reflect the importance given on the one hand to the amount of change (that is, adaptation and drift seen in morphology and genetics), and on the other to the sequence of branching events regardless of the amount of change that has occurred (Fig. 5.2). In an ideal situation, of course, the two should reflect the same thing, but sometimes this does not occur, and so priority has to be given to one or the other – branching sequence or quanta of change. Out of this arise different schools of classificatory theory.

In very general terms there are two basic approaches to classification (Fig. 5.3). One is generally referred to as phenetic. This approach looks at as many features as possible, and constructs relationships on the basis of overall similarity and difference. It is closely associated with numerical taxonomy, which uses traits such as the sizes of particular bones to determine relationships between taxa. Because similarity does not always reflect evolutionary relationships based on descent (the aim of classification being to identify such relationships), due to convergence, phenetic relationships may not always be accurate. The second approach is known as cladistics (also called phylogenetic systematics), which emphasizes only phylogeny and branching

**FIGURE 5.2 The process of evolution and classification:** Biological classification arises out of the process of divergence that occurs in evolution, and should reflect what has happened in history. However, the process of evolution involves the direction of change, the sequence of change, and the amount of change, and these can produce conflicting classificatory priorities.

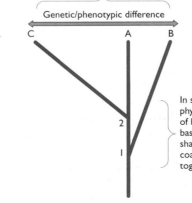

In some classificatory systems (e.g. phenetics) greater emphasis might be placed on the amount of evolutionary change that has occurred. In this case A and B might be grouped together.

Genetic/phenotypic difference

C          A     B

Time

2

1

In some classificatory systems, especially phylogenetics and cladistics, the sequence of branching events is the key or only basis for classification. In this case A and C share a more recent branching point (or coalescence) and so would be grouped together.

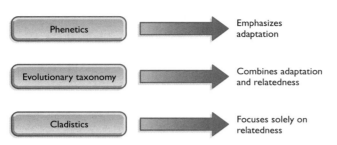

**FIGURE 5.3**
**Approaches to classification:** Different methods of describing relationships among organisms effectively emphasize different aspects of the world. For instance, by concentrating on characteristics that reflect genetic relatedness, cladistics produces an evolutionary tree. In contrast, phenetics measures all aspects of similarity among organisms and therefore emphasizes similarities in adaptation.

How do we measure evolutionary closeness?

sequence. To deduce branching sequence it seeks to identify and use in the classification only those traits which can be identified as having been derived from common ancestors, and therefore reflect evolutionary relatedness.

If evolution proceeded at regular rates, so that after branching two lineages diverged steadily in terms of morphological adaptations, then the phenetic pattern would be identical to the phylogenetic pattern. This generally does not happen. Sometimes a new lineage will diverge quickly, accumulating many evolutionary novelties that put a great morphological distance between it and its sister species; sometimes a new lineage will remain almost identical to its sister species over vast periods of time, with the morphological distance remaining minimal while genetic distance increases. As a consequence of these different tempos of evolution, phenetics will sometimes yield a different pattern from that produced by cladistic analysis.

These two approaches are to some extent extremes in a continuum of ways of trying to build phylogenies.[125] The choice of a classification system is a matter of philosophy and what question is addressed: should the grouping be developed according to overall morphological similarity, which emphasizes adaptation? Or should it reflect relatedness? Which is the more "natural" system? Proponents of phenetics claim that their analysis is completely objective and completely repeatable, and therefore will reflect meaningful patterns in nature. Cladists argue that the phylogenetic hierarchy (branching sequence) is the only important reality, whether we discover it or not. Only one pattern of phylogenetic branching exists – the path that evolution actually followed. The challenge is being able to infer that pattern from the morphology and other evidence, such as genetics.

Since the 1960s there has been a shift away from phenetics toward cladistics, and this is the technique that is now used most often in paleoanthropology and other branches of evolutionary biology. Its terminology and concepts are often quite hard to grasp, and so the next section focuses on these and some of the problems that arise from a cladistic approach.

## From homology to cladistic concepts

Although there are important differences between phenetics and cladistics, nonetheless they both proceed from the same basic principles. Fundamental to any evolutionary classification is the concept of homology or the "homologue."[32,126] At its simplest level, this means that in any comparison we must compare like with like (Fig. 5.4). If we measure the length of a skull in a baboon and a horse, we must make sure that the measurements are identical in terms of the parts of the anatomy they

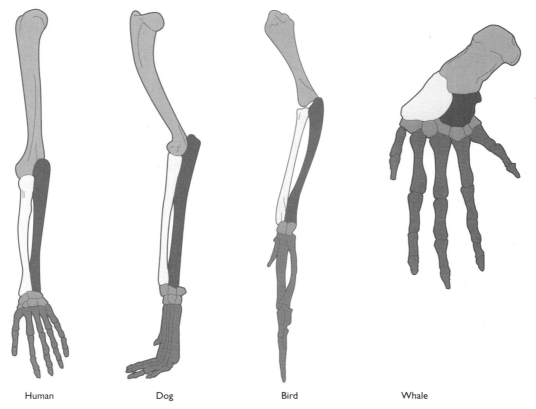

Human          Dog          Bird          Whale

**FIGURE 5.4  The principle of homology:** The biological derivation relationship (shown by colors) of the various bones in the forelimbs of four vertebrates is known as homology and was one of Darwin's arguments in favor of evolution. By contrast, the wing of a bird and the wing of a butterfly, although they perform the same task, are not derived from the same structures: they are examples of analogy.

represent. The only basis on which that can be done is if the character or morphological landmarks being measured were present in a common ancestor, and therefore have what can be thought of as a single origin (Fig. 5.5). While determining what is or is not a homologue can be difficult in practice, it is an essential part of any classificatory system.

While all classificatory systems use homologies, the procedures vary. A vertebrate species' morphology is composed of a large suite of anatomical characters: shapes of bones, patterns of muscular attachments, skin color, and so on. Phenetics compares as wide a range of characters as possible between a group of species to produce multivariate cluster statistics, which are effectively an average of all such comparisons. The more characters that are included, the more objective the technique is said to be, automatically spitting out a phenetic hierarchy from the assembled cluster statistics. In fact, practitioners frequently must choose among several possible patterns, for all that is produced is a measure of statistical distance, not an actual phylogeny, which must in turn be inferred from the statistics.

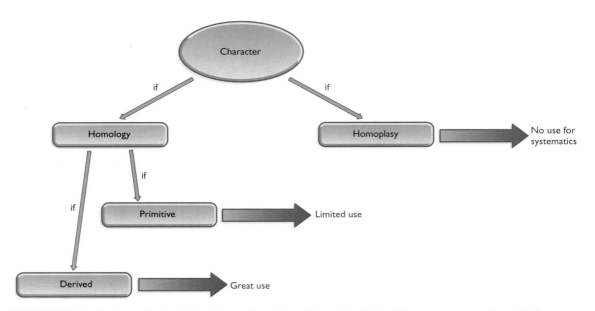

**FIGURE 5.5  Deducing relationships:** A genetic relationship can be deduced between two species only if homologous – not analogous – characters are used. Homologous characters come in two forms: primitive and derived. Primitive characters have limited use in deducing relationships because they occur in the ancestor of the group and therefore give no information about species within the group. Derived characters are the key to relationships because they occur in only some of the species under study and therefore can be used to differentiate within the group.

Pheneticists are using the total morphological pattern to determine evolutionary relationships, and although in practice they recognize that some traits are more important than others, it is the overall structure that is sought. However, one of the criticisms made of phenetics by cladists is that not all traits are equal in evolution. Starting from the common ground of homology, they argue that the only traits that can be used to classify organisms must be truly homologous – in other words, they must have been present in a common ancestor. From this point, though, they make a critical distinction which is fundamental to the method. Species share characteristics because they have a common ancestor, but simply adding up the number of homologous similarities does not provide a good guide to evolutionary relationships.

The key distinction that cladists make is between "primitive" traits and "derived" traits (Fig. 5.6). Primitive characters are those inherited from the ancestral stock for that group. For instance, baboons, chimpanzees, and humans all have nails on the ends of their fingers. These species are not uniquely linked by this character, however, because New World monkeys and all prosimians (lower primates) have fingernails as well. Fingernails are a characteristic feature of all primates. For baboons, chimpanzees, and humans, the possession of fingernails is therefore a primitive character with respect to primates. From the point of view of classification, counting

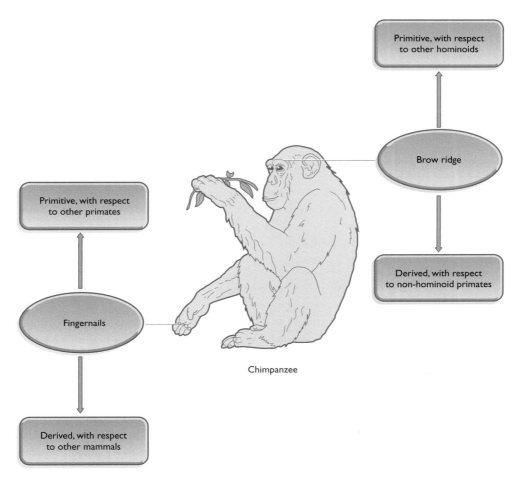

**FIGURE 5.6 Relative status of characters:** The state of a character depends on the reference point. For instance, for an ape fingernails are considered primitive in relation to other primates because all other primates have fingernails. Thus, fingernails would not serve to distinguish apes from, for example, monkeys. Fingernails are a derived character for primates as a whole, however, because no other mammals have them. Thus, fingernails serve to distinguish primates from other mammals. The second character illustrated here – brow ridges – is found only in hominoids, not in other primates, and is therefore derived for hominoids. This character distinguishes apes from monkeys. In a chimpanzee, however, brow ridges would be considered to be primitive with respect to other hominoids; that is, the character would not distinguish a chimpanzee from, for example, a gorilla.

this as one of the things that makes apes and humans uniquely related is unwarranted – because it is a trait that relates to much earlier evolution, it should be excluded from any consideration of the particular relationships between apes and humans.

On the other hand there are some traits that are shared uniquely among chimpanzees, humans, and Old World monkeys such as baboons, and are absent from New World monkeys and prosimians. One example of this is the loss of a premolar, so that there are only two premolars in the jaw.

In this case, these attributes represent shared derived characters for the Catarrhini (the infraorder that encompasses the Old World monkeys, apes, and humans as a group) with respect to primates. In contrast to the example of nails, this presence of only two premolars is a useful trait for distinguishing the relationships of the Catarrhini.

Obviously, the classification of homologous characters into primitive and derived is always relative to the level of the hierarchy being considered. For instance, although the possession of fingernails is a primitive character within the Catarrhini with respect to other primates, it is a derived character for primates as a whole: it distinguishes them from other mammals.

Needless to say, given that this is biological systematics, this method comes with a suite of specialized words which are essential for reading the literature. As described above, the key distinction is between primitive characters, which are known as plesiomorphies, and derived characters, which are known as apomorphies. If a trait is shared across a number of taxa, and defines it as a clade, then it is said to be synapomorphic, or in other words, a shared derived trait, as opposed to one that is unique to a single taxon (autapomorphic).

To infer a unique phylogenetic relationship among a group of species, one must identify derived characters or apomorphies – the evolutionary novelties that separate the species from their common ancestor. This idea, simply stated, is the principle behind cladistics. A collection of all species with shared derived characters that emerged from a single ancestral species is said to be a monophyletic group, or clade; a diagram indicating relationships is a cladogram (Fig. 5.7). Cladists reject groups which are not monophyletic – that is, are paraphyletic and polyphyletic groups – as unnatural groups. A paraphyletic group contains a subset of descendants from a single ancestor. If, for instance, only some descendants of the common ancestor diverged significantly from the original adaptation, the phenetic approach would recognize two groups; those that remain most

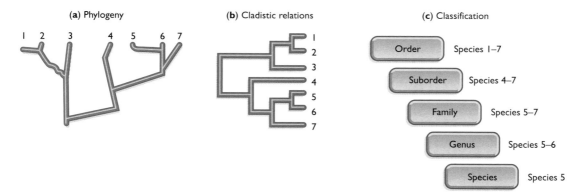

**FIGURE 5.7 Relation between phylogeny, cladistic classification, and Linnaean classification:**
(a) The phylogeny of seven species. (b) A cladistic classification. (c) The Linnaean classification for species 5.

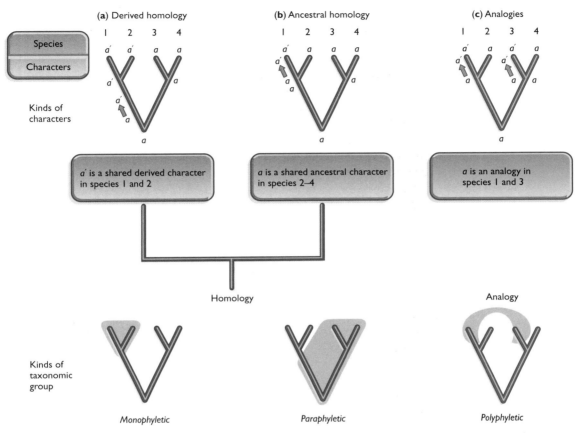

**FIGURE 5.8 Different types of groups:** In (a), the evolution of shared derived characters leads to the formation of a monophyletic group. When a species (or several species) diverges significantly from the ancestral status, it may be excluded, leaving a paraphyletic group (b). Convergent evolution may yield species with similar adaptations; these species may be encompassed within a polyphyletic group (c). Cladists recognize only monophyletic groups as being natural groups because they truly reflect phylogeny. Pheneticists accept the reality of both paraphyletic and polyphyletic groups because they reflect the results of evolution, or adaptation.

similar to the ancestral state form the paraphyletic group. Polyphyletic groups are those where different taxa may have different ancestors, implying convergence. Paraphyletic groups include the common ancestor, while polyphyletic groups do not, but only in monophyletic groups is the ancestor unique and universal (Fig. 5.8).

Is cladistics the best method for reconstructing evolution?

The theory of cladistics is often elegant and convincing, and its logic in terms of evolutionary processes is impeccable. It has many advantages over phenetics, not the least of which is that it is very explicit in the assumptions it makes, especially with regard to the fact that not all characters are equally useful for the purposes of classification. In practice the difference is not as great as the theory might suggest, as a good pheneticist would make judgments about the value of a character before including it in an analysis. Equally, a cladistic analysis also requires making biological judgments at the beginning.

For example, to assign a character to a species it is necessary to have made an initial judgment about what those species are – what are called the operational taxonomic units or OTUs – and to have placed them into at least some sort of phylogenetic position; in other words, for cladistics to work it is necessary to have some *a priori* knowledge of the relationships and units. Perhaps more importantly, it is also necessary to have a good hold on whether similarities are due to genuinely homologous traits, or to the same trait occurring more than once in evolution. This is the trap of convergent evolution, which is frequently difficult to detect.

Convergent evolution can operate on a gross level. For instance, it produced superficially similar "wolves" among placental and marsupial mammal populations that were separated for more than 100 million years – the Tasmanian wolf of Australia compared with the true wolf of Eurasia. It may also work on a discrete level, such as the details of dental and jaw anatomy – a consideration that is particularly pertinent to paleoanthropologists because dental and jaw parts are the most frequently recovered fossil specimens of hominoids. It is common for a particular cusp to be added or lost, or for teeth to become smaller or larger, in different lineages. Convergence results from the independent evolution of characters of similar function, and teeth and jaws are particularly susceptible to parallel adaptation. In cladistic terms these are referred to as homoplasies, and are in effect the characters that go against the reconstruction of a true phylogeny. In many studies it has been shown that in fact homoplasies occur with quite high frequencies – on average around one third of the characters may have evolved more than once in a set of organisms – which is one of the reasons that systematics is such a difficult procedure.[127,128,129]

## Cladistic practice

The cladistic approach was originally developed by the German systematist Willi Hennig in 1950,[130] and in recent years it has become the preferred approach for many researchers in paleoanthropology. Determining relationships between species involves a number of steps (Fig. 5.9):

1 The OTUs must be defined – for example, recognizing bonobos, chimpanzees and humans as separate species in an analysis. (In phenetics this is not necessary at the outset, as the basic units emerge out of the statistical properties of the populations examined.

2 The traits to be used must be defined. This step basically consists of listing features that are biologically meaningful and can be observed; these traits are such things as the size of the canine, whether the chin slopes forward or backwards, etc. While at one level this is a straightforward step, at another it is more complex, as the traits may be relatively arbitrary and may also covary. For example, if you take brain size as a trait

1 Define **operational taxonomic units** (OTUs): these may be species, populations, or higher taxonomic units.
2 Define **traits** that will be used for the analysis: these will normally be morphological characters such as teeth, tails, etc.
3 Define **states** that the traits can exhibit or possess: these will be, for example, the presence of a tail versus the absence of a tail. In this case there are just two states, but in many cases there may be several graded ones (absence of brow ridge, slight brow ridge, prominent brow ridge, etc.).
4 Define polarities for the states, or the order in which they are likely to change in evolution.
5 Define ancestral state or plesiomorphic character. This will be determined either by establishing it from an **outgroup** (another taxon that may be related to the sample taxa more distantly – e.g. if the sample constitutes hominoids, then the cercopithecoids might be used as an outgroup). Alternatively, developmental or ontogenetic stages can be used.
6 Assign values (states) for each trait to each of the taxa, thus creating a data matrix.
7 Construct **cladograms** representing the most parsimonious or likely pattern of change among the OTUs, and thus the most probable evolutionary history.
8 Evaluate the statistical validity of the set of trees and the path of state changes to see how particular characters have evolved.
9 Construct an evolutionary **phylogeny** and scenario derived from the cladograms.

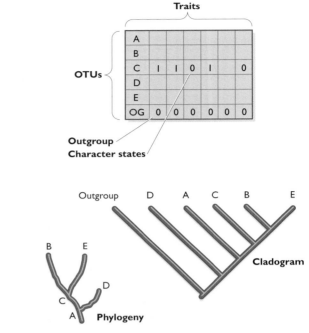

**FIGURE 5.9 Steps in cladistic analysis:** To carry out a cladistic analysis a number of distinct steps must be followed, from defining the basic taxonomic units to the interpretation of evolutionary history.

(large or small), and then also take cranial length, breadth, and height, this may appear as three traits, but obviously they are all closely related to each other; it is difficult for brain size to increase without an increase in at least one of the other three.

3 The traits having been defined, it is then necessary to define the states they can have. If we take canine size, for example, we could say that it has three possible states – small, medium, or large; for a tail we might recognize two states – present or absent. Here we again can see both the simplicity of the method, and hence its applicability very widely, and also its arbitrariness – how big does a canine have to be to be classified as being "large"? Early cladistic analyses, in contrast to phenetics, were largely qualitative in the definition of traits, but there has been great progress in developing more quantitative approaches and methods for defining character states.

4 Once the states have been defined, it is then necessary to define what are known as their polarities – that is, to indicate the direction of change from primitive or plesiomorphic to derived or apomorphic. There are normally two ways of doing this. The first is to use what is called an outgroup – that is, to go beyond the taxonomic units being studied, and look at the condition in the next most closely related species. For example, if we are interested in hominoids, which are characterized by

an absence of tails, we can look across the primates more generally, and indeed the mammals as a whole, and we can see that tails are widespread. From this observation we can determine that the ancestral or primitive state is "with tails" and thus the polarity goes from "with tails" to "without tails." On the whole this system works very well, but there are problems that arise, either because it is not clear what the outgroup might be, or because homoplasies are common and so the ancestral state is obscure. The other method for determining polarities is to use developmental patterns – that is, to consider traits that occur at an earlier embryological stage as more ancestral.

**5** Once the polarities are established, it is then possible to assign values to all the OTUs for each trait.

The matrix that is produced by these steps is the datum from which a cladistic analysis is then carried out. Basically this is done using a series of simple rules for building evolutionary trees (or cladograms); it is, for any reasonable dataset, an extremely long procedure and is dependent upon powerful computer programs. The programs look at all the possible trees that can be constructed between the taxa. If there are three taxa, then it is fairly simple and there are four possible trees. However, this number increases exponentially, and a tree with 50 species or OTUs will have $2.8 \times 10^{74}$ possible trees – considerably more trees than there are atoms in the universe. Thus in practice not all possible trees are scrutinized, and the powerful programs are the ones that rapidly exclude large numbers of unlikely trees.

There are two principles used for excluding trees and discovering what is hopefully the actual phylogenetic history. The first of these is known as parsimony. Briefly put, parsimony seeks the simplest explanation, with the belief that this path is the most likely to have been followed. Evolutionary change is inherently of low probability, so paths going from character state A to character state B are themselves inherently likely to be simple rather than complex. (Complex paths might involve, for example, reversals of evolutionary direction or the same trait evolving more than once.) In the context of phylogenetic analysis, the parsimony method looks for the tree (or trees) that uses the fewest changes to link the given species in an evolutionary hierarchy. A parsimony analysis produces a very large number of trees, and then ranks them in terms of the one that involves the fewest evolutionary changes (the shortest tree) and has the fewest reversals (a high consistency index). In practice, while it is often possible to find the shortest tree, that tree may actually exist in a forest of others that are not much longer, so that in the end it is, as with phenetics, a matter of statistical probability as to which is the actual evolutionary history (Fig. 5.10).

The second method is that of maximum likelihood. In this case, rather than looking for the shortest tree, the method attempts to assess the probability of any particular change occurring, and then pursues the route of evolutionary change that follows the path of maximum likelihood.

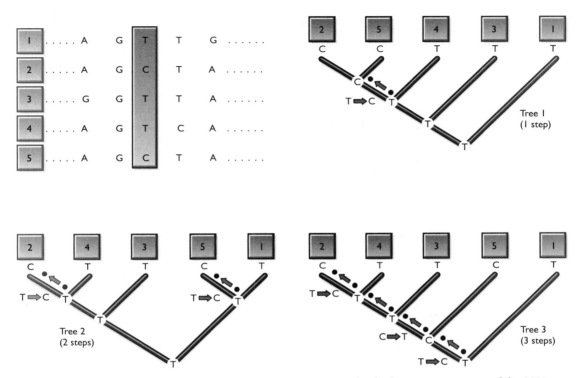

**FIGURE 5.10 The parsimony technique:** In this example of five individuals (1–5), we see part of the DNA sequence. By concentrating on position 3 in this instance, the parsimony technique seeks to find the tree with the lowest number of mutational steps to link all population members. Three trees are drawn here, with one, two, and three steps taken to link the five individuals. The parsimony technique would select tree 1 as the most likely relationship among the five individuals.

Computationally it is far more complex than parsimony analysis, and requires considerable *a priori* knowledge about the likelihood of any particular change. It has thus been used more for molecular data than morphological data, as the patterns of change in the former are easier to calculate.

Overall, cladistics has become the preferred method for reconstructing evolutionary relationships, and hence as a basis for systematics.[125] However, this is not to say that it is an approach without either theoretical or practical limitations, or that phenetics is not an appropriate method in some circumstances. Perhaps the most important understanding to emerge out of the debates about method has been the recognition that phylogenetic reconstruction is actually a process of hypothesis formulation – and cladistics is a powerful method for generating hypotheses. However, like all hypotheses, cladistic ones are there to be tested and refined, not set in stone.

Ultimately, all such hypothesis tests require an actual test of the conditions under which evolution works – difficult to achieve, because as we have emphasized, evolution is essentially a historical process. However,

Is parsimony a realistic assumption?

two approaches have been developed which do provide some sort of test. The first, using the power of modern computers, effectively simulates evolutionary events thousands of times, and thus produces a measure of the probability of particular events occurring, thus allowing the observed state of affairs to be compared with an expected one.[131]

The second testing avenue, that of using known phylogenies of living organisms, is more powerful, because it is based on real evolution. It is also more difficult to undertake. The work of Walter Fitch and William Atchley with inbred mice provides an example of this method (see next section); in fact, the two had embarked on the research explicitly as a way of testing analytical techniques. Recently another system has been developed, which overcomes some of the problems of the mouse system. It involves the bacteriophage T7, which has the practical advantage of a short generation time (measured in minutes rather than months) and the theoretical advantage of producing more realistic phylogenies (than the mice, for example). It is also easier to obtain extensive genetic information from phages than from mice. This ability not only helps in the phylogenetic analysis, but also gives insights into the details of change at the molecular level; aspects of this change can then be incorporated into simulation models. The researcher creates the phylogeny by separating phage colonies at selected times, eventually producing a series of lineages whose history is precisely known and offers a controlled test of phylogenetic methods.[132]

In a recent survey of these several tests of phylogenetic analysis, David Hillis and his colleagues at the University of Texas concluded that most methods currently being used, including parsimony, produce reasonably accurate reconstructions of known phylogenies.[133] This finding allows some confidence that the phylogenetic hypotheses that these methods develop from analyses of unknown phylogenies may be close to the historical truth.

## Cladistics and hominoid classification

How should apes be classified?

Cladistics has played a significant role in altering the systematics of the hominoids (humans and apes). Most cladistic analyses – using both morphological and molecular data – have in recent years supported a tree in which humans, chimpanzees, and gorillas are a monophyletic clade, and orangutans more distantly related to all of them.[134-9]

Now, suppose that this phylogenetic pattern is correct – and molecular data support this classification (Fig. 5.11). Surely it should be reflected in the formal classification, one might think. Traditionally, humans and their direct ancestors have been assigned to the family Hominidae, while the African apes and the orangutan occupy a separate family, the Pongidae. Such a grouping reflects overall morphological similarity, because humans have diverged dramatically from the apes; it ignores strict phylogeny, however, which groups humans with African apes and puts the orangutan separate.

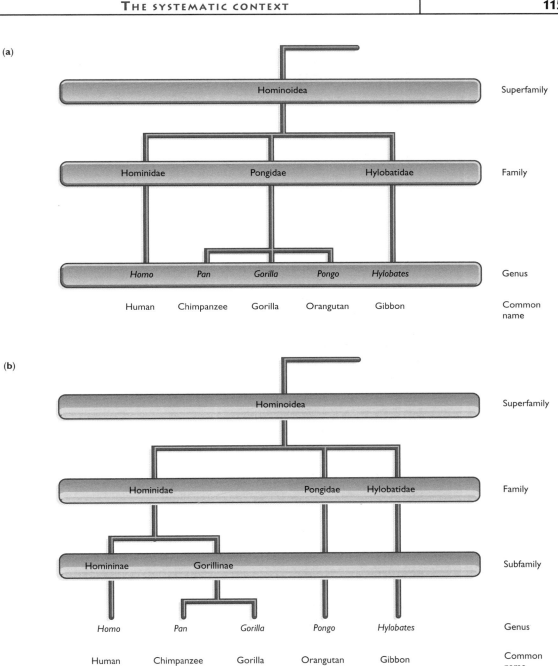

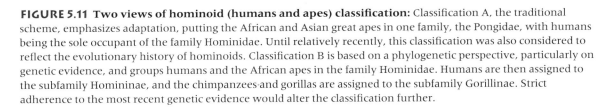

**FIGURE 5.11 Two views of hominoid (humans and apes) classification:** Classification A, the traditional scheme, emphasizes adaptation, putting the African and Asian great apes in one family, the Pongidae, with humans being the sole occupant of the family Hominidae. Until relatively recently, this classification was also considered to reflect the evolutionary history of hominoids. Classification B is based on a phylogenetic perspective, particularly on genetic evidence, and groups humans and the African apes in the family Hominidae. Humans are then assigned to the subfamily Homininae, and the chimpanzees and gorillas are assigned to the subfamily Gorillinae. Strict adherence to the most recent genetic evidence would alter the classification further.

If phylogeny is to be accurately reflected in classification, then one possibility is as follows. Hominidae would include the African apes and humans, with orangutans occupying the family Pongidae. Humans would be the sole occupant of the subfamily Homininae – hence the more general term "hominin" rather than the previously used "hominid." The gorilla and chimpanzee would occupy the subfamily Gorillinae.

Accurate in cladistic terms though this grouping may be, pheneticists and evolutionary taxonomists would demur. Classification should also reflect the very drastic ecological shift that has occurred in the hominin line compared with its ape cousins, they contend. According to this argument, maintaining family status for the apes and separate family status for humans is therefore appropriate. Indeed, as mentioned in chapter 1, Julian Huxley suggested in 1958 that, because our intellectual and psychological evolution so distances us from the rest of the animal world, humans might be envisioned as deserving an entire kingdom of their own: Psychozoa.[13]

If the different philosophies of classification routinely provoke debate, then the business of assigning the appropriate "rank" – genus, subfamily, family, and so on – among hominoids appears guaranteed to stoke the flames even further. Indeed, on the basis of the most recent molecular evidence, some researchers argue that the formal classification should reflect an even closer relationship between humans and chimpanzees than the one mentioned above.

## MOLECULAR SYSTEMATICS

**KEY QUESTION** How can molecular genetics be deployed to build phylogenies?

To this point, this chapter has focused mainly on the basic principles of classification, and these have been developed for the traditional database for reconstructing phylogeny – that is, morphological evidence. Genetic evidence has recently taken its place alongside morphology, however, creating the approach known as molecular systematics. Since the early 1980s this new approach has developed from a rare curiosity to become a powerful tool in modern evolutionary biology.[140] The reasons for its development are several. First, and most important, is the vast (and ever-growing) flood of molecular data from organisms of all kinds resulting from new techniques of molecular biology that allow access to the most fundamental of genetic information, the DNA sequence of genes. Second is the development of methods for analyzing these data. Third is the recently developed computing power necessary to conduct that analysis.

Not all genetic data used in molecular systematics take the form of DNA sequences, however (Fig. 5.12). Alternative forms of information include the following: comparison of immunological reactions of proteins;

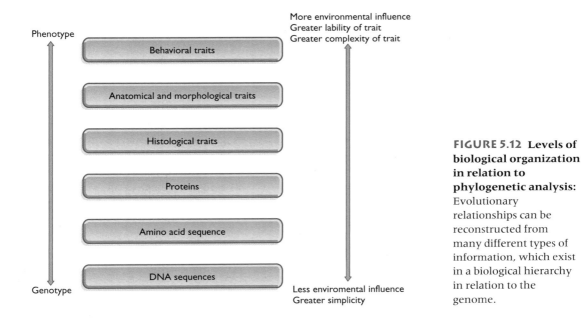

Phenotype

Behavioral traits

Anatomical and morphological traits

Histological traits

Proteins

Amino acid sequence

DNA sequences

Genotype

More environmental influence
Greater lability of trait
Greater complexity of trait

Less enviromental influence
Greater simplicity

**FIGURE 5.12 Levels of biological organization in relation to phylogenetic analysis:** Evolutionary relationships can be reconstructed from many different types of information, which exist in a biological hierarchy in relation to the genome.

comparison of electrical properties of proteins (gel electrophoresis); and DNA–DNA hybridization, which effectively compares the entire genetic complement of one species with that of another. All of these methods provide a measure of genetic distance between the species being compared. What is being measured, of course, is part of the phenotype, and in that sense is really another form of morphological data that can be used to reconstruct phylogenies. DNA sequence data (that is, the presence or absence of particular nucleotides at particular loci), on the other hand, do provide a very direct measure of genetic relatedness, and are thus thought by some to be a better source of information about evolutionary relationships, and easily susceptible to cladistic analysis, which is equivalent to the phenetic measure of overall similarity.

Is DNA the best source of phylogenetic information?

The principles underlying molecular systematics are exactly the same as those described above in relation to morphological data – that is, similarity among homologous systems denotes common ancestry. Molecular systematics relies on the fact that when two species diverge, mutations will accumulate independently in the DNA of the daughter lineages.[141] Scrutiny of similarities and differences among species' DNA therefore permits their evolutionary relationships to be inferred. In its early days, molecular systematics was perceived (by molecular biologists) as being inherently superior to traditional methods, for several reasons.

First, because molecular data are derived from the genes of a species, they were envisaged as carrying the fundamental record of evolutionary change. Second, molecular data were considered to be immune to the problem of convergence, for the following reason. Natural selection produces convergence, through adaptation to similar environmental

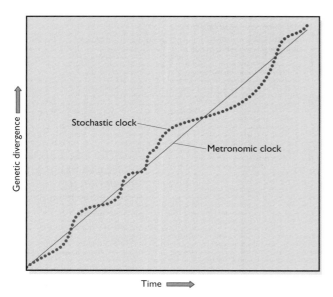

**FIGURE 5.13 The molecular clock:** If genetic mutation were to occur at a constant rate, then biologists would have access to a completely reliable, "metronomic" molecular clock. In fact, the rate of mutation for any particular region of DNA is likely to fluctuate through time, giving a "stochastic" molecular clock. By bringing together data on genetic divergence from different regions of DNA, it is possible in principle to average out these fluctuations, thereby providing a good, average clock. Because the technique of DNA hybridization effectively compares the entire DNA complement of two related species, fluctuations in mutation rate in different parts of the genome are automatically averaged out.

conditions. Because the majority of mutations that accumulate are selectively neutral, they remain invisible to natural selection. Convergence toward similar mutations in different lineages is therefore highly unlikely, except by chance. Third, molecular and morphological evolution were thought to proceed at very different tempos (the former always regular, the latter always erratic), which was assumed to imply that molecular data were more reliable.

Moreover, because genetic differences between lineages was suspected to proceed in a regular manner, the notion of a molecular evolutionary clock was developed.[142] The idea underlying the molecular clock is that the rate of mutational change for genetic material that is "silent" (that is, does not become expressed as proteins, and so is not subject to selection) is constant, and therefore differences between lineages accumulate at a steady rate (Fig. 5.13). If this is correct, then not only would it be possible to determine reliably the branching order of related species with genetic data, but one could also calculate when the lineages diverged from one another – that is, the branch length. Last, morphological features express complex and mostly unknown sets of genes and regulatory interactions among genes. In contrast, molecular data relate to much smaller and strictly defined sets of genes. According to proponents of molecular systematics, simplicity yields reliability.

A clear example of the utility of molecular phylogenetics had been known since the 1960s, in the form of the work of Morris Goodman (one of the pioneers of molecular evolution)[143] and, later, Allan Wilson and Vincent Sarich (a biochemist and an anthropologist at the University of California, Berkeley).[142] These researchers had demonstrated that the traditional classification of humans and apes was wrong in terms of evolutionary relationships of the group, and that the human clade had evolved a mere 5 million years ago, not the 15 million to 30 million years that anthropologists then held to be true[23] (Fig. 5.14).

One limitation of systematics in general is that its major goal – that of knowing a particular phylogeny – is impossible, because a particular conclusion cannot be tested. Phylogeny is history, and conclusions about it are merely hypotheses, not certainty. When molecular and morphological phylogeneticists reach their separate conclusions, they are therefore comparing their separate hypotheses, and neither group can be certain it is correct. If one could perform molecular and morphological phylogenetics on a known phylogeny, however, then a direct test of conclusions would be possible. Walter Fitch, then of the University of Southern California,

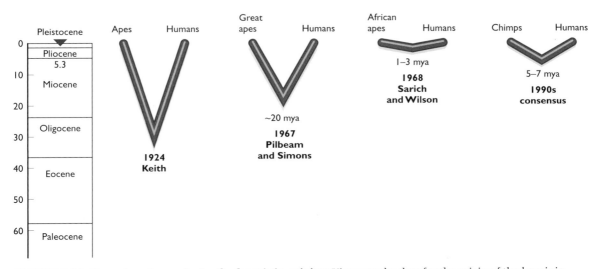

**FIGURE 5.14 Changing chronologies for hominin origins:** Views on the date for the origin of the hominin clade have changed dramatically over the last century. Early paleoanthropologists envisaged a very long chronology, and this view persisted into the 1970s, although there was some reduction in the time envisaged. The first applications of the molecular clock by Sarich, Wilson, and Goodman drastically reduced this. The current consensus, supported by both paleontology and molecular evolutionary genetics, suggests a date of between 5 million and 7 million years.

and William Atchley, then of the University of Wisconsin, realized that inbred strains of laboratory mice offered such a possibility.[132]

The phylogeny of these mice has been recorded over many years, as new strains were developed. The molecular data consisted of protein variants from 97 gene loci; the morphological data encompassed 10 measurements on the lower jaw on the mice at 10 weeks of age. Fitch and Atchley, using five different methods to analyze the two datasets, found that the molecular data accurately reconstructed the known phylogeny whereas the morphological data did not. They conceded that the phylogeny covered only 70 years – an extremely short period compared with typical phylogenies – and that the morphological data were quantitative (that is, widths and lengths) whereas morphologists typically use qualitative traits (that is, presence or absence of homologous features). Both these facts may work against producing an accurate phylogeny based on morphology. Despite these caveats, Fitch and Atchley concluded that "molecular data appear to be superior to morphological data for reconstructing phylogenies."

## Limitations of molecular systematics

Today, molecular approaches to systematics are recognized as less simple, and therefore less immediately reliable, than previously supposed. For instance, not all genes give precisely the sample phylogenetic pattern for the human/ape phylogeny. Moreover, when Fitch and Atchley recently extended their analysis of inbred mice to more gene loci, they found that while some genes gave the correct phylogeny, others did not.[144]

The reasons that molecular systematics is less straightforward than initially supposed are several. For instance, it is now recognized that the dynamics of mutation are highly complex and include the fact that not all regions of a gene or other regions of DNA are equally susceptible to change; indeed, some different regions are highly susceptible to similar kinds of change. For this reason, convergence can and does occur in DNA sequences.[141] Moreover, some mutation events may become hidden through "multiple hits." Imagine that a particular nucleotide position in a gene mutates early in the lineage's history. As time passes, other mutations will accumulate as well. If all subsequent mutations occur at different sites, a count of the mutations present will give an accurate record of the lineage's mutational history. If a later mutation occurs at a previously mutated site, however, then the count will be too low, giving an erroneous conclusion. The longer the time period under investigation, the greater a problem multiple hits become. Statistical methods are being developed to try to accommodate this factor.

Another potential confounding problem is that the degree of sequence divergence between the same gene in different lineages might not accurately measure the point at which the lineages diverged. The issue here relates to the potential difference between the species tree and the gene tree.[145] A species tree describes the evolutionary history of the species – that is, the true phylogeny. If all genes in two daughter species began to diverge only when the populations diverged, then the gene tree would be the same as the species tree. This, however, is not always the case. Genes often develop variants (polymorphisms) within a population, so that some individuals may possess one variant while other individuals carry the other. Once such variants become established, they begin to accumulate mutations independently.

Suppose a polymorphism of a gene X arose in a species A some 4 million years ago, giving variant X in some individuals and variant X1 in others (Fig. 5.15). Suppose, too, that allopatric populations became established 2 million years later,

**FIGURE 5.15 Species trees and gene trees:** A gene X in a species A undergoes polymorphism, producing variants X and X1, which then continue to accumulate differences between them. A speciation event occurs later, producing species B and C. Through various circumstances, gene variant X predominates in species B while variant X1 predominates in species C. A comparison of the differences between X and X1 would overestimate the time at which the daughter species B and C diverged. In other words, the gene tree is older than the species tree.

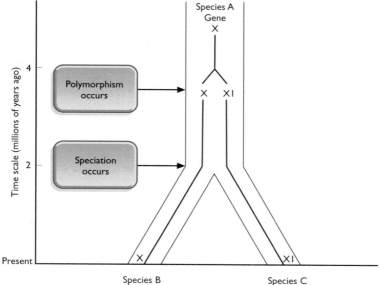

with X remaining in the parent population while X1 appeared exclusively in the newly isolated population. Such a situation can lead to speciation. Now suppose that the modern populations of the descendant species are subjected to molecular systematics analysis, using gene X. Calculations based on the nucleotide differences between X and X1 would indicate that the two daughter species diverged 4 million years ago, when their separate sequences would have begun to diverge (this is the gene tree). In fact, the species did not begin to diverge until 2 million years ago (as the species tree reveals). In general, therefore, when the gene tree/species tree problem arises, the divergence date inferred from the molecular data will be too old.

This example assumes, of course, that molecular data can be used to calculate time since divergence, on the basis of the molecular clock concept. It was once assumed that mutations accumulated at a regular rate in all genes, in all lineages, and at all times in lineages' histories. Some genes might mutate at higher rates than others because of functional constraints – globin genes mutate at a higher rate than histone genes, for example. In fact, a series of clocks would operate, each ticking regularly but at different rates.

Again, though, this assumption turns out to be too simplistic. It has now been established that some genes in some lineages at some points in their evolutionary history do indeed accumulate mutations in a clocklike manner. Differences arise, however, in mutation rates in the same gene between lineages, as well as differences in rates in a single gene within a single lineage at different points in its history. The notion of a global clock is therefore no longer tenable. The existence of local clocks is, nevertheless, a reality, and they have great utility. (The accumulation of mutations is not strictly clocklike, as a metronome; rather it is a stochastic clock with some fluctuations in rate, but with an average that looks regular over time.) Researchers must determine whether their gene of interest is behaving in a clocklike manner, using the relative rate test, before they can proceed to measure branch lengths in phylogenies (Fig. 5.16).

**FIGURE 5.16 The relative rate test:** The diagram represents two evolutionary events. At 1, a split occurred, leading to species C and a second lineage. The second lineage then split at node 2, leading to species A and B. According to the rate test, if the average rate of genetic divergence is the same in all lineages, then the genetic distance from species A to species C (dotted line) should be the same as the genetic distance from species B to species C (dashed line). If gene mutation slowed down in lineage B, then the B-to-C genetic distance would be shorter than the distance from A to C.

## Homology of gene sequences

The issue of homology of gene sequences is by no means a simple one. Indeed, in some ways it is more complex than for anatomical features. First, the term "homology" should be restricted to common ancestry, while "similarity" should be constrained to likeness of sequences. Unfortunately, when molecular biologists first began reading gene sequences,

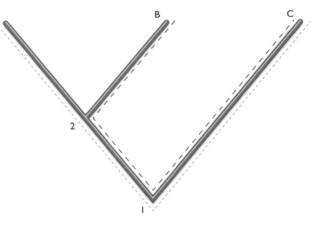

they frequently used "homology" when they actually meant "similarity." Next, it should be recognized that seeking homology between gene sequences is much more difficult than may be imagined. Two species that have recently diverged may display no sequence differences at all, or perhaps just a few. After a very long period of divergence, the two sequences will accumulate differences; indeed, they may become hardly recognizable as the same gene. This situation arises because the gene sequence, and with it the amino acid sequence of the protein for which it codes, can change substantially, while still retaining the functional three-dimensional structure of the product protein. This structure is the one that is visible to selection. Thus, even genes of little sequence similarity can be homologous.

Homology of genes is further complicated because in higher organisms many genes duplicate, producing gene families; each member of the family represents a slight variant on the original sequence. The discovery of gene families prompted molecular biologists to coin new terms for homology, which take this complexity into account. For genes in which duplication has not occurred, and that therefore exist as single copies, the term "orthology" was introduced for purposes of comparing them between related species; this term is functionally equivalent to traditional morphological homology. The term "paralogy" is used to describe members of a gene family, such as the globin family, within a species.

Thus, the terms "orthology" and "paralogy" refer to aspects of homology that arise from a gene's history among related species or within a species. A third form of homology, also restricted to molecular systematics, is xenology. Although species' genetic packages are largely constrained to vertical inheritance from generation to generation, remaining isolated from the genes of other species, occasionally this barrier breaks down. Specifically, genes may be transferred horizontally between species, often as passengers on certain viruses, and become incorporated into the host's genome. Although common in micro-organisms, horizontal transfer is rare among higher animals, but is not unknown. A clue to xenology surfaces when a gene in two distantly related species appears surprisingly similar in sequence, in the context of significant dissimilarity of other genes.

Other complications may arise as well, such as the fact that most genes in higher organisms exist as coding sequences, or exons, interspersed with non-coding sequences, or introns. A gene may be assembled by bringing together individual exons from other, existing genes, in a process known as exon shuffling. This text will not go into detail about this process, but will merely note that this exon shuffling leads to partial homology: a gene A, which consists of a selection of exons from genes X, Y, and Z, for example, will be partially homologous to these genes. Partial homology does not occur with morphological characters.

## Morphology and molecules compared

Paradoxically, one of the advantages of morphological systematics stems from the erratic nature of the tempo of morphological evolution. An important feature of evolution is adaptive radiation, which occurs when a new group diversifies at its establishment, yielding many lineages with unique features that subsequently may change little. If such a radiation occurred deep in evolutionary history, a clocklike accumulation of genetic mutations would be unable to track the details of the brief burst of change, for the following reasons. A slow rate of mutation in DNA sequences would leave the event unrecorded. DNA sequences that change rapidly, on the other hand, would capture such change, but this information would be overwritten to the point of illegibility by subsequent mutation. By contrast, the morphological changes that accompany the radiation would, in principle, persist in the lineages' subsequent history, preserving the event for comparative morphologists to discern. The rapid radiation of placental mammals 100 million years ago, prior to the end-Cretaceous extinction, is a good example of this type of development.

What affects the rate of molecular evolution?

A major advantage of molecular phylogenetics is the potential extent of information it can evaluate, which at the limit is equal to the entire genome (in humans, for example, the genome includes 3 billion nucleotides). Morphological characters necessarily represent only a subset of this information. Moreover, because different sectors of the genome accumulate mutations at variable rates, genetic methods offer access to both ancient divergences (with slow-changing DNA, such as ribosomal DNA) and recent events (with fast-changing DNA, such as mitochondrial DNA). Morphological information cannot encompass this range of evolutionary history. It is also powerless to discern evolutionary history in cases involving limited morphology, such as in the early divergence of microorganisms nearly 3 billion years ago.

## Molecular systematics and human evolution

The premise underlying this chapter is that an important issue in evolutionary biology is that of determining evolutionary relationships – answering the classic question of who is related to whom in an evolutionary tree. In fact this question has in many ways dominated much of the research into human evolution. However, as we have seen, answering this question relies on having good methods for building biological classifications, and these are not simple. They depend upon an understanding of evolutionary process as well as complex methods of measurement and analysis. Into this complexity has been thrown the added dimension of molecular genetics. While this has added considerably to our knowledge of the evolutionary history, including that of the hominins, nonetheless it is becoming increasingly clear that interpreting molecular information in an evolutionary context is also full of difficulties. The net result is that while we are undoubtedly gaining enormously in terms of knowledge, the complexity – especially computational and statistical – of the methods being used has increased enormously. The days when phylogenies could be scribbled on the back of an envelope are gone, and the technical demands of cladistic methods have replaced them. What remains the case, regardless of the data used, though, is that any phylogenetic reconstruction is only as good as the methods on which it is based, and all classifications and phylogenies remain hypotheses subject to constant testing with new data.

Molecular systematics has been important in three areas of human prehistory. Its application to the issue of the origin of the hominin clade has already been mentioned. A second area is in the origin of modern humans, while the third relates to the diversification of modern human populations and events such as the timing of human colonization of the Americas, discussed later in this book.

## BEYOND THE FACTS
# WHAT'S IN A NAME?

*The issue: formal Latin names (Linnaean names) are an essential part of "doing biology." Ironically, the Linnaean system was designed to facilitate communication between scientists, but has become a byword for pedantic obscurity. Does it matter if humans are hominins or hominids? Is it just a matter of convention, or do the names tell us something about what has happened in biology? Should we be more relaxed about biological names, or more strict?*

One of the problems of studying human evolution is the plethora of Latin names. This problem is compounded by the fact that many of them are very similar – hominoid, hominid, hominin – and all of them in some way appear to be fancy ways of saying "human." To add insult to injury, just when one name becomes established as meaning one thing – for example, "hominid" to describe all the species on the lineage leading to humans – it is replaced by another meaning exactly the same thing – namely "hominin."

The technical basis for these differences lies in the fact that the different endings of the words simply reflects common usage across the whole of biological systematics to show degrees of relatedness. To call humans and humans alone "hominoids" would imply that we belong to our own superfamily, and thus are very different from any other apelike form. To call humans "hominins" is to give humans only a subfamily distinction, and so recognize a closer relationship to the other apes. One could even go to further extremes, and place humans and chimpanzees, for example, in the same genus (as Jared Diamond has implied),[14] or at the other extreme, place them in their own kingdom, the Psychozoan, as Julian Huxley proposed.[13]

As is discussed later in the text, the current consensus is to use the term "hominin," certainly implying a closer relationship between humans and other apes than was the case in the late 1990s. But does this change in terminology make a difference? One answer would be no; the relationships remain broadly the same, and greater accuracy has only been bought at the expense of greater confusion. After all, a hominid in a 1980 textbook is clearly something that is closer to humans than to apes, whereas in a 2000 textbook a hominid could easily be a chimpanzee or a gorilla. If, for example, new data were to emerge showing that the differences were greater than was thought, then would it be right to return to the word "hominid," causing even greater confusion?

This is the dilemma facing systematicists, not just in human evolution, but across biology as a whole. The key issues are whether terminology should faithfully reflect the order of evolutionary branching events or the degree of evolution that has occurred after the branching events (humans are, after all, very different from any other species), and whether the power of the Linnaean system lies in its stability or it should respond to changing evidence. There are no simple answers to these questions, not least because behind them is the fact that these higher taxonomic levels are arbitrary, in the sense that they are not underpinned by biological mechanisms.

# CHAPTER 6

# HUMAN EVOLUTION IN COMPARATIVE PERSPECTIVE

**FIGURE 6.1**
**Determinants of evolutionary structure:** The outcome of selection is the result of the interaction between the demands of the environment, the nature of the existing adaptations (phylogenetic heritage), and the effects of general rules governing ecology.

Thus far we have looked at the framework for human evolution in terms of the development and nature of evolutionary theory; in terms of the way fossils, the key evidence for human evolution, are formed and dated; and in terms of the way in which we place humans into a phylogenetic context. The reason for the last of these stems from the fact that it is impossible to study human evolution in isolation – to know what is unique and special about humans, we need to know also what they share with other groups of animals. This chapter will therefore explore the principles relating to viewing human evolution in a comparative perspective. In looking at this comparative perspective we need to consider two elements that contribute to it. The first of these is drawn from our phylogenetic position, the fact that humans and other hominins are primates, and so share many characteristics (synapomorphies in the cladistic terminology of the previous chapter) with other primates. The second is the rules governing energy, size, and shape in biological systems (Fig. 6.1). In a sense these two comparative approaches represent the two overarching elements that determine evolutionary trajectories – phylogenetic heritage and adaptation.

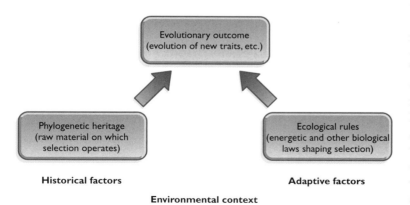

Evolutionary outcome
(evolution of new traits, etc.)

Phylogenetic heritage
(raw material on which selection operates)

Ecological rules
(energetic and other biological laws shaping selection)

**Historical factors**　　　　　　　　　　**Adaptive factors**

**Environmental context**

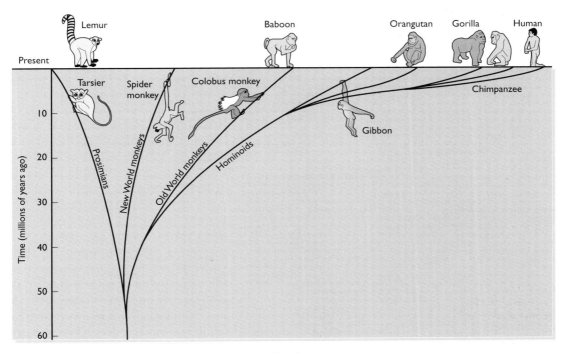

**FIGURE 6.2 Primate family tree** (Courtesy of John Fleagle.)

## PRIMATE HERITAGE

*Homo sapiens* is one of approximately 200 species of living primate, which collectively constitute the order Primates[146,147,148] (Fig. 6.2). (There are 22 living orders in the class Mammalia, which includes the bats, rodents, carnivores, elephants, and marsupials.) Just as we, as individuals, inherit many resemblances from our parents but also are shaped by our own experiences, so it is with species within an order. Each species inherits a set of anatomical and behavioral features that characterize the order as a whole, but each species is also unique, reflecting its own evolutionary history.

> **KEY QUESTION** What light does primate evolution throw on the probable phylogenetic heritage of hominins?

Matt Cartmill, of Duke University, says of anthropology that: "Providing a historical account of how and why human beings got to be the way they are is probably the most important service to humanity that our profession can perform."[149] An understanding of our primate heritage provides the starting point for writing that historical account. In this chapter we will consider what it is to be a primate, in terms of anatomy and behavior.

The roots of primate studies lie in anatomy, for it was comparative studies of human and ape anatomy that first gave insights into many

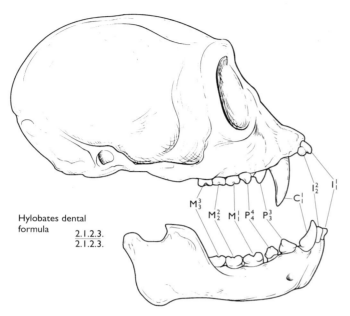

Hylobates dental
formula
2.1.2.3.
2.1.2.3.

**FIGURE 6.3 Primate dentition:** Teeth are particularly important in the reconstruction of primate phylogeny, for two reasons. First, their extreme hardness means that they are the most common item recovered from the fossil record, and hence provide a disproportionate amount of information about fossil species. Second, teeth give very clear information about dietary habits because the shape is strongly influenced by the type of food eaten. By convention, dental formula is written as shown in the diagram. This species (a siamang) possesses two incisors, one canine, two premolars, and three molars (a common scheme in higher primates). (Courtesy of John Fleagle.)

aspects of human evolution (Fig. 6.3). Although anatomy remains central to the subject, the study of primates – primatology – has undergone important changes in recent years for three reasons. First, ecological research has been thoroughly incorporated into primate studies. As a result, primate biology can be interpreted within a more complete ecological context. Second, the science of sociobiology has enabled a keener insight into the evolution of social behavior. And primates, if nothing else, are highly social animals. And third, primatology has become more integrated with evolutionary biology generally, such that approaches such as life-history theory allow much stronger comparative inferences to be made. Modern primatology therefore promises to serve as the focus of some of the most serious intellectual challenges of behavioral ecology.

Modern primates can be classified into four broad types (Fig. 6.4), although these are not necessarily monophyletic in the sense discussed in the previous chapter:

♦ prosimians, which include lemurs, lorises, and bushbabies, in addition to tarsiers;
♦ New World monkeys, such as the marmosets, spider monkeys, and howler monkeys;
♦ Old World monkeys, such as macaques, baboons, and colobus monkeys; and
♦ hominoids, which comprise apes and humans.

Among these four groups the prosimians constitute one major branch, and the other three constitute another, collectively referred to as the anthropoids. Although these two represent the two major clades of primate evolution in terms of traditional classification, they are not in fact monophyletic. Recent work, both molecular and anatomical, has shown that the tarsier is closer to the anthropoids than the other prosimians in terms of branching sequence.[150,151,152] Consequently modern classification has as the two major clades the strepsirhines (lorises and lemurs) and haplorhines (tarsiers, New World monkeys, Old World monkeys, apes, and humans).

As a whole, primates now occur only in the tropics and subtropics, and are not found in Europe, Australia, or North America (Fig. 6.5). There are strong biogeographical patterns to primate distribution – for example, the lemurs are found only in Madagascar, and the division

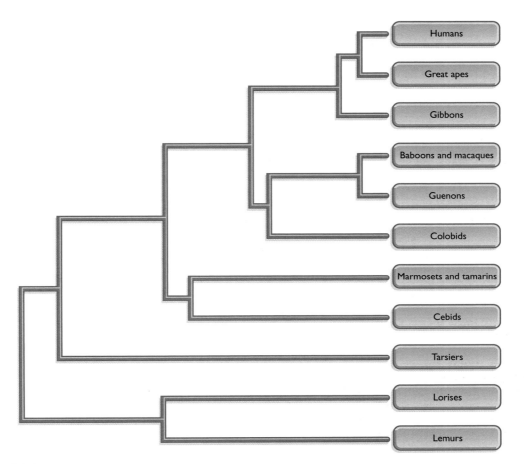

FIGURE 6.4 Major types of primate

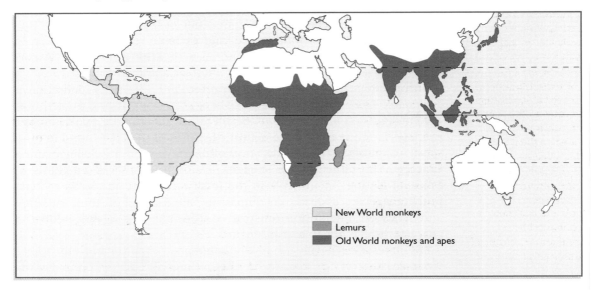

FIGURE 6.5 Geographical distribution of living primates

between the Old and New Worlds represents the major divide among the anthropoids.

What are the main types of primate?

Modern primate species constitute an extraordinarily varied order, in terms of both morphology and behavior. Some species are among the most generalized and primitive of all mammals, while others display specializations not seen in other mammalian orders. Nevertheless, primate bodies are generally primitive in the sense that they retain many basic mammalian characteristics. True, some have lost tails and others have developed large brains. None, however, has turned hands into wings (as bats have), or reduced fingers and toes to single digits (as horses have), or lost limbs altogether (as baleen whales have, being without hindlimbs), or transformed its dentition into something that no self-respecting primate would put into its mouth (as the baleen whales have, with their hairlike combs designed for filtering tiny prey from water).

Modern primates vary enormously in size, ranging from the diminutive mouse lemur, which weighs in at 80 grams, to the male gorilla, at more than 2000 times the mouse lemur's size.[146] Whatever their size, primates are quintessentially animals of the tropics. Although different primate species occupy every major type of tropical environment – from rainforest to woodland, shrubland, savannah, and semidesert scrub – 80% of them are creatures of the rainforest. Several Old World monkeys and one ape – the mountain gorilla – live in temperate and even subalpine zones. Among primates, *Homo sapiens* is unique in ranging so widely geographically and in tolerating so extreme a variety of environments.[153]

## Definition of primates

Although humans have clearly departed from our primate roots in colonizing so broad a range of habitats, many of the characteristics that we often envision as separating us from other primates – such as habitual upright walking, great intelligence, and extreme sociality – are actually extensions of, rather than discontinuities with, what it means to be a primate. We should therefore ask: what is it to be a primate?

Surprisingly, this question, which essentially asks for a definition of "primate," has proved difficult to answer concisely. "It has, in fact, been a common theme throughout the literature on primate evolution that primates lack any clear-cut diagnostic features of the kind found in other species of placental mammals," notes Robert Martin, of the University of Chicago. The difficulty, he suggests, stems from an overemphasis on "skeletal features identifiable in the fossil record." If one looks at living primate species instead, encompassing all aspects of their anatomy and behavior, a definition constructed from universal or near-universal characteristics is possible, says Martin.

"Primates are typically arboreal inhabitants of tropical and sub-tropical forest ecosystems," begins Martin's definition.[154,147] It goes on to describe features of hand and foot anatomy, overall style of locomotion, visual abilities, intelligence, aspects of reproductive anatomy, life-history factors (such as longevity and reproductive strategy), and dental architecture. The definition generally depicts species that have a rather special niche in the world. As Rockefeller University anthropologist Alison Jolly has noted, "If there is an essence of being a primate, it is the progressive evolution of intelligence as a way of life." In fact, the history of primatology reflects this difficulty; on the one hand there have been those, such as Martin, who have searched for unique and universal characteristics of the primates (autapomorphies) to define the group, and others, such as the eminent British researcher Sir Wilfrid Le Gros Clark[155] and the British anatomist John Napier,[156] who focused on identifying general trends. Autapomorphies are unsatisfactory as they often reflect only relatively trivial elements of the order, while the latter group of researchers has been criticized for viewing primates as stepping stones towards the human condition. This difficulty is, however, perhaps informative about the overall nature of primate adaptations. Some of the key components of Martin's definition are described below.

Primate hands and feet have the ability to grasp because they are equipped with opposable thumbs and opposable great toes. Humans are an exception, as the human foot has lost its grasping function in favor of forming a "platform" adapted to habitual upright walking. In higher primates, fingers and toes have nails, not claws; and finger and toe pads are broad and ridged, which aids in preventing slippage on arboreal supports and in enhancing touch sensitivity.

Primate locomotion is hindlimb-dominated, whether it consists of vertical clinging and leaping (various small species), quadrupedal walking (monkeys and the African great apes), or brachiation (all apes) (Fig. 6.6). In each case, the center of gravity of the body is located near the hindlimbs, which produces the typical diagonal gait (forelimb preceding hindlimb on each side). It also means that the body is frequently held in a relatively vertical position, making the transition to habitual bipedalism in humans a less dramatic anatomical shift than is often imagined.

Vision is greatly emphasized in primates, while the olfactory (smell) sense is diminished. In all primates, the two eyes have come to the front of the head, producing stereoscopic vision, to a greater extent than in other mammals. Although some primates (the diurnal species) have color vision, this character does not discriminate the order from many other vertebrate groups. The shifting of the eyes from the side of the head to the front, combined with the diminution of olfaction, produces a shorter snout; this character is accompanied by a reduction in the number of incisor and premolar teeth from the ancestral condition of 3 incisors, 1 canine, 4 premolars, and 3 molars (denoted as 3.1.4.3) to a maximum of 2.1.3.3. (Prosimians and New World monkeys demonstrate this latter pattern, whereas Old World monkeys and hominoids have one fewer premolar.)

Why are primates difficult to define?

**FIGURE 6.6 Modes of primate locomotion:** The monkey (top right) walks quadrupedally, while the gibbon (top left) is an adept brachiator (it swings from branch to branch like a pendulum). The orangutan (mid-left) is also adept in the trees, but as a four-handed climber. The gorilla (bottom left), like the chimpanzee, is a knuckle-walker (it supports its weight through the forelimbs on the knuckles of the hand rather than using a flat hand as the monkey does). The tarsier (foreground) moves by vertical clinging and leaping. The hominid (right) is a fully committed biped. Note the grasping hands and forward-pointing eyes characteristic of primates. (Courtesy of John Gurche/Maitland Edey.)

Partly because of the emphasis on vision, primate brains are larger than those found in other mammalian orders. This increase also reflects a greater "intelligence." In this character, the lemurs, lorises, and other prosimians are, however, less well endowed than monkeys and apes. Tied to this enhanced encephalization is a shift in a series of life-history factors: animals with large brains for their body size tend to have a greater longevity and a low potential reproductive output. For instance, primate gestation is long relative to maternal body size, litters are small (usually one), and offspring precocious; age at first reproduction is late, and interbirth interval is long. "Primates are, in short, adapted for slow reproductive turnover," observes Martin.

If we think of humans as animals with particular physical and behavioral habits, this discursive definition describes us as well, apart from the fact that we do not live in trees. For instance, throwing is a skill dependent upon the mobility of the shoulder joint that is part of the primate anatomical repertoire. Hindlimb-dominated locomotion, grasping and touch-sensitive hands, stereoscopic vision, and intelligence – all are required in that activity, and all are general characteristics of primates. More historically, when hominins first began making stone tools, they were not "inventing culture" in the sense that the phrase is often used, but merely applying primate manipulative skills to a new task. Although it is true that even by primate standards *Homo sapiens* is particularly well endowed mentally, our generous encephalization merely represents an extension of just another primate trait.

Later we will return to some of these and other themes, particularly the issue of life-history strategy and brain size. In this chapter, we will address the question of how primates arrived at their current form – that is, how a small, ancestral mammal species developed the above suite of characteristics, and how this provides the heritage upon which human traits are based.

## Theories of the origin of the primate adaptation

The first systematic attempt to account for the differences between primates and other mammals was made by T. H. Huxley in his 1863 book, *Evidences as to Man's Place in Nature*.[2] Early in the twentieth century, the British anatomists Grafton Elliot Smith and Frederic Wood Jones continued this quest.[15] Ancestral primates and, by extrapolation, humans were different from other mammals, they argued, because of adaptation to life in the trees – hence the arboreal hypothesis of primate origins. Grasping hands and feet provided a superior mode of locomotion, according to these scientists, while vision was a more acute sensory system than olfaction in among the leaves and branches.

What was the basic primate adaptation?

The arboreal theory was modified and extended in the 1950s by Wilfrid Le Gros Clark,[155] and remained the orthodox view until the 1970s, when Cartmill proposed an alternative, using the biologists' most powerful tool – comparative analysis.[156,157] He noted that "The arboreal theory was open to the most obvious objection that most arboreal mammals – opossums, tree shrews, palm civets, squirrels, and so on – lack the short face, close-set eyes, reduced olfactory apparatus, and large brains that arboreal life supposedly favored . . . If progressive adaptation to living in trees transformed a tree shrew-like ancestor into a higher primate, then primate-like traits must be better adapted to arboreal locomotion and foraging than their antecedents."[157] In other words, if primates are truly the ultimate in adaptation to arboreal life, you would expect that they would be more skillful aloft than other arboreal creatures. "This expectation is not borne out by studies of arboreal nonprimates," he noted. Squirrels, for instance, do exceedingly well with divergent eyes, a long snout, and no grasping hands and feet, often displaying superior arboreal skills to those of primates. "Clearly, successful arboreal existence is possible without primate-like adaptations," concluded Cartmill.

If the close-set eyes and grasping hands and feet were an adaptation to something other than arboreality, what was it? Once again Cartmill used the comparative approach to find an answer, one that formed the basis of the visual predation hypothesis. Boldly put, the hypothesis states that the suite of primate characteristics represents an adaptation by a small arboreal mammal to stalking insect prey, which are captured in the hands.

Cartmill sought individual elements of the primate suite in a range of other species. For instance, chameleons have grasping hindfeet, which they use to steady themselves when approaching insect prey on slender branches. Some South American opossums show similar behavior, capturing their prey by hand or mouth. And, of course, the convergence of the eyes is found in many predatory animals that need to be able to judge distance accurately, such as cats, owls, and hawks.

"Most of the distinctive primate characteristics can thus be explained as convergence with chameleons and small bush-dwelling marsupials (in the hands and feet) or with cats (in the visual apparatus)," concluded Cartmill. "This implies that the last common ancestor would have subsisted much as modern tarsiers, the mouse lemur, and some lorises do today." These species should not be considered "living fossils" because, like humans, they are also the products of 60 million years of evolution. It is simply that their ecological niche resembles the niche occupied by their ancestors.

Cartmill's visual predation hypothesis has more recently been challenged by American primatologist Robert Sussman.[159] He points out that many primate species locate their prey by smell or hearing, so that visual predation by itself is not sufficient to explain this suite of primate adaptations. He also argues that the earliest primates evolved at a time when flowering plants were in the midst of an evolutionary diversification. Grasping hands and feet would have enabled small primate species to move with agility in terminal branches rich with fruit; keen visual acuity would allow fine discrimination of small food items. Sussman's hypothesis is obviously similar in some ways to the earlier arboreal one, and shows, despite the obvious power of Cartmill's hypothesis, there is still much to learn about the adaptive basis for the origins of the primate.

The nature of the diet is obviously important for theories of primate origins, but it is important to note an important fact about this. Living primates do not follow a single "primate diet." Insects, gums, fruit, leaves, eggs, and even other primates – all are found on the menu of one primate species or another, and most species regularly consume items from two or more of these categories. The key factor that determines what any individual species will principally subsist on is body size. Small species have high energy requirements per unit of body weight (because of a high relative metabolic rate), and they therefore require food in small, rich packets. Leaves, for instance, are simply too bulky and require too much digestive processing to satisfy small primates. Because of their reduced relative energy demands, large species have the luxury of being able to subsist on bulky, low-quality resources, which are usually more abundant. From the small to the large species, the preferred foods shift, roughly speaking, from insects and gums, to fruit, to leaves. Thus trends in primate evolution are strongly related to dietary requirements and correlates.

A good deal of variation upon this basic equation exists, however. As Cambridge University primatologist Alison Richard points out, "Almost all primates, regardless of size, meet part of their energy requirements with fruit, which provides a ready source of simple sugars." What sets the basic equation, she says, is "how they make up the difference in energy and how they meet their protein requirements." This is where body size is crucial, and why, for instance, the bushbaby's staple is insects and the gorilla's is leaves.

## The origin and evolution of primates

The overall evolutionary pattern of primates remains still unsettled (Fig. 6.7). Some kind (or kinds) of species ancestral to all primates survived the mass extinction 65 million years ago that spelled the end of the Age of Reptiles, with the dinosaurs being the most notorious of the extinctions. Soon into the subsequent Age of Mammals, "primates of modern aspect" appear, approximately 50 million years ago, beginning an adaptive radiation that included an increase in range of body size and a concomitant broadening of diet. The 200 modern species represent the remains of that adaptive radiation, which, in total, probably gave rise to some 6000 species.[154]

What is the pattern of early primate evolution?

The known fossil record provides only the briefest of glimpses of this radiation, a sketchy outline at best; somewhere between 60 and 180 fossil primate species can be recognized. In the past some researchers considered the earliest primate group to be the Plesiadapiformes, the best-known specimen of which was *Purgatorius*, found a century ago in Montana and later at several other sites. The plesiadapiforms constituted a successful group living in the Paleocene and early Eocene (55 to 65 million years ago) of North America and Europe, amounting to some 25 genera and 75 species. The range of body size was considerable, stretching from 20 grams to more than 3 kilograms. Most members of the group were probably insectivores. Their supposed phylogenetic link with later primates is rather limited, resting on the primate-like structure of the cheek teeth and ear structure. In other respects the plesiadapiforms are somewhat specialized; for example, they possess large anterior teeth and three or fewer premolars.

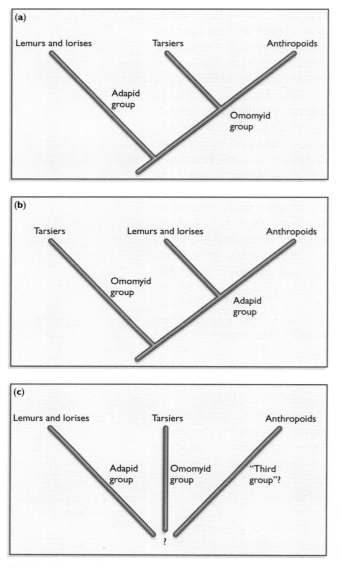

**FIGURE 6.7 Three views of primate evolution:** A good deal of uncertainty exists over the pattern of primate evolution. Until recently most opinion was divided between schemes (a) and (b), which show differences over the origin of anthropoids. A third view (c) has also been proposed, which postulates a third, early group of primates that was ancestral to modern anthropoids. On the basis of the most recently discovered fossil evidence, however, scheme (a) is now most strongly supported.

(Many of the earliest prosimians have four premolars.) For these reasons, it is probable that the plesiadapiforms were not ancestral to prosimians, but possibly formed a sister group in the primate clade.

Although identifying the first primate has been elusive, it is well established that by the Eocene there were two major groups of primates, both of them strepsirhines – the Omomyidae and the Adapidae. A flurry of discoveries in the 1990s threw light on the early evolution of these groups, although they have not allowed us to pinpoint the origins of the primates as a whole. Indeed, some have argued that the origins of the primates go back considerably further, and should be sought in the Cretaceous.[147]

For instance, in 1990, French researchers announced the discovery in Morocco of a collection of 10 undoubtedly primate cheek teeth, which were described as a new species, *Altiatlasius koulchii,* dated to the Late Paleocene.[160] The species, which is estimated to have weighed less than 100 grams, is thought to belong to the Omomyidae. A North American discovery, consisting of a relatively rare cache of fossil skulls, is also said to be an omomyid, of the species *Shoshonius cooperi,* which lived a little more than 50 million years ago.[161] The omomyids – tiny, nocturnal, fruit-eating species – are considered to be ancestral to tarsiers, while the Adapidae – diurnal folivores, frugivores, and insectivores – were bigger and are putative ancestors of lemurs. Although these two large and geographically widely dispersed families now seem well accepted as the earliest known primates, the question of their origin persists, if they are not derived from the plesiadapiforms.

One of the most spectacular discoveries, announced in 1994, included five new types of early primate, of both omomyid and adapid affinities, at the Shanghuang site in southeastern China. The diversity of species at this site exceeds that found in all of the rest of Asia and in well-documented sites in Europe and North America. One of the most interesting finds involved teeth that are virtually identical to those of modern tarsiers. Thomas Huxley speculated that the anatomical range of the lower-to-higher primates in today's world gives a window into the group's evolutionary history. The Chinese find indeed implies the modern tarsier might be a "living fossil." (Not literally, of course, but the group simply has not changed much since its origin.)

Uncertainty has long swirled around the evolutionary root of the suborder Anthropoidea (monkeys, apes, and humans). Some anthropologists have argued that its origin lies within the adapids; others have favored the omomyids.[146,162] Both schemes put the origin of anthropoids close to 35 million years ago. A recently developed argument suggests that neither group is ancestral to anthropoids, but that a third group existed. *Algeripithecus minutus*, discovered in Algeria and reported in May 1992, is suggested to be a specimen of this group.[146] The Shanghuang fossils provide support for the omomyid affinity with anthropoids, however. The dental formula of one specimen, *Eosimias*, is what would be expected of an ancestral anthropoid (hence its name, which means "dawn ape"). *Eosimias* is more closely linked to omomyids than to adapids, thus forging a link with the ancestral tarsier group. It now seems likely that modern tarsiers and modern anthropoids shared a specific common ancestor. If correct, this pattern of primate evolution would put the origin of anthropoids closer to 50 million years rather than the 35 million years ago that was previously believed.

The earliest known fossil of the superfamily Hominoidea, which includes all living and extinct species of humans and apes, is some 20 million years old; it was found in Africa (see chapter 8).

## Primate radiations

While the detailed phylogenetic relations of the early primates are obscure, the general pattern is relatively clear (Fig. 6.8). If we return to the four basic types of primate – the prosimians, New World monkeys, Old World monkeys, and hominoids – we can see that they represent a series of successive evolutionary radiations occurring at particular times and places (Fig. 6.9). The prosimians underwent the first of these radiations during the Eocene, and at a time of major changes in world biogeography. The prosimians, for many millions of years, were the only representatives of the primates, although the fossil evidence increasingly suggests that the first anthropoids (monkeys) did appear alongside them in the Eocene, but remained rare and isolated.

Adaptive radiations – the pattern of primate evolution?

The evolution of the anthropoids becomes the dominant evolutionary trend of the Oligocene. The first forms are known in Asia, but they are also known at a slightly later date in Africa. These appear to include the

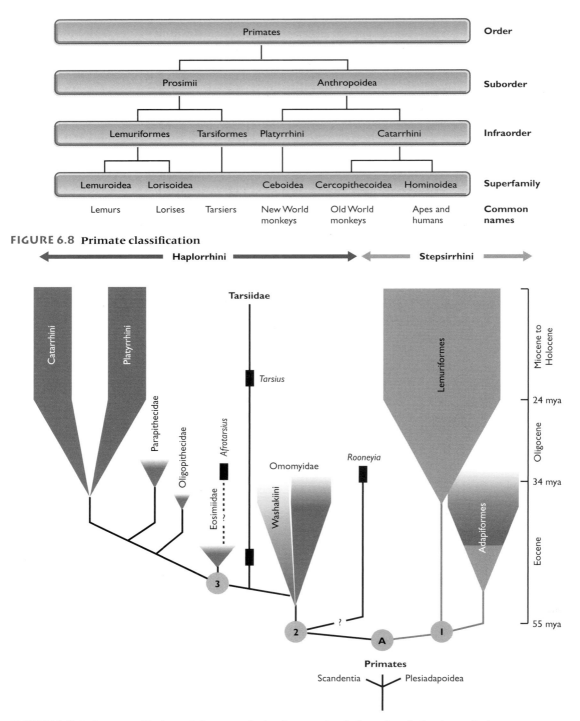

FIGURE 6.8  Primate classification

FIGURE 6.9  Primate radiations: Primate evolution has consisted of a series of adaptive radiations. (Adapted from Kay. Ross and Williams/*Science*.)

ancestors of both the Old and New World monkeys. The origins of the latter have long been a matter of dispute. Some have argued that they spread from the Old to the New World via Asia and North America, and then crossed the oceans that separated the two parts of the Americas. The problem with this hypothesis is the absence of any fossil evidence for anthropoids in North America at any time. Increasingly support has shifted to an alternative model, whereby the ancestors of the living Old and New World monkeys and apes are thought to have been in Africa during the Oligocene, and from there a founding population crossed the Atlantic (at that time not as wide as today), and thus established an isolated platyrrhine (New World monkey) lineage (see Fig. 6.8). This model fits the actual fossil evidence rather better, and is supported by similarities across a range of non-primate groups between Africa and South America. However, some researchers have questioned the feasibility of such a long-distance rafting event as crossing the Atlantic.[163]

However they arrived in South America, once established they underwent a major process of diversification, resulting in a remarkably diverse group today, showing both many unique forms such as the marmosets and tamarins, and the distinctive prehensile tails of the spider monkeys, as well as convergences with the monkeys of the Old World.

In the Old World, as will be discussed in more detail in chapter 8, the cercopithecoids (Old World monkeys) and the hominoids also diversified, across a time during which the prosimians became increasingly rare. The ancestors of the living superfamilies were probably present in the Oligocene, but it was in the Miocene that the early apes became abundant, while the monkeys remained relatively rare. The diversification of the living apes, the ancestors of humans, and the cercopithecoid monkeys underwent further diversifications during the last 10 million years or less.

## Primate evolution and human phylogenetic heritage

In one respect it can be argued that the deeper the origins of the primates go, and the more generalized their characteristics, the less there is to be learnt about the features derived from phylogeny that will have shaped human evolution. However, the overall pattern of primate evolution does provide some insights. Perhaps the most important of these, as indicated earlier, is that over the course of primate evolution there is a sustained pattern of increased brain size, intelligence, behavioral flexibility, and social complexity. Many of the anatomical traits observed and inferred in the fossil record are correlates of this overall pattern. In this context, humans can be said, by and large, to have extended the pattern of primate evolution, rather than to have set if off on another course. Another important point to note is that humans, along with other primates, have mostly

What do humans share with other primates?

retained the basic anatomical and physiological structure found in the order as a whole, and more specifically within the primates. Differences between humans and other primates are ones of degree, not kind. Certainly there are some specializations that are unique among the primates – bipedalism and relative hairlessness being two important ones – but in other ways humans are no more specialized than other primates. In one sense, therefore, it may be possible to say that one important element of the phylogenetic heritage of humans is its lack of specialization.

## THE COMPARATIVE PERSPECTIVE

**KEY QUESTION** Are there general rules governing the adaptations of animals (their size, shape, behavior, etc.) that may apply to humans when viewed comparatively?

Humans are unique, but that uniqueness is the product of evolutionary processes. That was the theme of chapter 1, and included the way in which different scientists over the years have grappled with how to explain that uniqueness. However, all species are unique, for that is what makes them separate species – hence the title of an earlier book by Foley, *Another Unique Species*.[30] While anthropology can perhaps, on account of the special nature of humans, get away with having its own discipline, it is not possible to separate out the evolutionary biology of each lineage, and treat it in isolation. On the contrary, the great power of evolutionary biology comes from looking at problems in comparative perspective. Why are plants different from animals? Why are elephants large and mice small? Why are the geographical ranges of carnivores larger than those of herbivores? In other words, the biology of the elephant is made more interesting, and more light is thrown on it, if it is compared with a mouse. From this perspective has come a great wealth of insight into patterns and processes in evolution and biology – for example, the patterns of body-size variation in relation to climate, or the determinants of the number of species on an island, and so on.

For the most part human evolution has remained relatively isolated from the growth of comparative evolutionary biology – attention has been too closely focused on human uniqueness. However, this has changed in recent years because of the power of comparative biology as a whole, which has encouraged a number of new approaches.[164] Sociobiology and socioecology (the study of variation in social organization), for example, have been one strong impetus, examining the interactions between social behavior and ecological conditions.[165,166] Another has been life-history theory – the recognition that there is variation in patterns of development that is strongly related to demography and to adaptation.[167,168] Yet another is ecomorphology, the patterns of variation in the shape and size of organisms in relation to the environment.[169] Each of these attempts to use interspecific patterns to generate general rules governing adaptation, and so to get at why any particular species has the form and the behavior that it does.

Many of these models are extremely powerful, and make excellent predictions of, for example, the amount of energy, in relation to body size, that an animal requires. In other ways it is often found that certain animals do not fit those models – they represent outliers or anomalies, and so are apparent exceptions to the rules. Where this occurs – and humans might be such a case in some contexts (for example, "expected" brain size) – then the models have highlighted unique elements that require special explanations. Thus, even when the models fail, they still throw light on evolutionary history.

## Evolutionary rules

Comparative approaches in essence go back to Darwin and Wallace, but it was really with the spread of their ideas to other biologists that scientists started proposing "evolutionary laws." Some, such as Bergmann's rule and Allen's rule (see next section) have stood the test of time. Others, such as Dollo's law (that evolution is irreversible), have faded and been disproved. Some of the rules are very grand and all embracing – for example, Kleiber's law of energy (see next section), while others are relatively limited in application. Major breakthroughs came in the second half of the twentieth century, with the development of island biogeography by Harvard biologist E. O. Wilson and Princeton ecologist R. McArthur, who showed how evolutionary and ecological diversity could be predicted from a few basic parameters.[170] This was also the start of a more mathematical and quantitative approach to evolution, for comparison depended upon being able to plot species in relation to each other; hence they had to be reduced to numerical values (Fig. 6.10). Perhaps the most striking development at that time was the r-K strategy theory (see the section below on "Bodies, brains, and energy"), which proved to be very powerful in predicting the characteristics of taxa on the basis of their reproductive potential.[171,172]

Are there "laws" of evolution?

The 1960s saw this approach expanded to include behavior, especially social behavior. Work by Crook and Gartlan[173,174] established the methods for making socioecological comparisons – that is, looking for patterns of social behavior in relation to resource availability, an approach extended to primates by Cambridge ecologist Tim Clutton-Brock and Oxford zoologist Paul Harvey.[166] To this suite of models was added life-history theory, developed originally by Oxford ornithologist and theoretical biologist David Lack,[175] who showed that features such as brain size and reproductive rate were not only strongly related to each other, but were also directly related to body size; and as we shall see in the next section, body size is a central element in all adaptive biology.

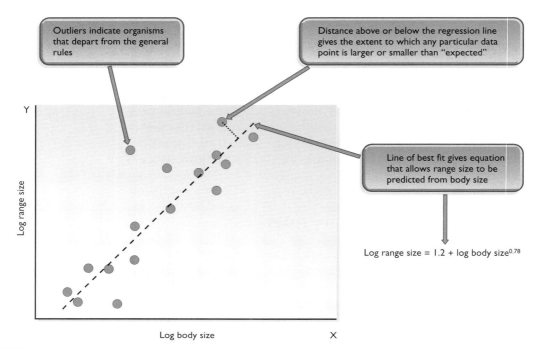

**FIGURE 6.10 The comparative perspective:** General rules in biology derive from a comparative perspective. For example, if two biological parameters are plotted against each other for a number of species, the relationship between them can be observed and one parameter (for example, range size) can be predicted from the other. Deviations from the predicted values provide important insights into unique adaptations.

The second reason for the growth of the comparative perspective in human evolution has been a greater tendency among anthropologists to treat humans as "just another animal," and certainly for them to see the fossil forerunners of modern humans as part of a suite of primate forms. A less anthropocentric perspective on humanity has led to an acceptance that humans should fit the patterns of evolutionary biology more generally. And if they do not, then it is only by comparison that it will be possible to discover where and when they deviate from the expected.[30]

A further factor in this change to a more comparative perspective is the growth of the fossil record for human evolution. In the days when the number of fossils could be counted on two hands, and when the number of species recognized was only two or three, each progressing inevitably toward the human condition, it was not really necessary to focus too much on the comparative element in human evolution. This was even more the case when it was thought that the ancestors of humans had diverged from the other apes as long ago as 20 million years. The massive growth in the recognition of fossil human diversity has meant that modern humans are now just one among many hominins – inviting comparison and a comparative approach.[176,177,178]

In the 1980s major critiques of this comparative approach were mounted, on the grounds that the work emerging was not statistically valid. The key point of any comparative analysis was to plot two variables on a graph and to see if there was a correlation, and if so, what sort of correlation. For example, body size is strongly correlated with home-range size (the size of area an animal needs to forage): the larger the animal, the larger the home-range size (Fig. 6.10). However, the basis of any statistical test is that the data points should be independent of each other – that is, that they should have at least the potential of varying in ways not directly related to other data points. Two evolutionary biologists from Oxford University, Paul Harvey and Mark Pagel, argued that in the comparisons being made there was no such independence, because the animals being compared were related to each other by common ancestry.[164] What they were suggesting was that if two species of baboons shared a particular relationship for body size and brain size, this was not because they were subject to the same adaptive laws, but because they had both inherited them from a common ancestor. This was in fact a statistical problem that had been recognized in the nineteenth century by one of the founding fathers of anthropology, Francis Galton.

How does common ancestry influence evolutionary comparisons?

However, rather than abandoning comparative approaches, Harvey and Pagel and others went on to develop what is now called "the comparative method in evolutionary biology," which takes the themes of the earlier comparative work, but analyses them in the context of what is known about phylogenetic history. In other words, these researchers developed methods which could take into account both that evolution is a process whereby laws of adaptive biology do occur, and that they do so in the context of a historical process. These models have themselves evolved over recent years, and are proving to be extremely useful.

Modern paleoanthropology is thus much more integrated into evolutionary biology as a whole, and is concerned with applying general rules – and indeed developing them – rather than just reconstructing the evolution of a lineage in isolation. In the next section we will look at how a comparative approach and the presence of general biological "rules" can throw light on what is a distinctly unique event – the evolution of humans.

# BODIES, SIZE, AND SHAPE

## Size and shape in evolution

Size is generally considered to be one of the most important factors in evolution – indeed, the relationship between size and

**KEY QUESTION** How do size and shape reflect adaptation to the environment?

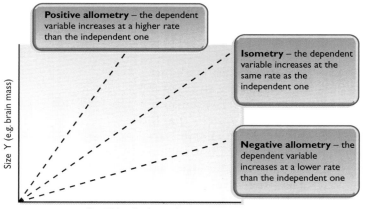

Size Y (e.g. brain mass)

**Positive allometry** – the dependent variable increases at a higher rate than the independent one

**Isometry** – the dependent variable increases at the same rate as the independent one

**Negative allometry** – the dependent variable increases at a lower rate than the independent one

Size X (e.g. body mass)

**FIGURE 6.11 Isometric and allometric relationships:**
Allometry refers to the way in which shape (or some other parameter) changes in relation to size. The graph shows three possible relationships between two biological variables in relation to size.

Why is allometry important?

evolution has been discussed for more than 100 years. Being large or small is part of an organism's adaptation (to reach resources, to fight conspecifics (members of the same species), to hide from predators, etc.), but it also correlates strongly with many other evolutionary patterns. Furthermore, larger animals have relatively lower energy requirements (Kleiber's curve), such that there is a form of economy of scale in nature; larger animals have larger geographical ranges; and larger animals have relatively smaller brains. The relationships of size are critical for understanding how adaptation actually works.[179]

Where size has no effect, then it can be said that there is an isometric relationship. In other words, if a particular organ or limb simply gets bigger at the same rate as the organism as a whole, then the relationship is isometric and size itself is not an important factor. However, for a large number of biological relationships this is not the case, and there are strong scaling effects.[180] For example, as body mass increases the energy required for metabolism increases at only three quarters of the same rate. This is referred to as an allometric relationship (Fig. 6.11). Allometric relationships have been shown to be of considerable interest in evolutionary biology, and in particular affect the shape of organisms, including humans. In particular, limb size and shape have an allometric relationship with body size, and are strongly influenced by factors such as climate. These allometric relationships – part of the "evolutionary rules" discussed above – have provided important insights into human evolution and adaptation.

It is striking that human populations in different parts of the world vary significantly in their body form, suggesting, among other things, an adaptation to different climates. An understanding of anatomical adaptation of many animal species to different climates has a long history, with two specific "rules" relating to this issue. Bergmann's rule, published in 1847, states that in a geographically widespread species, populations in warmer parts of the range will be smaller-bodied than those in colder parts of the range. Allen's rule, published in 1877, states that populations of a geographically widespread species living in warm regions will have longer extremities than those living in cold regions. Allen's rule applies to animals and plants alike, so the nature of the extremity depends on the species concerned (Fig. 6.12). In the case of humans, the most important extremities in this context are the limbs.

**(a)**

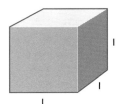

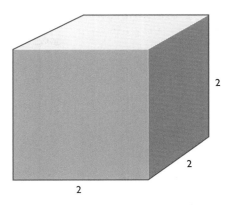

Mass = 1
Surface area = 6
**Surface area/mass = 6**

Mass = 8
Surface area = 24
**Surface area/mass = 3**

**(b)**

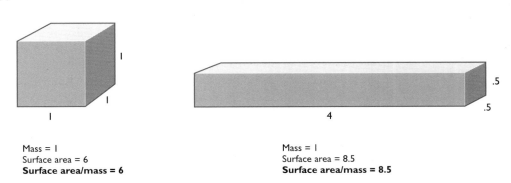

Mass = 1
Surface area = 6
**Surface area/mass = 6**

Mass = 1
Surface area = 8.5
**Surface area/mass = 8.5**

**FIGURE 6.12 Allen and Bergmann's rules:**

**(A) Geometric basis of Bergmann's rule:** An increase in size decreases the ratio of surface area to mass; in humans, this relationship is reflected in the breadth of the trunk.

**(B) Geometric basis of Allen's rule:** An elongated shape increases the ratio of surface area to mass; in humans, this relationship is reflected in limb length.

## Principles of climatic adaptation in human populations

Despite the long pedigree of Bergmann's and Allen's ecogeographical rules, anthropologists were slow to apply them to human variation. Interest in this relationship emerged in the 1950s and 1960s, when climate began to be recognized as an important influence in determining anatomical differences among different geographical populations. For instance, the bodily differences between the tall, thin Nilotics at the equator and the short, bulky Eskimos in the Arctic came to be viewed as a direct

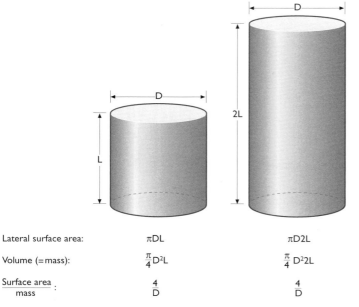

| | | |
|---|---|---|
| Lateral surface area: | $\pi DL$ | $\pi D2L$ |
| Volume (=mass): | $\frac{\pi}{4}D^2L$ | $\frac{\pi}{4}D^22L$ |
| $\frac{\text{Surface area}}{\text{mass}}$ : | $\frac{4}{D}$ | $\frac{4}{D}$ |

**FIGURE 6.13 The cylindrical model of body shape:** An increase in the length (L) of the trunk has no effect on the ratio of surface area to body mass. (Courtesy of C. B. Ruff.)

reflection of optimal strategies for balancing heat production and dissipation at different latitudes with different prevailing climates.

In recent years, Christopher Ruff, of Johns Hopkins University, has been bringing together the study of ancient and modern human variation in relation to climate.[181,182] For his analyses, Ruff views the human body as a cylinder, the diameter of which represents the width of the body, or, more specifically, the width of the pelvis; the length of the cylinder represents trunk length (Fig. 6.13). The link between anatomy and climate relates to thermoregulation, or the balance between heat produced and the ability to dissipate it. This relationship translates to the ratio of the surface area to the volume of the cylinder, or body mass. In hot climates, a high ratio – that is, a large surface area relative to body mass – facilitates heat loss. In cold climates, a low ratio – that is, a small surface area relative to body mass – allows heat retention. Simple geometry shows that the ratio of surface area to body mass is high when the cylinder is narrow, and low when it is wide. This finding forms the basis of Bergmann's rule.

A strong prediction flows from this analysis: people living at low latitudes will have narrow bodies and a linear stature, while those at high latitudes will have wide bodies and a relatively bulky stature. When Ruff surveyed 71 populations around the globe, he found that the prediction was sustained very well (Fig. 6.14). He also discovered that Allen's rule applies convincingly, with tropical people having longer, thinner limbs, which maximizes heat loss, while people at high latitudes have shorter limbs. This difference in limb proportions enhances the linear look of tropical people and the stocky appearance of high-latitude populations. A comparison of the tall Nilotic people of Africa with the relatively stocky Eskimos in the northernmost latitudes of North America illustrates this difference very clearly.

Body width represents the key variable, even though tropical people also tend to be linear. A further step of simple geometry shows that linearity is not a necessary feature of low-latitude populations. The ratio of the surface area to body mass in a cylindrical model of a certain width is not altered by changing its length, as Fig. 6.13 above shows. Peoples who live in similar climatic zones will have the same body width, no matter how tall or short they are, because they have the same surface area to body mass ratios (Fig. 6.15). This fact is revealed in a comparison of

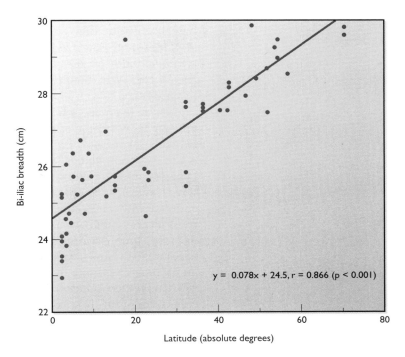

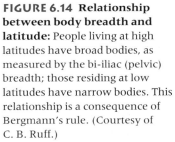

**FIGURE 6.14 Relationship between body breadth and latitude:** People living at high latitudes have broad bodies, as measured by the bi-iliac (pelvic) breadth; those residing at low latitudes have narrow bodies. This relationship is a consequence of Bergmann's rule. (Courtesy of C. B. Ruff.)

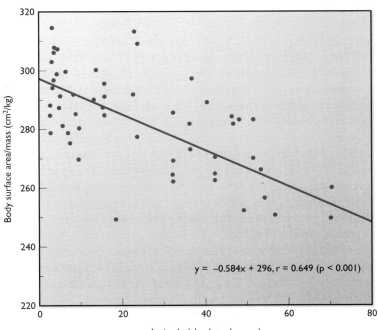

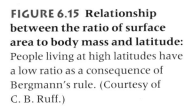

**FIGURE 6.15 Relationship between the ratio of surface area to body mass and latitude:** People living at high latitudes have a low ratio as a consequence of Bergmann's rule. (Courtesy of C. B. Ruff.)

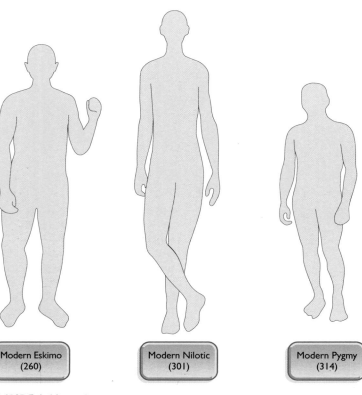

Modern Eskimo
(260)

Modern Nilotic
(301)

Modern Pygmy
(314)

**FIGURE 6.16 Body outlines of modern populations:** Figures below the outlines give the surface area to body mass ratio (cm²/kg). Note the broad body and short stature of the Eskimo, and a low ratio; the Nilotic body is narrow and linear, with a high ratio. The Pygmy has the same body breadth as the Nilotic and a similar ratio. (Courtesy of C. B. Ruff.)

What determines human body shape?

Nilotic people, whose average height is more than six feet, and Mbuti Pygmies, who are two feet shorter on average (Fig. 6.16). Why the difference in stature?

The answer is related to efficiency of heat dissipation. Humans rely heavily on sweating to cool their bodies. Nilotics live in open environments, where sweating is efficient; in contrast, Mbuti pygmies, like most Pygmy populations, live in moist, humid forests, where the air is still and sweating is an inefficient cooling mechanism. Under these environmental conditions, the best strategy is to limit the amount of heat generated during physical exertion, which is achieved by reducing the volume of the cylinder. With the width of the cylinder remaining constant, this requirement implies a reduction of its length – in other words, reduced stature.

This insight may have implications for the lifestyles of both the Nariokotome (Turkana) boy (*Homo erectus* individual) and Lucy, an individual of an early hominin species (see chapter 9), whose differences in stature are similar to the differences observed between Nilotics and Pygmies. Despite their varying statures, Lucy and the Nariokotome boy exhibited very similar body widths, comparable with the width of modern tropical populations. This observation makes sense because, living in east Africa as they did, they were exposed to a tropical climate (albeit more than a million years apart). Ruff speculates that, like the Nilotics of today, the Nariokotome boy and his fellow *Homo ergaster* people lived an active life in open environments. Lucy and her companions, by contrast, inhabited more closed, forested environments, comparable with the environment of modern Pygmies (Fig. 6.17).[183]

Climatic adaptation of body form can also be seen in Neanderthals, who lived in Europe between 250,000 and 34,000 years ago – a time when, for the most part, the Pleistocene ice age still held the continent in its grip. The frigid conditions under which the Neanderthals evolved are reflected in their wide bodies and their relatively short limbs, characteristics comparable to those seen in modern Eskimos.[184]

# Complications flowing from population migration

Although a population's location in the world may be stamped in the anatomy of its members' bodies, complications sometimes arise. For instance, the correlation of body width to latitude is less strong in New World populations than it is in the Old World. Some odd paradoxes exist, too, such as the Pecos Pueblo American Indians of New Mexico. Although these people experienced a warm climate, they had an average body width only slightly less than that of Eskimos, but their limb proportions were not vastly different from those of equatorial populations. The original colonizers of the Americas came from northeastern Asia and would have had a cold-adapted body shape. As they moved south through the New World, populations would have begun to adapt to local conditions, but only gradually. Moreover, the difference in the body width/latitude equation between the Old and New Worlds, and the chimeric form of, for instance, the Pecos Pueblo Amerindians, suggests a difference in speed of adaptation of different parts of the body. Although limb proportion can change relatively quickly, alteration of body width occurs much more slowly. Moreover, variation in skin pigmentation with latitude is not as great in the New World as that seen in the Old World, again indicating a slow rate of local adaptation. Amerindians have been in the New World for at least 10,000 and perhaps as long as 30,000 years (see chapter 19).

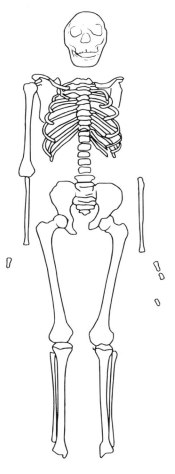

KNM-WT 15000
Nariokotome boy
(307)

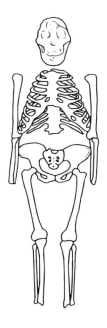

AL 288-1
Lucy
(320)

**FIGURE 6.17 Body proportions in fossil humans:** Despite their different statures, the Nariokotome (Turkana) boy and Lucy have very similar body breadths and surface area to body mass ratios. (Courtesy of C. B. Ruff.)

Population movements such as the colonization of the Americas occurred many times in human prehistory, and they inevitably muddy what might otherwise be a clear relationship between body shape and climate, and its change through time. It is not that they undermine the importance of adaptive rules such as Bergmann's, but that they act as a reminder that evolutionary change occurs in particular historical contexts, and that it takes time for selection to operate. One important example relates to the issue of the origin of modern humans. Many anthropologists believe that anatomically modern *Homo sapiens* evolved from a small population in Africa, perhaps as much as 200,000 years ago, and then spread into the rest of the Old World, reaching western Europe only 45,000

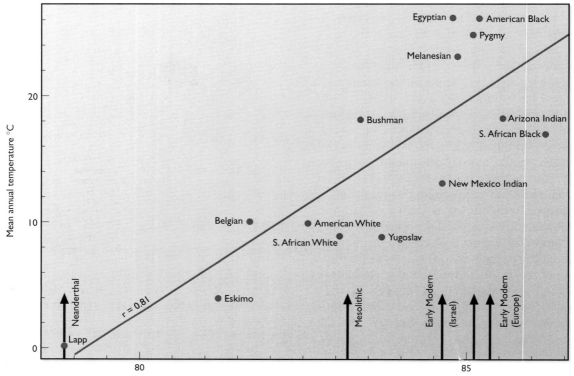

**FIGURE 6.18 Body proportions in early modern humans and Neanderthals:** Limb proportions of modern humans in relation to temperature are shown in the graph. The limb proportions of early modern humans and Neanderthals are indicated by arrows, and suggest that the former were tropical in origin, while the latter were cold-adapted. (Courtesy of C. Stringer.)

to 35,000 years ago. If true, the African origin of anatomically modern humans would be reflected in their body and limb proportions (Fig. 6.18). Indeed, such populations do show this trend, as these tall, long-limbed people entered lands located at a latitude more conducive to wide bodies and short limbs. Erik Trinkaus, of Washington University, has promoted this climatic hypothesis since the early 1980s.[184]

A competing hypothesis also attempts to explain the long lower-limb proportions of the early modern human populations of Europe (Fig. 6.19). It relates to the mobility of these populations compared with the established Neanderthals. According to the mobility hypothesis, early modern Europeans were more active than Neanderthals, covering more territory in their daily foraging. A long stride length would be beneficial for such a subsistence pattern; and a long stride length is best achieved with long lower limbs. Milford Wolpoff, of the University of Michigan, and David Frayer, of the University of Kansas, have been prominent proponents of the mobility hypothesis.

A comparative test of the competing hypotheses was recently conducted by Trenton Holliday at Tulane University.[185] If a connection between mobility and lower limb length explains the observations on early modern humans in Europe, then the same connection should be evident in modern hunter-gatherer populations. A survey of 19 such populations showed no such connection, which argues against the mobility hypothesis. Although the survey did not produce strong support for the climatic hypothesis, Holliday concludes that it remains the most likely explanation. Very probably, several factors were involved, including adaptation to climate.

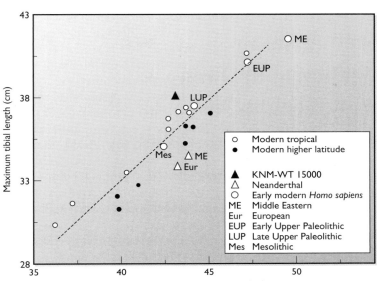

**FIGURE 6.19  Tibia length relative to femur length in modern and prehistoric populations:** A short tibia relative to the femur indicates a relatively short leg, an indication of cold adaptation. The dotted line represents the division between modern tropical and high-latitude populations. Tropical populations fall above the line, indicating a long leg; high-latitude populations fall below it, indicating a short leg. European and Middle Eastern Neanderthals had relatively short legs; early African *Homo* (KNM-WT 15,000, the Nariokotome (Turkana) boy) had relatively long legs. The early modern human populations are intermediate between tropical and high-latitude populations, but tend to the tropical end. (Courtesy of C. B. Ruff.)

# Changes in recent human populations

We now turn to changes in body form of humans through time. Many anthropologists agree that from early in the *Homo* lineage, some 2 million years ago, to the appearance of archaic *Homo sapiens*, 300,000 or 400,000 years ago, robusticity steadily increases before finally reaching a plateau. In this case, we are talking about people having thick skulls and heavily muscled limbs. (Brain size increased from approximately 900 cc to more than 1400 cc in this time.) These people were immensely strong, reflecting their arduous subsistence pattern. Early anatomically modern humans, who appeared 200,000 years ago, were significantly less robust than archaic *sapiens*, but much more so than people today. (As mentioned above, the early anatomically modern people in Europe were also more linear, because of their African origin.) The robusticity of early moderns decreased gradually over a long period, and then dramatically so after the end of the ice age, 10,000 years ago, but not in all populations. Australian Aborigines, Patagonians, and Fuegans, for instance, are still relatively robust in their skull and skeletal anatomy.[186] Where it occurred, the loss of robusticity occurred principally between 10,000 and 5000 years ago, then halted. Reductions in brain size (to 1300 cc), size of teeth and jaws, and overall stature followed similar patterns, but to different degrees.[187]

For instance, in his studies of Australian populations, Peter Brown, of the University of New England, Armidale, found the following changes in the five millennia after the ice age: tooth reduction, 4.5%; facial size reduction, 6–12%; brain-size reduction, 9.5%; and stature reduction, 7%.[188] Where data exist in other parts of the world, such as in Europe and southeast and west Asia, similar changes are observed, although paleoanthropologists disagree on whether, for instance, significant brain shrinkage began as early as 30,000 years ago or only 10,000 years ago. Whatever the details of the timing of events in these later stages, it seems irrefutable that, until the nutritional effects of the last century or so kicked in, modern people were comparatively small on the human evolutionary stage.

What overall pattern held, beginning with the increase in robusticity until archaic *sapiens* arrived? Subsistence was strenuous in those days, as our ancestors plied a life of hunting and gathering with only rudimentary technology to aid them. Muscles, not missiles, were their weapons. Other explanations for this trend have been suggested, too. For instance, Foley suggested that people became stronger because they were embroiled in increasing conflict between neighboring groups.[189] The conflicts arose because the groups were dominated by bands of males, probably closely related, who sought to appropriate the plentiful resources in their area, including females from other groups.

How and why have humans changed biologically in the recent past?

Why, then, did robusticity decline with the origin of anatomically modern humans, and continue to diminish for tens of millennia? Not because these humans changed their social structure and became more peaceable, but because technological inventions usurped the role previously played by sheer strength. One key invention involved projectiles, spears in which stone points were hafted onto wooden shafts. Stone tools became more versatile, which perhaps buffered people from some bare-hands contact with their environment. And people were smarter, too, indicating that guile rather than brawn might have filled the larder.

Loring Brace, of the University of Michigan, has long been a proponent of technology, or culture, as an important force in diminishing human robusticity.[190] Eventually, food preparation, through cooking, took the pressure off teeth, which became smaller as a result. This development emerged at different times in different parts of the world. But, as Peter Brown observes of the Australians, tooth reduction can occur even in the absence of food preparation, so other forces must be operating as well. Perhaps, whatever the reason underlying the change, it is important to remember that, as with any such shift, it is a question of costs and benefits. Large body size endows individuals with certain benefits, principally strength, but there are costs as well, associated with growth and energetics; so when the benefits of being large and robust disappear, then selection will favor smaller individuals.

Many dramatic changes transpired with the end of the Ice Age, not least of which was the disappearance of plentiful game, some of it very large.

Gone were the mammoth and mastodon, for instance. Foley suggests that this reduction of resources forced recourse to one of two subsistence strategies. The first was food production, or agriculture; the second was a shift to a different kind of social structure in hunting and gathering bands. Although we now think of agriculture as producing plentiful food, early food production was a hazardous venture, with many lean times. The archeological record shows that nutritional stress was rife for early farmers – a sure way of keeping body size small. In the second strategy, because male hunters were unable to monopolize food resources to the same degree as their ancestors had done, they were unable to monopolize many females as mates (a practice known as polygyny). As a result, less aggressive competition for females occurred among males, and therefore less of a premium was placed on raw strength. Thus, males became smaller because they did not need to fight as much.

Nutritional stress is a popular explanation of human shrinkage for many anthropologists, not because of the loss of the megafauna of the ice age but because of a booming human population. Limitations on resources often lead to reduced body size, says Christopher Stringer, of the Natural History Museum, London, as is seen in the dwarfing of species on islands.[191] According to Robert Martin, of the University of Chicago, the stress results from a shift to early weaning, a strategy that boosts reproductive output in the face of the competition associated with increased populations. Early weaning inevitably leads to a reduction in brain size, though not, says Martin, necessarily to a reduction in body size. This viewpoint separates Martin from most anthropologists, who argue that reductions of brain size in recent history simply followed in the path of body-size reductions.

The advent of agriculture was once viewed as the universal change in human culture that produced a universal change in human physique. As more was learned about the shift from foraging to food production, however, this notion appeared less likely to be the answer. Agriculture was developed at different times in different parts of the world, and in some places not at all. In Australia and parts of the Americas, for instance, people were still hunting and gathering as they had been for tens of millennia, and yet the pattern of body-size reduction still applied. The one change that applies everywhere, of course, is the increase in global temperature associated with the end of the Pleistocene. That fact alone is sufficient to force serious consideration of climate as the causative agent. Moreover, body-size reduction has occurred in many non-human animals in this same period, in Australia, Israel, and Indonesia, for example. Wherever researchers look (and so far not many places have been analyzed), the same phenomenon is found in non-human animals.

Body structure could change in several ways in the face of increased ambient temperature. For example, it might reduce body mass, which in humans translates to reduced stature. Militating against this hypothesis as a general explanation, however, are two observations. First, changes in body form almost certainly began before the end of the Ice Age, so warming cannot be adduced as an explanation. Second, the Pleistocene was punctuated by several warm interglacials – one around half a million years ago, and a second between 125,000 and 75,000 years ago – and there is no obvious shift in stature to accompany these global changes in temperature.

The changes in body size observable in human evolution over the relatively recent past are complex and do not offer themselves to simple explanations. That complexity – different timings in different populations, and different trends in geographically separated populations – is a useful reminder that evolutionary change is made up not of broad general patterns, but of local patterns of adaptation occurring under conditions of particular environment, and with specific historical population movements. All evolutionary change is likely to be of this form; it is the relatively short time scale and the richness of the fossil record that make it possible to observe this occurring in the evolution of modern human diversity. However, in seeking explanations it is important to remember that what we are observing is the playing out of general rules of adaptation in particular conditions associated with human evolution.

## BODIES, BRAINS, AND ENERGY

**KEY QUESTION** How does the size of an animal influence its growth, development, and reproduction, and thus its evolutionary strategy?

The last section introduced size as an important factor in shape and discussed this in relation to climatic adaptation. In this section we pursue another aspect – the impact of size, of both brains and bodies, on life-history variables and behavioral ecology. Life-history variables are those factors that describe how individuals of a species proceed from infancy through maturity to death, and the strategies involved in producing offspring (Fig. 6.20). We will see why hominins, with their large body size, have many more options open to them in terms of diet, foraging range, sociality, expanded brain capacity, and so on than, for example, the diminutive mouse lemur.

In 1978, Princeton ecologist Henry Horn[192] encapsulated the range of potential ecological options by posing the following set of questions:

> In the game of life an animal stakes its offspring against a more or less capricious environment. The game is won if offspring live to play another round. What is an appropriate tactical strategy for winning this game? How many offspring are needed? At what age should they be born? Should they be born in one large batch or spread out over a long lifespan? Should the offspring in a particular batch be few and tough or many and flimsy? Should parents lavish care on their offspring? Should parents lavish care on themselves to survive and breed again? Should the young grow up as a family, or should they be broadcast over the landscape at an early age to seek their fortunes independently?

In responding to these challenges, the animal kingdom as a whole has come up with a vast spectrum of strategies, ranging from species (oysters, for instance) that produce millions of offspring in a lifetime, upon which no parental care is lav-

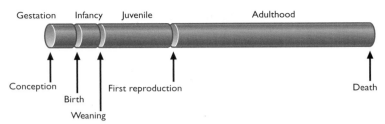

ished, to species (such as elephants) that produce just a handful of offspring in a lifetime, each born singly and becoming the object of intense and extensive parental care. In the first case, the potential reproductive output of a single individual is enormous, though typically curtailed by environmental attrition; in the second case, it is small.

**FIGURE 6.20 Life-history variables:** The life of any individual is a series of stages from conception to death. Timing and stages of life influence and are influenced by body size, metabolic rate, brain size, number of offspring, and mortality. Life-history theory looks at how the evolutionary context has set the timing and duration of these for particular species.

## r- and K-selection

In seeking an answer to Horn's question about how best to play the game of life, biologists have developed some powerful predictive models based on demography and ecology. A few simple parameters, it turns out, go a long way to showing when it pays to have lots of offspring and when it pays to just have a few, and what the consequences are.

Imagine an island being colonized for the first time by a pair of individuals of a species. Assuming there is a niche for them, what will happen to them? The answer is that they will reproduce and the population will increase. In fact the rate of population increase will be exponential, for in the absence of any serious limitation on their survival (it is, after all, an empty island), and its rate will be determined by the maximum rate at which they can produce offspring. The rate can be described by a simple equation:

$$R = rN$$

where $R$ = the rate of population growth, $r$ = the maximum fecundity of an individual, and $N$ = the number of breeding individuals. The rate of exponential growth is determined by the size of $N$ and $r$.

However, we can ask what will happen to this population over the longer term. As the population increases, at a more and more rapid rate, it will start to overcrowd the island, and to eat into the resource base on which it depends. It will have reached its carrying capacity (Fig. 6.21). Now the population will not increase, and its growth will be described not just by the parameters $r$ and $N$, but by the carrying capacity, expressed as $K$ in this equation:

$$R = rN\left(\frac{1 - N}{K}\right)$$

What is the difference between an r- and a K-selected species?

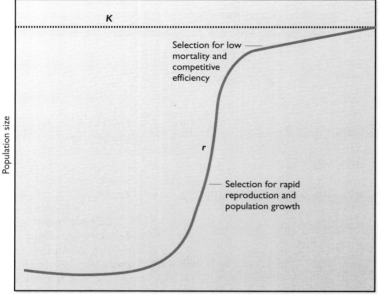

K

Selection for low
mortality and
competitive
efficiency

r

Selection for rapid
reproduction and
population growth

Population size

Time

**FIGURE 6.21 r- and K-selection:** The principles of r- and K-selection are drawn from the way in which a population may increase in relation to resource availability. Early in time, when a population is very small, it will increase exponentially in relation to its maximum intrinsic reproductive rate (r-selected); when the population is large and close to carrying capacity (K), it will be selected for its ability to compete effectively (K-selected).

What this means is that the growth of a population is determined by two things: its intrinsic rate of reproduction and the availability of food. Which of these two is the most important depends on the context. When the population is well below carrying capacity, it is the reproductive rate, r, that is important; at carrying capacity, it is the environment, or rather the organism's ability to survive in a competitive environment, that is critical.

From this simple model come some important evolutionary insights. Where resources are not the limiting factor, such as in very unstable and unpredictable environments, then the most successful strategy is to have a high reproductive rate and to produce many offspring. This is described as an "r-selected" strategy, reflecting the role of reproductive rate in the equations. Where resources are limiting, then producing large number of offspring is counterproductive – they will simply die – and it is a better strategy to have few offspring that are well adapted to the resource base and can compete well. This is known as "K-selected," again from the equations.[172]

In other words, from a simple consideration of the way in which populations grow under different ecological circumstances we have an insight that helps answer Horn's question of what is the best strategy. It depends on the circumstances, and some species will be "r-selected" or "r strategists" and have high reproductive and mortality rates, while others will be "K-selected" or "K strategists," having a few high-quality offspring. Of course, it is important to remember that this is a simple model: many other factors are involved in determining life-history strategies, and the r–K model is a continuum one – no species is purely r-selected or K-selected. However, these terms are widely used to describe the link between ecology and the evolution of life-history characteristics (Fig. 6.22). What is more, there are a number of correlates of the r–K model (Fig. 6.23), and these provide insights into primate and human evolution.

## Primates as large mammals

By their nature, mammals are constrained in the range of life-history patterns open to them: mammalian mothers are limited in the number of

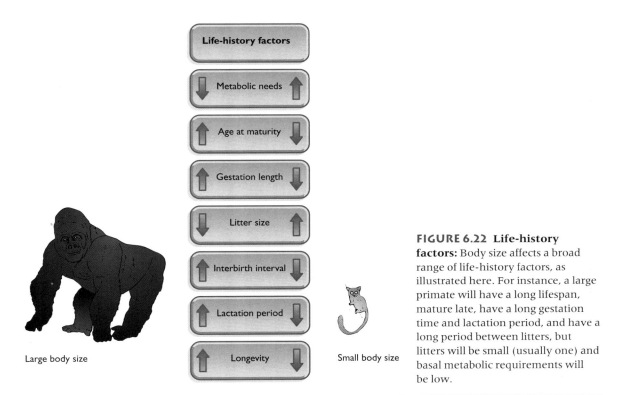

**FIGURE 6.22 Life-history factors:** Body size affects a broad range of life-history factors, as illustrated here. For instance, a large primate will have a long lifespan, mature late, have a long gestation time and lactation period, and have a long period between litters, but litters will be small (usually one) and basal metabolic requirements will be low.

| FACTOR | R-SELECTION | K-SELECTION |
|---|---|---|
| Climate | Variable and/or unpredictable; uncertain | Fairly constant and/or predictable; more certain |
| Mortality | Often catastrophic, nondirected, density independent | More directed, density dependent |
| Survivorship | High juvenile mortality | More constant mortality |
| Population size | Variable in time, nonequilibrium; usually well below carrying capacity of the environment; unsaturated communities or portions thereof; ecological vacuums; recolonization each year | Fairly constant in time, equilibrium; at or near carrying capacity of the environment; saturated communities; no recolonization necessary |
| Intra- and interspecific competition | Variable, often lax | Usually keen |
| Selection favors | ♦ Rapid development<br>♦ High maximal rate of increase, $r_{max}$<br>♦ Early reproduction<br>♦ Small body size<br>♦ Single reproduction<br>♦ Many small offspring | ♦ Slower development<br>♦ Greater competitive ability<br>♦ Delayed reproduction<br>♦ Larger body size<br>♦ Repeated reproduction<br>♦ Fewer, larger progeny |
| Length of life | Short, usually less than 1 year | Longer, usually more than 1 year |
| Leads to | Productivity | Efficiency |

**FIGURE 6.23 Characteristics of *r*- and *K*-selection:** *r*-selected species (such as oysters) live high-risk lives and are more affected by external factors than by competition from within the population. *K*-selected species pursue low-risk strategies in which intraspecies competition is an important factor in success. Primates as a whole, and apes and humans in particular, are *K*-selected.

**FIGURE 6.24 A difference in body sizes:** The gorilla and the mouse lemur represent the largest and the smallest of the primates, with the females of the species weighing 75 kilograms and 80 grams, respectively. Such differences in body size have many implications for a species' social and behavioral ecology. One of the most dramatic involves potential reproductive output: the female mouse lemur can grow to maturity and, theoretically, leave 11 million descendants in the time it takes a female gorilla to produce a single offspring.

offspring that can be carried successfully through gestation and suckling. Nevertheless, potential reproductive output can be relatively high if more than one litter is raised each year and the lifetime lasts several years.

In the order Primates, potential reproductive output is low compared with that of mammals as a whole, with litters being restricted in the vast majority of species to a single offspring. In the parlance of population biology, primates are said to be *K*-selected. Of all primates, humans are the most extremely *K*-selected species.

Within the overall Primate order, however, a wide range of life-history patterns exists, as biologists Paul Harvey, Robert Martin, and Tim Clutton-Brock have pointed out:

Adult female mouse lemurs [the smallest species of primate] can probably produce one or two litters of two or three offspring each year, and the young can be parents themselves within a year of their own birth [Fig. 6.24]. On the other hand, adult female gorillas [the largest species of primate] produce a single offspring every 4 or 5 years, and the young do not breed until they are about 10 years old.[193]

In terms of potential reproductive output, the female mouse lemur (which weighs 80 grams) can leave 10 million descendants in the time it takes the female gorilla (weighing 75 kilograms) to produce just one. "Such differences between species have presumably evolved as adaptations for exploiting different ecological niches," note Harvey and his colleagues. "Each niche is associated with a particular optimum body size, dictated in part by an animal's ability to garner and process available food supplies."

Success in simple Darwinian terms is often measured in the currency of reproductive output, which is determined by a series of interrelated life-history factors. These factors include age at maturity, length of gestation, litter size, duration of lactation period, interbirth interval, and lifespan.

Some species live "fast" lives – during their short lifespan, they mature early, produce large litters after a short gestation period, and wean early. The result is a large potential reproductive output. Other species live "slow" lives – during their long lifespan, they mature late, produce small litters (a single offspring) after a long gestation period, and wean late. Thus, their potential reproductive output is small.

As it happens, the best predictor as to whether a species lives "fast" or "slow" is its body size. Small species live fast lives, while large species live slow lives. As potential reproductive output is highest in species that experience fast lives, it might seem that all species would be small. That some species are large implies that a bigger body size provides some benefits that offset the reduced potential reproductive output.

Such benefits might include (for a carnivore) a different spectrum of prey species or (for a potential prey) better anti-predator defenses. Another potential benefit of increased body size is the ability to subsist on poorer-quality food resources. Basal energy demands increase as the 0.75 power of body weight; in other words, as body weight increases, the basal energy requirement per kilogram of body weight decreases, a relationship known as Kleiber's curve. This concept explains why mouse lemurs must feed on energy-rich insects and gums, for instance, while gorillas can subsist on energy-poor foliage. A further potential benefit of increased body size is improved thermoregulatory efficiency.

The generally close relationship between body size and the value of the various life-history factors is the outcome of certain basic geometric and bioenergetic constraints – the basis for allometry, discussed in the previous section. Any particular body size increase is associated with a more or less predictable change in, for example, gestation length, and age at maturity. For each life-history variable, therefore, a log/log plot against body size produces a straight line, with a particular exponent that describes the relationship (0.75 for basal energy needs, 0.37 for interbirth interval in primates, 0.56 for weaning age, and so on). In effect, such plots remove body size from species comparisons and allow us to assess the significance of particular organs – for example, brain size – after body size is taken into account. This examination amounts to analyzing how far particular features depart from predictions based on body size.

*What are life-history variables?*

If basic engineering constraints were all that underpinned life-history factors, then every species would be directly equivalent with every other species when body weight was taken into account. That is, all figures for each life-history variable would fall on the appropriate straight lines. In fact, individual figures often fall above or below the line, indicating a good deal of life-history variation. This variation reveals an individual species' (or, more usually, a group of related species') adaptive strategy.

Researchers now know that, in addition to body size, brain size is also highly correlated with certain life-history factors, in some cases much more so than is body size.[194]

## Altricial and precocial strategies

Among mammals as a whole, a key dichotomy exists in developmental strategy that has important implications for life-history measures: the

altricial/precocial dichotomy. Altricial species produce extremely immature young that are unable to feed or care for themselves. The young of
precocial species, on the other hand, are relatively mature and can fend for
themselves to a certain degree.

Life-history factors critically associated with altriciality and precociality
include gestation length. In altricial species, gestation is short and neonatal
brain size is small. Gestation in precocial species is relatively long, and
neonatal brain size is large. There is, however, no consistent difference in
adult brain size between altricial and precocial species. Primates as a group
are precocial with the exception of *Homo sapiens*, which has developed a
secondary altriciality and an unusually large brain.

In addition to the distinction between fast and slow lives based on
absolute body size, some species' lives may be fast or slow for their body
sizes. Such deviations have traditionally been explained in terms of classic
*r*- and *K*-selection theory outlined above. According to this theory, environments that are unstable in terms of food supply (that is, are subject
to booms and busts) encourage *r*-selection: fast lives, with high potential
reproductive output. Alternatively, stable environments (which are close
to carrying capacity and in which competition is therefore keen) favor
*K*-selection: slow lives with low potential reproductive output and high
competitive efficiency.

As mentioned earlier, primates are close to the *K*-selection end of
the spectrum among mammals as a whole, but some primates are less
*K*-selected than others. For instance, Caroline Ross of the Roehampton
Institute, London, has shown that, when body size is taken into account,
primate species that live in unpredictable environments have higher
potential reproductive output than species residing in more stable
environments.[195]

A second factor that influences whether a species might live relatively fast or slow for its body size has been identified by Paul Harvey and
Daniel Promislow.[194] In a survey of 48 mammal species, the two found
that "those species with higher rates of mortality than expected had
shorter gestation lengths, smaller neonates, larger litters, as well as earlier
ages at weaning and maturity." In other words, species that suffer high
natural rates of mortality live fast. "The reason is that species with higher
rates of mortality are less likely to survive to the following breeding season
and will therefore be selected to pay the higher costs associated with the
earlier reproduction."

Again, does the very slow life lived by *Homo sapiens* imply evolution
from an ancestor that experienced very low levels of mortality?

Given that most mammals measure less than 32 centimeters in length,
hominins – even the early, small species – must be classified as large mammals. One of the earliest known hominin species, *Australopithecus afarensis*,
stood in the general range of 1 meter (females) to 1.7 meters (males)
tall, and weighed some 30 to 65 kilograms. These general proportions
persisted until approximately 1.5 million years ago and the evolution of

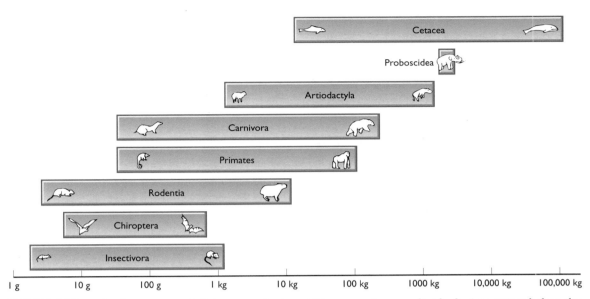

**FIGURE 6.25 Body size compared:** Primates are in the middle range of mammalian body sizes. Nevertheless, the biology of hominoids is the biology of large mammals. Most mammalian species are small and are concentrated in Rodentia, Chiroptera, and Insectivora.

*Homo erectus*, which stood close to 1.8 meters (with a much reduced difference between males and females).

## Predictions for early hominin species

With a knowledge of these general body proportions and the estimates of brain size, it becomes possible to predict various life-history factors for the early hominin species, given what is known of the only extant hominin, *Homo sapiens*. Surely, hominins lived slow lives in terms of life-history variables, with a vastly increased brain capacity eventually distorting some of them.

*What are human life-history characteristics?*

In addition, we can identify several behavioral ecology traits that would be associated with large body size (Fig. 6.25). For instance, dietary scope could be broad; day and home ranges could be large; mobility could be high; predator–prey relations would be shifted from those of smaller primates; thermoregulatory efficiency would be improved; sociality would be extended; and enhanced encephalization would be energetically possible.[30]

In sum, studies of life-history strategies have identified body size, brain size, environmental variability, and mortality rates as being crucial to the rate at which a species will live. Much of human evolution may therefore be explained in terms of a large hominoid exploiting a relatively stable food supply, its stability perhaps being enhanced by virtue of its breadth.

Technology may eventually have contributed to this stability by permitting more efficient exploitation of meat and certain plant foods, thus broadening the diet even further. A reduction in mortality, perhaps through improved anti-predator defense, would further encourage a "slow" life-history strategy. The selection pressure leading to increased body size remains to be identified. Furthermore, because these models, drawn from using a comparative approach, allow us to have *expectations*, we can then have a means of exploring a key issue in human evolution: when do humans deviate from the general models? When, for example, do humans' brains become larger than would be expected on the basis of their body size as a typical primate?

## BEYOND THE FACTS
# IMPROBABLE BIOLOGY

*The issue: much of science – and all of statistics – is concerned with probability. Much of what we see in biology makes sense, and can be said to be highly probable. However, there are many issues in evolution where no matter how hard one tries, the hypotheses still seem very unlikely. How can we come to terms with the apparently improbable nature of some aspects of evolution? Indeed, the appearance of humans may count as one such event. Does the vast amount of time involved in evolution make a difference?*

Both Darwin and Wallace, the co-founders of modern evolutionary theory, recognized the importance of distributions as evidence for evolution. Wallace wrote a vast monograph on the *Geographical Distribution of Animals*, which is still a major contribution to biogeography, and one of the major barriers to distribution – the break between southeast Asia and Australasia – still bears his name as the Wallace line. For the most part, related animals occur in geographical proximity to each other, and barriers such as oceans, mountains, etc. act as the divisions between major biomes. Where interesting anomalies are found, such as the presence of marsupials in both South America and Australia, this is understandable in terms of the plate-tectonic history of the continents,

for at the time the marsupials evolved, the two areas were part of a single continent.

However, there are some distributions which apparently defy the normal rules. The New World monkeys are one such example. Their closest relatives are in Africa and Asia, from which they have been isolated for many millions of years. The continent closest to them, North America, lacks any possible ancestral forms. The only conclusions to be drawn are that they must have got there as a founding population by drifting across approximately 900 km of sea (there may at that time have been some intermediate islands), or else they dispersed from Africa (where forms similar to early platyrrhines are known) and had a distribution across the whole of Asia and North America, and yet left no fossils (there are abundant fossils of other groups at this time).

Following the dictum of Sherlock Holmes, in this case we can rule out a large number of completely impossible hypotheses, but we are left with two apparently deeply improbable ones: a miraculous voyage or massive lost geographical distribution. Certainly there is far more circumstantial evidence in support of the African route than the Asian one, but there are still problems. What are the chances of (1) a sufficiently large group to act as a founding population being washed off the African coast; (2) their being on a sufficiently large raft of "land" (that is, a mat of vegetation washed into the sea, as has been observed); (3) their having enough food and water for the length of the journey; and (4) their ending up beached on an appropriate part of South America?

At first sight it might be simplest to say that it is really virtually impossible, but against that it is worth bearing in mind the lengths of time involved. The window of opportunity for such an event, it could be argued, may have been over 10 million years. If events (1) and (2) occurred on average once a year, then in fact there are 10 million such potential events, and only one of them has to be successful – that is, the event may have a probability of only one in 10 million, but if there is enough time, then it becomes quite probable. In other words, the length of time involved in evolution is such that what seems vastly improbable is in fact within reasonable statistical probability. Although there is a strong adaptive element in the direction of evolution, as is evidenced by the convergence between Old and New World Monkeys, nonetheless there must always be chance events.

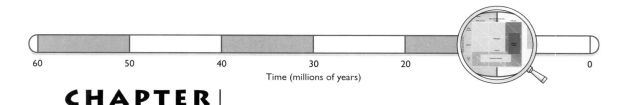
# CHAPTER 7

# RECONSTRUCTING BEHAVIOR

## BODIES, BEHAVIOR, AND SOCIAL STRUCTURE

**KEY QUESTION** Given that anthropoid primates are highly social, what patterns of social organization are to be found among them, and what determines social structure?

The great majority of primate species are social animals, living in groups that range from 2 to 200 individuals. Whatever the size of the group, it serves as the focus of many important biological activities, including foraging for food, finding mates, raising offspring, and defending against predators.[196] The group is also the center of intense social interaction that has little apparent direct bearing on the practicalities of life: in the human sphere we would call it socializing, the making and breaking of friendships and alliances. The size, composition, and activity of a group define what is usually meant by a species' social organization.

Animal behavior is a far more variable characteristic than, for instance, anatomy or physiology. Consequently, an order such as the Primates will display an astonishingly wide range of social organization, in which even closely related species may carry out their daily social lives in very different ways. We saw in the previous section that body size can have a powerful influence on many aspects of a species' way of life, but social organization is not one of them. Even if we consider just the apes – the largest of the non-human primates – the array of social organization found in the species is as great as among the primates as a whole.

As highly social creatures ourselves, we may find it odd to ask "Why should animals live in groups?" This problem is, in fact, a very good biological question because gregariousness carries many costs. Across the animal kingdom as a whole, social living is quite rare, and indeed their intense sociality is one of the key characteristics of the anthropoid primates.[197] For instance, a lone individual does not have to share its food with another individual, but competition for all resources characterizes a group. A lone individual is not exposed to diseases that flourish in communities, which provide a viable host pool for pathogens. A lone individual is much less conspicuous to predators than is a group of individuals, and so on. Clearly, as most primates do live in groups, the benefits must outweigh the costs.

This chapter will discuss current thinking about the benefits – that is, the causes – of living in groups. It will also examine some of the consequences of group living – not the costs mentioned, but the ways in which individuals might adapt behaviorally and anatomically to different types of social structures. Although social organization is highly variable, we will discover that there is a pattern to that variation, and like anatomical structures, it is shaped in evolution by both adaptive need and phylogenetic heritage.

*Why are animals sometimes social?*

## Social organization in apes

To obtain a feel for some of the details of social organization and the range to be found among primates, we will first survey the social lives of the apes: gibbon (and siamang), orangutan, chimpanzee, bonobo, and gorilla.[198,199] It should be remembered that the apes are just a small part of the primates as a whole, and that across the order there is a considerable variety of social organization, but it is an interesting evolutionary fact that the apes do show a considerable diversity of social organization (Fig. 7.1).

Gibbons and siamangs, the smallest of the apes (sometimes called the lesser apes), live in forests in southeast Asia. The basic social structure of these highly acrobatic, arboreal creatures is very similar, consisting of a monogamous mating pair plus as many as three dependent offspring. Gibbons are territorial, and eat a diet of fruit and leaves. On reaching maturity, the offspring leave the natal group and eventually establish one of their own by pairing with another young adult of the opposite sex. Mature males and females have essentially the same body size. Gibbons provide a good example of a tendency toward long-term monogamy.

The other Asian ape, the orangutan, is much larger than the gibbon and pursues a very different lifestyle, although it is also highly arboreal. The core of its social organization is a single mature female and her dependent offspring. The mother and offspring occupy a fairly well-defined home range, which usually overlaps with that of one or more other mature females and their offspring. In contrast, males are rather solitary creatures, each occupying a large territory that usually contains the home ranges of several mature females with whom he will mate. Males, which are about twice the size of females, actively defend their territories against incursion by other males. Although in some senses the orangutan may be thought of as essentially asocial (that is, the only social bonds are between mother and offspring), another way of looking at it is that the mating system is

**FIGURE 7.1**

**Hominoid social organization:** The range of social organizations among the apes matches that found among anthropoids as a whole. Gibbons are monogamous, with no size difference between males and females. Single male orangutans defend a group of females (and their offspring), distributed over a large area; this organization is a variant of unimale polygyny sometimes known as exploded polygyny. Single male gorillas also exert control over a group of females (and their offspring), but the females are in a smaller area; this system is another example of unimale polygyny. In chimpanzees, several related males cooperate to defend a group of widely distributed females (and their offspring); this system provides an example of multimale polygyny or polygynandry.

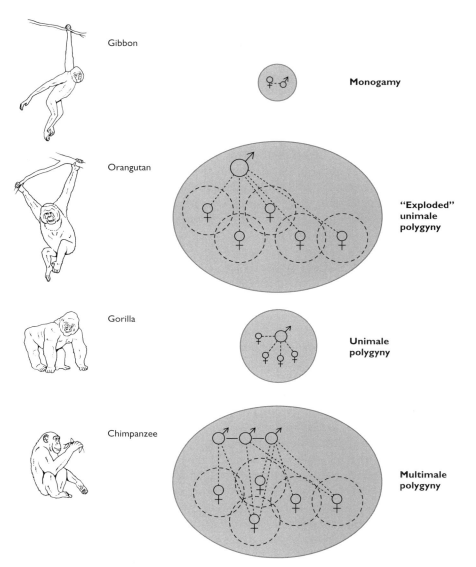

therefore one of a loosely organized harem, with one male mating with several females (technically known as unimale polygyny).

Gorillas, the largest of the apes, live in the forests of central and west Africa. These animals follow a mating system similar to that of the orangutan – unimale polygyny – although their ecology and organization are distinctly different. Predominantly terrestrial animals that live on low-quality herbage found in abundant but widely dispersed patches, gorillas live in close groups composed of from 2 to 20 individuals. The adult male – the silverback – has sole mating access to the mature females, whose immature offspring also live in the group. Mature males compete for control of the group. Nevertheless, a female, usually a young adult, will sometimes transfer from one group to another, seemingly as a matter of free choice. New groups may be established when a lone silverback begins to attract transferring females. As with orangutans, male gorillas are twice the size of females. The similarity with orangutans is that this is another case of unimale polygyny, but in this case the harem is not scattered over a large area, but lives in close proximity to each other.

Chimpanzees, which are terrestrial and arboreal omnivores, live in rather loose communities composed of between 15 and 100 individuals, representing a mixture of mature males and mature females and their offspring. Unlike savannah baboons, which live in close, cohesive troops of mature males, related females, and their offspring, sometimes numbering 200 individuals in total, chimpanzee communities are maintained by occasional contact between males and females. On a day-to-day basis the community breaks up into smaller foraging groups, before recombining – hence the description of chimpanzee social system as a fission–fusion one. The core of chimpanzee social life is a female with her offspring; these units are often found by themselves but sometimes link up with other females and their offspring. Each female maintains a core area, which usually overlaps with that of one or more other females. By contrast with orangutans, single chimpanzee males do not maintain exclusive control of a group of female home ranges. Instead, a group of males defends the community range against the males of neighboring communities. Mating in chimpanzee communities is promiscuous, with each oestrus female copulating with several males. The social organization is therefore known as polygynandry or multimale polygyny.

How do apes vary in their social structure?

A key feature of chimpanzee social organization is that, unlike in the general pattern of cercopithecoid (Old World monkey) societies, males remain in their natal group while young adult females transfer to other communities. As a result, the adult males that are cooperating to defend their community are usually closely related to one another. Adult male chimpanzees are typically 25 to 30% larger than females.

Among the apes, then, one finds monogamy, unimale polygyny, and multimale polygyny. (Polyandry – one female having exclusive access to several mature males – which is rare in mammals generally, is absent here.) This spectrum of social organization raises questions about several aspects of group living. For example, how big will a social group be? What is the ratio of adult females to adult males? Among which sex is there the greater degree of relatedness? What difference arises in the size of males and females?

## Causes of social organization

The fact that such a rich array of social organizations exists among primates as a whole, and among the apes in particular, surely indicates that a rather complex set of processes underlies them. For each species, some kind of interaction must take place between its basic phylogenetic heritage – its anatomy and physiology – and key factors in the environment. Thus, different species will probably react differently to the same environmental factors, creating at least part of the observed diversity. What else plays a part?

"There is no consensus as to how primate social organization evolves," Wrangham observes, "but a variety of reasons suggest that ecological pressures bear the principal responsibility for species differences in social behavior."[200] Indeed, since the 1960s ecological influences have been a popular source of explanations. As Wrangham explains, the problem is that "we do not know exactly what the relevant ecological pressures are, or which aspects of social life they most directly affect, or how."

One of the most frequently advanced explanations of the benefits of group living has been defense against predation. Even though it may be more conspicuous than a lone individual, a group can be more vigilant (more pairs of eyes and ears) and more challenging (more sets of teeth). Effective defense against predators has been observed in many group-living species of primate.[201]

It is certainly true that terrestrial species, which face greater risk from predators than arboreal animals, live in larger groups and commonly include more males in the group; in addition, the males in such species frequently are equipped with large, dangerous canine teeth. For each of these factors, however, one can advance equally plausible explanations of their origin that have nothing to do with protection against predation. So, it is possible that terrestrial primates evolved these characteristics for these other reasons; once evolved, the properties proved highly effective in mitigating the threat of predation. Protection against predation may to some degree be a consequence, not the primary cause, of group living.

Food distribution has also been suggested as a trigger of social organization. Groups might be more efficient than individuals at discovering discrete patches of food, for instance, or, where food patches are defensible by territorial species, the patch size will then influence the optimum group size. Wrangham has proposed a theory of social organization that includes food distribution as a key influence, but the focus of this model differs from that of earlier ideas.

Wrangham's model examines the evolutionary context of male and female behavior, and proposes that "it is selection pressures on female behavior which ultimately determine the effect of ecological variables on social systems." In other words, whatever ecological setting a species might occupy, the behavior of females is fundamental to the social system that evolves within it.

The starting point for Wrangham's model is that male and female reproductive success – the key measure of Darwinian fitness – is determined by very different things. Basically, for females reproduction is energetically expensive and time-consuming. Each offspring requires a long period of development, whereas for the male the investment is relatively little –

*Is social structure shaped by ecology?*

sperm comes cheap (Fig. 7.2). One consequence of this is that females have relatively little scope for varying their reproductive output except by ensuring the survival of their offspring. Access to mature males is not usually a limiting factor, whereas access to food resources most certainly is. Most male primates, along with 97% of all male mammals, bestow no parental care on their offspring. As a result, their reproductive success is determined by successful access to mature females. From this comes the fundamental principle of Wrangham's model – female reproductive success is determined by access to food, male reproductive success by access to females.

In the great majority of primate societies, females remain in their natal group while males transfer. Any explanation of why primates should form social groups at all must also explain this asymmetry. Attempts to correlate different types of habitat with the tendency to form different types of social groups fail to satisfy this criterion. Wrangham's model does offer an explanation, as follows.

If food generally comes in patches that can support only one female and her offspring, then females will forage alone, as orangutan and chimpanzee females do much of the time (Fig. 7.3). Food that comes in larger, defensible patches can, however, support several mature females and their offspring (Fig. 7.4). Sharing a food resource also brings an element of competition into the group, which leads to loss of time and energy through aggressive encounters. Wrangham suggests that the costs of competition within a group are balanced against the benefits of cooperating with group members to outcompete other groups for access to food patches. Cooperation is most beneficial when it occurs among relatives: helping kin is like helping yourself, because they share your genes.

Thus, when a species exploits food resources that come in discrete, defensible patches, multifemale social groups will evolve in which the females are closely related to one another. In anthropology, such groups are known as matrilocal; with non-human primates, a better term is "female kin-bonded." Where do the males fit in? If patches of food resources are relatively densely distributed, allowing a group of females to defend them all and exercise territoriality, extra males are somewhat extraneous and a unimale social system usually forms. If, however,

♂ : Access to mature females

♀ : Access to food resources

**FIGURE 7.2 Different reproductive strategies:** For a female primate, the variable that determines ultimate reproductive success is access to food resources. By contrast, a male's reproductive success is limited by his access to mature females. This difference critically influences the overall social structure of primate societies.

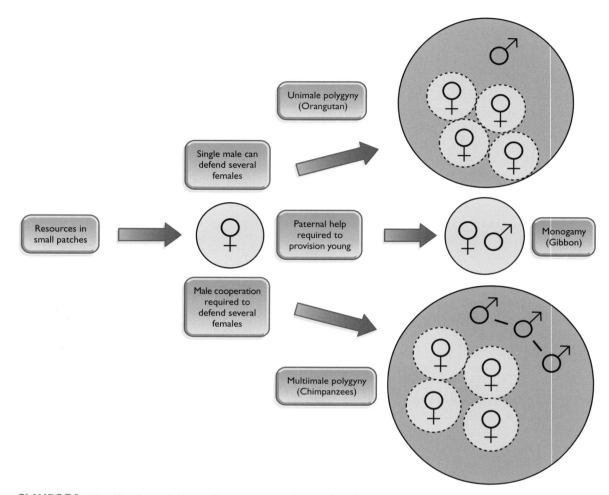

**FIGURE 7.3  Distribution with small resource patches:** When food exists in patches too small to support more than one mature individual, females will forage singly (with their offspring). If a male can defend a "community" of lone females, unimale polygyny will result, as observed with the orangutan. If a male can defend only one female, or if paternal help is required in raising offspring, monogamy will result, as found with the gibbon. If a community of females can be defended only by several males, then a group of related males will defend a number of unrelated females, as observed in chimpanzees.

territoriality is not possible and increased group size does not create major problems, several adult males can be accommodated. Indeed, extra males can prove useful in the occasional competitive encounters with other groups. In such a situation, some kind of multimale system would form.

In non-female-bonded systems, such as those of the chimpanzee and orangutan, where food does not come in defensible patches and females are mostly alone, the distribution of males depends on whether they can defend a community range alone or need the cooperation of other males. For orangutans, community defense by a single male is feasible, but for

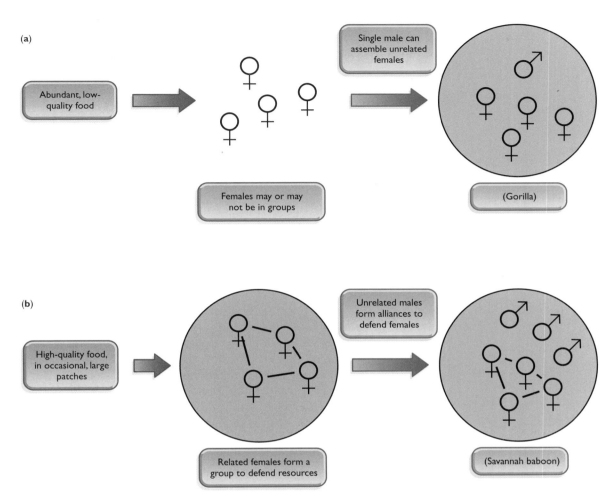

**FIGURE 7.4 Distribution with larger resource patches:** (a) When low-quality food is widely distributed, females may forage alone or in groups (in which the individuals are unrelated). A male may be able to assemble a harem, as does the gorilla. (b) When high-quality food occurs in large but scarce patches, related females will form a group to defend them. Alliances among unrelated males may form to defend the females from other males, as in savannah baboons.

chimpanzees, cooperation is essential. Again, cooperation is most effective among relatives. Thus, chimpanzees have evolved a multimale social system in which females, not males, transfer to other groups on reaching maturity.

Given these underlying influences, says Wrangham, several predictions can be made in terms of behaviors within and between groups. For instance, intense social interaction – grooming and so on – is expected among females in female-bonded groups, but is less frequent in non-female-bonded groups. Aggression within female-bonded groups should arise over access to food resources, and females should play a very active role in the encounters. By contrast, aggression within non-female-bonded groups should relate to access to females, and males should be the principal aggressors. These predictions appear to have some support.

Another possible consequence of primate sociality is body size – specifically, the difference between males and females, known as sexual dimorphism in body size.[202–5] Male primates often must compete with other males for access to breeding females, and the bigger their body size, the more likely they are to succeed. Natural selection in species in which such male–male competition occurs is likely to lead to increased male body size. Other factors that might be important in such encounters – canine teeth, for example – may also become exaggerated in males.

What is the effect of body size on ecology and behavior?

In monogamous species, in which competition between males is low or absent, males and females are typically the same size. In addition, all species characterized by significant sexual dimorphism exhibit some degree of polygyny. Enlarged canines are also found in polygynous species. The equation is not simple, however, because no direct correlation exists between the degree of polygyny and the degree of body-size dimorphism. Species in which males typically control harems of, for example, 10 females do not necessarily display greater body-size dimorphism than species in which males control harems of two females.

Although the notion that body-size dimorphism represents the outcome of competition among males for access to females is popular among biologists, other explanations are also possible. The simplest is that males are large and aggressively equipped so as to provide effective protection against predators. Once again, the problem of circularity arises here. Another suggestion is that males and females assume different sizes as a way of exploiting different food resources, thus avoiding direct resource competition.

Robert Martin of the University of Chicago adds an important note of caution to this discussion, noting that perhaps our explanations have been too male-oriented in seeking to explain why the male size has increased. Instead, he suggests, perhaps the size difference indicates that the females have become smaller. "Smaller females may breed earlier," he notes; "selection for earlier breeding might explain the development of sexual dimorphism in at least some mammalian species."[202]

Even though many aspects of the interaction of species and their different environments remain to be fully worked out, one thing is clear: the complete social behavior of a species is the outcome of a mix of causes and consequences of individuals coming together to coexist in groups.

## Primate socioecology

Whatever the details of primate social behavior might be, the main point to emphasize is that social behavior is not an ephemeral "add on" to the biology of any species. Social behavior is deeply ingrained in primate evolution – primates are social specialists where bats are flying specialists – and the comparative approach adopted here makes it clear that the type of social organization found in any species is related to its ecology and the

Polygynous social system
Dimorphic canines

Monogamous social system
Monomorphic canines

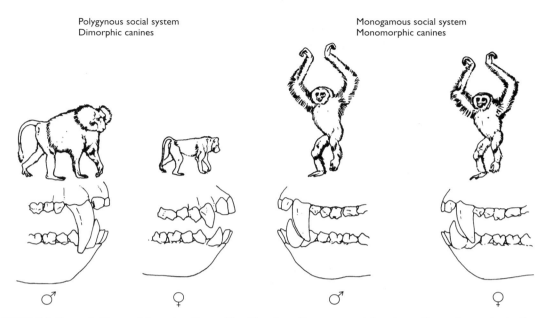

♂                    ♀                    ♂                    ♀

FIGURE 7.5 **Sexual dimorphism, teeth, and bodies:** In polygynous social systems, the males are typically larger than females, in terms of both body size and canine teeth, as illustrated here for baboons. By contrast, in monogamous species, body size and canine size are usually very similar between the sexes, as illustrated here for gibbons. (Courtesy of John Fleagle/Academic Press.)

resources on which it depends – whether they are dispersed or clumped, high-quality or low-quality.[165, 206–8] From this flows both the morphological adaptations – body size, tooth size, gut size, pattern of locomotion – and the social behavior – monogamous or polygynous, male kin-bonded or female kin-bonded. When looking at the evolution of humans we can observe that many of their key adaptations relate to social organization and behavior, but the comparative perspective acts as a reminder that this is just as much a part of the evolutionary process as the evolution of their physical characteristics, and indeed is directly related to those physical and ecological traits (Fig. 7.5).

## NON-HUMAN MODELS OF EARLY HOMININ BEHAVIOR

Some scholars argue that evolution has carried humans so far away from the rest of the animal world that little can be learned about ourselves by looking at non-human primates. There is, of course, a gap between the behavioral repertoire of *Homo sapiens* and that of

**KEY QUESTION** What can be inferred about the social behavior of extinct hominins from the patterns of social organization among non-human primates?

even our closest genetic relatives, the African apes. However, that gap is not an inherent one, intrinsic to humans, but is a product of the disappearance or extinction of many species of hominin. The gap must have been filled to some extent by the behavioral repertoires of our ancestors, the various extinct species of hominin.[209] It is therefore legitimate to use the information available about the behavioral ecology of living primates to infer something about the behavior of our forebears.

There are two major goals of attempts to use living non-human primates as part of the study of human evolution. One goal is to get a handle on what can be thought of as the starting point of human evolution. Put into a more technical framework, one objective is to reconstruct the characteristics, particularly the social organization, of the last common ancestor between humans and the apes, especially chimpanzees. The other goal is to use an understanding of primate socioecology to reconstruct as best we can the behavior of particular extinct hominins – australopithecines or Neanderthals, for example. This second objective has become both more pressing and more difficult as more and more species of fossil hominin have been discovered, and so we expect greater variation and nuances in their behavioral repertoire. If the experience of observing the behavior of modern ape species teaches us a lesson, it is that we can expect different forms of social organization among different hominin species. In either case, we are looking for means of answering questions on which the fossil record itself is largely silent, or can provide only supplementary evidence: did they live in groups, and, if so, what was their size? What was the ratio of mature males to mature females?

## Three approaches to models

What are the strengths and weaknesses of different analogue models?

There are several ways in which modern primates can be used to model the lives of the extinct species.[210,211] First, one can identify a living species that appears to match some basic hominin characteristics and then seek lessons about behavior that might transfer to hominins. Second, guided by phylogeny, one can consider only the living African apes and humans and seek commonalities of behavior that might therefore have been present in a common ape/hominin ancestor. Third, one can look at the pattern of variation across a range of species, and see how the environment conditions behavior and social organization, and then apply these general principles to the specific conditions of particular hominin groups. Each of these approaches has strengths and weaknesses.

## Primate models

The first of these three models – the specific primate model – is the longest-established approach. Several different species have been offered as the most appropriate model at certain times (Fig. 7.6), including the

**FIGURE 7.6  A catalog of candidates:** Several different species have been nominated as instructive models for early hominin evolution. Here we see the pygmy chimpanzee (*top left*), the common chimpanzee (*top right*), the savannah baboon, and the lion (a social carnivore).

savannah baboon, the common chimpanzee, and, most recently, the pygmy chimpanzee.

Although the baboon is a monkey, not an ape, and is therefore genetically related to hominins only rather distantly, baboons are attractive as models for early hominins because they share a similar habitat: bushland savannah.[212,213] Living in troops with as many as 200 individuals, the savannah baboon offers a striking picture of the social life of our forebears. A troop consists of mature females (often related to one another) and their offspring, and many mature males (unrelated to one another). The males are larger than the females and are equipped with impressively threatening canines. In other words, baboons operate within a multimale, female kin-bonded social organization. It was argued that these characteristics – larger social groups, complex social organization, and sexually differentiated roles – were the product of living in a savannah environment, and as this was where it was posited ancient hominins had evolved, then these characteristics would have occurred among these groups. An even more specific baboon model, the so-called "seed-eating hypothesis," developed by New York University anthropologist Clifford Jolly, used the gelada baboon, which was a grass-eating specialist, as a model for the evolution of many human characteristics.[214]

The chimpanzee has also been proposed as a model for the last common ancestor and the early hominins, and for good reason: it is our closest genetic relative.[199,215,216] Many aspects of chimpanzee behavior have been seen as the basis for significant features in human evolution – hunting, stone-tool use, an ability to use symbols, and complex social behaviors. One problem with the chimpanzee, as with all specific models, is the trap of the present: just as extinct species are likely to be unique anatomically and not represent some slight variant of a living species, so the behavior of extinct species is also likely to be unique. When, for instance, a chimpanzee model is proffered, "an ape–human dichotomy is created," says Richard Potts, an anthropologist at the Smithsonian Institution (Washington, DC). "The problem with placing early hominids along a chimp–human continuum is that it precludes considering unique adaptations off that continuum."

Potts points out that the dentition of the early hominin genus *Australopithecus* – large, thickly capped cheek teeth set in robust jaws – resembles that of neither chimpanzees nor humans. "Thus, in this aspect of dental anatomy *Australopithecus* did not fall on the proposed continuum," notes Potts.

The most recent entry into the primate model stakes is the pygmy chimpanzee, or bonobo, proposed in 1978 by Adrienne Zihlman, John Cronin, Vincent Sarich, and Douglas Cramer.[215] Randall Susman, of New York University at Stony Brook, is also a proponent.[216] The pygmy chimpanzee, found only in a small area in the Democratic Republic of the Congo, is strikingly similar in overall body proportions to the early hominin species *Australopithecus afarensis*. Although bonobos and chimpanzees share many characteristics, there are some subtle differences. Bonobo society is more female-centered and less overtly aggressive, and in it sex is a substitute for aggression. Bonobos engage in sex in every possible partner combination, often face-to-face, usually as a way of reducing tension in the group. They also lack subsistence technology altogether. This species, *Pan paniscus*, may thus represent a very different model for early hominins. Given that the two species of *Pan* are equally closely related to humans, it is difficult to determine which is the more appropriate model – one where intergroup violence and cannibalism are well-demonstrated behaviors (*Pan troglodytes*), or one where sex is used for social purposes other than reproduction (*Pan paniscus*).

Primates have generally been used in these species-based models, but there is no reason why such analogies should be confined to our own order. Social carnivores – lions and hyenas – have been used as a model, given the suggestion that a switch to a more hunting-based way of life was an important evolutionary trigger for hominins.[217] More controversially, Elaine Morgan, on the basis of an earlier idea of British zoologist Alistair Hardy's, has used similarities between marine mammals and humans (hairlessness, subcutaneous fat, descended larynx, for example) to suggest that ancient hominins had their origin in an aquatic way of life.[218]

## Phylogenetic models

The second approach – phylogenetic comparison – is considerably more conservative, seeking only to identify basic shared behavioral characteristics among humans and other primates, especially African apes. The approach was developed in the late 1980s, and is effectively an application of the cladistic methodology described in chapter 5, extended in this case to behavioral rather than morphological or genetic data.[219–22] The aim is to look for characteristics that are shared across lineages, and therefore possibly infer from them what may have been present in the common ancestors. The rationale, as explained by Wrangham, is as follows: "If [a behavior] occurs in all four species, it is likely (though not certain) to have occurred in the common ancestor because otherwise it must have evolved independently at least twice. If the four species differ with respect to a particular behavior, nothing can be said about the common ancestor."

Wrangham examined 14 different behavioral traits – such as social group structure, male–female interactions, intergroup aggression, and so on.[219] He found eight traits to be common to gorillas, the two chimpanzee species, and humans; six traits were not shared (Fig. 7.7). On this basis Wrangham infers that the common ancestor of hominins and African apes "had closed social networks, hostile male-dominated intergroup relationships with stalk-and-attack interactions, female exogamy and no alliance bonds between females, and males having sexual relationships with more than one female."

| METHOD | PHYLOGENETIC COMPARISON | CHIMPANZEE MODEL | CHIMPANZEE MODEL | BEHAVIORAL ECOLOGY | BEHAVIORAL ECOLOGY |
|---|---|---|---|---|---|
| SPECIES | COMMON ANCESTOR | A LATE PREHOMININ | THE EARLIEST HOMININ | THE EARLIEST HOMININ | AN EARLY HOMININ |
| 1 Closed social network | Yes | – | Yes | Yes | – |
| 2 Party composition | ? | Unstable | Unstable | Stable | Unstable |
| 3 Females sometimes alone | ? | Yes | Yes | No | Yes |
| 4 Males sometimes alone | Yes | Yes | Yes | No | Yes |
| 5 Female exogamy | Yes | – | Yes | No | – |
| 6 Female alliances | No | – | No | Yes | – |
| 7 Male endogamy | ? | Often | Yes | No | Yes |
| 8 Male alliances | ? | – | Yes | – | – |
| 9 Males have single mates | No | No | No | No | Yes |
| 10 Length of sexual relationships | ? | Short | Short | Short | Long |
| 11 Hostile relations between groups | Yes | No | Yes | – | – |
| 12 Males active in intergroup encounters | Yes | – | Yes | – | – |
| 13 Stalking and attacking | Yes | – | – | – | – |
| 14 Territorial defense | ? | – | ? | – | – |

**FIGURE 7.7 Ancestral social organization:** Using different models, it is possible to determine those aspects of behavior that might have appeared in an ancestral species. In the phylogenetic comparison, each of the 14 questions asks if a particular aspect of behavior exists in *all* modern African apes. If it does, then this same behavior likely also appeared in the common ancestor with hominins. (Courtesy of Richard Wrangham.)

This ancestral suite, as Wrangham calls it, is merely a foundation upon which past social behavior can be constructed. But, for instance, it does seem to preclude the suggestion made in 1981 by Owen Lovejoy, an anatomist and anthropologist at Kent State University, that the then earliest known hominin, *Australopithecus afarensis*, was monogamous and non-hostile.[223]

The approach can be extended further, beyond just the issue of the behavior of the last common ancestor with the two chimpanzee species. By looking at what traits are shared at deeper and deeper phylogenetic levels – that is, with catarrhines, anthropoids, etc. – it is possible to get some idea of when certain traits may have evolved. For example, sociality is a trait that is found universally across the anthropoid primates, suggesting that the mechanisms promoting and permitting individuals to tolerate and support each other were in place tens of millions of years ago.[224] The phylogenetic approach has become more and more popular, and has been used to explore the evolution of many traits across the primates – patterns of residence, advertisement and concealment of ovulation, etc.

**Within-sex relationships**

**Between-sex relationships**

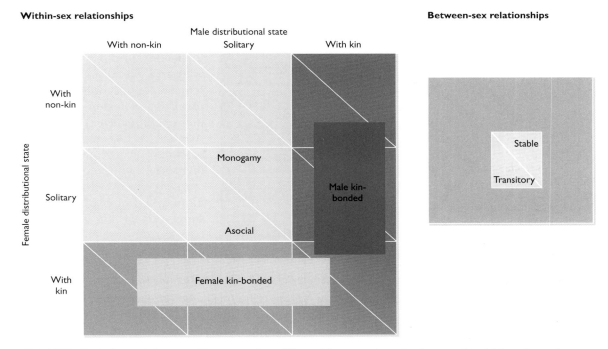

FIGURE 7.8 **Finite social space:** The range of possible social systems is set by the ways in which males and females associate with each other and with their own sex – thus the idea of a finite social space. Into this world of social diversity can be put the range of primate and human social organization.[221]

## Behavioral ecology models

The third approach – reconstructing social organization from first principles of behavioral ecology – is the newest and most promising. The technique seeks to establish the range of social structures that might have been available to hominin ancestors, and then determine how these structures might be altered in the face of changing environments.

The basis of the analysis is the recognition of phylogenetic constraints in ecological context. Just as ancestral anatomy limits the paths of subsequent evolution, so too does ancestral social structure (Fig. 7.8). For instance, evolving from a multimale, non-female-bonded organization to a multimale, female-bonded structure is highly unlikely, because the intermediate steps would be inappropriate under prevailing conditions. In other words, only certain evolutionary pathways are available for ecologically driven shifts in social organization. Thus, if you know where an ancestral species began among the many possible social structures, you can predict the nature of ecologically driven social change, because you know the available pathways.[221]

The phylogenetic context for hominins is, of course, the apes – particularly the African apes. The social structures found among the apes vary greatly,

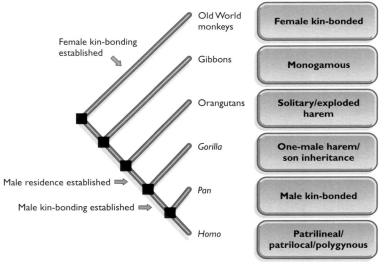

**FIGURE 7.9**
**The phylogeny of hominoid social organization:** Although there is considerable diversity in the nature of hominoid social systems, there is a phylogenetic pattern relating to an increase in the level of kin-related males remaining resident in their natal groups, and thus the development of male kin-bonding. (Courtesy of Robert Foley.)

ranging from solitary individuals among orangutans, through monogamous families in gibbons, to single-male units with small numbers of unrelated females among gorillas, to complex fission–fusion communities of chimpanzees. However, two striking facts that provide a basis for reconstructing hominin social evolution are that, in marked contrast with Old World monkeys, none of the apes show female kin-bonding or have a core of related females, and the African apes in contrast show a tendency towards male kin-bonding (Fig. 7.9).

Foley has proposed a model of the social evolution of the hominoids.[209] It can be suggested, on ecological grounds, that the most likely social structure in species ancestral to African apes and hominins is relatively gorilla-like (Fig. 7.10). Toward the end of the Miocene, approximately 10 million years ago, a steadily cooling climate was reducing forest cover. A drier, more diverse habitat developed, especially in east Africa, which created a patchy distribution of food resources. Such an ecological shift would favor the evolution of a chimp-like social structure: communities of dispersed females and their offspring, with genetically related males defending the community against males from other groups.

The emergence of the hominins can be seen as part of the African hominoid radiation, with this clade exhibiting increasingly strong male kin alliances under certain ecological conditions. As long as the cooling persisted, the ecological shift would continue. Such environments appear to promote, among other things, larger group size among primates – partly as a response to the greater threat of predation, partly due to the effects of resources being more patchily distributed. Given the evolutionary pathways available under the model, the larger social groups are more likely to be built upon the male kin alliances rather than related females.

Given this background, the most probable social organization for the early australopithecines consists of mixed sex groups, with males linked by a network of kinship. Females, forced to forage over larger areas to find dispersed and seasonally limited food and to aggregate in the face of some predation, would be expected to form stable associations with either specific males within the alliance, or the entire alliance of males.

Within the hominin species of 3 to 1 million years ago there developed a degree of morphological diversity, presumably reflecting adaptation to

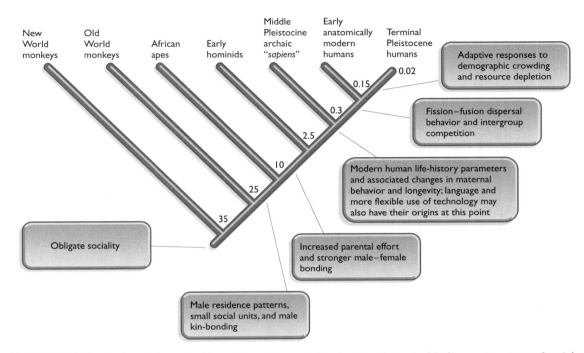

**FIGURE 7.10 Evolution of hominoid and hominin social behavior:** Using the phylogenetic pattern of social behavior of hominoids, linked to the principles of socioecology discussed in the text, it is possible to hypothesize about the evolution of social organization among extinct hominins. The numbers show the approximate age of the bifurcating points (evolutionary divergence of the clodes) in millions of years. (Courtesy of Robert Foley.)

different patterns of subsistence. At one extreme, the robust australopithecines apparently exploited a diet of coarse, low-quality plant foods. Such foods tend to occur in large, widely dispersed patches. The expected effect on the ancestral hominin socioecology would have been to weaken male kin bonds within a less structured, large or fluid group. More competition between males would develop, presumably accompanied by dimorphism in body size.

*How does primate social behavior relate to phylogeny?*

At the other extreme, *Homo erectus* produced adaptations including increased brain size and much reduced dental apparatus. Faced with the same problem of subsisting in tropical savannah environments – that is, maintenance of a constant food supply in the face of seasonality – this hominin species adopted a strategy different from that of the robust australopithecines. Instead of exploiting low-quality food resources, its members increased their consumption of meat, a patchily distributed but high-quality resource.

While the causes of meat-eating may be ecological, the consequences for hominins would have been distributional and social.[221] One consequence would be greatly increased home and day ranges, which would complicate direct defense of females by males. The predicted response would be resource defense through territorial exclusion, and given the size of the area involved, this would involve alliances of males rather than individual defense. Such a pattern of behavior would enhance male kin associations as a means of coping with high levels of intergroup competition and interactions.

In addition to changes wrought by this subsistence strategy, *Homo* would face another key change: the consequences of brain enlargement. Producing and rearing large-brained offspring is energetically expensive. At some point it would have become too expensive for the mother to provide for the offspring by herself, necessitating paternal involvement. The effect would be to increase the frequency, intensity, and stability of male–female associations. Is this point the beginning of the nuclear family, so much a part of Western society? No, because the nuclear family is actually rather uncommon among human societies; an analysis of social structure variation among modern human societies shows that most permit polygamy, and where polygamy is not the case, serial monogamy rather than strict monogamy is often the situation.

One question posed by these models is how they might be tested. While they make predictions about the past, these cannot be directly observed. The key lies in looking for evidence that may be consistent with the predictions, or alternatively not fit them. In terms of the fossil record a number of observations can be made. For example, some of the changes associated with the genus *Homo* are predicated upon an ecological shift to higher-quality resources, and so the fossil and archeological record can be examined to see if there is evidence for this at the critical time, or indeed at other times.

A second approach is to look at levels of sexual dimorphism. A comparative examination of patterns across the primates shows that there is a relationship between the levels of sexual dimorphism and the type of social organization – monogamous primates, for example, exhibit relatively little sexual dimorphism. This can be exploited in the fossil record; size differences within species of hominin can be used to estimate the levels of sexual dimorphism. What this shows is that in fact there is a high level of sexual dimorphism among the hominins in general, often exceeding that of chimpanzees and approaching that of gorillas. This certainly suggests that monogamy was not the social organization of the earliest hominins. There is, however, a decline in levels of sexual dimorphism across the course of the evolution of *Homo*.

A further test is to look at what might be described as the scars of evolution to be found in living humans. Sexual dimorphism in human populations is one example. The 20% body-size dimorphism in modern humans would indicate a degree of male–male competition in our recent past, not monogamy (Fig. 7.11). And the fact that more resources are devoted to male fetuses than female fetuses, thus giving them a higher birth weight, is also consistent with male–male competition. One further factor is the size of the male's testis, an indicator of subtle competition among males in multimale groups. For instance, chimpanzees live in promiscuous,

What determines levels of sexual dimorphism in primates?

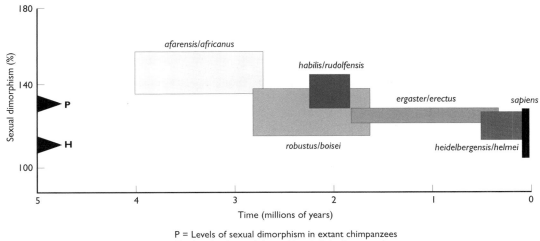

**FIGURE 7.11 Patterns of sexual dimorphism among the hominins:** Sexual dimorphism may reflect social and mating behavior. It can to some extent be estimated for extinct hominins from size distributions in the fossil record. These estimates suggest that the early hominins were markedly dimorphic, and that this decreased over the course of the evolution of *Homo*. (Courtesy of Robert Foley.)

multimale groups. One way that an individual male might outcompete his fellows is to produce more sperm in his ejaculate. Gorillas and orangutans do not face this kind of competition, and consequently they have small testes.[225,226]

What of *Homo sapiens*? Human testes are small as well, apparently ruling out competition in promiscuous, multimale groups. Monogamy also appears to be eliminated, leaving a form of unimale polygyny. But, as Robert Martin and Robert May commented recently, "these biological antecedents are today often overlain by extremely powerful socioeconomic determinants."[227]

There are alternative models that have been proposed, ranging from Lovejoy's one based on monogamy and provisioning (working from analogies with birds)[223] to others involving female kin-bonding.[228] Currently there is no clear consensus, and given the limited tests possible, there is unlikely to be for some time. The point of presenting these models here is less to establish a firm factual basis for early hominin social behavior than to illustrate the principles involved, and to show that a comparative approach, which throws up "evolutionary rules" even for social behavior, is an important means of gaining access to the past (Fig. 7.12).

**FIGURE 7.12
Contributions to social organization:** A species' social structure will be determined by the outcome of interaction between its phylogenetic heritage – body size, and so on – and the environment in which it lives. Species with different phylogenetic constraints may therefore exhibit different social structures under the same environmental conditions.

## JAWS AND TEETH

**KEY QUESTION** How does the way in which teeth develop, erupt, and wear down tell us about the behavior of extinct hominins?

At first sight it might seem odd to include a section on jaws and teeth in a chapter on reconstructing the behavior of extinct hominins, for there is little as anatomical as a tooth. The reason for this inclusion, however, is that these are the parts of the skeleton that are most likely to be preserved, and they also contain an enormous amount of information about the lives and deaths of animals in the past. If behavior is preserved anywhere, it is in the teeth, and so reconstruction of behavior demands a knowledge of what can be inferred from dentition. Much of what is said later in this book will have been derived from dental evidence.

Jaws – particularly lower jaws – and teeth are by far the most common elements recovered from the fossil record. The reason is that, compared with much of the rest of the skeleton, jaws and teeth are very dense (and teeth very tough), which increases the likelihood that they will survive long enough to become fossilized.

Because jaws usually serve as an animal's principal food-processing machine, the nature of a species' dentition can yield important clues about its mode of subsistence and behavior. Overall, however, the dental apparatus is evolutionarily rather conservative, with dramatic changes rarely appearing. For instance, human and ape dentition retains roughly the basic hominoid pattern established more than 20 million years ago.[229] Moreover, different species facing similar selection pressures related to their feeding habits may evolve superficially similar dental characteristics, as we shall see, for example, in the matter of enamel thickness. Similar sets of jaws and teeth may therefore arise in species with very different biological repertoires.

Any understanding of human evolution is going to be considerably dependent upon information gained from jaws and teeth; as the most durable material they are the most common fossil, but they are also extremely informative about the life history, diet, and phylogenetic position of extinct species. In this chapter we will examine four facets of hominoid dentition: the overall structure of jaws and teeth; the pattern of eruption; the characteristics of tooth enamel; and the indications of diet that are to be found in microwear patterns on tooth surfaces.

## Basic anatomy

Perhaps the most obvious trend in the structure of the primate jaw (and face) throughout evolution is its shortening from front to back and its deepening from top to bottom, going from the pointed snout of the tarsier to the flat face of *Homo sapiens* (Fig. 7.13). Structurally, this change involved the progressive tucking of the jaws under the brain case, which steadily reduced the angle of the lower jaw bone (mandible) until it reached the virtual "L" shape seen in humans. Functionally, the change involved a shift from an "insect trap" in prosimians to a "grinding machine" in hominoids.

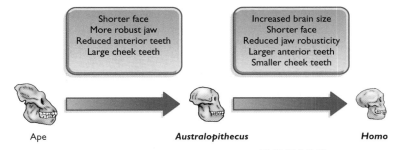

Shorter face
More robust jaw
Reduced anterior teeth
Large cheek teeth

Increased brain size
Shorter face
Reduced jaw robusticity
Larger anterior teeth
Smaller cheek teeth

Ape

*Australopithecus*

*Homo*

**FIGURE 7.13**
**Evolutionary trends in dentition:** The transition from ape to *Australopithecus* and from *Australopithecus* to *Homo* involved some changes that were continuous and others that were not. For instance, the face became increasingly shorter throughout hominin evolution, while robusticity of the jaw first increased and then decreased. The combined increase in cheek tooth size and decrease in anterior tooth size that occurred between apes and *Australopithecus* was also reversed with the advent of *Homo*.

Grinding efficiency increases as the distance between the pivot of the jaw and the tooth row decreases, with hominins being closest to this position.

The primitive dental pattern for anthropoids includes (in a half-jaw) two incisors, one canine, three premolars, and three molars, giving a total of 36 teeth. This pattern is seen in modern-day New World anthropoids, while Old World anthropoids possess two premolars (not three), giving them a total of 32 teeth. Overall, the modern ape jaw is rather rectangular in shape, while the human jaw more closely resembles an arc. One of the most striking differences, however, is that apes' conical and somewhat blade-shaped canine teeth are very large and project far beyond the level of the tooth row; in these animals, males' canines are substantially larger than those found in females, an aspect of sexual dimorphism with significant behavioral consequences (Fig. 7.14).

When an ape closes its jaws, the large canines are accommodated in gaps (diastemata) in the tooth rows: between the incisor and canine in the upper jaw, and between the canine and first premolar in the lower jaw. As a result of the canines' large size, an ape's jaw is effectively "locked" when closed, with side-to-side movement being impossible. By contrast, human canines – in both males and females – are small and barely extend beyond the level of the tooth row. As a result, the tooth rows have no diastemata, and a side-to-side "milling" motion is possible, which further increases grinding efficiency. The upper incisors of apes are large and spatula-like, which is a frugivore adaptation. In contrast, human upper incisors are smaller and more vertical, and, with the small, flat canines, they form a slicing row with the lower teeth.

The single-cusped first premolar of apes is highly characteristic, particularly the lower premolar against which the huge upper canine slides. Ape molar teeth are larger than the premolars and include high, conical cusps. In humans, the two premolars assume the same shape and have become somewhat "molarized." The molars themselves are large and relatively flat, with low, rounded cusps – characteristics that are extremely exaggerated in some of the earlier hominins (Fig. 7.15).

What are the functions of different teeth?

The hominin dental package as a whole can therefore be regarded as an extension of a trend toward a more effective grinding adaptation. In some of the earliest known hominins – *Ardipithecus ramidus* and *Australopithecus*

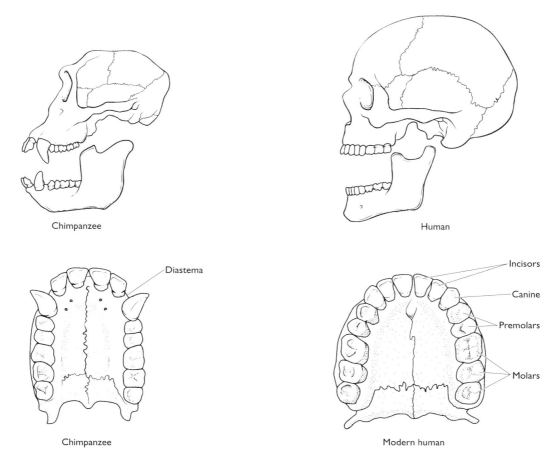

**FIGURE 7.14 Jaws and teeth:** Note the longer jaw and more projecting face in the chimpanzee, the protruding incisors, and large canines.

*anamensis* from over 4 million years ago – the dentition remains strikingly apelike, with a significant degree of sexual dimorphism. Within 2 million years, however, the canines in several hominin species have become smaller and flattened, looking very much like incisors (Fig. 7.16).

## Eruption patterns

The pattern of eruption of permanent teeth in modern apes and humans is distinctive, as is its overall timing. Recently anthropologists have debated this aspect of hominoid dentition, specifically asking how early hominins fit into this picture. Were they more like humans or more like apes? Although the issue remains to be fully resolved, indications are that until rather late in hominin history, dental development was in many ways rather apelike, particularly in its overall timing.

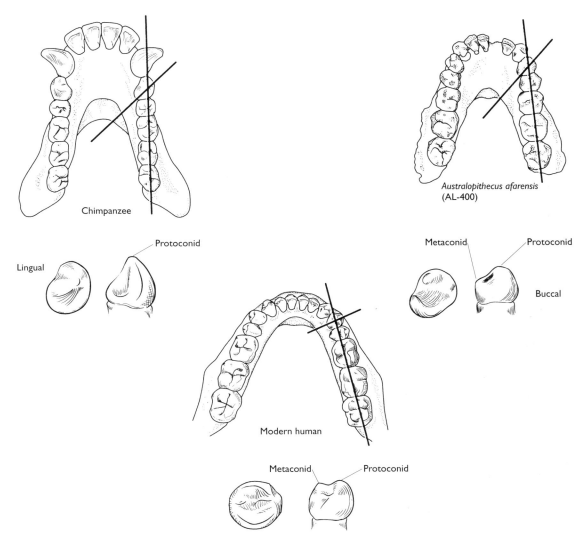

**FIGURE 7.15 Early hominin dentition:** The first premolar in apes is characteristic in having one cusp (protoconid); in humans, the tooth has two cusps (the protoconid and metaconid). (Lingual = the side toward the tongue; buccal = the side toward the cheek.) In apes, the axis of the premolar in relation to the tooth row is more acute than in modern humans. In *Australopithecus afarensis*, an early hominine, the tooth is intermediate in shape between humanlike and apelike, but its axis resembles that seen in apes.

The ape tooth eruption pattern is:

M1 → I1 → I2 → M2 → P3 → P4 → C → M3

The corresponding human pattern is:

M1 → I1 → I2 → P3 → C → P4 → M2 → M3

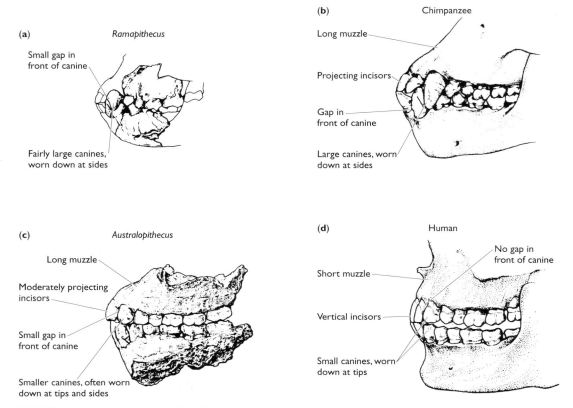

**FIGURE 7.16 Tooth characteristics:** This diagram shows some of the major characteristics in (a) a Miocene ape, (b) a chimpanzee, (c) *Australopithecus afarensis*, and (d) *Homo sapiens*. (From *Our Fossils Ourselves*, courtesy of the British Museum Natural History.)

where M = molar, I = incisor, P = premolar and C = canine. The principal difference, therefore, is that in apes the canine erupts after the second molar, while in humans it precedes the second molar. Associated with the prolonged period of infancy in humans is an elongation of the time over which the teeth erupt. The three molars appear at approximately 3.3, 6.6, and 10.5 years in apes, whereas the ages are 6, 12, and 18 years in humans.

Thus, a human jaw in which the first molar has recently erupted indicates that the individual was roughly 6 years old. An ape's jaw with the first molar just erupted would indicate an individual a little more than 3 years old. The question is: how old is an early hominin jaw in this state? Is it 3 years old or 6 years old? As it happened, the first australopithecine to be discovered – the Taung child, *Australopithecus africanus* – had just reached this state of development.

University of Michigan anthropologist Holly Smith recently analyzed tooth eruption patterns in a series of fossil hominins and concluded that most of the early species were distinctly ape-like.[230] For *Homo erectus*, which lived from 1.9 million until approximately 400,000 years ago, her results implied that early members of this species showed a pattern that was intermediate between humanlike and apelike. For instance, in 1985 a remarkably complete skeleton of *Homo erectus* (denoted KNM-WT 15000) was discovered on the west side of Lake Turkana, Kenya. The individual was a youth whose second molar was in the process of erupting. A human pattern of development would imply an age of 11 or 12 years when he died, while an ape pattern would give 7 years. In fact, Smith's analysis suggests that he was probably 9 years old.[231] The fully human pattern of dental development did not evolve until late in *Homo erectus*'s existence.

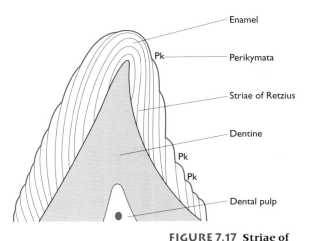

**FIGURE 7.17 Striae of Retzius:** These bands that appear on the surface of the tooth enamel are the product of the way in which the enamel develops and can be used to calculate the absolute age of a fossil hominin. The striae emerge on the enamel surface as shallow ripples called perikymata (pk). (Courtesy of T. Bromage and C. Dean.)

Smith's conclusion has been challenged by University of Pennsylvania anthropologist Alan Mann, who a decade earlier had proposed that all hominins followed the human pattern of development. Nevertheless, Smith's position received support in late 1987, when Glenn Conroy and Michael Vannier of Washington University School of Medicine published results of their computed tomography (CT) analysis of the Taung child's skull. The two were able to "see" the unerupted teeth within the jaw bone, and consequently concluded that the teeth would have emerged in an apelike pattern.[232]

The debate has been extended further by two researchers at University College London, who claim to be able to determine the exact age of a tooth by counting the number of lines – striae of Retzius – within the enamel (Fig. 7.17). Although this technique is not universally accepted, the two researchers, Timothy Bromage and Christopher Dean, believe that the lines represent weekly increments, thus giving an anthropological equivalent of tree rings, which measure yearly increments.

When Bromage and Dean applied their technique to a series of australopithecine and early *Homo* fossils,[233] they obtained ages that were between one-half and two-thirds of what would be inferred if a human standard of dental development had been applied (Fig. 7.18). If they and Smith are correct, then hominins followed a distinctly apelike pattern of dental development until relatively recently in evolutionary history.[234] This concept has important implications for the period of infant care. Once that is prolonged, which becomes necessary when significant postnatal brain growth takes place, then social life becomes greatly intensified. The dental evidence indicates that this prolongation may have begun with *Homo erectus*, which is in accord with data on increased brain size.[235]

What can be learnt from dental development?

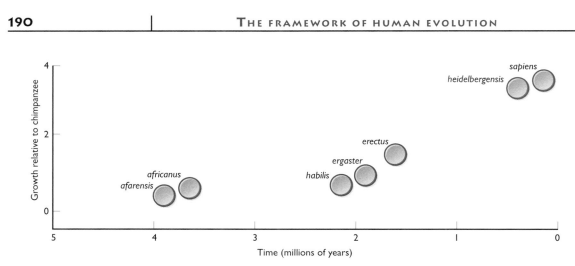

**FIGURE 7.18 Rates of maturation in hominins:** Applying the histological technique developed by Bromage and Dean, it is possible to estimate the rates of development of fossil hominins relative to both humans and living apes. The evidence suggests that australopithecines had life-history strategies more like those of the apes than like those of humans.

## Enamel thickness

The relative thickness of enamel on cheek teeth has played an important role in anthropology, not least because Elwyn Simons interpreted *Ramapithecus* as being an early hominin through identification of this character (see chapter 8). Modern humans carry a thick enamel coat on their teeth, whereas the African apes exhibit thin enamel (in the orangutan, the enamel layer is of intermediate thickness). Until the 1994 discovery of *Ardipithecus ramidus* changed the picture, all known fossil hominins also possessed thick enamel. Thick enamel was therefore assumed to be a shared character for the African hominoid clade. Thin enamel was seen as an adaptation to fruit eating, while thick enamel was envisioned as an adaptive response to processing tougher plant foods.[236]

The evolution of thin and thick enamel followed a complex path throughout hominoid history.[237] Thin enamel appears to be a primitive character for the hominoid clade as a whole, but thick enamel has arisen several times independently during the history of the group. What about the African hominoid clade? As already indicated, thick enamel was traditionally considered to be a characteristic of this clade, with the chimpanzee and gorilla having reverted to a primitive state of thin enamel. The most recent analysis of enamel formation in hominoids and a re-evaluation of late Miocene hominoids in Africa have turned this view around, however. It now seems likely that the common ancestor of modern African hominoids had thin enamel, that the earliest hominins also possessed thin enamel (with thick enamel developing only later in the clade's history), and that chimpanzees and gorillas represent the primitive state of the group, not a reversal. The thick enamel of later hominins and, for instance, the late Miocene ape *Sivapithecus* reflects independent evolution, not homology.

## Toothwear patterns

The surface of tooth enamel bears the brunt of an animal's primary contact with its food, and to some extent at least a signature of that contact is left behind. Using a scanning electron microscope, Alan Walker of Pennsylvania University has produced images of a range of characteristic toothwear patterns: for grazers, browsers, frugivores, bone-crunching carnivores, and so on.[238] The teeth of grazers, for instance, are etched with fine lines that are produced by contact with tough silica inclusions (phytolyths) in grasses; browsers' teeth are smoothly worn, as are those of fruit-eaters; scavengers' teeth are often deeply marked as a result of bone crushing. In a series of comparisons, all early hominins appear to fit into the frugivore category, along with modern chimpanzees and orangutans. This pattern entails a rather smooth enamel surface into which are etched a few pits and scratches.

A major shift occurs, however, with *Homo erectus*, whose enamel is heavily pitted and scratched. Such a pattern resembles a cross between a hyena (a bone-crunching carnivore) and a pig (a rooting omnivore). Although it is not yet possible to interpret precisely the implications for the *Homo erectus* diet, it is significant that toothwear patterns indicate some sort of abrupt change in hominin activities at this point in history – perhaps significant brain expansion, reduction in body-size dimorphism, systematic tool making, use of fire, or migration out of Africa.

In spite of their limitations, then, teeth clearly have the ability to yield information about hominin history that goes far beyond simply what went down our ancestors' throats. Teeth are central to understanding the behavior and biology of any species, and in many cases they are the only source of such information.

## BEYOND THE FACTS
## DOES BEHAVIOR EVOLVE?

*The issue: evolutionary biology is generally concerned with changing shapes and structures – anatomy and morphology – but it is often the behavior of animals that actually determines their survival. However, is behavior – especially complex social and cultural behavior – subject to the same principles of evolution as anatomy?*

When the principles of evolution were developed by Darwin and many of his successors, it was done in terms of the morphological attributes of the biological world. Evolution was primarily about a bigger beak here and a more brightly coloured tail there. Genetics, as it developed, showed the way in which such morphologies might be rooted in genes, and therefore heritable from one generation to another, and fixed in any one particular individual. Each Galapogos finch had the particular beak it had inherited from its parents, and that beak had been shaped over generations by natural selection. Outside creationist circles, there is little that is controversial about any of this.

However, much of an animal's survivorship is dependent not so much upon what it looks like as what it does – in other words, upon its behavior. For many creatures it can be argued that their behavior is as much a product of their genes as are their anatomy and physiology – indeed, their behavior is driven and constrained by their genes. Classic examples where it has been possible to show a direct relationship between behavior and genes include the hygienic behavior of certain bees, or the imprinting that occurs in the behavioral development of birds. Again, this would be accepted by most evolutionary biologists and others.

The behavior of more complex animals, such as primates and humans, on the other hand, is clearly not what can be called simple instinct, but rather is learnt and developed by experience. Such behaviors might vary from individual to individual, and from population to population, and indeed would change over the course of a lifetime. In humans in particular, behaviors can be picked up and discarded, influenced by those around as well as by parents. The point here is that this sort of behavior is unlikely to be the product of simple genetic instruction. If this is the case, how can it be said to be under the influence of natural selection, and to be the product of Darwinian evolution? This is particularly the case where culture and social behavior are considered.

Intense controversy has raged over this issue. Some have argued that without a clear-cut genetic basis, social and cultural behavior cannot be said to be part of evolution, and that explanations for such phenomena as human courtship or social kinship structures are beyond evolutionary explanation. Much of social anthropology might take this position, seeing biology and evolution as providing only the most basic substrate for human behavior. Others have argued, however, that behavior is as much subject to selection as is morphology, and that the mind, social organization, and cultural variants are all the end products of the evolutionary process. This may come about either because there is indeed a genetic basis, even if we have yet to establish it, or else because evolution can operate on heritable characteristics (and much behavior is heritable) even in the absence of a genetic basis. Given that so much of what is important in human evolution is indeed behavioral, it is important to consider what are or are not the limits of evolutionary processes. Behaviour and culture are the ground on which those limits should be set.

# PART 2

# EARLY HOMININ EVOLUTION

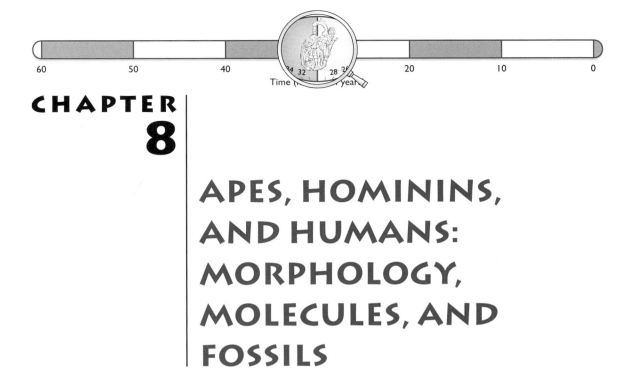

# CHAPTER 8

# APES, HOMININS, AND HUMANS: MORPHOLOGY, MOLECULES, AND FOSSILS

The superfamily Hominoidea (colloquially, hominoids) includes all living and extinct ape and human (hominin) species. This chapter will address the relationships among living hominoids and their formal classification, the timing of the evolutionary divergence between the human and ape lineages, and the probable anatomical characteristics of the ancestor of humans common to both humans and apes. We will examine our knowledge of extinct ape species and their possible relationship to living hominoids.

Until relatively recently, all hominoid systematics relied on comparative anatomy of living species and of fragmentary specimens from the rather sparse fossil record. Beginning in the early 1970s and with great intensity since the early 1990s, molecular data have been exploited to tackle the same questions asked by anatomists. Strikingly different answers have emerged from these two perspectives, partly because of the difficulties involved in inferring evolutionary relationships from anatomy, particularly when the available evidence is so fragmentary, and partly because the methodology for doing so had not been fully developed. With a few qualifications (as discussed later in this chapter), molecular systematics has had the greatest influence in shaping our current views of hominoid relations and classification. If the molecular perspective is indeed the correct one, it raises questions about why the same picture was slow to emerge from morphological analyses, and why that picture is so muddy compared with the apparent clarity provided by the molecular view.

We will begin with an outline of the development of ideas of hominoid relations; we will then incorporate the impact of the molecular perspective and its development. As Morris Goodman, one of the pioneers of molecular evolution, has recently observed,[239] this history has involved two paradigm shifts: one in the philosophy of systematics, with the increasing dominance of genealogical rather than adaptive interpretations, and one in how we view ourselves, with the gap between humans and apes becoming ever narrower.

## MORPHOLOGY AND MOLECULES: A HISTORY OF CONFLICT

### Morphological interpretations

**KEY QUESTION** Do molecular genetics and comparative anatomy provide the same insights into the origins of hominins?

The systematics of modern hominoids is, put simply, a question of which species are evolutionarily more closely related to which other species, and how this relationship should be reflected in both species' classification. When Carolus Linnaeus published his *Systema Naturae* in 1758, first-hand knowledge of apes was at best sketchy. Nevertheless, in coining the term "Primates" (that is, the principal order of mammals), he included apes and humans in that group. Not all naturalists of the time (the late eighteenth century) accepted the idea of so intimate a relationship between humans and apes, however. For instance, although the French naturalist Buffon recognized echoes of similarity between apes and humans (based on his knowledge of the chimpanzee), he declared that the distance between man and brute must be infinite. Similarly, the German naturalist Blumenbach considered the gap to be vast, rejecting the validity of the Primate order. Humans, he suggested, should be the sole occupants of the order Bimanus, while the rest of the primates should be grouped into a separate order, Quadrumana. His suggested classification gained wide popularity that lasted for a century. In 1868, for instance, the British paleontologist Sir Richard Owen increased the human/ape distinction by elevating Bimanus to the level of a subclass, Archencephala.

Dissenting voices were heard, however, the most prominent of them from Thomas Henry Huxley.[2] As noted in chapter 1, in his 1863 book, *Evidences as to Man's Place in Nature*, Huxley not only recognized a close affinity between humans and apes, but also concluded that humans are closer to African apes than to Asian apes. Darwin accepted this idea, and used it to speculate in his 1871 *The Descent of Man* that humans therefore probably evolved in Africa (land of our closest relatives), not Asia.[19]

The first primate fossils with hominoid-like features were discovered (in Europe) in the early nineteenth century, and other discoveries followed sporadically through the subsequent decades. The most important of these finds were assigned to the genus *Dryopithecus*, the first specimen of which was uncovered in southwest France in 1856. Various *Dryopithecus* species were considered to be ancestral to modern apes – particularly the gorilla, chimpanzee, and possibly orangutan. A second genus, *Palaeosimia*, was thought by some scholars to be ancestral to the orangutan, while a third lineage, *Sivapithecus*, was said by some to form the basis of the human lineage. For the most part, these inferences were drawn from assumed homologous similarities in jaw and tooth anatomy.

The evolutionary interpretation of these fossil apes had an important impact on perspectives on living hominoid relationships. For instance, writing in the 1920s, the British anatomist Sir Arthur Keith saw human origins as arising within an apelike stock; he considered the *Dryopithecus* group to be ancestral to all three living great apes. This group diverged from the human lineage in the middle Miocene, he concluded, which in today's terms would mean some 15 to 20 million years ago. In this scheme, the chimpanzee, gorilla, and orangutan would be closely related to one another, and all equally distant from humans. (Note that this differs from Huxley's idea of a close relationship to the African apes.) Although the American paleontologist Henry Fairfield Osborn, a contemporary of Keith's, agreed with the overall shape of the evolutionary tree, he considered the divergence between humans and apes to have been even more ancient.

We have in these early workers the basis for two ways of looking at hominoid relationships. On the one hand there is the idea that the apes as a whole represent one (monophyletic) group, with humans as the separate lineage (the sister clade, in modern terminology); and on the other hand there is the idea that human origins are to be found amongst the great apes, which are not monophyletic, and in particular among the African apes. The first of these models presupposes a more ancient split and origin, the second a more recent one.

For most of the twentieth century the "ancient split" model was dominant. For example, it was developed by the Swiss anatomist Adolph Schultz[240] using embryological evidence (in the 1930s) and on comparative anatomical grounds by the British anatomist Sir Wilfrid Le Gros Clark[21] (in the 1930s through 1950s). However, although Schultz and, particularly, Le Gros Clark were highly influential, their view was not unanimously adopted. For instance, William King Gregory, Osborn's contemporary, inferred a human/African ape affinity from dental evidence. Sherwood Washburn, of the University of California, Berkeley, later came to the same conclusion principally on the basis of an analysis of the postcranial skeleton (in the 1960s), and the great evolutionary biologist Ernst Mayr, writing in 1950, came to the same conclusion[25] (Fig. 8.1).

By the 1960s, therefore, although no universal consensus had been reached, strong support existed for two points: (1) various fossil apes could be identified as being ancestral to modern apes, and (2) the divergence between this group and the human lineage occurred in ancient times. The implication was that the great apes were one another's closest relatives, and all were equally distant from humans. Although several species of

*What is the morphological pattern of hominoid diversity?*

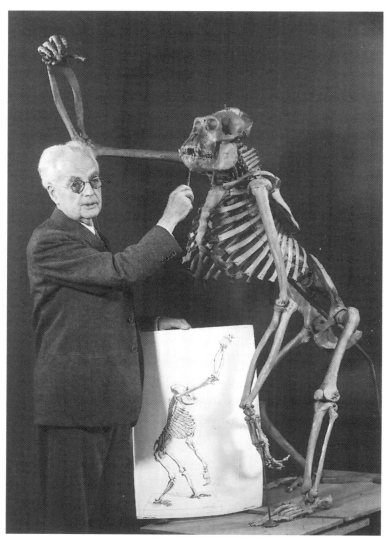

(a)

**FIGURE 8.1**

**Supporters of an African ape–human affinity:** William King Gregory (a), Sherwood Washburn (b), and Ernst Mayr were three scientists who championed the view that humans may have had a close relationship with the African apes.

hominin had been discovered by this time (of the genus *Australopithecus*), they were thought to have lived relatively late in the lineage (dating at between 1 and 3 million years old). Missing was a good candidate for an early member of the human lineage, which was expected to be at least 15 million years old.

Such a candidate was proffered in the early 1960s by Elwyn Simons, then at Yale University. Simons analyzed upper-jaw and dental characteristics of a specimen of *Ramapithecus* that had been found by a Yale doctoral student, G. Edward Lewis (Fig. 8.2), three decades earlier in the Siwalik Hills of India. Simons identified *Ramapithecus* as an early hominin on the basis of the reconstructed shape of the jaw (parabolic rather than U-shaped) (Fig. 8.3), the shape of the teeth, and the thick layer of enamel on the cheek teeth. Moreover, Simons judged the divergence of apes and humans to have occurred in the Oligocene, some 30 million years ago.[241]

Simons was joined in this view by David Pilbeam (Fig. 8.4), now of Harvard University, and together they developed a rather detailed evolutionary scenario of Miocene apes and the beginning of the hominin and modern ape lineages (Fig. 8.5).[242] To this Asian group was added an Africa version – *Kenyapithecus* – which Louis Leakey claimed was associated with stone tools, thus raising the possibility of its being cultural and bipedal at a very early date.[243] This view was widely, but not universally, accepted. However, its period of development coincided with a new approach, the application of molecular genetic data to hominoid systematics. This work was to draw the field of hominoid relationships back to the views of Huxley, and a more recent and African relationship for the first hominins.

## Molecular studies

The term "molecular anthropology" was coined in 1962 by Emile Zuckerkandl, who, with Linus Pauling (both were pioneers in this field), invented the notion of using molecular evidence to uncover evolutionary histories. At the time, Zuckerkandl had already discerned a hint of what was to unfold in the science when he compared enzymic digests of proteins from humans, gorillas, chimpanzees, and orangutans. First Morris Goodman, of Wayne State University, and then Allan Wilson and Vincent Sarich, of the University of California, Berkeley, actually went on to establish the new field of research.[24,142] They used the immunological reactions of certain blood proteins to measure genetic distances among the living hominoids. In the early 1960s, Goodman established the human/African ape affinity, while in the late 1960s Wilson and Sarich used the genetic distances to identify times of divergence between the ape and human lineages.

**(b)**

**FIGURE 8.1** (*cont'd*)

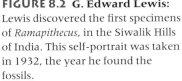

**FIGURE 8.2  G. Edward Lewis:** Lewis discovered the first specimens of *Ramapithecus*, in the Siwalik Hills of India. This self-portrait was taken in 1932, the year he found the fossils.

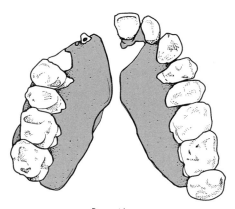

*Ramapithecus*

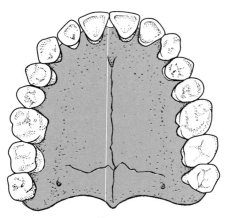

Human

**FIGURE 8.3**
*Ramapithecus*
**reconstructed
(wrongly):** In the
original reconstruction
of the two fragments
of upper jaw (maxilla) of
Lewis's *Ramapithecus*
specimen, the shape
appeared to be
humanlike. This reason
partly explains why the
Miocene ape was
thought to be an early
hominin.

**FIGURE 8.4  David Pilbeam (left)
and Elwyn Simons:** Seen here at a
1982 Rome conference on human
origins, Simons, later joined by
Pilbeam, was a powerful influence
as the proponent of *Ramapithecus*
as a putative early hominin. By
the 1980s, Pilbeam had become
convinced by molecular evidence
that their earlier position was
wrong; Simons later agreed.

As with all such calculations, Wilson and Sarich calibrated their molecular clock using known (or assumed) divergence times derived from the fossil record. They applied the then-accepted divergence time of Old World monkeys (superfamily Cercopithecoidea) and Hominoidea of 30 million years ago. According to their research, the genetic distance between humans and African apes was one-sixth of that between living African hominoids and Old World monkeys. This finding implied that African apes and humans diverged 5 million years ago (one-sixth of 30 million years) (Fig. 8.6).

In the decades since this first calculation of human/ape divergence based on molecular data, many different techniques have been applied to the problem, including electrophoresis of proteins, amino acid sequencing of proteins, restriction enzyme mapping of various types of DNA, sequencing of mitochondrial and nuclear DNA, and DNA–DNA hybridization. Although their results are by no means unanimous, the great majority of these techniques have supported the human/ African ape linkage and have yielded a divergence time near 5 million years, and probably not greater than 6 million. This finding is generally in good accord with the known fossil record.

Much controversy surrounded this work, and not all disagreements pitted molecular biologists against morphologists. For instance, considerable debate surrounded the issue of the rate at which genetic change in the hominoid lineages accumulated. Supporters of the molecular clock (such as Wilson) argued that the rate was constant and universal. Others (such as Goodman) believed that accumulation rates could change over time and in different lineages. Indeed, Goodman initially attributed some of the surprisingly small genetic distance between humans and African apes to a slowdown in the clock. A slowdown could, of course, affect calculations of divergence times: a small genetic distance might disguise a long evolutionary separation. By now, fluctuations in the clock's rate in general have been accepted, and a slowdown among hominoids in particular. Nevertheless, as long as such fluctuations are taken into account, it remains possible to use genetic data for calculations of divergence times via local clocks. For instance, using extensive DNA sequences of certain globin genes, Goodman (previously a critic of the clock) and his colleagues recently calculated the human/chimpanzee divergence as 5.9 million years.[136,239]

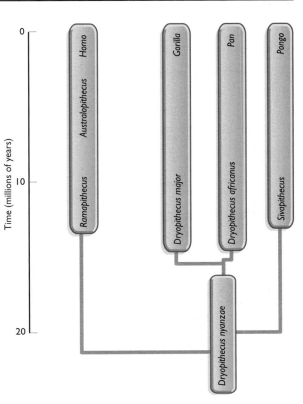

**FIGURE 8.5 Hominoid evolution – the view in the 1970s:** The view in the 1970s, originally proposed by Simons and Pilbeam, saw a long chronology for the divergence of hominins and apes, an Asian origin for hominins, and close affinities between Miocene apes and living descendants.

How has molecular genetics influenced the reconstruction of hominoid phylogeny?

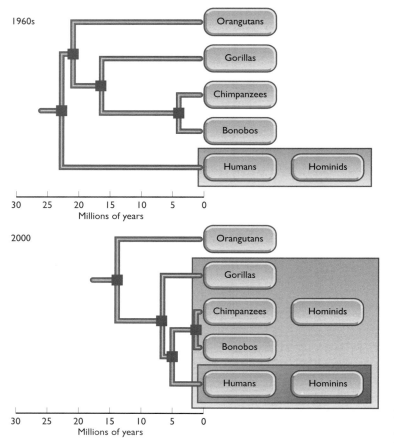

During the first two decades of molecular anthropology, the vast majority of work agreed on two things: the reality of a human/African ape affinity and an inability to be certain of the exact relationships between chimpanzees, humans, and gorillas – the so-called trichotomy where the sequence of branching cannot be determined. The latter factor implied that either the trichotomy was real (that is, there was a three-way lineage split, rather than two sequential branching events) or the techniques were not sensitive enough to detect what might be rather short branches in a tree with two divergence points. In the mid-1980s, evidence began to build in favor of a tree with two divergence points: the separation of the gorilla, followed later by a human/chimpanzee split. During the subsequent decade, most molecular datasets of various types supported the same pattern. Cladistic analysis requires specific characters (not genetic distance); in this context, it means gene sequences. Of 10 such independent datasets collected to date, eight support a human/chimpanzee link, two a chimpanzee/gorilla link, and none a human/gorilla link. (Humans are known to share 98.3% identity in nuclear, non-coding DNA sequence and more than 99.5% identity in nuclear coding sequences, or genes.)

**FIGURE 8.6 Hominoid relationships – molecular and morphological views of the 1970s:** When the molecular approach was first developed, it presented a radically different picture of hominoid relationships and evolution than the preferred one of many paleontologists.

Molecular phylogenetics involves several potentially confounding complications. The gene-tree/species-tree problem is particularly relevant here. It can yield a phylogenetic pattern of the sort now heavily supported, even though the evolutionary reality is a simple trichotomy, as Jeffrey Rogers,[244] of the Southwest Foundation for Biomedical Research in San Antonio, Texas, has argued. The problem involves the evolution of gene polymorphisms within an ancestral species and the subsequent unequal distribution or extinction of variants in descendant species. A thought experiment will illustrate the point (Fig. 8.7).

Imagine that an ancestral species possessed a gene A. Now imagine that a variant of the gene, A', arose 10 million years ago, making the gene polymorphic. Individuals in the population of the common ancestor may now possess two copies of variant A (that is, be homozygous for A), two copies of variant A′ (homozygous for A′), or one copy of each variant (hetero-zygous). Suppose that 5 million years ago the ancestral species split into three daughter spe-cies, X, Y, and Z. In the popula-tion that leads to X, the variant A′ is lost, leaving just A. In the population that leads to Z, vari-ant A is lost, leaving just A′. A comparison of the sequences of this gene in species X and Z would indicate that they diverged 10 million years ago, despite the fact the speciation event occurred only 5 million years ago. This erroneous dating, based on conflation of so-called gene trees and species trees, would follow from the gene polymorphism.

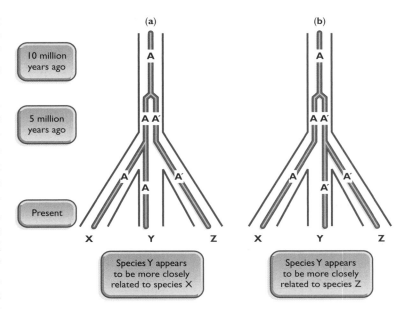

**FIGURE 8.7 Gene trees versus species trees:** Gene polymorphism in an ancestral species followed by differential sorting of variants can lead to erroneous conclusions, in both the timing of divergence and the relationship among descendant species. (a) Genetic analysis would make species Y look more closely related to species X than to species Z. (b) Y looks more closely related to Z than to X. The reality is a trichotomy. (See text for details.)

What about species Y? If its population lost variant A, a comparison of all three species would imply that Y is more closely related to species Z than to species X; similarly, if Y lost variant A′, it would appear to be more closely related to species X than to species Z. In fact, all three species are equally related.

As this model indicates, for ancestral species possessing many highly polymorphic genes, no simple, single picture will emerge in a comparison of its descendants' genes. This complexity, suggests Rogers, explains the mixed data for the hominoids, and he states that a trichotomy is the most likely pattern. Others, such as Jon Marks, agree. The hominoid history, he contends, is effectively a three-way split.

It is true that the gene-tree/species-tree problem can lead to an errone-ously old divergence date. It is also true that the problem can yield a pat-tern of two divergences apparently separated in time whereas the reality is a trichotomy. How is hominoid history to be assessed, given the data to hand?

The processes involved are stochastic, in terms of the timing of the origin of polymorphisms and the subsequent sorting of variants. As a result, many datasets are required to test hypotheses. The fact that so many datasets point to a similar divergence time for the inferred human/ chimpanzee split provides some confidence in that date, unless all genes just happened to have produced polymorphisms at the same time in the ancestral species prior to speciation – an unlikely event.

The same principle can be applied to the putative two-divergence pat-tern, as Maryellen Ruvolo, of Harvard University, has argued. Given the stochastic nature of the sorting of variants, there is a one-third probability of genetic data implying a human/chimpanzee alliance and a two-thirds probability of seeing chimpanzee/gorilla or human/gorilla alliances. Statistically speaking, Ruvolo calculates, the probability of eight human/chimpanzee alliances emerging from 10 datasets as a matter of chance is close to one in 3000. In other words, the observed pattern is very likely to reflect history rather than being a statistical quirk.[245,246]

Over time there has been a gradual firming up of the evidence; few would now question the monophyly of the African apes and humans, or the divergence in the Later Miocene, and there is general agreement that the sequence of branching is most likely to have been gorilla first, then humans and chimpanzees. This is a considerable advance on the often heated debates of the 1970s.

## A shift in interpretation: morphology

During the 1980s, while the molecular techniques were emerging and being developed, opinion concerning the morphological evidence moved much closer to the genetic perspective (particularly the human/African ape affinity, but also the identity of the earliest hominin), for several reasons. First in importance was the weight of the molecular evidence itself, but a second factor was derived from morphological evidence. There were three elements to this: the discovery of new fossils relating to the great apes; the use of cladistics to analyse the evolutionary relationships of living apes and humans; and reinterpretations of the characteristics of the earliest hominoids, and thus the characteristics of the ancestors of both the great apes and humans.

Turning to the first of these, the discovery of new and more complete *Sivapithecus* specimens (cranial and facial regions) in Turkey and Pakistan (Fig. 8.8) led to two conclusions: (1) the facial elements in particular suggested a much closer relationship with the Asian great ape, the orangutan; and (2) a clear link was established between *Ramapithecus* (the putative early hominin) and *Sivapithecus*. If *Ramapithecus* was truly part of a group ancestral to the modern orangutan, then it could not be a hominin, because hominins are more closely related to African apes. This change of view was very much due to the fact that for the first time there were relatively complete fossil remains, whereas earlier views were based on fragmentary pieces, especially teeth and jaws.

With *Ramapithecus* removed as a putative early hominin, the known fossil record of hominins (early *Australopithecus*) was accepted as a reliable guide to the origin of the clade; that is, it was likely to have occurred closer to 5 million years ago than the previously favored 15-plus million years. Early *Australopithecus* was also seen as sharing a number of features with chimpanzees (Fig. 8.9). The timing of morphological and molecular chronologies became much closer. The central issue then became the

How have morphological and molecular approaches become integrated?

evolutionary relationships (and history) among chimpanzees, gorillas, and humans.

The African apes share many anatomical similarities, particularly in their forelimbs, which show adaptations to their knuckle-walking mode of locomotion, and in their dentition, which has a thin layer of enamel on the cheek teeth. Modern humans and their (then known) extinct relatives exhibit thick enamel, as do many fossil apes. In several cladistic analyses of living hominoids (by, for example, Lawrence Martin of the State University of New York at Stony Brook and Peter Andrews of the Natural History Museum, London),[247] the shared limb anatomy and dental features of African apes were judged to be derived characters that linked chimpanzees and gorillas as a separate clade from humans. Under this scheme, humans were seen as having diverged first from the hominoid lineage, with gorillas and chimpanzees sharing a common ancestor in which knuckle-walking and thin tooth enamel evolved. A second scheme – the trichotomy discussed above in which African apes and humans diverged simultaneously from a common ancestor – was said to be possible, though less likely (Fig. 8.10).

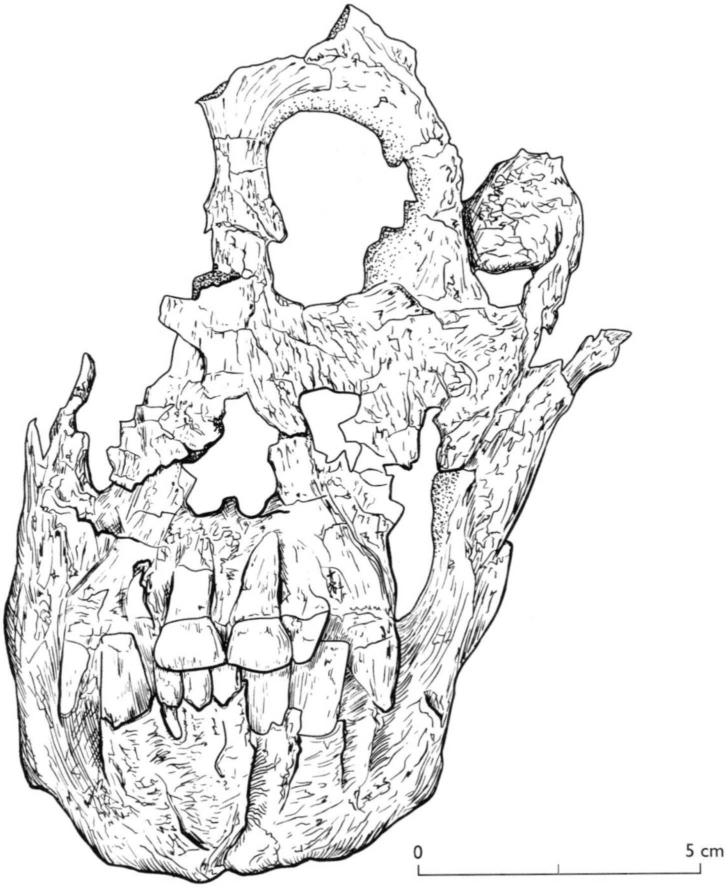

**FIGURE 8.8** *Sivapithecus*: This 8-million-year-old specimen of *Sivapithecus indicus* was found in 1980 in the Potwar Plateau in Pakistan, by David Pilbeam and S. M. Ibrahim Shah (of the Geological Survey of Pakistan). The animal was approximately the same size as a chimpanzee but had the facial morphology of an orangutan; it ate soft fruit (detected from the toothwear pattern) and was probably mainly arboreal. The species' close similarity to *Ramapithecus* effectively removed the latter from contention as the earliest known hominin.

The Martin/Andrews view of human/African ape affinity won wide support, although different views were expressed as well. For instance, one cladistic analysis grouped the orangutan with the African apes in a clade separate from humans, while another identified an African ape clade and a human/orangutan clade. In this plethora of morphological analyses, only one, published in 1986 by Colin Groves of the Australian National University, concluded (weakly) that humans and chimpanzees are one another's closest relatives;[248] this assessment was based on forelimb anatomy, particularly the wrist. That is, gorillas were suggested to have

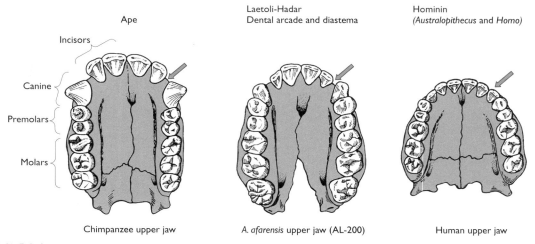

**FIGURE 8.9  Palate and tooth anatomy:** In apes, the jaw is U-shaped; in modern humans and later extinct hominins, it is parabolic. The jaws of early hominins such as *Australopithecus afarensis* are somewhat intermediate in shape. Ape incisors are large and spatulate; a gap, the diastema, separates the second incisors from the large canine; the premolars and molars have high cusps. In humans, the incisors are small; no diastema appears; the canines are small; the premolars and molars have low cusps. In *Australopithecus* species, the incisors are larger than in modern humans, as are the canines; a diastema is sometimes present in early species; the premolars and molars are large with low cusps. The very earliest hominin species are more chimplike in their dentition. (Courtesy of Luba Gudz.)

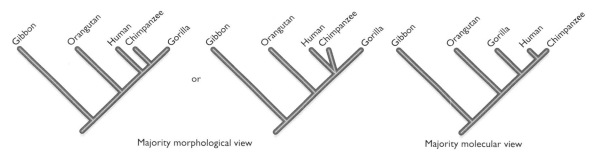

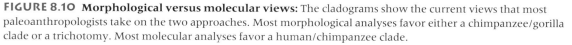

**FIGURE 8.10  Morphological versus molecular views:** The cladograms show the current views that most paleoanthropologists take on the two approaches. Most morphological analyses favor either a chimpanzee/gorilla clade or a trichotomy. Most molecular analyses favor a human/chimpanzee clade.

diverged first from the hominoid ancestor, with humans and chimpanzees sharing a common ancestor from which they later diverged.

Morphologists resisted this interpretation because of the many anatomical similarities between gorillas and chimpanzees, which were assumed to be shared derived characters. If the human/chimpanzee association was indeed correct, then morphologists faced awkward puzzles. For instance, the many striking anatomical similarities of gorillas and chimpanzees must be explained either as homoplasies (independent, parallel evolution), which seems unlikely, or as shared primitive characters that were present in the common ancestor of apes and humans. Furthermore, why have the homologous features that reveal the human/chimpanzee link been so hard to find? Groves was the lone voice in identifying any at all. Recently, however, analyses of fossil and living hominoids have added further evidence related to this point.

There have also been changes in analyses of the earlier hominoids, and so what to expect early apes to look like. For instance, David Begun, of the University of Toronto, compared cranial and dental features in the Miocene ape *Dryopithecus*, an early member of the hominin clade, and living hominoids. He concluded that many characters in gorillas once considered to be derived are actually primitive, and that humans, chimpanzees, and australopithecines share several characters that are derived for the group as a whole.[249] This finding links humans and chimpanzees as one another's closest relatives.

Groves, in collaboration with Australian and US researchers, conducted further extensive cladistic analyses on a large suite of cranial and postcranial anatomy (264 morphological characters in 18 primate taxa, using maximum parsimony analysis) (Fig. 8.11). Like his earlier study, this (more detailed) analysis supports a human/chimpanzee association.[250] A phylogenetic tree with a hominoid trichotomy is much less parsimonious, requiring 15 additional evolutionary steps. Note, however, that the use of subsets (rather than the entire suite) of morphological data in such analyses can yield different phylogenetic results. The selection of which characters should be analyzed is therefore critically important.

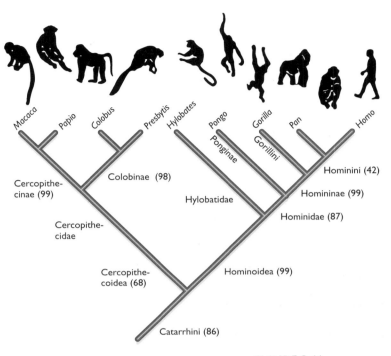

**FIGURE 8.11**
**Cladogram of catarrhine relations:** This analysis of 264 morphological characters leads to a chimpanzee/human association as the most parsimonious tree; a tree with a hominoid trichotomy is less parsimonious. This study is one of very few morphological analyses that identifies a chimpanzees and humans as one other's closest relatives. (Courtesy of Shoshani et al.)

## Nature of the hominin ancestor

The research in both paleontology and genetics since the 1970s has brought the study of human origins to an interesting position that is very distant from that of Darwin and Huxley. For them, starting out on the problem of human evolution, it was a question of finding some place to begin, and very much working in the dark. Beyond the overall idea of humans as primates, and the general theory of evolution, there were few

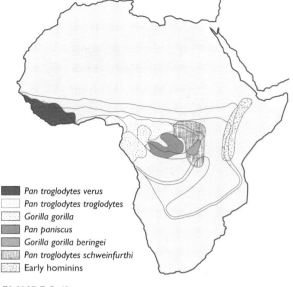

Pan troglodytes verus
Pan troglodytes troglodytes
Gorilla gorilla
Pan paniscus
Gorilla gorilla beringei
Pan troglodytes schweinfurthi
Early hominins

**FIGURE 8.12**
**Expecting the hominins:** As both Wallace and Darwin noted, geographical distributions can be used to infer past evolutionary events. The distribution of the extant African apes – the closest relatives of humans – can be seen as a guide to where hominins may have evolved. The current distribution of early hominin fossils perhaps indicates that they were the eastern and more arid range of African ape distributions. The lines indicate current five- and six-month dry season distribution. (Courtesy of Robert Foley.)

expectations. Now, after a century and a half of work on comparative anatomy and genetics, and with a reasonable fossil record of the apes, it is possible to have some sound expectations of where and when the first hominin should appear, and what it might look like.

First, given the close relationship between humans and the African apes, especially the chimpanzee, it is most probable that hominin origins lie in Africa rather than elsewhere in the world. More specifically, one might, given the current distribution of the African apes, expect that to be broadly equatorial, rather than in one of the continental extremes (Fig. 8.12). Second, the molecular and the fossil evidence leads to the expectation that the last common ancestor should have occurred at some point in the later Miocene. A broad level of confidence would place this between 10 and 4 million years ago; the most probable bracketing of the time would be between 7 and 5 million years ago. During that period, in Africa, we should expect to find taxa that approximate either to the last common ancestor, or to the first ancestral forms of the living human and chimpanzee lineages, or to forms that are equally related to both lineages, but may not be directly ancestral to either.

Third, it is also possible to have some concrete ideas about the appearance of the last common ancestor. The classic view has been to assume that this ancestor was relatively generalized. This arises from the idea that hominins diverged from the stem of the other great apes prior to their divergence into orangutans, gorillas, and chimpanzees. As the close relationship with chimpanzees has been established, the last common ancestor has been seen as still generalized but also more chimplike (Fig. 8.13). It has been widely believed to have been intermediate in size between the gibbon and chimpanzee; it is imagined to have been principally (but not exclusively) arboreal and to have incorporated a significant amount of bipedalism in posture and locomotion, both in trees and on the ground. The ancestor is thought to have lacked the anatomical specialization of the African great apes (such as in the forelimbs and axial skeleton – that is, the vertebrae and ribs) that relate to knuckle-walking. The cranium would have been prognathic (protruding), as is seen in fossil and living apes. And because the cheek teeth in many fossil apes and (until recently) all known hominins are both large and covered with a thick enamel layer, the common ancestor has been assumed to fit this pattern. African apes, for example, have thin enamel, a presumed shared derived character.

David Pilbeam has proposed an alternative hypothesis, one influenced in part by the phylogeny suggested by the molecular data. Because

Adaptations to bipedal locomotion

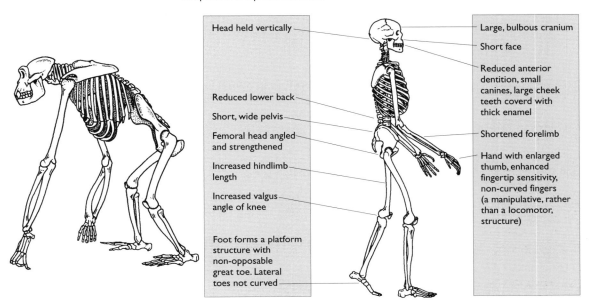

Head held vertically

Reduced lower back

Short, wide pelvis

Femoral head angled and strengthened

Increased hindlimb length

Increased valgus angle of knee

Foot forms a platform structure with non-opposable great toe. Lateral toes not curved

Large, bulbous cranium

Short face

Reduced anterior dentition, small canines, large cheek teeth coverd with thick enamel

Shortened forelimb

Hand with enlarged thumb, enhanced fingertip sensitivity, non-curved fingers (a manipulative, rather than a locomotor, structure)

**FIGURE 8.13  Ape and human anatomy:** The ape (left) is adapted to a form of quadrupedalism known as knuckle-walking, which is seen only in chimpanzees and gorillas. Rather than support the forelimb on the palm of the hand (like most primates) or the palmar surface of the fingers (like baboons), the African apes support it on the dorsal surface of the third and fourth digits of their curled hands. The wrist and elbow anatomy is adapted so as to "lock," thus providing a firm support for the body weight. Human bipedalism (right) involves a number of anatomical differences from that seen in quadrupedalism, as indicated. Anthropologists are divided over whether the common ancestor of humans and African apes was a knuckle-walker.

humans and chimpanzees are one another's closest relatives, and because chimpanzees and gorillas share so many anatomical features, the common ancestor is likely to have been rather chimplike, says Pilbeam. (Such a pattern is more parsimonious than one involving parallel evolution of knuckle-walking in separate gorilla and chimpanzee lineages.) This proposed pattern would include a degree of knuckle-walking and thin-enamelled teeth. The hominin lineage has lost many of these features, partly through its adaptation to bipedal locomotion and a change in diet. The recent discovery of a 4.5-million-year-old hominin, *Ardipithecus ramidus*, bolsters this view. This species is chimp-like in some aspects of its dentition, including possessing thin enamel, and in its postcranial anatomy.

What was the last common ancestor like?

The suggestion of a chimplike ancestor has been resisted in the past and continues to inspire controversy because it would require "reversal" in the direction of evolution, particularly in the configuration of the vertebral column. For instance, African apes have four lumbar vertebrae, early hominins (as seen in two specimens of *Australopithecus africanus* and one *Homo erectus*) have six (presumably as an adaptation to bipedalism), and modern humans have five. An evolutionary progression along these lines would therefore involve an increase from four to six lumbar vertebrae, followed by a decrease to five. Anatomists consider such a progression as evolutionarily difficult, or at least unparsimonious. Pilbeam adduces new insights into the genetics of embryological development to argue that such transitions are actually achieved rather easily. For instance, experimental modification in the timing of expression of certain genes that control development (homeobox genes) in mice readily changes the number of lumbar vertebrae that develop.

If the common ancestor was actually chimplike, discerning the identity of a chimplike fossil from, for example, 6 million years ago would pose significant challenges – as the new *Sahelanthropus tchadensis* discovery shows (see chapter 9).[251] Such a specimen might be the common ancestor, but it might also be an early member of the modern chimp lineage.

## Classification of hominoids

The story of the impact of molecular approaches on studies of human origins is one that has been repeated across the whole of evolutionary biology. Evolutionary relationships that had been well established by classical morphological studies have been overthrown by the new genetics; in some cases relationships that had been difficult to resolve have now been widely accepted.

One effect of these changes has been that many names have been revised within the Linnaean classificatory system. This change has not been at the most fundamental level of the biological classification system – the species – but at the higher levels – genera, families, and so on. Basically there are two issues involved in relation to humans and apes.

The first of these is that the traditional classification reflected the anatomists' interpretations, and these tended to recognize relatively great differences between humans and the other apes. This was partly a simple reflection of the biological facts: humans have a skeleton that has been radically altered from that of other primates in response to their bipedal gait, and their brains and behavior are also significantly different from those of apes. However, it also reflected the more subjective perspective that humans were something new and different in the evolution of the primates.

How should hominoids be classified?

As previously mentioned, hominoid relationships as inferred from molecular data coincide with the view expressed more than a century ago, with the added detail of the human/chimpanzee affinity. In contrast, the traditional classification of hominoids continues to reflect the view that dominated in the intermediate time. That is, it claims that the superfamily Hominoidea was divided into three families: Hylobatidae (gibbons and siamangs), Pongidae (orangutans, gorillas, and chimpanzees), and Hominidae (humans) (Fig. 8.14).

However, we now know, on the basis of both molecular data and a re-examination of the classical morphological information, that this classification does not reflect genealogy (an important principle in systematics), for it links together orangutans, chimpanzees, and gorillas in the Pongidae, despite the fact that they are not the most closely related. The correct genealogical solution would be to have three families, with the Hominidae including both humans and the African apes. The Hylobatidae family would remain intact, while orangutans would represent the sole occupants of the Pongidae. However, this solution, proposed by Morris Goodman, runs counter to the principle that family differences should represent significant grade shifts (Fig. 8.15). Goodman's model places humans and chimpanzees in the same family, despite the very great differences in adaptation, while separating at family level the chimpanzee and orangutan, which share many adaptive features.

These difficulties arise because there is really only a poor fit between the levels of genetic differentiation, which reflects the sequence of branching events, and the amount of evolutionary change at the phenotypic level. If the genealogy principle were adhered to, given the very small amount of genetic change among the hominoids, it is probable that all apes and humans should be placed in a single family – the Hominidae. If the grade principle were adhered to, then the family difference at least would be retained.

There is no perfect way to resolve these conflicting demands, and it is always important to remember that there is an arbitrary element to all classificatory systems. To some people classification matters enormously, while to others the debate about nomenclature can seem reminiscent of medieval theologians disputing how many angels can fit on the head of a

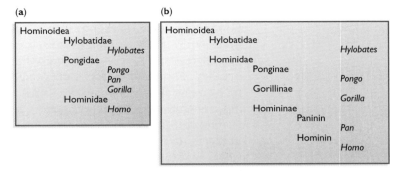

(a)

Hominoidea
  Hylobatidae
    *Hylobates*
  Pongidae
    *Pongo*
    *Pan*
    *Gorilla*
  Hominidae
    *Homo*

(b)

Hominoidea
  Hylobatidae
                *Hylobates*
  Hominidae
    Ponginae
                *Pongo*
    Gorillinae
                *Gorilla*
    Homininae
      Paninin
                *Pan*
      Hominin
                *Homo*

**FIGURE 8.14 Alternative classifications of the Hominoidea:** (a) This shows the traditional classification of hominoids. (b) This shows the classification used in this text for humans and their ancestors, based on current molecular and morphological consensus. Although the traditional scheme has now been abandoned by most people, there is no clear consensus as to the preferred alternative.

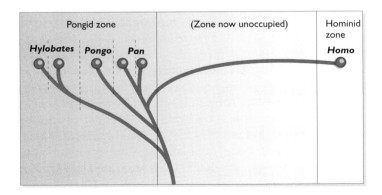

**FIGURE 8.15 Hominoid adaptations and classification:** This diagram by George Gaylord Simpson expresses his rationale for supporting the traditional hominoid classification, in which all the great apes are members of a single family, the Pongidae. During evolution, hominines shifted their adaption to a very non-apelike pattern. The "unoccupied zone" would have been filled by now extinct hominoids and hominins.

pin. The best outcome must be one that is faithful to the principle of monophyly while retaining some level of common sense concerning the fact that humans are different (see chapter 1) and some stability.

Until recently, everyone knew what was meant by a "hominid" – it included living and extinct human species. Now, some see it as including humans and African apes; to others, it means humans, African apes, and orangutans; and to still others, it signifies humans, African apes, orangutans, and Asian lesser apes. There is, however, a general consensus emerging that the term "hominid" does not denote just humans and their extinct relatives, but also the African apes at least. In turn, the term "hominin" (Homininae, subfamily level) or "hominin" (Hominini, tribe level), is now used where hominid was previously.

## EVOLUTION OF THE CATARRHINES: THE CONTEXT OF HOMININ ORIGIN

**KEY QUESTION** How does the overall primate fossil record throw light on the origin of the hominins?

The Hominoidea (apes and humans) is one of two superfamilies that constitute the infraorder Catarrhini; the second superfamily is the Cercopithecoidea (Old World monkeys). The infraorders Cartarrhini and Platyrrhini (New World monkeys) together constitute the suborder Anthropoidea, or anthropoids, often called the higher primates. This section will describe current thinking about the evolutionary history of anthropoids, and particularly the hominoids, including relationships between fossil and extant species.

Despite a decade of important discoveries, the catarrhine fossil record remains frustratingly sparse. Ironically, one impact of these discoveries has been that anthropologists' ideas about the evolutionary history of the group have become less certain. This is due to a change in the way fossils are interpreted. In the past the tendency was to use the available evidence, however tenuous, to link fossil specimens with living groups; thus *Pliopithecus* was the ancestor of the gibbons, *Proconsul major* was the ancestor of the gorilla, and so on. This view seemed reasonable, as the past species must obviously have been the precursors of the living ones. However, it has increasingly been realized that biodiversity in the past cannot simply be seen in the light of present biodiversity, and that many of the fossil species known were not ancestral to living forms, but represented separate evolutionary events. The result of this change of attitude is that there is less certainty concerning the origins of living groups such as the hominoids and cercopithecoids, but on the other hand a greater understanding of the general patterns. Today, the history of the group is more typically described as a series of adaptive radiations scattered through time, with the origin of most radiations being uncertain.

The uncertainty partly stems from the fact that the fossil discoveries highlight just how sparse the record really is. In addition, many of the new finds cannot be fitted into the inferred evolutionary patterns. More important, however, is the difficulty involved in selecting anatomical characters that may be taken as reliable indicators of evolutionary relatedness. Many similarities that were once assumed to be shared derived characters, for instance, are now suspected to be homoplasies; and others that were assumed to be derived may actually be primitive.

One general difficulty is that, although living hominoids are usually defined in relation to postcranial anatomy (adaptations to a suspensory habit), phylogenies are often based on cranial and dental characters. This trend has arisen for a very practical reason: the latter elements (particularly teeth) are more durable and therefore more common in the fossil record. In addition, they are widely assumed to be superior to postcranial characters for phylogenetic reconstruction, although this assumption is not necessarily valid. When phylogenetic analysis using one set of characters produces answers different from analysis of another set, as it often does, there is a very real – and difficult – question of which character set is truly superior.

## Some general patterns

Three key points stand out in any review of the evolution of the catarrhines. First, the fossil record of the group generally does not overlap with the geographic areas where catarrhines are most abundant today (Fig. 8.16). The early fossil record is concentrated in North Africa and Eurasia, with some specimens found in east and southern Africa. Modern Old World monkeys and apes are most abundant in the forests of sub-Saharan Africa and southeast Asia. This pattern may reflect real biogeographic changes in the history of the group, or it may partly result from a biased fossil record: forest habitats are generally poor environments for fossil preservation.[251]

**FIGURE 8.16 Features of the catarrhine fossil record:** A major lesson to be inferred from the main features of the catarrhine fossil record is that the present is not always a direct key to the past.

Second, among living catarrhines, Old World monkeys are both more abundant and more diverse than apes (Fig. 8.17). Some 15 genera and 65 species of Old World monkey exist, compared with five genera (*Pan, Gorilla, Pongo, Hylobates,* and *Homo*) and two dozen species of hominoid (a dozen of these are members of the *Hylobates* group, or gibbons). In earlier times, precisely the opposite situation prevailed, with apes being more abundant and more diverse than monkeys.[252,253]

Third, the early apes were not merely primitive versions of the species we know today.

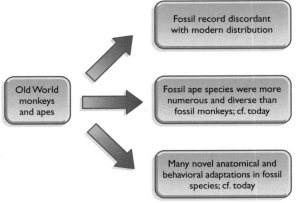

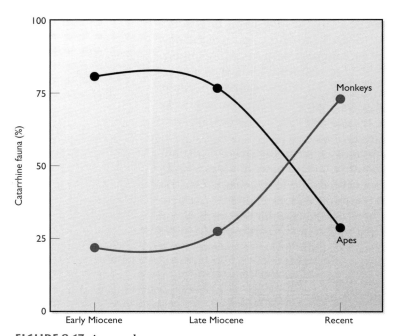

What is the pattern of
anthropoid evolution?

**FIGURE 8.17 Ape and
monkey diversity:**
Apes are relatively rare
now, and monkeys
relatively common. In
the past the situation was
the opposite of this.
(Redrawn from
P. Andrews.)

They combined various sorts of characters: some apelike, some monkeylike, and some unknown in modern large primates. In fact, most early fossil apes are apelike only in their dentition, while much of the post-cranial skeleton was monkeylike. Consequently, they are often referred to as "dental apes." Such anatomical novelties probably caused the early apes to be behaviorally distinct as well, as measured in terms of locomotor patterns and dietary activities. In addition, this variation makes it much more difficult to predict the appearance and behavior of ancestral species, including the ancestor of the human lineage.

The general pattern of anthropoid evolution is as follows: anthropoids are thought to have originated approximately 50 million years ago. Historically, on the basis of a very scant fossil record, Africa seemed the most likely source of origin, but recent finds have lent support to Asia. The earliest haplorrhines, *Bahinia* and *Amphipithecus,* are known from Asia, making it a distinct possibility that the anthropoids also arose there (see next section).[162]

However, the most abundant early fossil evidence of anthropoids is found in north Africa, at the early Oligocene sites of the Fayum Depression, Egypt, where specimens range in age from 37 million to 31 million years.[254] Some of the species found at these sites are thought to represent a time prior to the division between platyrrhines and catarrhines. The present fossil evidence strongly indicates an African origin for both the catarrhines and the hominoids, the latter occurring some 25 million years ago. Approximately 18 million years ago, hominoids dispersed into Eurasia (following the joining of the continents through continental drift) and underwent a subsequent adaptive radiation there. The later Miocene saw the extinction of many of the more archaic hominoids, shifts in biogeographical distribution, and the radiation of the cercopithecoids. Within this general pattern, which does help explain the origins of the major groups, there is little evidence directly associated with particular species of extant hominoid; indeed, in all likelihood it is only for the orangutan that we have any direct fossil evidence for ancestry. No ancestor has yet been identified for the African apes and humans, which diverged 5 million to 6 million years ago in Africa (Fig. 8.18).

# Early anthropoids

*Algeripithecus minutus*, a small primate that lived in north Africa perhaps as long as 50 million years ago (early Eocene), holds uncertain claim to be the earliest known anthropoid. It exhibits some anthropoid cranial characteristics but is otherwise rather primitive for primates. A little younger is the newly named Chinese genus *Eosimias* (dawn ape). A small creature (weighing between 2.5 and 3 ounces), *Eosimias* also possesses some dental characteristics of living hominoids but is prosimian in all other respects. Both species bear some resemblance to a possible basal anthropoid, but their status would be strengthened by more extensive fossil evidence. The presence of other Eocene anthropoid species, *Amphipithecus* and *Pondaungia*, in Burma is taken by some to imply an Asian – not African – origin of anthropoids.

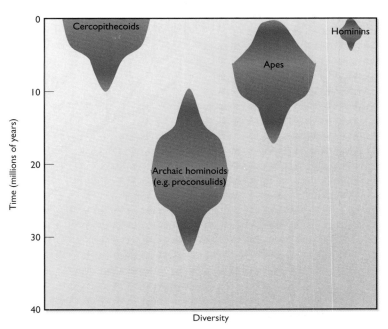

**FIGURE 8.18** The general pattern of catarrhine evolution: The fossil record shows a pattern of diversification and radiation of the catarrhines.

However, wherever their origins may lie, it is clear that once established and diversifying they are predominantly or exclusively an African lineage. The evidence for this comes from the late Eocene/early Oligocene, at the Fayum Depression, where Elwyn Simons, of the Duke University Primate Center, started working in the early 1960s. Currently one of the driest places on Earth, the region was covered with tropical forest bordering an inland sea 35 million years ago. The rich fauna and flora were typical of tropical forest and swamp ecosystems. Simons and his colleagues have recovered fossils of 11 anthropoid species, from beds dated at 37 million to 31 million years ago (Fig. 8.19).[254] The species may be assigned to two groups, the parapithecids and propliopithecids.

Parapithecids, which include *Qatrania*, *Serapia*, *Algeripithecus*, and *Apidium*, were small, marmoset-sized anthropoids that were mostly leaf-eaters. Like earlier putative anthropoids, the parapithecids exhibited a mix of anthropoid and prosimian features. They also possessed the New World monkey dental formula: two incisors, one canine, three premolars, and three molars on each side of the upper and lower jaws. (Catarrhines, by contrast, have only two premolars.) The New World dental structure may therefore have been primitive for all anthropoids. The Parapithecidae are not thought to have been ancestral to any later anthropoids.

The Propliopithecidae include *Propliopithecus*, *Catopithecus*, and *Aegyptopithecus*, the largest of the Fayum anthropoids (males weigh as

**FIGURE 8.19 Fayum primates:** The Fayum primates give one of the few glimpses into early anthropoid diversity. (Courtesy of John Fleagle.)

much as 6 kg). Tooth structure indicates that members of this group were principally fruit-eaters. There is marked sexual dimorphism of body size, implying social systems in which males competed for females in some kind of polygynous structure (see chapter 7). The 1995 announcement of 37-million-year-old cranial and dental specimens of *Catopithecus browni* makes the species the earliest known undisputed anthropoid. The origin of this group (and the parapithecids) cannot be directly linked with known earlier Eocene primates, however. Although the group cannot be definitively identified as being ancestral to later cercopithecoids and hominoids, many workers believe that the group likely played this role.

Although *Aegyptopithecus zeuxis* was once considered to be an early ape, it is now presumed to antedate the evolutionary divergence of the anthropoid stock into Old World monkeys and apes. *Aegyptopithecus*, or something like it, may therefore represent the basal catarrhine stock. Some authorities consider it possible that a species akin to *Aegyptopithecus* and its contemporaries might represent the basic anthropoid condition prior to the split between Old World and New World anthropoids. *Aegyptopithecus* was probably a generalized arboreal quadruped and the closest living behavioral analogue would be *Alouatta*, the howler monkey of the New World.[255]

## The earliest hominoids

Hominoid fossils are known throughout much of the Miocene in Africa and Eurasia, with the earliest specimens of a species of *Proconsul* (dated at approximately 22 million years) coming from Africa, the likely region of origin for the clade. Although claims have been put forth for an even earlier *Proconsul* specimen, at 26 million years, their validity cannot be established because of the absence of reliably diagnostic parts. In any case, the clade apparently originated some time between 31 million and 22 million years ago.[256]

Hominoids underwent several adaptive radiations, producing a great abundance and variety of species that followed lifestyles that would not be considered typical of modern apes. *Proconsul* itself comprised several species, including one as small as a gibbon and another the size of a female gorilla. Miocene hominoids were creatures of tropical and subtropical forests. Climate change – the result of global cooling and local tectonic activity – greatly reduced hominoid habitat through the Late Miocene in the Old World and was probably responsible for the drop in the diversity of hominoids. Cercopithecoid diversity increased in parallel with this change, and many monkey species came to occupy niches previously filled by hominoids.

The fossil hominoids as a whole can be considered in terms of two informal groups: those that clearly show hominoid traits but lack the derived and specialized characteristics that distinguish the living apes (called here "primitive apes"), and those that can at least partially be placed within the radiations of the modern apes (called here "modern apes"). Broadly speaking the former belong to the Early and Middle Miocene, and die out by the Late Miocene, while the latter appear from the Middle Miocene and continue through to the present day.

The basal hominoids, from which the primitive apes are derived, are best represented by *Proconsul*, a genus from the Early Miocene of east Africa (Fig. 8.20) that was originally discovered and described by Louis and Mary Leakey. *Proconsul* fossils have been found at several sites in Kenya, and this species is probably the best-known Miocene ape. In its cranial and dental features, *Proconsul* is judged to be primitive; the thin enamel layer on its cheek teeth apparently reflects a non-hominoid origin. The brain was relatively large, and the increased surface area of the molars and broadening of the incisors imply a commitment to a more frugivorous diet. In its postcranial skeleton, *Proconsul* displays a mix of ape and monkey features. For instance, although it had no tail (like an ape), its thorax was narrow and deep, a characteristic seen in pronograde (body horizontal to the ground) monkeylike locomotion rather than orthograde (body more vertical to the ground) apelike locomotion. "In the forelimb skeleton, the shoulder and elbow region are remarkably apelike," notes

**FIGURE 8.20**
*Proconsul africanus*: This reconstruction is based on fossils found prior to 1959 (light shading) by Mary Leakey and in 1980, among the Nairobi Museum collections, by Alan Walker and Martin Pickford (dark shading). This individual, a young female that lived approximately 18 million years ago, has characteristics of both modern monkeys (in its long trunk and arm and hand bones) and modern apes (in its shoulder, elbow, cranial, and dental characteristics). (Courtesy of Alan Walker.)

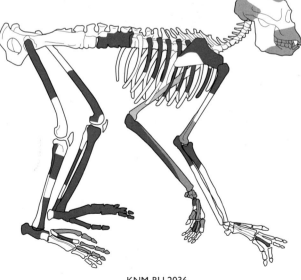

KNM-RU 2036

Alan Walker of Pennsylvania State University, "but the arm and hand bones look more like those of some monkeys. In the hindlimb the reverse is true: the foot and lower leg bones are very apelike while the hip region looks less so." This unique combination of ape and monkey characters probably conferred a unique mode of quadrupedal locomotion, but *Proconsul* would have been more like a monkey than like the forelimb-dominated ape in terms of posture and locomotion.[257]

Interestingly, the hand had a large, opposable thumb, which makes *Proconsul* more like humans than either monkeys or apes. This feature suggests that *Proconsul* might have had considerable manipulative skills, perhaps including making and using simple tools, such as stripped twigs.

Although *Proconsul* is the earliest, best-known, and probably most basal of the primitive apes, discoveries in recent years have shown that it is far from alone. Indeed, it is also important to note that within the genus *Proconsul* a number of species have been recognized. Among the other Miocene hominoids that fall into this "primitive" group are *Afropithecus, Turkanapithecus, Morotopithecus,* and *Micropithecus.* What characterizes all these specimens is that they are clearly apes in their facial and dental anatomy, and yet lack the specializations, especially of the postcranium, that characterize the living apes. The specimens are also found only in Africa, or, as in the case of *Heliopithecus,* which is known from Saudi Arabia, in the immediately adjacent parts of the world. Slightly later one would also include fossils such as *Kenyapithecus* and the recently discovered *Equatorius,* both known from east Africa, and *Otavipithecus,* discovered in Namibia.

## Later hominoids

*What is the evidence for the evolution of hominoids?*

The earliest hominoid species have (so far) been recovered from African sites, indicating an African origin of the clade. However, from around 17 million years ago (the beginning of the Middle Miocene), hominoids are discovered on other continents – Asia and Europe. These forms also, in contrast to those known from Africa, show more modern features and can be linked to the living apes to some extent (Fig. 8.21).

The best-known of these is also one of the earliest – *Sivapithecus* – with specimens found in Pakistan, India, Nepal, and possibly Turkey, dating between 13 million and 8 million years ago. *Sivapithecus,* within which has also been included the material from *Ramapithecus* that was discussed earlier, is a relatively large-bodied ape, with a large, robust face, and thick tooth enamel. Its cranial morphology is strongly reminiscent of the orangutan, which is why it has been considered to represent the first fossil evidence for the evolution of an ape lineage that is still extant.

If *Sivapithecus* is indeed ancestral to the modern orangutan, then the facial and palatal similarities in the two species are shared derived characters. Recently, questions have arisen about this ancestral relationship,

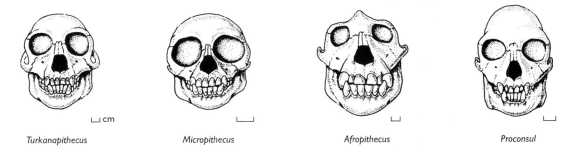

Turkanapithecus          Micropithecus          Afropithecus          Proconsul

**FIGURE 8.21 Early Miocene apes:** The range of Miocene hominoids is considerable, and they are widely dispersed in space and time. (Courtesy of John Fleagle/Academic Press.)

based partly on differences in dental and mandibular anatomy, but more particularly on certain postcranial evidence, including arm and leg bones. The postcranial evidence implies that *Sivapithecus* had a narrow, monkey-like archaic trunk. Today's orangutans and the living African apes share many similarities in their trunk anatomy. If *Sivapithecus* is a unique orangutan ancestor, then these similarities must have arisen independently (that is, as homoplasies) because the Asian fossil ape would have already diverged from the common hominoid stock. However, David Pilbeam, for example, believes it more parsimonious to view the facial and palatal similarities as homoplasies and the postcranial similarities as shared derived characters in the common ancestor of living hominoids, thus removing *Sivapithecus* from orangutan ancestry.[258]

Also known from Eurasia are a number of other apes, which, although not directly related to modern apes, nonetheless are clearly similar to *Sivapithecus*, and hence belong to the same clade. These include *Dryopithecus*, found in Europe, and one of the earliest genera that enjoyed a widespread radiation and dated from the same time period. *Dryopithecus* was the first fossil ape to be discovered, with a specimen located at a site in France in 1856. The presence of Miocene hominoids in Eurasia reflects faunal migrations (and subsequent adaptive radiations) from Africa after the continents joined through tectonic action, 18 million years ago. (Contact had been intermittent in earlier times.)

The January 1996 announcement of the discovery of an extraordinary partial skeleton of *Dryopithecus laietanus* from the site of Can Llobateres in Spain greatly increases our understanding of the species' postcranial anatomy and locomotor pattern, but it does not solve its phylogenetic affiliation. In a paper published in the journal *Nature*, Salvador Moyà-Solà and Meike Köhler describe the newly discovered postcranial material as reflecting more suspensory adaptation and orthograde posture (similar to living apes) than are seen in any Miocene ape, with the possible exception of *Oreopithecus*.[259,260] For instance, the lumbar vertebrae are proportionally shorter than in monkeys and most Miocene apes; the arms are powerful and capable of a wide range of movement; the hand is large and adapted for powerful grasping. The ratio of arm length to leg length (intermembral index) is larger than in living African apes and similar to that in the orangutan. The Spanish species is dated at 9.5 million years, indicating that the postcranial adaptations of living apes might have evolved by that date, depending on the still unsettled evolutionary relationship between *Dryopithecus* and the living apes.

Other hominoids that belong to this group include *Ouranopithecus* (Greece), *Lufengpithecus*, and *Ankarapithecus* (a Turkish species, dated at 9.8 million years). The first two lived approximately 8 million years ago. *Ouranopithecus* (also called *Graecopithecus*), which has extremely thick enamel, has been suggested by various researchers to be ancestral to hominins, African hominoids, and orangutans, reflecting the many different interpretations that are possible with the material available and the criteria employed.[261] *Ankarapithecus*, details of which were published late in 1996, exhibited a mix of gorillalike and orangutanlike features in its cranial anatomy. A very large hominoid, *Gigantopithecus*, lived in Pakistan and China between 8 million and less than half a million years ago. It had large, thickly enamelled molar teeth, stood as high as 2.7 m tall, and weighed as much as 300 kg, making it the biggest hominoid ever.

In 1984, the discovery of a maxilla (approximately 8 million years old) was reported from Samburu in northern Kenya; this specimen displays features not seen in other hominoids. So far unnamed, the Samburu fossil has been said by David Pilbeam and Peter Andrews, of the National History Museum in London, to be possibly ancestral to the gorilla. More recently, however, Pilbeam has noted that "It could be related to almost any of the African lineages, but is likely unrelated to any."

*Oreopithecus*, the first specimens of which were found in the late nineteenth century, lived approximately 8 million years ago. Its dentition represents a mix of primitive and derived characters (but not like those of living hominoids); its trunk was short and the thorax broad, with long arms and short legs. Its elbow joints resembled those of modern apes. Its evolutionary relationships are unknown. Recent analyses by Moyà-Solà and Köhler have suggested that *Oreopithecus*, although not a hominin, might nonetheless have been bipedal, indicating that this trait might have arisen more than once in the course of hominoid evolution.[262] *Oreopithecus*, known from Italy, was apparently adapted to living in very swamplike conditions, and this may have been the basis for its unique postcranial anatomy.

## Where are the monkeys?

This discussion of the Miocene fossil evidence has focused on the apes, and this might be thought to reflect a bias of interest toward the closest relatives of humans. However, this is not the case. While the fossil record shows a relatively abundant record of apes, that of the cercopithecoid monkeys is very poor. The earliest possible monkey is known from Uganda, dated to around 25 million years ago (*Victoriapithecus*). It is probable that this genus and some related forms are part of the cercopithecoid lineage prior to the divergence of the cercopithecids and colobids. The earliest unambiguous evidence comes from much later. From perhaps 15 million years ago there is evidence in both Asia and Africa for monkeys, and both the colobids and cercopithecids are present. However, what is striking is that until the terminal Miocene such finds are relatively rare, and there is little evidence for a very extensive radiation. Rather, the radiation of the cercopithecoids appears to have occurred more dramatically during the Pliocene.

It is possible that the dearth of fossil monkeys represents the bias of the fossil record, but it must also be considered that this pattern – diverse apes and scarce monkeys, the complete opposite of today's pattern – may signify some important evolutionary information. To consider this, we need to examine the overall pattern of catarrhine evolution in the context of climatic and environmental change.

## Evolutionary ecology of the catarrhines

We can characterize the evolution of the catarrhines as a series of adaptive radiations (see Fig. 8.18 above). Although, as was clear from the previous sections of this chapter, the phylogenetic relationships of these groups are very hard to determine, the general pattern seems to be relatively well established. The radiations start with the evolution of the basal catarrhines, the platyrrhines, and the primitive anthropoids of Africa. This occurred in Africa during the Oligocene, and from this group, the platyrrhines survive only from the descendants that made it to South America, the early catarrhines as hominoids and cercopithecoids, and the others are extinct. The second major radiation was of the primitive hominoids, represented by *Proconsul, Afropithecus*, etc. This was a specifically African event, which is worth considering briefly. At this time Africa was isolated from Eurasia; the landbridge, formed by the Arabian peninsula and Sinai, did not exist. Although there was some faunal exchange between Africa and Eurasia, prior to the Middle Miocene we can, in zoologist Jonathan Kingdon's terms, describe the continent as "island Africa."[263] Climatically it was a period of relative warmth, with moist equatorial rainforests widely distributed over large areas. It was a period when the opportunities to adapt to an arboreal, fruit-eating way of life were considerable, and the early, primitive hominoids represented the animals that did so.

The isolation of Africa ended in the period around 18 million years ago. As discussed in chapter 3, tectonic movements played an important part in shaping evolutionary history. At the end of the early Miocene the Tethys Sea, the name given to the area now occupied by the Mediterranean Sea, was closed off at its eastern end. Prior to that, there had been open, ocean

What ecological factors shaped catarrhine evolution?

water connecting the Atlantic and the Indian Ocean. This closure had major climatic effects, but it also resulted in the formation of a landbridge between Eurasia and Africa. Across this landbridge many lineages of animal dispersed, in both directions. Into Africa went the ancestors of the big cats and the antelopes; across into Eurasia went the hominoids.[264]

This event led to the major adaptive radiations of the Middle Miocene. *Sivapithecus*, *Gigantopithecus*, *Dryopithecus*, *Oreopithecus*, etc. can all be considered as the product of this faunal exchange, although whether they were truly a monophyletic radiation, or the result of several dispersal events, remains to be established. Eurasia offered a wider range of biomes, and the radiation reflects this, with the very large-bodied *Gigantopithecus*, the swamp-dwelling *Oreopithecus*, etc. The result was the remarkably diverse set of apes that thrived through the Middle Miocene across Eurasia.

However, this was a period of climatic change as well as tectonic change – indeed the two are related to a considerable extent; the closure of the eastern end of the Tethys Sea resulted in the end of the flow of water between the Atlantic and Indian Oceans, and the buildup of a temperature differential between them; and in southern Asia the formation of a contact between the Indian subcontinent and the mainland, leading to the uplift and mountain-building that produced the Himalayas. This new mountain range played an important role in changing the climate that led to the monsoonal system of the Indian ocean and the increased seasonality of the tropics, which had important consequences for human evolution.

This climatic change – toward colder and drier climates – had a major effect on the pattern of catarrhine evolution. The cooler temperatures led to less rainfall and a break-up of the great rainforests, with more open woodland and grassland spreading more widely. In these more demanding environments, especially in Africa, the primitive apes began to disappear, and by around 9 million years they were all extinct. In the higher latitudes of Eurasia the cooler climate also led to a contraction of range and diversity, and the extinction of many lineages. Indeed, by 2 million years ago, among the greater apes, only *Gigantopithecus* and the orangutan were still extant, and now it is only the latter that survives. The lesser apes of Asia, however, did survive, and flourished among the islands of southeast Asia, becoming what are now gibbons.

While these climatic changes of the Middle and Late Miocene were detrimental for the hominoids, for the cercopithecoids the story is very different. As we have seen, they were rare in the extreme during the earlier parts of the Miocene. During the later Miocene, however, they become increasingly common and widespread in both Asia and Africa, and all the major forms – baboons, colobines, cercopithecines, etc. – radiate. In contrast to the apes, these found more wooded and less forested conditions suitable, favoring their greater opportunism, their smaller body size and higher reproductive rates, and their greater dietary breadth. Peter Andrews has suggested that one way in which monkeys were able to survive the changing conditions of the last 15 million years, where the apes could not, was in their ability to digest coarser vegetation and less ripe fruits – a great advantage in drier and more seasonal environments.[253]

Whatever the ultimate cause, the Miocene is thus a very dynamic phase of catarrhine evolution. In terms of diversity, we see an increase in hominoid species and the disappearance of the archaic catarrhines. As the Miocene continues, that diversity reaches a peak, and then declines, such that the living apes are just a relict of former richness. At the same time, monkeys become increasingly common. Both patterns are responses to climatic and tectonic changes.

Evolutionary dynamism can be seen not just in the changing patterns of diversity, but also in the nature of the adaptations. Early catarrhines are almost exclusively arboreal, quadrupedal, and frugivorous, as well as being relatively small. Over the course of the later Cenozoic they become more and more terrestrial, more folivorous, and among the apes less quadrupedal. Evolutionary changes are reflected in adaptive changes.[146]

It is also important to note that this is the context in which hominin origins are to be found (Fig. 8.22). Hominins are part of the Hominoidea and Catarrhini, and they do not emerge in isolation, but as part of this general trend. It is perhaps relevant to observe that the evolutionary trends among the catarrhines are toward terrestriality, and that hominins are extreme terrestrial specialists (see chapter 9 on bipedalism); and that their evolution occurs in the context of increasingly open environments,

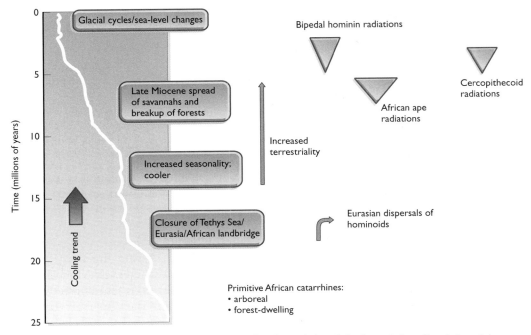

**FIGURE 8.22 Evolutionary and climatic context for the origin of the hominins:** Hominin origins occur in the context of the overall patterns of evolution among the catarrhines in general.

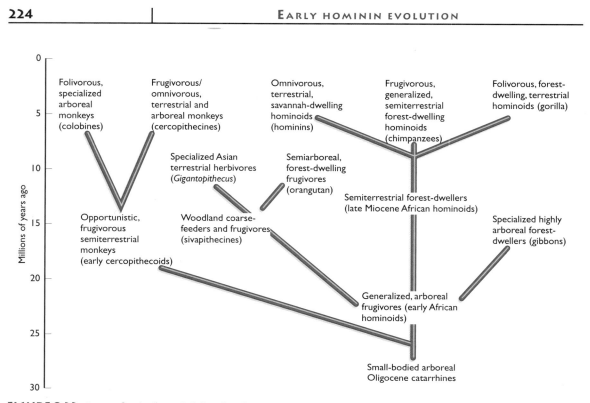

**FIGURE 8.23 An ecological model for the divergence of the catarrhines:** From a common ancestor, the catarrhines have diverged broadly, each pursuing a different solution to the environmental challenges of the Miocene. (Courtesy of Robert Foley).

and involves a change of diet (Fig. 8.23). Rather than being an aberrant part of the evolutionary story, humans are an extreme end of a general trend, responding in their own way to the pressures faced by all hominoids and cercopithecoids.[28]

## Out of Asia

There is one final question to be considered in relation to the evolution of the Hominoidea that bears directly on hominin origins. In the first part of this chapter we saw that the molecular and the morphological evidence pointed to hominins being closely related to African apes. In this chapter we have seen that the hominoids (apes) do have an African origin, and that there are apes in Africa during the Early and Middle Miocene. It has long been thought that they then evolve into the common ancestors of the living African apes and humans within Africa, while the Asian apes (orangutans) are descended from the dispersals that led to the sivapithecines.

The problem with this model is that there is little fossil evidence to support it. After the period of 10 million years there is little evidence for apes

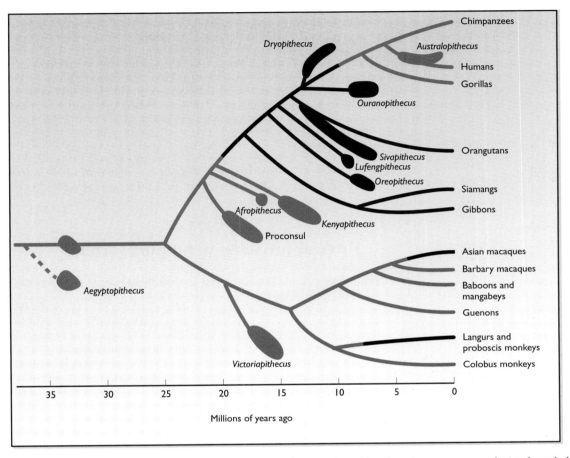

**FIGURE 8.24 Out of Asia:** Recent attempts to integrate the genetic and fossil evidence on ape evolution have led to the suggestion that the extant African apes and the common ancestor with hominins may have derived from a Miocene Asian form that had migrated back into Africa at the end of the Miocene. Dark shading indicates Asian forms; light indicates African. (Courtesy of B. Stewart and T. Disotell.)

in Africa, and for the earlier periods the apes known belong to the primitive groups. The closest affinities of the living African ape/human clade actually lie with the Eurasian apes (Fig. 8.24). This observation has led to a radical view, proposed by Beth Stewart at SUNY and Todd Disotell at New York University.[265] They have proposed that the "African apes" in fact have a Eurasian origin. Somewhere around 7 million years ago there was a further faunal interchange between Eurasia and Africa. Part of that, these researchers have claimed, involved a dispersal of some of the Asian apes (which were far more "modern" or "derived" than their African counterparts, such as *Kenyapithecus*) back into Africa, and that rather than there being evolutionary continuity from groups such as *Afropithecus* into the chimpanzee and gorilla, there was a replacement event. If this is true – and several scientists have questioned the model[266] – then it means

Did "African" apes originate in Africa?

that the common ancestor of humans and African apes was itself an Asian lineage which came into Africa, and then diversified, in response to the changing climates of Africa during the later Miocene, into the lineages that we now think of as so innately African – chimpanzees, gorillas, and humans.

If this is the case, then one hominoid fossil in particular takes on a special interest. This is *Ouranopithecus*.[261] Known from the Late Miocene of Europe, this group is the most derived of the Eurasian hominoids, and so has been seen as perhaps having a close relationship to the African apes and hominins (and lacking the specializations of *Pongo*). If this interpretation is correct, then perhaps it is evidence for the distribution of a group of later Eurasian apes on the perimeters of Africa, and thus the source lineage from which the hominins and African apes are derived.

## Three approaches to hominin origins

In this chapter we have looked at three lines of evidence that have a bearing on the first hominins. One line is the genetics of the living apes and humans, which has shown unambiguously that humans and African apes are a closely related and monophyletic group. The second line is the comparative phenotypes of humans and other primates, which show that humans, while distinctive, are nonetheless clearly great apes, and most probably more similar to African apes than the orangutan. It is also the case that the differences lie in the very large brain and the bipedal adaptations of humans. The third line of evidence is that of the fossil record and its associated climatic context, which show that the hominoids, the group to which humans belong, evolved over the course of the Miocene, principally in Africa during the earlier phases, and then more widely across Eurasia.

These lines of evidence can be used to predict where and when the first hominins should appear in the fossil record, and to some extent what they might be expected to look like. What they suggest is that the first hominins should be found at some point in the later Miocene or earlier Pliocene (that is, between 8 million and 4 million years ago), in Africa, most probably a more central part of Africa, and that they should either look like a fairly generalized modern ape or have characteristics shared with the African apes and with chimpanzees in particular. These predictions can now be tested against the fossil record, and to the place and time of hominin origins can be added insights into their adaptive character.

# BEYOND THE FACTS
# SUCCESS AND FAILURE IN EVOLUTION

*The issue: the simplest view of evolution is that it is all about reproductive success – the more offspring you leave, then the more successful you are. While this may work at the level of particular individuals and populations, it is not clear whether it is possible to talk about evolutionary success more generally and across longer spans of time.*

Within evolutionary theory there is little doubt as to what constitutes evolutionary success. According to Darwinian ideas, success is measured simply in terms of reproductive output. The most successful individual is the one that has the most surviving offspring, and the most successful species is presumably the one that is most abundant, and so on. It is this that leads to the view that bacteria are highly successful in evolutionary terms, for example.

However, when we want to consider the bigger picture in evolution, this simple reproductive accountancy approach breaks down; what is the point of comparing the reproductive output of bacteria and elephants, when they have such different intrinsic reproductive rates? Alternative measures of success seem to be required.

The problem is that once one moves away from the theoretically firm ground of Darwinian selection, then one is mired in the more subjective swamps of macroevolutionary measures. A number of alternatives do suggest themselves, but each in turn raises as many difficulties as it solves.

One possibility is longevity of the lineage. One often hears, for example, that the dinosaurs were far more evolutionarily successful than humans because they had been on the planet for hundreds of millions of years, compared to the few hundred thousand of *Homo sapiens*. However, this is not comparing like with like, for the dinosaurs consist of a whole class whereas humans are a single species. The true comparison would be either to compare mammals and dinosaurs (there's not much in it in terms of longevity, especially as mammals are still going strong), or else to compare humans with a particular dinosaur species (some are long-lived and some short-lived).

Another measure is diversity: successful lineages proliferate and form multiple species, whereas unsuccessful ones remain alone. The poor old, long-lived, unchanging coelacanth would be unsuccessful in this light, despite having being around for hundreds of millions of years. In contrast, the guenons (the cercopithecine monkeys of Africa) have speciated many times, and so in that sense are the most successful primate group.

There is no simple answer to these questions, and that is at least partly because the issues about evolutionary success depend entirely upon the question being asked. When we look at the overall pattern of catarrhine evolution during the last 30 million years we can see several different types of success. In one sense, the hominoids have been a successful group because they have persisted throughout that whole period; in another they have clearly been in decline since about 10 million years ago. The monkeys, on the other hand, are currently extremely successful in terms of diversity, but represent a radiation that has really only taken off in the last 5 million years. And humans, within the hominoid stem, are clearly highly successful in terms of reproductive success – at 6 billion we are probably three orders of magnitude more successful than any other primate species has ever been – but currently exist as only a single species. Each of these measures can be informative about the evolutionary pressures faced by lineages over very long periods of time.

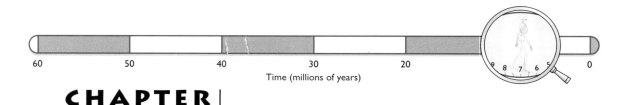

# CHAPTER 9

# SEARCHING FOR THE FIRST HOMININS

## THE EARLIEST HOMININS

**KEY QUESTION** What is the fossil evidence for the earliest representatives of the hominin fossil record?

Genetic evidence implies that the hominin clade arose between 5 million and 7 million years ago. Fossil evidence of the clade is plentiful late in its history, but becomes progressively sparser toward its origin, particularly earlier than 4 million years ago. This early hominin fossil record consists of some relatively well-established and well-known taxa, such as *Australopithecus afarensis* and *Australopithecus africanus*, as well as some material which has been known for many years from earlier times, but which has been too fragmentary to provide more than a few insights into possibilities. However, since the early 1990s there have been a number of spectacular finds which have radically transformed our understanding of the first hominins, although they still leave many doubts and uncertainties.

## The moving history of the missing link

There are few concepts of human evolution more widely known than that of the missing link. This was a term that arose in the nineteenth century to express the idea that "out there" should lie a crucial fossil that would provide the link between humans and apes, and thus the evidence to prove that Darwin's ideas about the descent of humans were essentially right. To creationists it has been an essential part of their armament, expressing the idea that such evidence is indeed missing. To evolutionary biologists it has been something of an embarrassment, as by definition if it is found it can no longer be missing! However, despite this, the idea does express in crude terms the idea of the first hominin – the earliest specimen to show features uniquely associated with humans.

Defined like this, the missing link has in fact changed continuously as the fossil record has become better and better known (Fig. 9.1).[16] When Darwin first published *The Origin of Species*, there was no recognized fossil record as such, and so the concept of a missing link was indeed quite appropriate. The first type of fossil hominin to fill this gap was the Neanderthals, known from Europe, and dis-

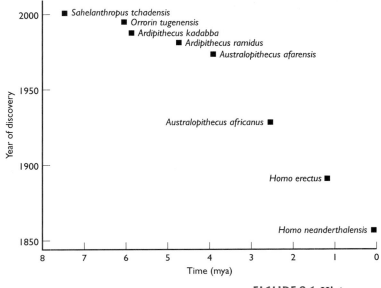

**FIGURE 9.1 History of the "missing link":** As new fossils have been discovered the nature and date of the earliest hominin have changed, especially since the early 1990s.

covered during the middle part of the nineteenth century. These are now known to have been late in the overall story, and to have been advanced compared to other species, and so the controversy over their brutishness and primitive character now seems strange.

The next fossil to step into the frame of the missing link was that of Eugene Dubois' discovery, in 1892, in Java of *Pithecanthropus erectus* – what became known as Java Man. This was certainly more "apelike" than the Neanderthals, but did nonetheless look fairly humanlike – and indeed has been shown postcranially to be very modern in overall characteristics. However, *Pithecanthropus* played an important role, not just in having the right name ("pithecanthropus" means "apeman"), but also by pushing back the chronology of human origins.[16]

The next missing link was something of a diversion. As discussed in chapter 1, the claim that fossils from Piltdown in Sussex, England, discovered during the early part of the last century, were the best evidence for the link between humans and apes led to one of the most prolonged errors in biology. The Piltdown fossils showed a pattern of a very modern, humanlike brain and an apelike jaw. This is the reverse of the pattern seen in *Pithecanthropus*, and indeed in later finds. *Eoanthropus dawsoni*, as Piltdown was known, was anatomically a perfect match for the missing link, largely or course because it was a faked hybrid between an ape jaw and a modern human skull. In the long run, new finds overturned this candidate, although it was not until the 1950s that it was actually shown to be a deliberate fake.[20,267]

In 1924 Raymond Dart found the first of the African australopithecines that have dominated early hominin research ever since. In complete contrast to Piltdown, the Taung child (*Australopithecus africanus*) had a rather

How have fossil discoveries changed views on hominin origins?

humanlike dentition and a small brain. Although this led to its hominin status being challenged, this and other finds established over the next 50 years or more the pattern of early hominin origins.[15]

Between 1924 and 1975 there were many new fossil discoveries, and many of these came to be looked on as "missing links." These include creatures such as *Ramapithecus* and *Zinjanthropus*, but while these had and still have a place in human evolution, they were either not early enough to be candidates for the earliest hominin, or their hominin status was questioned. Usually, though, it is a case of something better or earlier coming along. The series discoveries that challenged the Taung *Australopithecus africanus* child was made in the 1970s in Ethiopia and Tanzania, and the find became known as *Australopithecus afarensis*.[268] Although challenged (as usual) on first announcement, this taxon soon became accepted as the earliest known hominin, and as such the current version of the missing link.

For a considerable period of time *A. afarensis* was accepted as being the closest to the last common ancestor, and indeed was increasingly interpreted as possessing many apelike features. This situation has, however, changed rapidly since the early 1990s. First, in 1994 Tim White and his Ethiopian colleagues discovered a new genus and species of possible hominin in the Middle Awash, Ethiopia, dated to nearly 4.5 million years ago, which appeared to possess some – but very few – hominin traits. This species, *Ardipithecus ramidus*, thus became the earliest known hominin.[269] Furthermore, other finds from Kenya – *Australopithecus anamensis* – indicated not only that *A. afarensis* was not the oldest hominin, but that it was not alone at the time it was thought to have lived.[270]

Then, in 2000, a French-led team working in Kenya announced the existence of a yet older possible hominin – over 6 million years – which they have called *Orrorin tugenensis*.[271] And finally, in 2002, another French-led team pushed back the dates for the earliest potential hominin when it announced the discovery of a hominin-like creature in Chad, further to the north and west in Africa than any previous find of this nature.[251] The specimen, *Sahelanthropus tchadensis*, dated to between 6 million and 7 million years, pushes the potential first hominin back to the very limits of where and when a hominin might be expected to be found. The suite of 1990s and 2000s discoveries is beginning to throw light not just on what the "missing link" might be, but also on the context in which various apelike-homininlike forms evolved.

Thus the latest finds have pushed back the earliest claimed hominin to over 6 million years ago. It is striking that over a century and a half the "missing link" has moved from Neanderthals – now known to be around 150,000 to 50,000 years old only – to a very apelike creature over 7 million years old. It is also striking that there is now a good accord between what is "predicted" on the basis of the current distribution of apes and their phylogenetic relationships, and the actual first evidence for hominins.

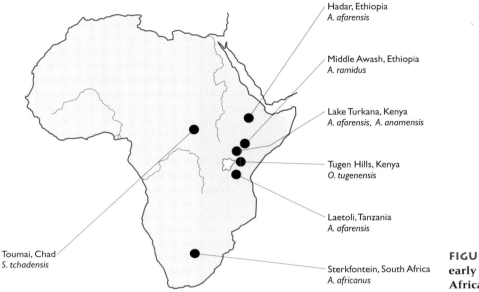

Hadar, Ethiopia
*A. afarensis*

Middle Awash, Ethiopia
*A. ramidus*

Lake Turkana, Kenya
*A. afarensis, A. anamensis*

Tugen Hills, Kenya
*O. tugenensis*

Laetoli, Tanzania
*A. afarensis*

Toumai, Chad
*S. tchadensis*

Sterkfontein, South Africa
*A. africanus*

**FIGURE 9.2** Sites of early hominins in Africa

## Major sites of early hominin fossils: south Africa

As can be seen from the history of the earliest hominins, the story has moved from Europe to Africa, and it is in Africa that human evolutionary history is principally written (Fig. 9.2). The sites of east and south Africa are thus of considerable importance. Raymond Dart, an Australian anatomist at the University of the Witwatersrand, Johannesburg, South Africa, discovered the first australopithecine in November 1924 and published his interpretation of it in the journal *Nature* in February 1925. The fossil, which had been collected by workers at a lime quarry at Taung, southwest of Johannesburg, was that of an immature apelike individual. It comprised the face, part of the cranium, the complete lower jaw, and a brain endocast, formed when sand inside the skull hardened to rock, recording the shape of the brain. An expert in neuroanatomy, Dart considered the brain to have a humanlike rather than apelike configuration; he also noted that the foramen magnum was placed centrally in the basicranium, as it is in humans, and not toward the rear, as is the case in apes (Fig. 9.3). Moreover, the canine teeth were small – a humanlike character. He concluded that the creature was a biped and was therefore a primitive form of human.[272]

Because the fossil's jaws included a complete set of deciduous teeth, with the first molar beginning to erupt, Dart stated that the specimen, often called the Taung child, died when it was approximately 7 years of age. This age was based on a humanlike pattern of development; as recent

**FIGURE 9.3 The Taung child:** Raymond Dart's discovery of a fossil at Taung, which he named *Australopithecus africanus*, set the way for a century of major discoveries in Africa. (Courtesy of Peter Kain and Richard Leakey.)

work has shown, however, australopithecines follow an apelike pattern of development, which makes the Taung child three years old at its death. In his *Nature* paper, Dart declared that the Taung child was an earlier form of human, and named it *Australopithecus africanus* (southern ape from Africa).

The anthropological community roundly rejected Dart's claim, stating that the Taung child was far too apelike to be considered part of the human family. (Another factor, not usually articulated, was that the Taung child was in the "wrong" place. At the time, Asia – not Africa – was considered to have been the birthplace of humanity.) No further discoveries have been made at the Taung quarry.

A decade later, Robert Broom, a Scottish paleontologist, joined Dart in Johannesburg and initiated new searches for early human fossils. He soon recovered a braincase of an adult *A. africanus* from another lime quarry, Sterkfontein, near Johannesburg. This find was important, because the childhood status of the Taung fossil made taxonomic interpretation more difficult. Sterkfontein became a rich source of hominin fossils, including cranial and postcranial elements (including an associated pelvis, femur, spine, and ribs). Prospecting continues there today.

As shown by the Sterkfontein specimens, *A. africanus* appeared to be apelike in having a protruding face and small brain, but had distinctly unapelike dentition, including small canines and large, flat molars. A bipedal posture was again indicated by the central position of the foramen magnum, and by the anatomy of the spine, pelvis, and femur.

While continuing his work at Sterkfontein, Broom found two new hominin-bearing cave sites near Johannesburg, Kromdraai and

Swartkrans.[16] Although similar in many ways to *A. africanus*, the hominins from these two caves were more robustly built. For instance, their jaws and teeth were larger, and a sagittal crest was located atop the cranium, providing anchor points for large muscles that powered the jaw. Broom coined a flurry of different species names for his new discoveries, although these labels were later simplified to a single name, *Australopithecus robustus*. (Some scholars consider that the more robust species was sufficiently different from the smaller, gracile species to warrant a different genus name, *Paranthropus*. The more conservative *Australopithecus* will be used here, but keep in mind that a lively debate persists over which genus name is more appropriate.)

In the late 1940s, Broom joined Dart at another cave site, Makapansgaat, found northeast of Johannesburg. The hominins found at that location were gracile, similar to the original finds at Taung and Sterkfontein. Dart developed the hypothesis that the australopithecines were aggressive, "killer apes" (because many of the crania were damaged), and that they used animal bones as weapons, an activity he termed "osteodontokeratic culture."[273] The killer-ape hypothesis has since been abandoned because the damage to the crania was discovered to have occurred after death, in the complex taphonomy of caves.[110] Dart may have been correct in suggesting that the australopithecines used animal bones as tools – not as weapons, however, but as digging sticks for unearthing nutritious bulbs and tubers. Robert Brain, of the Transvaal Museum, Pretoria, has found stick-shaped animal bones in the Swartkrans cave that carry abrasions that could have been caused by their use as digging sticks.[274] Simple pebble tools were also found at Sterkfontein in the late 1950s and later at Swartkrans.

Despite the rich haul of australopithecine fossils that (primarily) Broom and Dart recovered from the cave sites, the anthropological community remained unconvinced of australopithecines' hominin status for many years. This view changed only after the eminent British paleontologist Sir Wilfrid Le Gros Clark visited Johannesburg in January 1947, where he made a detailed study of the extensive fossil collection, which then contained elements from more than 30 individuals. A month later he wrote a paper in *Nature* stating that Dart and Broom's interpretations had been "entirely correct in all essential details." With this statement, the anthropological community recognized that early human ancestors were indeed extremely apelike and that the cradle of humankind was Africa, as Darwin had predicted 76 years earlier.

Dating the south African hominins has proved difficult, because their cave context is not appropriate for radiometric dating (Fig. 9.4). A combination of paleomagnetic dating and faunal correlation has yielded ranges of 3.5 million to 2.5 million years for the gracile australopithecines and 2 million to 1 million years for the robust species. A recent reassessment of the ecology of the australopithecines indicates, for instance, that the habitat at Makapansgaat consisted of a mixture of forest and thick bush, rather

What are the major south African hominin sites?

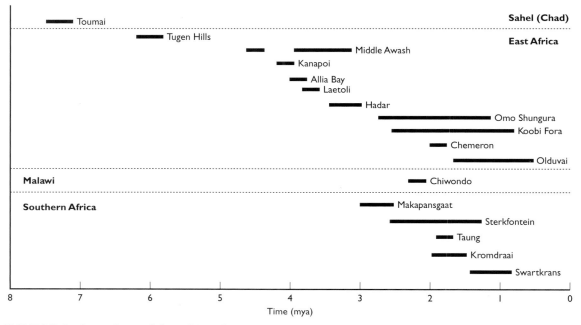

**FIGURE 9.4 Chronology of the African hominin sites:** Dating is key to understanding human evolution. Although there are still uncertainties, the broad spectrum of early hominin sites has been dated. The dates in east Africa, where potassium/argon methods can be used, are better than those in south Africa.

than the open savannah once assumed to have prevailed in the area. At the caves near to Johannesburg, the habitat was more open.

Modern paleoanthropology was therefore established in south Africa, where at least two species of australopithecine thrived early on, in coexistence with early species of *Homo*. Work continues at those sites, and some others such as Drimolen and Gladysvale, and is making important contributions to our understanding of human evolution.

## Major early hominin discoveries: east Africa

The first hominin discovery in east Africa was made in mid-1959 at Olduvai Gorge in Tanzania, when Mary Leakey found a skull (but no lower jaw) that was similar to the robust australopithecines of South Africa. The skull, which was reconstructed from hundreds of fragments, was even more robust than the southern species, however, having a large face, strong sagittal crest, and massive molar teeth. Louis Leakey had been visiting the gorge since the early 1930s, searching for hominins. He had found pebble tools and handaxes, and his wife undertook detailed analysis of them, establishing the foundation of east African paleolithic archeology. Almost three decades were to pass before Mary Leakey found what her husband initially took to be the gorge's tool maker. Because of the differences between the Olduvai hominin and those discovered in south Africa, Louis Leakey gave it the name of a new genus species, *Zinjanthropus boisei*. (Colloquially, the fossil was known simply as Zinj, or, to Mary Leakey, Dear Boy – for good reason, as it ended three decades of fruitless search.) The formal name was later changed to *Australopithecus boisei*.

The age of the Olduvai fossil was soon established as 1.75 million years via the first application of radiometric dating (potassium/argon) in anthropology.[95] It marked an important milestone in the science.

Soon after Zinj was unearthed, parts of a second hominin were found. This fossil proved much more gracile and possessed a larger brain and smaller teeth. Leakey decided that this specimen – not Zinj – had been the tool maker and that it was an early species of *Homo*, which he named *Homo habilis* in 1964.[275]

What are the major hominin sites in east Africa?

Once the hominin drought at Olduvai Gorge had been broken, discoveries of hominin fossils were frequently made in the years that followed. They included more of the robust australopithecine, more *H. habilis*, and a later species of *Homo, Homo erectus*; no unequivocal specimens of *A. africanus* were found, however. Louis Leakey died in 1972, and Mary Leakey continued working at the gorge until the mid-1980s, when she retired. Since then work has continued, and at least one important hominin fossil has been recovered, a partial skeleton of an early *Homo* (OH 62). To the south of Olduvai important work has also continued at the site of Laetoli. This covers an earlier period, stretching back to about 3.8 million years ago, and has yielded important early australopithecine fossils (*A. afarensis*) and footprint tracks.

The Leakeys' work at Olduvai Gorge helped establish east Africa as an important source of early hominins, but it was their son, Richard, who built on that foundation and made the region pre-eminent in paleoanthropology. Richard Leakey had taken part in an international expedition to the lower Omo Valley, Ethiopia, in 1967, which eventually unearthed two relatively modern skulls, many isolated teeth, and a robust but toothless partial mandible, which was given the species name *Australopithecus aethiopicus*. After leaving the Omo expedition at the end of its first year, Leakey began exploring the fossil-rich sediments on the east side of Lake Turkana, in northern Kenya.[276]

In his first full season of prospecting in 1969, Leakey found a complete, intact skull of *A. boisei*, which is known under its museum acquisition number of KNM-ER 406. This find initiated an almost uninterrupted period of discovery, which continues today under the direction of Leakey's wife, Meave. One of the most famous finds, made in 1972, was a cranium of an early species of *Homo*, known as KNM-ER 1470. Another important fossil was a cranium of *Homo ergaster*, KNM-ER 3733, which was found in the same sedimentary layer as the robust 406 skull. The extreme anatomical differences between the two contemporary skulls scuttled the single-species hypothesis.

Since the early 1980s, collections have also been made on the west side of Lake Turkana. These finds include a complete cranium of a 2.6-million-year-old robust australopithecine, which some term *Australopithecus aethiopicus*, and a partial skeleton of a *Homo erectus* youth. Stone tools have also been uncovered at many sites around Lake Turkana, including some that date to 2.5 million years – among the oldest tools known anywhere.

Because the sediments around Lake Turkana are interleaved with volcanic tuffs, the fossils of the region can now be securely dated. The collection shows the coexistence of several hominin species (*Australopithecus* and *Homo*) between 3 million and 2 million years ago,[276] but no unequivocal *A. africanus*. Many consider the latter to be an exclusively South African species, with *A. robustus* and *A. boisei* being geographical variants of the robust form. Another australopithecine species, *A. anamensis*, has been found recently on the western side of Lake Turkana. The sediments around Lake Turkana have also now yielded another fossil species dated at around 3.5 million years old, namely *Kenyanthropus platyops*.[277]

In addition to Kenya and Tanzania, Ethiopia has yielded many important fossils, particularly for the earlier period. Work was originally carried out along the sediments of the River Omo, to the north of Lake Turkana, by teams from France and the United States. The finds from the Omo have seldom been spectacular, but are important because they show a long and continuous record which can provide excellent chronological control and also evidence for paleoenvironmental and climatic community.[278]

The Hadar region, to the east, was again explored by French and American scientists, and resulted in one of the most publicized of finds, that of "Lucy," more formally a 3-million-year-old australopithecine. In addition to Lucy, the Hadar has provided a large sample of important specimens, including a set of fossils which may all belong to one community.[268]

Finally, since the early 1990s, Tim White and colleagues have been working in the Middle Awash region of Ethiopia, and have during that time uncovered some spectacular finds that have thrown light on the whole range of human evolution – from the earliest forms known as *Ardipithecus ramidus*,[269] to a later australopithecine called *Australopithecus garhi*,[279] to later representatives of Middle Pleistocene *Homo*.[280] It is clear that the whole region of the northeastern part of Africa is rich in such fossils, for they have also been recovered from Eritrea (the Buia specimen).[281]

Until very recently it appeared that hominins could only be found in east and south Africa. However, in the last decade or so this situation has changed. In the region lying between east and south Africa – and part of the Rift Valley system – fossils dated to over 2 million years have been found in Malawi. In the other direction, australopithecines and even earlier hominins are now also known to the northwest, from the area of Lake Chad. The later of the specimens have been called *Australopithecus bahrelghazali*,[282] and are around 2.3 million years old, contemporary with *Australopithecus afarensis*, while the former are currently claimed to be the oldest hominins, at over 7 million years old, *Sahelanthropus tchadensis*.[251]

No undisputed australopithecine fossil has been found outside the African continent. In fact, most scholars agree that hominins did not leave Africa until approximately 2 million years ago, when *Homo erectus* expanded its range to include Eurasia.

What is the first hominin?

## The first hominins: current contenders – *Orrorin, Ardipithecus,* and *Sahelanthropus*

As earlier chapters showed, we can expect the first hominins to appear in Africa somewhere toward the end of the Miocene, or between 8 million and 5 million years ago. This expectation is based on current biogeography, the systematics of the living apes, and molecular genetics. It is perhaps particularly satisfactory that current evidence fits these expectations very well.

While *Australopithecus afarensis* remains an important pointer to hominin origins, it has been replaced recently by other fossils which may give a more accurate insight into the first hominins. One of these significant new discoveries was announced in the year 2000. It was discovered in the Tugen Hills in central Kenya by Martin Pickford and Brigitte Senut, two anthropologists based in Paris. Although the nature of the discovery is mired in political controversy, there seems to be little doubt that what they have found is a significant addition to the hominin story. Dated to around 6 million years, the fossil has been called *Orrorin tugenensis.* The material is still relatively rare, and has only been published in provisional form (Fig. 9.5). There is relatively little cranial material, so diagnosis is based on the more intractable dentition and post-crania. The femora are the critical evidence, as it is claimed that they show bipedal adaptations. These are indicated by the angle of the femur head, a spherical femoral head that

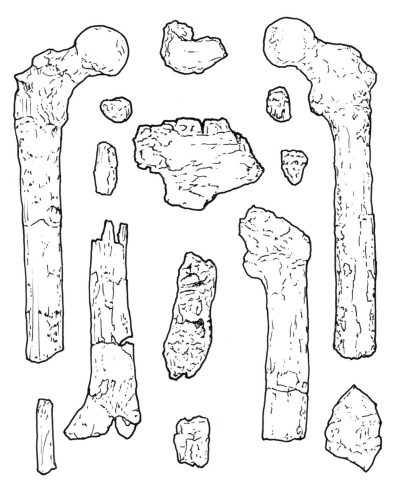

**FIGURE 9.5** *Orrorin tugenensis*: This taxon, discovered in 2000, is dated to approximately 6 million years ago, and may represent one of the earliest hominins. (Redrawn from M. Pickford and B. Senut.)

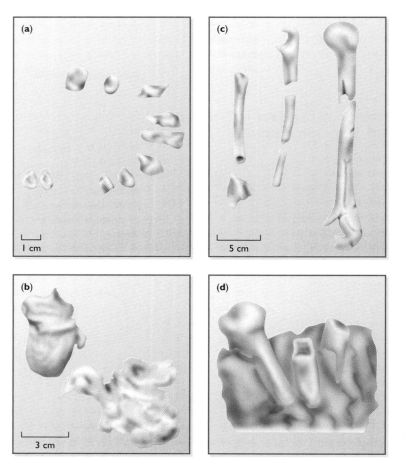

**FIGURE 9.6**

*Ardipithecus ramidus*: Discovered in 1994, and dated to around 4.4 million years ago, *Ardipithecus ramidus* represents a form very close to the common ancestor with chimpanzees. (From White et al./*Nature*, with permission.)

is rotated anteriorly, and an intertro-chanteric groove that runs from a small and moderately deep trochanteric fossa to just above the lesser trochanter. The forelimb is represented by the humerus and phalange, and these are similar to those of *A. afarensis* and of apes, thus perhaps indicating some arboreal activity and adaptation. The teeth are in contrast relatively apelike, with large canines that hone the first premolar (P4) and large incisors, but with thickened tooth enamel, a hominin characteristic.

Another contender for the earliest hominin is *Ardipithecus ramidus*. In 1994, Tim White and two colleagues, Gen Suwa and Berhane Asfaw, published details of 17 hominin fossils they found in 4.4-million-year-old deposits at Aramis, in the Middle Awash region of Ethiopia. These specimens include part of a child's mandible, some isolated teeth, a fragment of basicranium, and three bones of a left arm of a single individual (Fig. 9.6). The dentition is more primitive (that is, more apelike) than that seen in *afarensis*, with narrower molar teeth capped with thin enamel, unlike the condition in all other known hominins; the canines are larger, but not as large as in living apes. The arm exhibits both apelike and non-apelike features, from which, White and his colleagues conclude, the mode of locomotion cannot confidently be determined. Nevertheless, the position of the foramen magnum, through which the spinal cord passes in the basicranium, suggests that the creature employed some sort of bipedal posture. The mix of hominin and ape features persuaded White and his colleagues that the Aramis fossil was a very early hominin, and the group named it *Australopithecus ramidus* ("ramid" means "root" in the Afar language of the region).

In an unusual move, seven months after their initial publication in *Nature*, White and his colleagues published a note in the same journal changing the genus name to *Ardipithecus* ("ardi" means "ground floor" in the Afar language). The authors noted that *Ardipithecus* was probably the sister taxon of *Australopithecus*; in other words, both species derived from an as yet unknown common ancestor. No substantive reason was given for the change, but many observers believe that it may be based on an early analysis of a partial skeleton of *ramidus* that was discovered at the end of 1994. Preliminary indications are that it might reveal a more primitive, chimplike morphology – hence the change of genus name. In a commentary, Bernard Wood raised the possibility that *ramidus* might even be an ape, not a hominin. Nevertheless, the very earliest hominin is expected to be apelike (possibly chimplike) in many ways, particularly in dentition. Recently, earlier material has been described as a subspecies of *A. ramidus* – *A. ramidus kadabba*, close to 5.0 million years. The perspective from which early hominins should be assessed is that of Miocene apes, not later hominins.

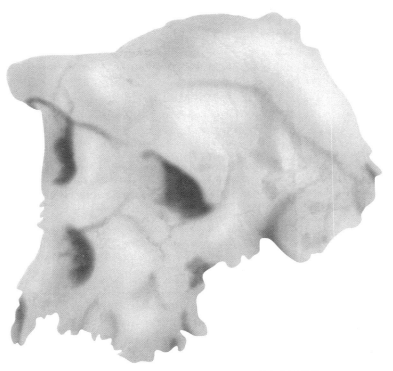

**FIGURE 9.7**
*Sahelanthropus tchadensis:* One of the recent discoveries of "the earliest hominin" comes from Chad, and extends both the possible age of the hominin lineage to between 6 million and 7 million years, and the possible geographical range to more northerly parts of Africa. (From Brunet et al./*Nature*, with permission.)

To the questions raised by the discoveries of *Ardipithecus* and *Orrorin* must be added the further complexities of the discovery announced in 2002 – that of *Sahelanthropus tchadensis* (Fig. 9.7). From Chad, French anthropologist Michel Brunet and colleagues recovered the skull of what may be the earliest hominin, dated to between 6 million and 7 million years ago. The skull has a number of features which link it to later hominins, in particular the relatively short face (not prognathic like an ape), the reduced canines, and the thick tooth enamel. There is also some indication that the creature was capable of an upright, bipedal stance. However, there are other aspects which are very apelike, notably the small, rounded cranium and brain (comparable to a chimpanzee's), and the heavy brow ridge. This mosaic led the team to the conclusion that this was on a lineage closer to the hominins and after the separation from the common ancestor of hominins and chipanzees. Although its precise phylogenetic position is unclear, it indicates that there were trends toward the hominin condition in parts of Africa during the later Miocene, and that the origins of the hominin lineage as part of an ape radiation are likely to be more complex (and perhaps earlier) than previously supposed.

Taken together, *Sahelanthropus*, *Orrorin*, and *Ardipithecus* greatly add to our knowledge of the diversity and nature of the first hominins and their close evolutionary relations. The fact that they have all been placed in separate and new genera gives some idea of how different they are from other hominins and hominoids. All of them show hominin characteristics, but the overall suite is variable – *Ardipithecus* is homininlike in having a reduced canine, but primitive in having thin enamel; *Orrorin*, on the other hand, has thick tooth enamel. Furthermore, they show considerable differences one from another. These differences do not conform to a simple chronological trend, for in many ways *Ardipithecus* is the more apelike, and yet is the youngest of the early genera. What they all share in common, which does throw light on the nature of the context in which hominins evolved, is an apelike brain case. Their faces, however, vary considerably. *Sahelanthropus*, for example, which is the earliest, has a face that is reminiscent of many later hominins.

At this stage you do not know with any precision what their relationships to each other and to later hominins are. Perhaps the key point to remember about them is that their chronology and geographical location are entirely consistent with the expectations of molecular genetics, and that they indicate the presence of at least a trend toward hominins if not the hominins themselves. At present they do not confirm unequivocally that bipedalism was a significant part of early hominin adaptation, but they do suggest that encephalization is not part of the early shift to the hominin clade.

In summary, this is a confusing but exciting time for hominin origins investigations. What we are probably getting a glimpse of, for the first time, is the earliest radiations of the African apes, which would include the gorilla, chimpanzees, and later hominins, but which, on this evidence, may also have included yet greater diversity. Although this may be phylogenetically complex, it does suggest that the trends which characterize later hominin evolution may have been more widely spread across the African apes than would be inferred from the nature of the current extant species, chimpanzees, and gorillas. The Late Miocene of Africa was clearly a place of considerable hominoid diversity.

# BIPEDALISM

**KEY QUESTION** How and why did bipedalism evolve?

If increase in brain size is not the characteristic that marks out the early evolution of the hominins, then the alternative is that it is the evolution of upright, bipedal walking. Certainly, there is tentative evidence from *Ardipithecus* and from *Orrorin* that there may have been bipedal elements to the earliest hominins, and by the time *Australopithecus* is present in excess of 4 million years ago, there is unambiguous evidence that at least some hominins were bipedal.[283] Understanding the evolution of bipedalism is a key problem in early hominin evolution.

## Bipedalism and human evolution

"Human walking is a risky business," British anthropologist John Napier once remarked. "Without split-second timing man would fall flat on his face; in fact with each step he takes, he teeters on the edge of catastrophe." Many of the anatomical adaptations – skeletal and muscular – to bipedalism in hominins function to maintain this balance so as to avoid catastrophe.

Although *Homo sapiens* is not the only primate to walk on two feet – for instance, chimpanzees and gibbons may use this form of posture in certain environmental circumstances – no other primate does so habitually or with a striding gait. The rarity of habitual bipedalism among primates, and among mammals as a whole, has given rise to the assumption that it is inefficient and therefore unlikely to evolve. As a result, anthropologists have often sought "special" – that is, essentially human – explanations for the origin of bipedalism (Fig. 9.8).

Closely associated with this view is the insidiously seductive recognition that, once an ape is bipedal, its hands become "freed" – to carry things, such as food, and to manipulate things, such as tools and weapons. So powerful is this notion that it has often been difficult to escape the assumption that bipedalism evolved in order to free the hands.

Quite apart from its anthropocentric bias in which hominin origins are often interpreted as meaning human origins (see chapter 3), the freed-hands hypothesis exemplifies the danger of explaining the origin of a current structure or function in terms of its current utility. The human brain, after all, is unlikely to have evolved so that people might write symphonies or calculate baseball batting averages. Although it is possible that upright walking evolved because of the advantages of carrying or making things with emancipated hands, other explanations must be explored as well.

Human evolution is often cast in terms of four major novelties related to the basic hominoid adaptation: upright walking, reduction of anterior teeth and enlargement of cheek teeth, elaboration of culture, and a significant increase in brain size. As the current fossil and archeological records indicate, however, these novelties arose at separate intervals throughout hominin evolution. In other words, hominins show a pattern of mosaic evolution.

As we will explore in more detail in later chapters, dental change is complex and has a pattern which occurs both very early (anterior dentition) and very late (posterior dental reduction); stone tools appear around

Why is bipedalism central to human origins research?

**FIGURE 9.8**
**Hypothesized causes of bipedalism:** Perhaps the defining characteristic of hominins, bipedalism has inevitably long been the focus of speculation as to its evolutionary cause. Some of the main ideas are shown here.

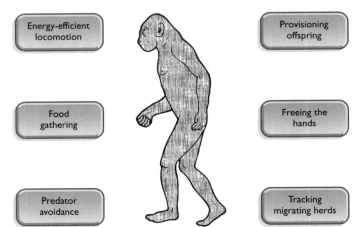

Energy-efficient locomotion

Provisioning offspring

Food gathering

Freeing the hands

Predator avoidance

Tracking migrating herds

2.5 million years ago (some 3 million to 4 million years after the molecularly estimated first hominin) and brain size only has any significant increase after 2 million years ago. Bipedalism, on the other hand, can be found in the fossil evidence, possibly of the very first hominins known, and certainly by more than 4 million years ago. If this is the case, then it can be claimed that bipedalism is the primary hominin adaptation, and that bipedal hominins existed with few other more humanlike traits.

In this chapter we will examine some of the mechanics of bipedalism, the ecological context in which it might have arisen, and the development of hypotheses that purport to account for its evolution.

## Biomechanics of bipedalism

The striding gait of human bipedalism involves the fluid flow of a series of actions – collectively, the swing phase and the stance phase – in which one leg alternates with the other. The leg in the swing phase pushes off using the power of the great toe, swings under the body in a slightly flexed position, and finally becomes extended as the foot again makes contact with the ground, first with the heel (the heel-strike). Once the heel-strike has occurred, the leg remains extended and provides support for the body – the stance phase – while the other leg goes through the swing phase, with the body continuing to move forward (Fig. 9.9).

Two key features differentiate human and chimpanzee bipedalism. First, chimpanzees are unable to extend their knee joints – to produce a straight leg – in the stance phase. Thus, muscular power must be exerted in order to support the body. Try standing with your knees slightly bent, and you'll get the idea. The human knee can be "locked" into the extended position during the stance phase, thereby minimizing the amount of muscular power needed to support the body. The constantly flexed position of the chimpanzee leg also means that no push-off and heel-strike occur in the swing phase.

Second, during each swing phase the center of gravity of the body must be shifted toward the supporting leg (otherwise one would fall over sideways). The tendency for the body to collapse toward the unsupported side is countered by contraction of the muscles (gluteal abductors) on the side of the hip that has entered the stance phase (Fig. 9.10). In humans, because of the inward-sloping angle of the thigh to the knee

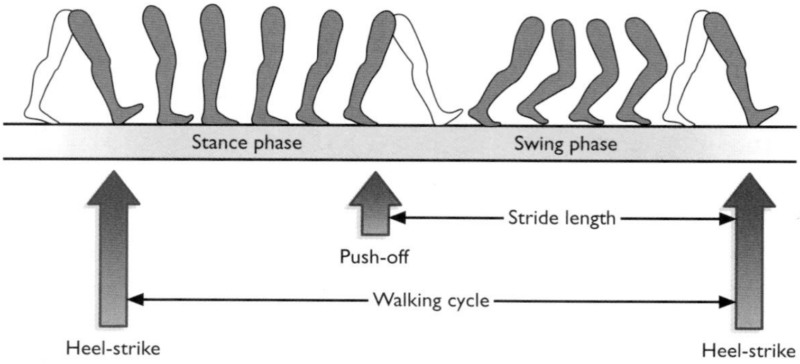

**FIGURE 9.9 Phases of bipedalism:** Upright walking in humans requires a fluid alternation between stance phase and swing phase activity for each leg. Key features are the push-off, using the great toe, at the beginning of the swing phase, and the heel-strike, at the beginning of the stance phase.

Stance phase    Swing phase

Push-off

Stride length

Walking cycle

Heel-strike    Heel-strike

(the valgus angle), the two feet at rest are normally placed very close to the midline of the body. Therefore, the body's center of gravity need not be shifted very far laterally back and forth during each phase of walking.

In chimpanzees, however, the femur (thigh bone) does not slope inward to the knee as much as in humans; thus, the feet are normally placed well apart. In addition, the chimpanzee's gluteal abductors are not highly developed. During bipedal walking the animal must therefore shift its upper body substantially from side to side with each step so as to bring the center of gravity over the weight-bearing leg. This characteristic waddling gait is exacerbated by the fact that the center of gravity in chimpanzees is higher, relative to the hip joint, than it is in humans.

Chimpanzee anatomy represents a compromise between an adaptation to tree climbing and terrestriality (mainly knuckle-walking). Modern human anatomy is a fully terrestrial adaptation, although the earliest hominins also demonstrated some arboreal adaptation. As we shall see later, these differences have implications for energetic efficiency.

Not surprisingly, the suite of anatomical adaptations that underlies human bipedalism is extensive, including the following characteristics (Figs 9.11–9.14):

♦ a curved lower spine;
♦ a shorter, broader pelvis and an angled femur, which are served by re-organized musculature;
♦ lengthened lower limbs and enlarged joint surface areas;
♦ an extensible knee joint;
♦ a platform foot in which the enlarged great toe is brought in line with the other toes; and
♦ a movement of the foramen magnum (through which the spinal cord enters the cranium) toward the center of the basicranium.

On the face of it, therefore, one might wonder how a quadrupedal ape might possibly have undergone the required evolutionary transformation to produce a fully committed biped.

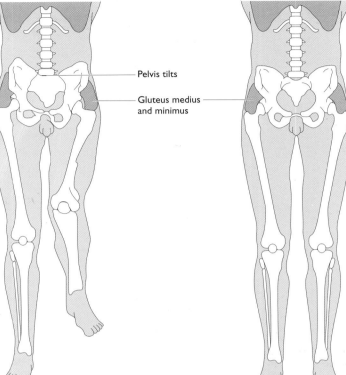

Pelvis tilts

Gluteus medius and minimus

**FIGURE 9.10 The pelvic tilt:** Gluteus medius and minimus muscles link the femur (thigh bone) with the pelvis. They contract on the side in stance phase, preventing a collapse toward the side of the unsupported limb. Nevertheless, the pelvis tilts during walking. (Courtesy of David Pilbeam.)

What are the adaptations of bipedalism?

# Ecological context of the origin of bipedalism

The nature of the evolutionary transformation that was required would, of course, have depended on the nature of the locomotor adaptation of the immediate hominin ancestor. The ancestor might have been a knuckle-walker, like the chimpanzee, or a species much more arboreally adapted. In any case, the quadrupedal to bipedal transformation is not as dramatic a shift as it might at first appear.

"All primates, with one or two possible exceptions, can sit upright, many can stand upright without any support from their arms, and some can walk upright," notes Napier. "In other words, we must view the human upright posture not so much as a unique hominin possession but as an expression of an ancient primate evolutionary trend. . . . The dominant motif of that trend has been an erect body."[156] That trend moved through vertical clinging and leaping (in prosimians), to quadrupedalism (in monkeys and apes), to brachiation (in apes). In other words, the transformation from ape to hominin did not involve the transformation of a true quadruped (such as a dog or a horse, for instance) into a true biped, a point that becomes important in calculating the evolutionary constraints that might have operated in the origin of hominins.

Until recent hominin fossil finds changed the picture, the habitat of the earliest hominins was envisioned as being relatively open woodland and even savannah. The discoveries of *Ardipithecus ramidus* and *Australopithecus anamensis* and analyses of their environmental context revealed, however, that these 4-million-year-old (or older) hominins died in wooded and perhaps even forested habitats. Although it is clear that for the most part hominins lived in environments that were not fully forested, it is also now apparent that they survived in a range of habitats with greater or lesser amounts of tree cover. The approach to bipedalism that is

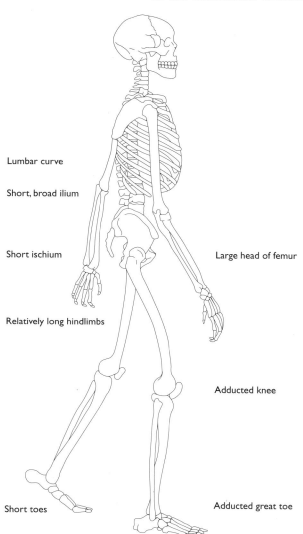

Lumbar curve

Short, broad ilium

Short ischium

Large head of femur

Relatively long hindlimbs

Adducted knee

Short toes

Adducted great toe

**FIGURE 9.11 Anatomical adaptations to bipedalism:** The principal adaptations involve a lumbar curve of the spine; a short, broad pelvis; and long hindlimbs. These characters bring the knees closer to the mid-line of the body (adduction) to form the valgus angle of the femur, and bring the great toe in line with the other toes (adduction).

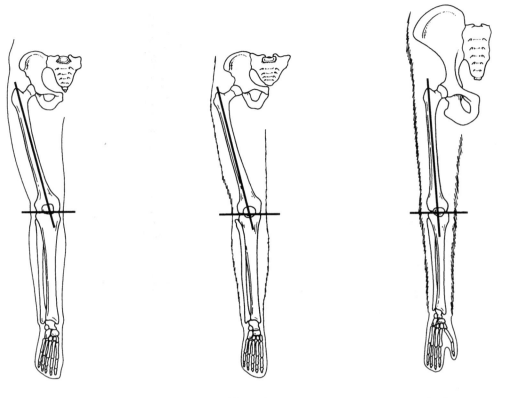

Human knee     *Afarensis* knee     Ape knee

**FIGURE 9.12 The valgus angles in humans, apes, and an early hominin:** The angle subtended by the femur at the knee, the valgus angle, is critical to bipedal locomotion. With the femur angled as in humans, the foot can be placed underneath the center of gravity while striding. An ape's femur is not angled in this way, causing the animal to "waddle" during bipedal locomotion. The valgus angle of *Australopithecus afarensis*, a 3-million-year-old (or older) hominin, is humanlike, indicating its commitment to bipedality. Also note the humanlike shape of the *afarensis* pelvis. (Courtesy of Luba Gudz.)

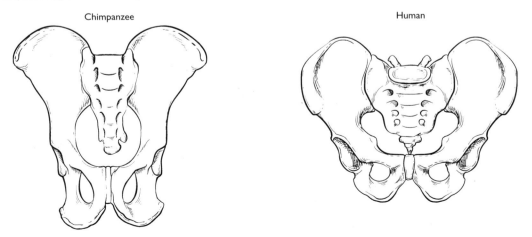

Chimpanzee           Human

**FIGURE 9.13 Pelvic anatomy:** In apes, the pelvis is long and narrow; in humans, it is short and broad.

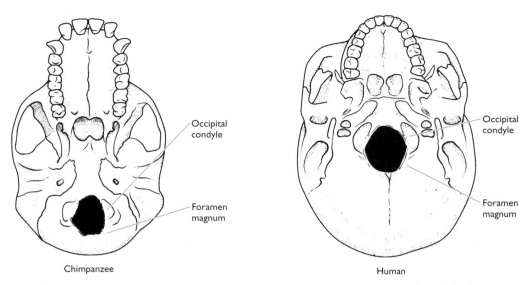

**FIGURE 9.14 The basicranium:** Because the skull is perched atop a vertical spine in a biped, the foramen magnum (through which the spinal cord enters the cranium) is located toward the center of the cranium; it is found toward the back in apes. The occipital condyles articulate with the first vertebra (atlas vertebra) of the axial spine.

required is one which considers under what circumstances it might have been advantageous for an ape (1) to stand and walk habitually on two legs, and (2) to give up its ability to brachiate efficiently when in the trees. That is what is meant by "the ecological context of bipedalism."

## Development of ideas on the origin of bipedalism

As we saw in chapter 1, Darwin essentially equated hominin origins with human origins, proposing an evolutionary package that included upright walking, material culture, modified dentition, and expanded intelligence. For Darwin – and many others who followed in his footsteps – upright walking was required for the elaboration of culture: "the hands and arms could hardly have become perfect enough to have manufactured weapons, or to have hurled stones and spears with true aim . . . so long as they were habitually used for locomotion and supporting the whole weight of the body."[19]

In the 1960s, this incipient "Man the Hunter" scenario found an added advantage in bipedalism: although humans are slower and less energy-efficient than quadrupeds when running at top speed, at a slow pace bipedalism allows for great stamina such as might be effective in tracking and killing a prey animal. Recently, with the challenging of the "Man the Hunter" image by "Man the Scavenger," it has been suggested that the endurance locomotion provided by bipedalism enabled the earliest hominins to follow in the wake of migrating herds, opportunistically scavenging the carcasses of the unfortunate young and the infirm old.

One problem arises with both these explanations: not only do stone tools that are required for wresting meat from carcasses apparently post-date hominin origins by as much as 3.5 million years, but also no indications of regular meat-eating have been found in the dentition of the earliest known hominins. In fact, evidence from microwear patterns on the surface of teeth shows that hominin diets remained predominantly vegetarian until approximately 2.5 million years ago, when cut marks are found on animal bones possibly associated with *Australopithecus garhi*.[279]

Other explanations offered for the origin of bipedalism have included the following:

♦ improved predator avoidance, as the biped would be able to see further across the "open plain" than the quadruped;
♦ increased thermoregulatory efficiency;
♦ display or warning;
♦ a shift in diet, such as seed eating or berry-picking; and
♦ carrying things.

The last explanation has been featured in two hypotheses in recent years: the "Woman the Gatherer" hypothesis, and the "Man the Provisioner" model.

The "Woman the Gatherer" hypothesis[284] marked an openly ideological challenge to the male-oriented "Man the Hunter" model.[285] Advanced initially in the early 1970s, the hypothesis shifted putative evolutionary novelties from males to females. "Plants, not meat, were major food items, and plants, not meat, were the focus for technological innovation and new social behaviors," explains Adrienne Zihlman, an anthropologist and the primary proponent of the "Woman the Gatherer" hypothesis.

The focal social ties envisaged in this scenario linked females and their offspring, with the males being rather peripheral; such a system would have been a continuum with social structures in most other large primates. The evolutionary novelty was that, living in a more open habitat than other large hominoids, hominin females had to travel substantial distances during foraging, sometimes using wooden tools to reach underground food items. They shared their food with infants, and often carried food and infants during foraging. Hence, bipedalism would have provided a selective advantage. The "Woman the Gatherer" hypothesis is more conservative than the "Man the Hunter" model, in that the first hominins are viewed as being basically apelike rather than already essentially human. Nevertheless, it focuses on the need to carry things: specifically, food for sharing within infants.

*Why did bipedalism evolve?*

Another hypothesis that focuses on the need to carry things is "Man the Provisioner," in which males gathered food and returned it to some kind of home base; there, the food was shared with females and offspring, specifically "his" female and offspring. Proposed in 1981 by Owen Lovejoy of Kent State University, this model envisages pair bonding and sexual fidelity between male/female couples, with the male providing an important part of the dietary resources.[223] Such a provisioning pattern would enable females to reproduce at shorter intervals, thus giving them a selective advantage over other large hominoids, which, says Lovejoy, were reproducing at a dangerously slow rate. The system would work only if a male could be reasonably certain that the infants he was helping to raise were his – hence the need for pair bonding and sexual fidelity.

Although it received widespread attention, Lovejoy's hypothesis was challenged. One line of criticism focused on the calculations that purported to show that large hominoids were at a reproductive disadvantage compared with humans. Another pointed out that if the first hominins were indeed monogamous, then the large degree of sexual dimorphism in body size seen in these creatures would be difficult to explain, given what is known about primate sexual systems and social structure.

Two other proposals focus on posture rather than locomotion. One, suggested by Nina Jablonski, examines typically hominoid threat displays, in which individuals stand bipedally in aggressive encounters: the taller you are, the more impressive (and therefore more likely to prevail in the conflict) you appear.[286] A second hypothesis, developed by Kevin Hunt, of Indiana University, derives from more than 600 hours of field observations of chimpanzees and their bipedal behavior, which included stationary feeding of fruits from bushes and low branches in small trees, and locomotion from one spot to another. Hunt recorded that 80% of bipedal behavior was related to stationary feeding; only 4% was observed during direct locomotion. He suggests, therefore, that the hominin bipedal adaptation was primarily a feeding adaptation; only later in hominin history did it become a specifically locomotor adaptation, he suggests.[287] One of the important elements of both Jablonski and Hunt's models is that they both suggest that selection to stand upright may have preceeded selection to move bipedally.

What all these models share is that they seek to explain bipedalism in terms of the indirect advantages it provides; in none of them, with the possible exception of the long-distance running one, do the advantages arise from what is made possible by bipedalism – with the hands freed, for example – rather than from the actual energetics of moving on two legs rather than four.

## Energetics of bipedalism: possible implication in its origin

A more direct and parsimonious – and therefore more scientifically attractive – explanation of bipedalism was proposed by Peter Rodman and Henry McHenry of the University of California at Davis in a 1980 publication.[288] Very simply, they suggest that bipedalism might have evolved not as part of a change in the nature of the diet or social structure, but merely as a result of a change in the distribution of existing dietary resources. Specifically, in the more open habitats of the Late Miocene, hominoid dietary resources became more thinly dispersed in some areas; the continued exploitation of these resources demanded a more energy-efficient mode of travel – hence the evolution of bipedalism. In this scenario, the evolution of bipedalism reflects the improved locomotor efficiency associated with foraging, and nothing else.

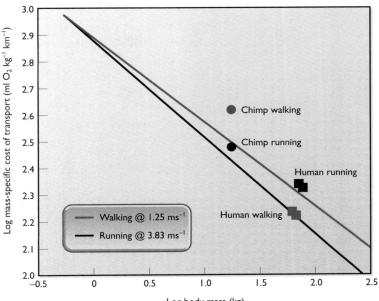

This proposal is based on a few simple points. First, although human bipedalism is less energy-efficient than conventional quadrupedalism at high speeds, it is more efficient at walking speeds (Fig. 9.15). Second, chimpanzees are roughly 50% less energy-efficient than conventional quadrupeds when walking on the ground, whether they employ knuckle-walking or move bipedally. Therefore, noted Rodman and McHenry, "there was no energetic Rubicon separating hominoid quadrupedal adaptation from hominin bipedalism."

For bipedalism to evolve among hominoids, only a selective advantage favoring improved energetic efficiency of locomotion was necessary. A more dispersed food resource could provide such a selection pressure. In other words, bipedalism was "an ape's way of living where an ape could not live," noted Rodman and McHenry. "It is not necessary to posit special reasons such as tools or carrying to explain the emergence of human bipedalism, although forelimbs free from locomotor function surely bestowed additional advantages to human walking."

Rodman and McHenry's hypothesis has recently been challenged on several counts, particularly by Karen Steudel of the University of Wisconsin.[289] She points out that this scenario implicitly assumes that the common ancestor of humans and African apes was a knuckle-walker, which many workers doubt. In addition, the postcranial skeleton of the early hominins differed from that of modern humans, specifically in including a significant degree of arboreal adaptation. The energy efficiency of bipedal locomotion in these creatures is therefore likely to have been lower than in modern humans, upon which the above energetic calculations were based. There is, concludes Steudel, "no reason to suppose that our quadrupedally adapted ancestors would have reaped energetic

**FIGURE 9.15**
**Energetics of locomotion:** The solid line represents the energy cost of running (at 3.83 meters per second) in mammals of different body size; the dotted line shows the cost of walking (at 1.25 meters per second). Note that chimpanzees are less efficient than other mammalian quadrupeds at both running and walking, while humans are less efficient at running but more efficient at walking.

Is bipedalism energetically advantageous?

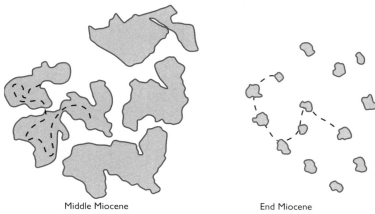
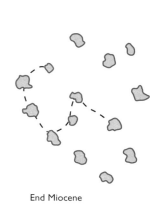

Middle Miocene          End Miocene

**FIGURE 9.16 Habitat change:** With the cooling and drying of the Late Miocene, once-continuous or near-continuous forest became fragmented in much of east Africa. Foraging for food resources would take place increasingly between patches rather than within them (as shown by the dotted line, left). This pattern of subsistence would demand significantly further daily travel (the dotted line at right), producing selection pressure for more efficient locomotion. Under such conditions, bipedalism might have evolved.

advantages when they shifted to an upright stance." Rodman and McHenry maintain that, although their hypothesis may have oversimplified the situation, it remains valid.

The energy efficiency hypothesis is accepted by some researchers. Lynne Isbell and Truman Young, of the University of California at Davis, are among these supporters and recently extended the evolutionary context to other African hominoids.[290] If, as the hypothesis argues, Miocene climate change made hominoid dietary resources less densely distributed (Fig. 9.16), then hominoids would have been forced to become more efficient in exploiting them. Isbell and Young accept that bipedalism represents one potential adaptation to this situation, which inevitably requires an increase in the daily travel distance while foraging for dispersed resources. A second strategy is to reduce the required daily travel distance, which is achieved by diminishing group size. (A large group requires more total food resources each day than a small group, and must therefore travel further to harvest it.) This strategy, argue Isbell and Young, was adopted by chimpanzees, which exhibit a fission–fusion group structure. As part of the researchers' argument, they cite field observations of gorillas and chimpanzees in Gabon, where the apes feed heavily on fruits. When these resources become scarce, gorillas maintain their group size, but switch their dietary emphasis to leaves. In contrast, chimpanzees continue to eat fruits, but forage in smaller groups or even alone. Isbell and Young's analysis is important because it puts hominin bipedalism within a general evolutionary ecology context of different behavioral adaptations by African hominoids to the same environmental circumstances.

## Thermoregulation and bipedalism

Other hypotheses for the evolution of bipedalism have recently been offered. For instance, Peter Wheeler, of Liverpool John Moores University, England, sees it as a possible thermoregulatory adaptation.[291,292] A bipedal posture reduces the area of body surface exposed to the sun while foraging (Fig. 9.17), particularly at midday; in this way, it minimizes the thermal stress with which the animal must cope through physiological cooling. Wheeler links the loss of body hair to bipedalism as a further way of reducing thermal stress, through more efficient cooling by sweating.

A 35-kg bipedal hominin, for example, would require 1.9 litres of water a day compared with 3.1 litres for a quadruped of the same weight. Moreover, the evolution of larger body size or tall stature, as occurred early in the *Homo* lineage (specifically, *Homo ergaster*, about 1.8 million years ago), increases locomotor and thermoregulatory efficiency, enabling the species to forage more widely for resources.

According to this model, therefore, it would pay in energetic terms for an ape that is living in an open environment where the solar radiation is high to be bipedal, in that it would allow foraging to continue through a larger portion of the day.

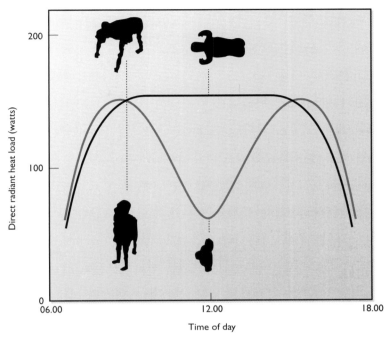

**FIGURE 9.17 A thermoregulatory model:** An upright stance means that a hominoid absorbs 60% less heat at midday than a hominoid in a quadrupedal stance.

## Time and energy:
## the ecology of a bipedal hominin

The question of the context is absolutely vital for understanding the evolution of bipedalism. A central element of evolutionary theory is that no adaptation is "absolute" – that is, perfect in all situations. Features are adaptive or not in particular contexts, and so the question is not whether bipedalism or quadrupedalism is the superior form of locomotion, but under which circumstances either might be – in other words, the costs and benefits of each.

The increasingly ecological approach that has been used in paleoanthropology in recent years provides us with a better means of looking at those circumstances. Time – that is, the number of hours in a day and how they are spent – is a critical part of that ecological approach. Robin Dunbar, of the University of Liverpool, has shown that baboon ecology is best understood by looking at the relationship between the way that an individual spends its time and the energetic costs.[293] Under more stressed ecological conditions, baoons had to spend more time feeding and travelling, and less time resting and socializing. This approach has been developed by Foley to apply to the costs and benefits of bipedalism, which are related not to general energy expenditure, but to how time is actually budgeted.[294]

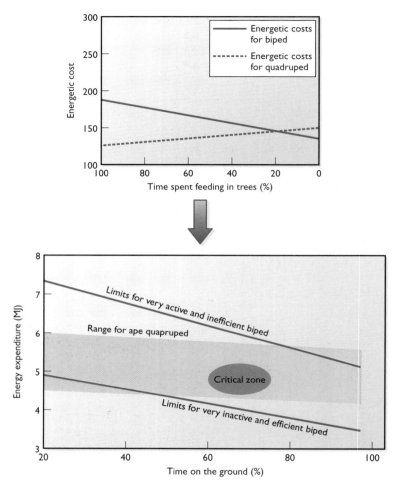

When this model was applied to the problem of bipedalism, using a simulation approach, a number of interesting insights emerged (Fig. 9.18). First, it was clear that as well as the benefits of bipedalism on the ground, it was also necessary to model the costs in relation to being less efficient in the trees, and therefore the critical element was the amount of time that had to be spent feeding in the trees. When this was modeled it emerged that as long as the "creature" (we can think of it as a proto-hominin, but of course in reality it is a set of numbers in a computer program) was spending at least 35–40% of its day feeding and traveling in the trees, then the costs of bipedalism exceeded the benefits. For bipedalism to evolve it was necessary to have a very significant component of terrestrial foraging (that is, 60%). However, another way of looking at it is that bipedalism can be advantageous – that is, the benefits exceed the costs – without a fully committed, 100% terrestrial way of life. A hominin could be spending up to 40% of its time in the trees and still benefit from being largely bipedal. This model can help make sense of the extent to which there is evidence for both bipedalism and relatively closed environments. It is the terrestrial component within

**FIGURE 9.18 Costs and benefits of climbing and walking:** The energetic costs of locomotion are an important part of why bipedalism may have evolved. While bipedalism is energetically efficient on the ground, it is more costly in the trees. For chimpanzee-style quadrupedalism the costs of arboreal movement are lower. In order to determine when bipedalism might have been beneficial, it is necessary to place the energetic costs in the context of time and daily activities. As can be seen from the upper graph, more time spent on the ground will lead to selection for bipedalism. Modeling the energetics in terms of daily time budgets suggests that the critical zone for transition to bipedalism is where about 60–70% of the activities are on the ground. One implication of this is that it would pay a hominin to be bipedal even when it is still feeding for considerable lengths of time in the trees. (Courtesy of Robert Foley.)

the environment that is more important than the presence or absence of trees per se.

The key element of the model is to consider the amount of time that would be spent feeding or traveling in the trees, for which bipedal adaptations would be a cost, versus the amount of time that would be needed to be spent on the ground, to allow for the advantages of bipedalism. Furthermore, these models fit nicely with the other observation that in environments that can be described as more seasonal – woodland, savannah, etc. – the resources are more dispersed, and so greater distances have to be traveled, and therefore there is a greater amount of time-related stress, and hence selection for energy efficiency.[295]

The problem of why and how bipedalism evolved remains an important and to some extent open one in early hominin evolution. Empirically it does seem that even if the very first hominins were not fully bipedal, and even if the australopithecines were less efficient in their bipedalism than *Homo*, nonetheless the switch to a more terrestrial and more bipedal way of life was among the most important adaptations of the early hominins. As we shall see in the next chapter, that bipedalism – hence the term "bipedal apes" – was the basis for the adaptive radiation in the Pliocene of the australopithecines, the best-known of the extinct hominin genera.

## BEYOND THE FACTS
# WHAT DOES IT MEAN TO BE A HOMININ? OR, WHAT DO YOU CALL A QUADRUPEDAL HOMININ?

*The issue: we generally define our evolutionary groups and lineages in terms of their end products – humans living today, for example. However, early in the evolution of a lineage it is not necessarily the case that all or indeed any of the features present at the end will have evolved. If this is the case, how would we recognize the first hominin?*

Imagine a physicist being asked if she would recognize a particular atomic structure, or a zoologist a type of ant. Obviously if they are any good, they would be able to make such a identification. Now ask a paleoanthropologist if he would recognize the first hominin. While most of us would like to say yes, in practice it might be far harder. The reason is that we have to have some agreement about what a hominin is, and what its characteristics might be.

In answer to the first of these questions, most people today would say that a hominin is a member of the sister clade of the lineage leading to *Pan*. Anything before the divergence of the two lineages could not be a hominin, and once the divergence has happened, then we would expect hominins to be identifiable.

But would they be? This depends on the nature of the evolutionary process involved. If the divergence is just the separation of two populations (divided by the Rift Valley, for example), and it was only much later that there was any actual change in the morphology, then a hominin would be indistinguishable in the fossil record from the last common ancestor. In other words, if the divergence did not involve any substantial evolutionary change, then a hominin would remain invisible in the fossil record for as long as it takes for drift or selection to produce novelties that we now associate with humans – and there is no reason why this might not have occurred only several million years later.

So the theoretical definition of a hominin does not necessarily help us know what the first one might look like. A more practical approach would be to say simply that a hominin can be recognized from its having hominin features. But what are they? Obviously the guide to that is to look at what humans have today – bipedalism, large brains, small faces, reduced canines, etc. But of course these may not all have evolved in the first hominin, or as a single package. This problem exercised the early anthropologists considerably, and, as we saw in chapter 1, there were those who considered bipedalism to be the first and key element, while others emphasized the expansion of the brain. Bipedalism is currently thought to be the most diagnostic and significant evidence for the presence of the hominin lineage. This is partly because the current evidence supports this position, and partly because there has been a disillusionment with brain-led models. It is also the case that bipedalism is such a complex adaptation that it has been argued that it is unlikely to have evolved more than once (that is, it cannot be homoplasic). It is worth remembering that change in

tooth size proved to be an unreliable guide for exactly this reason; hence the problems with *Ramapithecus*.

The current position is therefore that hominins are defined by bipedalism, and this is the feature to seek traces of in fossil material. But if it is not present, that would not, for the reasons outlined above, necessarily mean that the fossil is not a hominin, merely that we have yet to define what the earliest changes were. This is exactly the situation we find ourselves in with *Ardipithecus ramidus*, *Orrorin tugenensis*, and *Sahelanthropus tchadensis*. Furthermore, the new analyses of *Oreopithecus*, and the diversity of locomotor types found among the early hominins, do perhaps indicate that bipedalism may not be entirely unique to hominins, but may have evolved more than once, and possibly more than once in the hominins themselves – making the problem even more intractable.

# CHAPTER 10

# THE APELIKE HOMININS

## THE AUSTRALOPITHECINES

The recent discoveries of very early and very primitive hominins from the Pliocene and even Miocene are radically transforming our perspective on the origins of the hominins, although it is perhaps too soon to say how these finds are all going to be interpreted in the long run. However, they have already changed our view of the australopithecines. Prior to the discovery of *Ardipithecus, Orrorin*, and *Sahelanthropus*, all non-*Homo* taxa were placed in a single genus – *Australopithecus* – which was considered to be the earliest stem and radiation of the hominins. Such a position is no longer tenable, and the position of the australopithecines is currently undergoing a radical change. The purpose of this chapter is to give an overview of the australopithecines, as representatives of the best-known and most diverse early hominin genera.

> **KEY QUESTION** What is the role of the australopithecines in early hominin evolution?

## Australopithecines in outline

The term *Australopithecus* was coined by Raymond Dart when he discovered the Taung child in 1924. The type species of the genus is thus *Australopithecus africanus*, which is now considered to be one of the earlier forms, and confined to south Africa. *Australopithecus africanus* represents what have often been termed the "gracile" australopithecines, in contrast to the "robust" ones. These terms are slight misnomers, as there is not a clear size difference between them, and the robusticity, as will be described later in this chapter, is largely associated with the teeth and

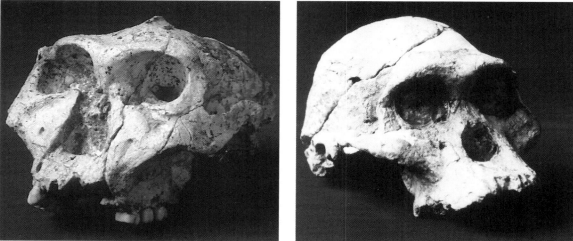

(a)                                                                                   (b)

**FIGURE 10.1 Two forms of australopithecine:** (left) The robust form of australopithecine, from Swartkrans; (right) the gracile form, from Sterkfontein. (Courtesy of Peter Kain and Richard Leakey.)

What are the australopithecines?

cranial musculature (Fig. 10.1). To some the robust australopithecines are sufficiently different to be placed in their own genus (*Paranthropus*), but for reasons that will be outlined here, a more conservative position is adopted in this book, and the "robust" forms are considered to be part of the genus *Australopithecus*.

By and large the more gracile australopithecines are earlier than the robust ones, and robusticity can be considered to be a later evolutionary trend or outcome. In the 1970s forms were discovered in Ethiopia and Tanzania which were similar to *Australopithecus africanus*, but earlier and more primitive. These were called *Australopithecus afarensis*, and are now among the best-known forms and considered to be a good model for a generalized early hominin. Other forms that have subsequently been discovered and broadly fall into this group include *Australopithecus bahrelghazali*, from Chad, and *Australopithecus garhi*, from Ethiopia. In the 1990s the earliest of the australopithecines yet discovered was announced, *Australopithecus anamensis* (Fig. 10.2).

The first of the robust australopithecines to be described was *Australopithecus robustus*, from sites in the Transvaal. Later a new and probably more robust form was discovered in east Africa, at Olduvai Gorge, and this is known as *Australopithecus boisei*. Finally, an earlier form was also discovered in Ethiopia and Kenya, and given the name *Australopithecus aethiopicus*.

Although there is some controversy about these names, with some people preferring less division of them, and others more (for example, dividing *Australopithecus robustus* into two species, the second being called *A. crassidens*), it is clear that the australopithecines, with about eight

### LAST COMMON ANCESTOR/EARLIEST HOMININS

SUPERSPECIES: *Ardipithecus ramidus, Orrorin tugenensis, Sahelanthropus tchadensis*

**General characteristics**: These taxa are generally considered to be very close to the point of divergence from the last common ancestor with *Pan*. Poorly known, but probably in the size range of a large chimpanzee (35–45 kg), they may or may not be bipedal. Overall they are probably frugivorous/omnivorous apes. Brain size and sexual dimorphism are unknown, but probably similar to *Pan*. These taxa perhaps indicate the geographical origin of Homininae in northeastern Africa.

**Variation**: The earliest hominins are highly variable (three genera), some showing indications of bipedalism, some possessing primitive traits.

### EARLY SAVANNAH BIPEDAL APES

SUPERSPECIES: *Australopithecus africanus*
SPECIES/SUBSPECIES: *A. anamensis, A. afarensis, A. africanus, A. bahrelghazali*

**General characteristics**: These are the early australopithecines. Although often described as gracile, these taxa are larger than chimpanzees, and mostly fall within the range 45–60 kg. In absolute terms brain size is between 400 and 550 cm$^3$. In relative terms, encephalization quotient (EQ – see chapter 17) is slightly above that of a chimpanzee (2.1), approximately 2.3 to 2.6. Facultatively bipedal; in general terms these australopithecines have relatively long arms, short legs, and large guts and chests, suggesting a mixed locomotion/positional behavior involving terrestrial and arboreal activities. Generally show a trend toward larger posterior teeth, with some anterior reduction. *Africanus* is often heavily megadontic. Tooth enamel is thick. On the basis of tooth morphology and wear, most of these are judged to have been frugivores, with elements of both coarser, lower-quality food and meat in the diet. Growth rates are apelike and rapid, with age of first reproduction probably similar to *Pan*. Probably highly sexually dimorphic, these species are best considered geographical and time-transgressive variants on the theme of African apes, less specialized than the later australopithecines.

**Variation**: *A. anamensis* and *A. afarensis* represent the earlier eastern forms, while *A. africanus* and *A. bahrelghazali* are slighly later southern and northwestern extensions of range and thus allopatric species. They exhibit considerable body size variation within and between species (*anamensis* (47–55 kg), *afarensis* (27–45 kg), *africanus* (30–43 kg). Posterior tooth size and wear in *africanus* overlap with those of some later australopithecines.

### LATER SAVANNAH BIPEDAL APES

SUPERSPECIES: *Australopithecus (Paranthropus) robustus*
SPECIES/SUBSPECIES: *A. robustus, A. crassidens, A. aethiopicus, A. boisei*

**General characteristics**: These are the so-called robust australopithecines or paranthropines. Their robustness is largely cranial, although they do tend to be slightly larger than the earlier forms.

**FIGURE 10.2 Summary of australopithecine and *Homo* diversity:** This table groups all the known hominins into broad ecological and chronological categories. The superspecies suggests a name if they are grouped together, but more reasonably they comprise several species.

Overall body size ranges from around 40 kg to over 80 kg, with an average around 50. There is some increase in brain size compared to other australopithecines, with an EQ between 2.2 and close to 3. They are bipedal, but still with relatively long forelimbs and shorter hindlimbs, and broad thoraxes. They have megadontic posterior dentition, with thick tooth enamel and highly reduced anterior dentition. All teeth show the effects of heavy wear and chewing, and have flat occlusal surfaces. Tooth wear and morphology indicate very coarse, small-object foods, probably high in grit and fiber, mostly plant foods, but likely to be eclectic on the basis of hominoid ancestry and including some meat. These species are highly sexually dimorphic across all taxa where known.

**Variation**: The robust australopithecines are all variants on a theme. *Boisei* is the most extreme in its megadonty, while the older *aethiopicus* possesses the smallest brain (410 cm³) and a projecting face. They may represent convergent evolutionary trends.

## EARLY INTELLIGENT AND OPPORTUNISTIC OMNIVORES

SUPERSPECIES: *Homo habilis*
SPECIES/SUBSPECIES: *H. habilis, H. rudolfensis, ? A. garhi, ? Kenyanthropus platyops*

**General characteristics**: Early *Homo* taxa show mixed features, in some ways similar to australopithecines but exhibiting larger brains and dental/facial reduction. Body size is very variable, but probably around 45–50 kg. Brain size exceeds 600 g, and the EQ estimates are close to 3.0. Early *Homo* is poorly known post-cranially, but some specimens indicate a body structure similar to australopithecines. Difficulties in taxonomic assignment make estimates of sexual dimorphism problematic, but it is likely to have been considerable. *Homo* is associated with the first stone tools, and may have been increasingly omnivorous. Growth patterns would be closer to apes than humans.

**Variation**: There are basically two forms – a smaller and more gracile australopithecine type (*habilis*), showing some brain enlargement and facial reduction, and a larger, more megadontic form with larger brain (*rudolfensis*). There is considerable doubt over their monophyly and affinities.

## LATER INTELLIGENT AND OPPORTUNISTIC OMNIVORES

SUPERSPECIES: *Homo erectus*
SPECIES/SUBSPECIES: *H. erectus, H. ergaster, H. heidelbergensis, H. antecessor*

**General characteristics**: Pleistocene *Homo* is generally larger than the Pliocene australopithecines and *Homo*, with brain sizes between 800 and 1200 cm³. EQs are greater than 3.0. Full bipedalism and linear body form are established, but with developed muscularity. Teeth are smaller, and technology extensive. Substantial hunting/meat eating may have been in place. Sexual dimorphism where known remains substantial. Growth patterns shifted toward the human condition.

**Variation**: Pleistocene *Homo* is very variable. There is a general temporal trend toward greater brain size, especially in African and European forms (*heidelbergensis*), and there is increased robusticity in the African lineage of *heidelbergensis*. The early African forms are much taller and more linear, with body mass above 50 kg. Later *erectus* are large and more robust (>60 kg) than *H. ergaster* (52–65 kg) *H. heidelbergensis* (55–80 kg) is often very large and robust (Bode, Petralona). Other differences may be local geographical ones.

**FIGURE 10.2** (*cont'd*)

**TECHNOLOGICAL COLONIZERS AND DOMINANT HERBIVORES**

SUPERSPECIES: *Homo sapiens*
SPECIES/SUBSPECIES: *H. helmei, H. neanderthalensis, H. sapiens*

**General characteristics**: These forms are generally large, sometimes greater than 60 kg, with larger cranial capacities well within the range of living humans. EQ is in excess of 5. There is full bipedalism, and a general loss of extreme cranial superstructures, with facial and dental reduction. Technology is much more complex (Mode 3, 4, and 5 – see text). Sexual dimorphism is reduced, and life-history parameters are likely to be close to or largely within the range of modern humans. Extreme habital tolerance appears to be characteristic, possibly associated with high levels of omnivory (hunting).

**Variation**: Variability in this form is quite marked across time. *Helmei* may be the early common ancestor to the later form, and so retains more primitive characters and is robust. *Neanderthalensis* is the most derived, with cold climate adaptation in terms of face size, body proportion, body mass (55–70 kg), and posture/locomotion. Early *sapiens* are large and robust, but become increasingly variable, gracile (35–70 kg), and widespread, with the most cultural and technological complexity. Sexual dimorphism is high in early forms of all taxa, but reduced in later *sapiens*.

**FIGURE 10.2** (*cont'd*)

species, are a diverse and important group. They need to be considered in terms of their adaptations (essentially bipedal in some form, but apelike in many other ways, hence the term "bipedal apes"), their evolutionary trends from "gracile" to "robust," their increasing dietary and dental specializations, and their relationships to later hominins as an adaptive radiation.

## The beginnings of *Australopithecus*

Despite the new finds of very early hominins, it is still the australopithecines that provide the best concrete evidence for the nature of the earliest hominins, and this is likely to remain the case as long as the published material on *Sahelanthropus, Ardipithecus*, and *Orrorin* remains so patchy. Although in historical terms the first australopithecine to be discovered was *A. africanus*, this has long since been supplanted by more recent discoveries, of which *A. anamensis* is now the earliest.

In August 1995, Meave Leakey (a paleontologist at the National Museums of Kenya), Alan Walker, and two colleagues published details of hominin fossils from two sites in northern Kenya, Kanapoi and Allia Bay, which they named *Australopithecus anamensis* ("anam" means "lake" in the local Turkana language).[270] The fossils (9 from Kanapoi and 12 from Allia Bay) include upper and lower jaws, cranial fragments, and the upper and lower parts of a leg bone (tibia). (In addition, the collection includes

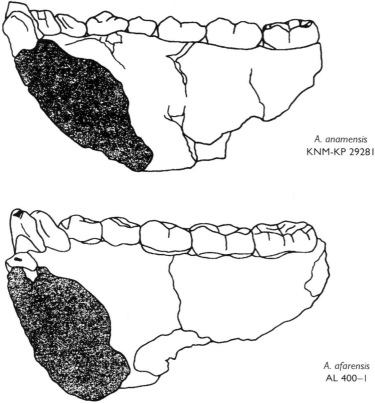

A. anamensis
KNM-KP 29281

A. afarensis
AL 400-1

**FIGURE 10.3**
*Australopithecus anamensis*: This is a recently discovered hominin from west Turkana, dated to over 4 million years old, and may represent the earliest australopithecine. (From Ward et al./*Evolutionary Anthropology*.)

a fragment of humerus that was found 30 years ago at the same site at Kanapoi.) The dentition is less apelike than in *ramidus*, having thick enamel on the molar teeth but relatively large canines. The tibia implies that *anamensis* was larger than *ramidus* and *afarensis*, with an estimated weight of 46 to 55 kilograms; its humanlike anatomy implies that *anamensis* was bipedal in posture and locomotion. Although distinct from *afarensis*, its discoverers claim that *anamensis* resembles the Laetoli fossils more than those found in the Hadar (see next section). The Kanapoi fossils have been dated at 4.2 million years and those at Allia Bay at 3.9 million years (Fig. 10.3).

The interpretation of the evolutionary relationships among these early hominins remains uncertain, but is focused principally on *ramidus* and *anamensis*. Some scholars, such as Tim White, of the University of California, Berkeley, have suggested an ancestor–descendant relationship, with *ramidus* being ancestral to *anamensis*, and *anamensis* being ancestral to *afarensis*.[269] White points out that the time range for the known fossils – 4.4 million years for *ramidus*, 4.2 million to 3.9 million years for *anamensis*, and 3.9 million to 2.9 million years for *afarensis* – is consistent with such a proposed lineage. Given that specimens of the new species remain rare, the time ranges are undoubtedly underestimates; extending them even modestly would very quickly cause an overlap that would invalidate the simple time-succession argument. In their paper announcing the discovery of *anamensis*, Leakey and Walker stated that the species might be ancestral to *afarensis*, but conceded the possibility of several species coexisting at this early period in hominin history, making firm phylogenetic reconstruction premature at this stage.[296]

The discovery of *bahrelghazali* further complicates the picture. Michel Brunet (of the University of Poitiers, France), David Pilbeam (of Harvard University), and their colleagues note that, because of differences between the newly named species and the recently discovered *Australopithecus anamensis* and *Ardipithecus ramidus*, *A. bahrelghazali* probably belongs to a clade that was separate from at least 4 million years ago and possibly longer.[282] Because it is more gracile than other hominins of the time, the authors say, this species may be related to the ancestry of *Homo*. If this is correct, a phylogeny of hominins that entails *afarensis* being ancestral to all later hominins is likely to oversimplify hominin evolution.

## The *Australopithecus afarensis* story

The Ethiopian hominin fossils were first found in the mid-1970s in the Hadar region of that country, by an international team led by Donald Johanson, of the Institute of Human Origins (IHO), Berkeley, and Maurice Taieb, a French paleontologist. The many hundreds of fossils recovered included mostly cranial and dental specimens (but no complete cranium) and postcranial elements. The most spectacular of these finds was the partial skeleton named Lucy (Fig. 10.4); in addition, remains of 13 individuals were found at a single site and were subsequently dubbed the First Family. It was clear from the start that some of the hominins were small while others were large.[268]

Work continued in the region until the early 1980s, but was then suspended for almost a decade. Recent prospecting in Hadar by Johanson and his colleagues at IHO, and in the nearby Middle Awash region by Tim White, has yielded many more fossil specimens, including the first complete cranium, details of which were published in 1993 and 1994.[297]

The Tanzanian fossils, which included a dozen complete or fragmented jaws, were found in the mid-1970s at the site of Laetoli, not far from the famous Olduvai Gorge, principally by Mary Leakey and Tim White. Unlike the Hadar hominins, the specimens found at Laetoli were mostly large. In addition to the fossils, a 30-yard-long humanlike footprint trail of three bipedal individuals was unearthed, which had been made (presumably by the Laetoli hominins) some 3.6 million years ago in a newly deposited layer of volcanic ash. The trail provides one of the more haunting relics of human prehistory, recording a few moments in the lives of three individuals, one of whom stopped briefly, turned to look eastward (possibly at the still-erupting distant volcano), and

**FIGURE 10.4 Skeleton of "Lucy":** This 40% complete skeleton, shown with her discoverer, Donald Johanson, in 1975, is one of the smallest specimens of *Australopithecus afarensis*. Her anatomy combines ape and human characteristics. Obviously adapted for considerable bipedalism, Lucy nevertheless had somewhat apelike limb proportions (short legs and long arms), and an apelike cranium. (Courtesy of the Cleveland Museum of Natural History.)

**FIGURE 10.5 Ancient trails:** The photograph shows a detail of hominin footprints at Laetoli, illustrating their very humanlike shape. (Courtesy of Andrew Hill.)

then continued onward (Fig. 10.5). The ash layer contains the prints of many other creatures as well, recording in graphic form something of the ecology of which the hominins were a part. No further discoveries have been made at Laetoli since the late 1970s.

The initial interpretations of the Hadar and Laetoli fossils were somewhat diverse. For instance, in their 1976 paper on the Laetoli hominins, Leakey, White, and their colleagues noted a "phylogenetic affinity to the genus *Homo*."[298] In the same year, Johanson and French anthropologist Maurice Taieb reported that the Hadar collection might include three hominin species, two australopithecines and one *Homo*.[299] After a collaborative analysis, however, Johanson and White concluded that the Hadar and Laetoli fossils represented a single hominin species, which they named *Australopithecus afarensis* in 1978.[300] It was the first major hominin species to be named since 1964, when *Homo habilis* was announced. As with *H. habilis*, the naming of *A. afarensis* generated considerable controversy among paleoanthropologists.

Johanson and White described *A. afarensis* as being much more primitive than other known hominins, with a strongly apelike appearance above the neck and a strongly humanlike form below the neck; as having extreme sexual dimorphism in body size (males larger than females); and as being ancestral to all later hominins. This position was challenged by some authorities, most notably by Richard Leakey and Alan Walker and by Yves Coppens and his colleagues in Paris. These critics argued that both the extreme size variation among the specimens and the many differences in detailed anatomy implied the existence of at least two, and perhaps more, species in the collections.

In recent years considerable effort has been made to calculate body weights of early hominins, with Henry McHenry, at the University of California, Davis, being a leader in this field.[301] His calculations give an average body weight of 30 kilograms for small (female?) *A. afarensis* individuals, and 45 kilograms for the larger (male?) individuals, with average heights of 105 and 151 centimeters, respectively. By comparison, average female/male body weights in the modern gorilla are 75 and 158 kilograms; for chimpanzees, 40 and 54 kilograms; and for *Homo sapiens*, 54 and 65 kilograms. This degree of putative sexual dimorphism in *A. afarensis* is consistent with a social structure that includes large, kin-related, multimale groups and unrelated females.

Although many anthropologists now accept the unity of the *afarensis* fossils as a single species, questions continue to emerge. For instance, a 1995 study of size range in the *afarensis* fossils by William Jungers and Brian Richmond, of the State University of New York (SUNY), Stony Brook, failed to give an unequivocal answer to the question of sexual dimorphism versus multiple species (Fig. 10.6).[302] Jungers and Richmond compared the inferred size range of various body parts in *afarensis* (as presently known) with that in chimpanzees, gorillas, and orangutans. They concluded that this range (for mandibles, upper arm bones, and thigh bones) is "either rare or not observed in human and chimpanzee samples, is rare to uncommon in orang-utans, and is unusual even in gorillas." The researchers concluded that this finding is consistent with either extreme sexual dimorphism in a single species or the presence of two species, noting that the evidence "does not permit a clear choice between these two alternatives." They point out, however, that the known size range in *afarensis* is likely to be smaller than the actual range because the fossils represent a sampling of the Pliocene population. For statistical reasons, that sampling will fall in the middle of the range, but will not indicate its full extent. If *afarensis* is indeed a single species, then the degree of sexual dimorphism was at least equal to the most extreme seen in living hominoids, and probably more so. Such dimorphism is inconsistent with a monogamous social structure, which has been proposed by some as fundamental to hominin social structure.

What does *Australopithecus afarensis* tell us about early hominin evolution?

Another 1995 study was less equivocal about the single-species status of *afarensis*. Peter Schmid and Martin Häusler, of the Anthropological Institute, Zurich, examined the pelvis of Lucy, who is a mature adult.[303] Using the extreme sexual dimorphism model, Lucy is identified as a female, as has been assumed (and as her name implies). Of 13 characters of the pelvis studied by Schmid and Häusler, however, only two were associated with females and the rest were considered male characters. They therefore concluded that Lucy is actually a male. If true, the extreme sexual dimorphism model is invalid, because Lucy is among the smallest of the Hadar hominins and a normal male would be much larger. According to Schmid and Häusler, the Hadar fossils are those of two species, not one. Owen Lovejoy, of Kent State University, who conducted the original reconstruction and study of Lucy's pelvis, rejects the Swiss researchers' analysis as flawed.

Finally, an analysis of the blood drainage pattern in the crania of hominins has also been taken to indicate the existence of multiple species prior to 3 million years ago.[304]

Countering these conclusions, both Johanson and his colleagues at IHO and White and his colleagues interpret the size diversity and anatomical uniformity of the fossils recently found at new sites in the Hadar and Middle Awash as "powerful support for the interpretation of *A. afarensis* as a single, ecologically diverse, sexually dimorphic, bipedal Pliocene primate species whose known range encompassed Ethiopia and Tanzania."

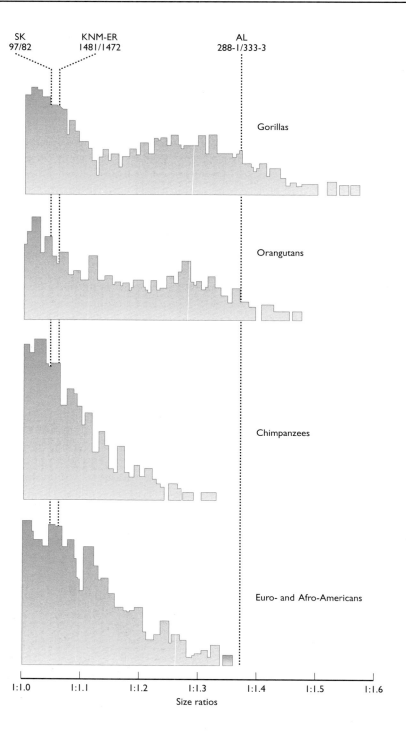

**FIGURE 10.6 Degrees of sexual dimorphism:** Richmond and Jungers (see text) compared the size ranges of various parts of the body in gorillas, orangutans, chimpanzees, and humans. These histograms show this measure for the size and shape of the top part of the femur among pairs of large and small individuals in populations. A comparison of a small *afarensis* (AL 288-1) and a large one (AL 333-3) gives a size ratio that lies at the extreme end of dimorphism in living hominoids. By contrast, similar pair-comparisons in early *Homo* (KNM-ER 1481 and 1472) and in robust australopithecines (SK 97 and 82) fall well within the range of living hominoids. (Adapted from Richmond and Jungers.)

# Anatomy of *Australopithecus afarensis*

Superficially, *A. afarensis* does appear to be essentially apelike above the neck and essentially humanlike below the neck. This form provides a good example of mosaic evolution, in which different parts of the body change at different rates and at different times. In fact, mosaicism is even more pervasive and detailed in this species because, throughout its postcranial skeleton, anatomy associated with bipedal locomotion has developed to different degrees in different places. One source of debate over this species concerns the interpretation of the various primitive aspects of the post-cranial anatomy: do they imply that, like most hominoids, *A. afarensis* spent a significant amount of time in the trees? Or were these primitive aspects of the anatomy simply genetic holdovers from an earlier adaptation, having no particular behavioral significance in *A. afarensis*? While individuals moved about on the ground, was their bipedalism significantly different from or essentially the same as that in modern humans?

In examining the biology of *A. afarensis* in more detail, we will look first at the cranial and dental anatomy and then return to functional inter-pretations of the postcranial skeleton.

The cranial capacity of *A. afarensis* ranges between 380 and 450 cm$^3$, or not much bigger than the 300 to 400 cm$^3$ range found in chimpanzees. The cranium itself is long, low, and distinctly similar to that of an ape, hav-ing a pronounced ridge (the nuchal crest) at the back of which were attached powerful neck muscles that balanced the head; the larger indi-viduals (males?) have a sagittal crest (a raised bony crest from front to back of the cranium). As in apes, the upper part of the *A. afarensis* face is small, while the lower part is large and protruding. The projecting (prognathous) lower face partly explains why powerful neck muscles are required to balance the head atop the ver-tebral column: in physical terms, this structure is a matter of moments (Fig. 10.7).

Many details of the under-side of the *A. afarensis* cranium (the basicranium) signify its hominin status, including the central positioning of the fora-men magnum, through which the spinal cord passes. The hominin status of *A. afarensis* is even more clearly seen in the jaws and teeth, however.

A comparison of a modern ape's dentition (the dentition of a chimpanzee, for example)

**FIGURE 10.7**
**Reconstruction of an**
***afarensis* cranium:**
The apelike features of *Australopithecus afarensis* are particularly evident in this cranium, which was constructed from fragments of several different crania. A relatively complete cranium was discovered in 1993, showing anatomy very much like this one. The increased robusticity in the jaws, the slightly enlarged cheek teeth, and reduced canines provide major clues to its hominin status. (Courtesy of the Cleveland Museum of Natural History.)

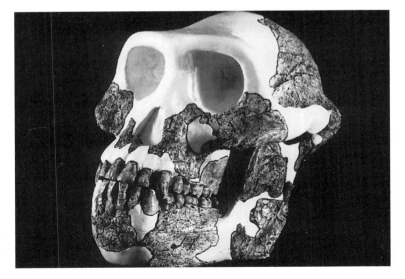

with that of modern humans reveals some striking differences. In most respects, *A. afarensis* is somewhat intermediate between these two patterns (Fig. 10.8). Although reduced, the canines are still large for the typical hominin, and significant sexual dimorphism is present; a diastema is required to accommodate each canine in the opposite jaw. In many individuals, the first premolar is distinctly apelike in having a single cusp, but the development of a second cusp can sometimes be discerned. Although the molars are characteristically hominin in overall pattern, they do not resemble the grinding millstones that are apparent later in the hominin lineage.

Clearly, an adaptive shift occurred with *A. afarensis*, one that looks dramatic by comparison with modern apes. The proper comparison, however, must involve some of the Miocene apes from which the hominin clade might have derived. From this viewpoint, *A. afarensis* really does exhibit an apelike appearance, in terms of both anatomy and diet. It would have been different from a behavioral standpoint, however, because *A. afarensis* habitually walked on two legs. For an insight into this novel adaptation, we move to the postcranial skeleton.

As we saw in chapter 9, the bipedal adaptation imprints itself in many different ways on the postcranial skeleton. The question is: how well does *A. afarensis* measure up as a biped? Functional analyses of various parts of the postcranial skeleton have been carried out by a large number of researchers, working in the United States, England, and France.

Owen Lovejoy collaborated with Donald Johanson and his colleagues to concentrate on the pelvis and lower limbs. The pelvis of *A. afarensis* is undoubtedly more like that of a hominin than that of an ape, being squatter and broader, but significant differences exist as well, such as the angle of the iliac blades (hip bones). These differences were not functionally significant in terms of achieving the balance required for bipedal locomotion, concluded Lovejoy.[305] Combined with the architecture of the femoral neck and the pronounced valgus angle of the knee, this character would permit a full, striding gait, essentially like that of modern humans in overall pattern if not in every detail. In other words, *A. afarensis* was said to be a fully committed terrestrial biped, with any apelike anatomy being genetic baggage and not functionally significant.

Meanwhile, other researchers began to see indications of arboreal adaptation in the *A. afarensis* anatomy. French researchers Christine Tardieu and Brigitte Senut studied the lower limb and upper limb, respectively, and inferred a degree of mobility that would be consistent with arboreality.[306] Russell Tuttle, of the University of Chicago, pointed out that the bones of the hands and feet were curved like those of an ape, which could be taken as indicative of climbing activity (Fig. 10.9). William Jungers reported that although *A. afarensis* arms are hominin in terms of length, its legs remains short, like those of an ape, which favors a climbing adaptation.[307] Examining certain *A. afarensis* wrist bones, Henry McHenry, of the University of California, Davis, concluded that the joint would have

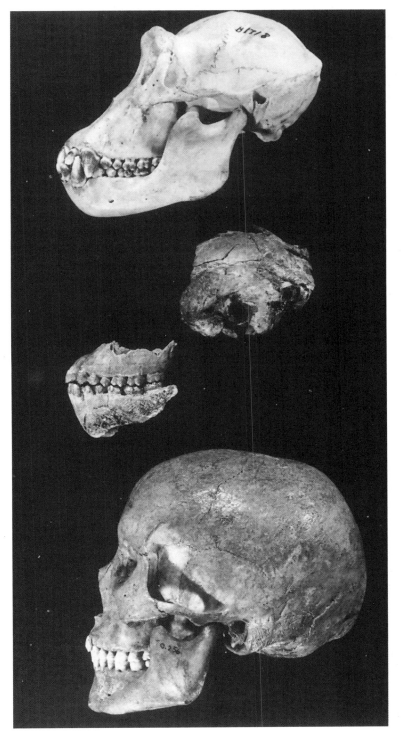

**FIGURE 10.8 Crania compared:** These profiles of chimpanzee, *afarensis*, and human crania show how very apelike these early hominins were. (Courtesy of the Cleveland Museum of Natural History.)

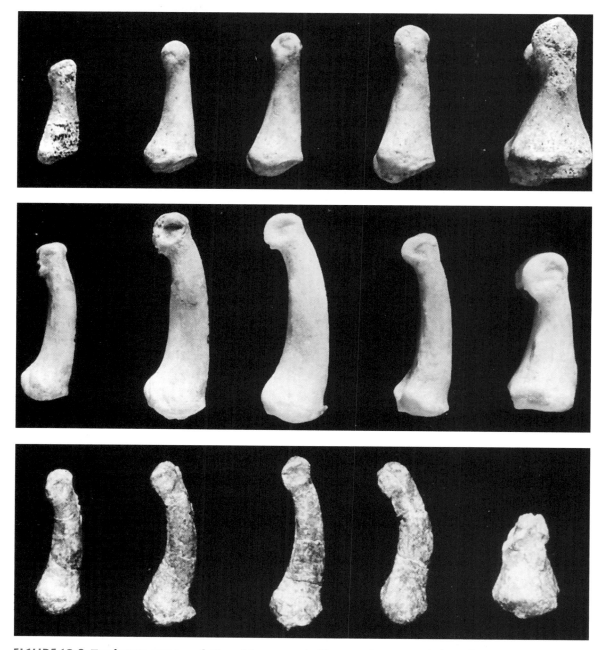

**FIGURE 10.9  Toe bones compared:** One of the anatomical features that some scholars take as an indication that *afarensis* was partly arboreal is the curved form of its toe bones. Upper row: modern humans; middle row: chimpanzees; bottom row: *A. afarensis*. (Courtesy of the Cleveland Museum of Natural History.)

been much more mobile than in modern humans, a character consistent with an arboreality.[308]

Following a more wide-ranging survey, Jungers, Jack Stern, and Randall Susman (all of SUNY, Stony Brook) argued that the full suite of postcranial anatomical adaptations indicated that, although *A. afarensis* was bipedal while on the ground, it spent a significant amount of time climbing trees, for sleeping, escaping predators, and foraging (Fig. 10.10).[309] Moreover,

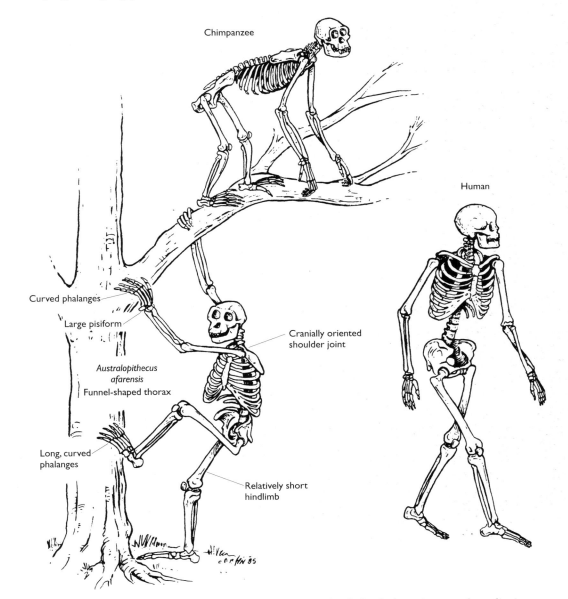

Chimpanzee

Human

Curved phalanges

Large pisiform

Cranially oriented shoulder joint

*Australopithecus afarensis*
Funnel-shaped thorax

Long, curved phalanges

Relatively short hindlimb

**FIGURE 10.10 Skeletons compared:** This diagram illustrates the skeletal adaptations to arboreality in *Australopithecus afarensis*. (Courtesy of John Fleagle/Academic Press.)

they concluded, while the animal was moving on the ground it could not achieve a full striding gait, as Lovejoy had argued, but instead adopted a bent-hip, bent-knee posture. Such a posture would clearly have important biomechanical and energetic implications for *A. afarensis*.

The differences of opinion in the *A. afarensis* locomotor debate stem partly from a lack of agreement over how to define the anatomy in certain instances and partly from differences in functional interpretation of other aspects of the anatomy. Over the years, however, the partially arboreal, bent-hip, bent-knee, bipedal locomotor posture, has been favored by including a 1995 study by Maurice Abitol, of the Jamaica Hospital, New York.

The key anatomical features cited in support of a partially arboreal adaptation include the following:
♦ curved hand and foot bones;
♦ great mobility in the wrist and ankle;
♦ a shoulder joint (the glenoid fossa) that is oriented toward the head more than in humans; and
♦ short hindlimbs.
Opponents of arboreal adaptation dispute the degree of mobility in the *A. afarensis* ankle, and cite the loss of the opposable great toe, which has become aligned with the other toes, a clear adaptation to bipedality (but see the discussion below).

Anatomical features that might imply a less than human style of bipedality are found in several parts of the body. For instance, although the forelimbs have assumed hominin proportions, thus improving weight distribution and balance required for bipedalism, the legs are short, as in an ape. Short legs mean short stride length. In addition, the foot is long relative to the leg, meaning that clearance could be achieved only by increasing knee flexion during walking (like trying to walk in oversized shoes).

The SUNY researchers and Abitol independently interpret the angle of the iliac blade of the pelvis in *A. afarensis* to imply a method of balance during bipedalism more like that of a chimpanzee than a human – that is, involving a bent hip. The SUNY group also claims that the lunate articular surface of the socket (the acetabulum) into which the head of the femur fits in the pelvis is less complete in *A. afarensis* than in modern humans. This incompleteness arises in a region that takes stress in humans when the fully extended hindlimb passes beneath the hip joint. Ergo, this kind of stride does not occur in *A. afarensis*.

Completing the case for a bent-hip, bent-knee walking posture is the suggestion by the SUNY researchers that the *A. afarensis* knee joint cannot lock in a fully flexed position, as it does in modern humans. The Kent State researchers dispute three points of this description of the anatomy, ultimately rejecting the functional interpretation. The shape of the joint surfaces of certain bones in the foot (the metatarsals) can be taken to imply a greater ability for flexion, which would be useful for climbing, and a poorly developed stability when in a push-off position. If *A. afarensis* did employ a bent-hip, bent-knee posture, then it would not have used the push-off step to the degree that occurs in the modern human striding gait.

Finally, Jungers has examined the size of hindlimb joints – particularly the femoral head – in modern apes, humans, and *A. afarensis*. The rationale was that distributing body weight on four limbs for most of the time – as chimpanzees and gorillas do, for instance – would not require the joint surfaces of the lower limbs to be as extensive, relatively speaking, as they must be if full weight was permanently balanced on the hindlimbs, as occurs in humans. Sure enough, humans have much larger femoral head surfaces than would an African ape of the same size. Although the femoral head surface in *A. afarensis* is larger than that of an ape of the same size, it does not even approach the human range. This finding leads Jungers to conclude that "the adaptation to terrestrial bipedalism in early hominins was far from complete and not functionally equivalent to the modern human condition."

*Was Australopithecus afarensis a biped?*

Another aspect of the postcranial anatomy worth noting in relation to the biology of *A. afarensis* is the structure of the hands. Although they have often been characterized as "surprisingly modern," they are actually rather apelike in manipulative capacity and overall curvature. For instance, the thumb is shorter than in the human hand, and the fingertips are much narrower. Human fingertips are broad, a trait related to the high degree of innervation required to perform fine manipulative tasks. It should be noted that the earliest stone tools recognized from the fossil record date to approximately 2.5 million years, which is post-*afarensis*.

Although bipedal in posture, *A. afarensis* retained several apelike aspects, particularly in body proportions. As can be seen in Fig. 10.10, its legs are relatively shorter and its arms relatively longer than in modern humans. In addition, as Peter Schmid (of the Anthropological Institute, Zurich) and Leslie Aiello (of University College London) have demonstrated independently, the shape of the trunk is apelike in being bulky relative to stature.[310]

Overall, then, *A. afarensis* anatomy – and presumably behavior – is somewhat intermediate between that of an ape and a human, a pattern that does not exist today. As a result, many researchers tend to see the adaptation as "unstable" – that is, as being in the process of striving toward the "perfection" of the human model. In fact, this unique anatomical and behavioral repertoire should be considered a stable evolutionary package. After all, the species persisted for at least 1 million years, from 3.9 million to 2.9 million years ago, and possibly even longer. The stability of its adaptation is reflected in the persistence of many anatomical aspects of the postcranium in *A. africanus*, a putative descendent of *A. afarensis*.

# Australopithecine diversity

We have looked at *A. afarensis* in detail here because it provides a good model for the more generalized australopithecines as a whole, it has been extensively studied, and the diversity of opinion illustrates the difficulty of interpreting complex and fragmentary fossil material. However, while *A. afarensis* is probably the best-known of the earlier australopithecines, it is now just one among many species of australopithecine.

All known australopithecines come from sub-Saharan Africa alone, and fall in date between over 4 million years to around 1 million years ago. They all share certain basic characteristics that place them within the genus *Australopithecus* (such as bipedal adaptations, thickened tooth enamel, small canine, small brains, and a tendency toward enlarged and flat molar teeth), but they also differ in minor ways one from another.

There are three forms that are known to date to the middle portion of the Pliocene – that is, older than 2.5 million years ago – and which have often been described as generalized or gracile australopithecines. As discussed earlier, *A. africanus* is known from sites in south Africa, and is dated to a little more than 3 million years to around 2.5 million years. This form is very similar in many ways to *A. afarensis*, although its face is somewhat less prognathic and its posterior teeth larger. Recently, new material has been recovered from one of the major south Africa sites, that of Sterkfontein, which is likely to have a major impact on our knowledge of the australopithecines. This is a partial skeleton, possibly older than 3.5 million years old.[311] Its foot has given it its nickname, Little Foot, and shows on the basis of the way that the metatarsals are angled together at their joints that this was a creature capable of grasping with its foot, and so likely to have been a competent arboreal primate. The rest of the skeleton is still being removed from its matrix, but is likely to reveal considerably more about these south African hominins, including the possibility that they represent another species of hominin.

*What is the pattern of australopithecine diversity?*

Because the history of australopithecine discoveries was, until recently, located exclusively in eastern or southern Africa, many anthropologists assumed that it reflected a real difference in the distribution of hominins and apes. That is, hominins were seen as being restricted to east of the Great Rift Valley, with apes remaining mainly in the west. The relatively continuous forest cover of central and western Africa was thought to provide an unsuitable habitat for hominins. At the end of 1995, however, this picture changed, with the announcement of the discovery of a hominin mandible in Chad, central Africa, which is 2500 kilometers west of the Rift Valley. The mandible, which has thick-enameled teeth, has been dated by faunal correlation to between 3 million and 3.5 million years old. Michel Brunet, David Pilbeam, and several colleagues initially described the jaw as being similar to that of *Australopithecus afarensis*.[282] On further study, however, they identified differences that signaled a different species, which they named *Australopithecus bahrelghazali* (Fig. 10.11). The discovery will prompt further prospecting east of the Rift Valley, even though the sparsity of suitable exposures makes such work difficult. Nevertheless, this single discovery breaks the traditional east–west divide, and implies that hominins were widespread in Africa, probably from the beginning. Such distribution often leads to geographically distinct populations and even different species. Inferring a phylogenetic pattern is therefore rather difficult.

The other recent addition to the australopithecine adaptive radiation is another discovery of Tim White's team in Ethiopia. Again, from the Middle Awash, he has described a new species dated to around 2.5 million years. This is similar to other australopithecines in terms of the shape of its cranium, the relative degree of prognathism, and the relationship between anterior and posterior dentition. However, it is also characterized by a relatively large skull, perhaps indicating the first evidence for increase in brain size, and a characteristically bipedal postcranium. The species has been given the name *Australopithecus garhi* (Fig. 10.12), and White and his colleagues have suggested that it may be close to the ancestry of the genus *Homo*.[279] This idea might be of interest in the context of the discovery close by of animal bone on which there are clear cutmarks made by stone tools.

All the australopithecines described so far are at least 2.5 million years old. From about this date the australopithecines show some distinctive trends and traits that have given rise to the name which describes them as a whole – the "robust" australopithecines. Compared to chimpanzees and to modern humans, australopithecines are characterized by a reduction in the size of the anterior dentition (the incisors and canines) and an increase in the size of the posterior dentition (especially the molars). This trend becomes very marked indeed in the later australopithecines, and these specimens have sometimes been referred to as "megadontic" or large-toothed. Along with these large teeth go other traits, such as massive jaws, heavy musculature, and crests along the top of their skulls (see below for a fuller discussion of their functional anatomy).[312]

These later robust australopithecines add considerably to the diversity of the group. The two best-known species are *Australopithecus robustus*, from south Africa, and *Australopithecus boisei*, from east Africa (Fig. 10.13). There is also an earlier species known from east Africa called *Australopithecus aethiopicus* (Fig. 10.14), which shares with the other species the larger back teeth, but also had a relatively prognathic face.

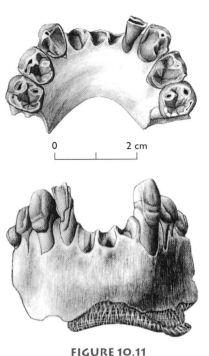

**FIGURE 10.11**
*Australopithecus bahrelghazali*: This newly discovered partial mandible from Chad, central Africa, is the first australopithecine to be found west of the Rift Valley, overturning the assumption that hominin habitat was restricted to areas east of the Rift Valley. The drawings show the top and front view of the mandible. (Courtesy of M. Brunet.)

# Australopithecine biology

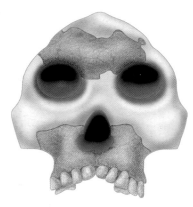

**FIGURE 10.12**
*Australopithecus garhi:*
This was discovered
in the Awash region
of Ethiopia, and has
characteristics that may
provide a link between
the australopithecines
and *Homo*. It was found
close to bones with
cutmarks on them,
perhaps suggestive of
tool use.

Like all early hominins, the australopithecines were essentially
bipedal apes with modified dentition. The hominin mode of
locomotion and dental apparatus are likely to have been adaptations to a habitat – and therefore diet – that differed from the
environments associated with apes. The later australopithecines
appear to have lived in a more open environmental setting – not
the open plains of traditional stories, but bushland and woodland
savannah. Food was probably located in widely scattered patches
and, judging from the structure of these species' teeth and jaws,
appears to have required more grinding than an ape's diet.

Traditionally, robust australopithecines were considered to have
eaten an exclusively vegetarian diet, while the gracile species included
some meat. This assumption was based on the larger jaws and molar teeth
of the robust species (Fig. 10.15). Recent work on the strontium-to-
calcium ratio in the bones of south African robusts, however, appears to
imply that they, too, ate some meat.[115]

Analysis of microwear patterns of australopithecine teeth gives some
insight into diet as well. For instance, using scanning electron microscopy,
Alan Walker found that the microwear pattern in robust australopithecines resembled that of chimpanzees and orangutans, both of which
eat various forms of fruit. More recently Frederick Grine, of the State

**FIGURE 10.13 Robust
australopithecine at Lake
Turkey:** Richard Leakey
found this intact cranium of
*Australopithecus boisei* (KNM-ER 406)
on the first major season of work on
the east side of Lake Turkana.
(Courtesy of Peter Kain and Richard
Leakey.)

University of New York at Stony Brook, and Richard Kay, of Duke University, concluded that the robust species consumed foods that were tougher than those eaten by the gracile species. The difference, they suggested, matches that found between the modern-day spider monkey, which eats fleshy fruits, and the bearded saki, which lives on seeds encased in a tough covering. Such a dietary difference is consistent with evidence that robust australopithecines lived in drier habitats, where soft fruits and leaves would be rare.[313]

Judging from the fossil record, australopithecines were as common on the landscape as other large, open-country primates (specifically, baboons). Thus, foraging strategies of hominins and baboons would not have differed dramatically. If australopithecines had been significant carnivores, for example, their population density would have been much lower than that of the principally vegetarian baboons.

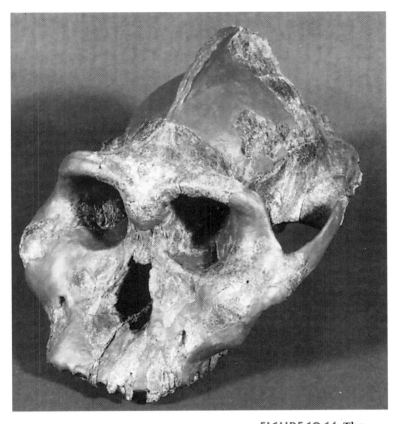

**FIGURE 10.14 The black skull:** Found by Alan Walker in 1984, the skull shows extreme features of australopithecine robusticity, but is dated at 2.6 million years. It is considered by some to be a member of *Australopithecus aethiopicus*. (Courtesy of Alan Walker.)

## Australopithecine anatomy

We will now examine aspects of the australopithecines' anatomy that qualify them as hominins, and the differences that arise between the gracile and robust forms. This survey will include the teeth, jaws, and cranium, the pelvis and associated locomotor skeleton, and the hands. In each case we will discuss the functional implications of the anatomy.

The terms "gracile" and "robust" appear to imply substantial anatomical differences between the two forms, with one being small and delicately built and the other exhibiting a larger and generally more massive form. In recent years, however, scholars have come to realize that the difference between the two forms lies mainly in the dental and facial adaptations to chewing: the robust forms have larger grinding teeth, more robust jaws, and more bulky chewing muscles and muscle attachments.

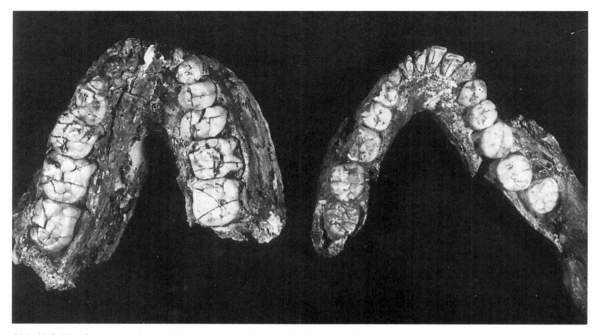

**FIGURE 10.15** Comparison of lower jaws of *Australopithecus robustus* and *A. africanus*

Recent body weight and stature estimates for australopithecines are as follows:[301]

♦ *A. africanus*: 41 kilograms for males and 30 kilograms for females, with statures of 138 and 115 centimeters, respectively;
♦ *A. robustus*: 40 kilograms for males and 32 kilograms for females, with statures of 132 and 110 centimeters, respectively; and
♦ *A. boisei*: 49 kilograms for males and 34 kilograms for females, with statures of 137 and 124 centimeters, respectively.

Estimates of brain size, which are based on a small number of specimens, typically give the robust species an edge over their gracile cousins. In fact, both are very close to 500 cm³. When body size is taken into account, brain size among all australopithecines is slightly, but perhaps not significantly, larger than in living great apes.

The teeth, jaw, and cranial anatomy is really one functional complex. As we have seen, the hominin dental adaptation can be described in general as moving in the direction of producing a grinding machine (Fig. 10.16). The two forms of australopithecine differ in that the robust species have taken this adaptation to an extreme, having enormous, flat molars and relatively small, blade-like incisors and canines.

This exaggeration of the hominin dental adaptation is most extreme in the robust australopithecine group. For instance, all hominins have a tooth row that is tucked under the face more than in apes, giving them a less projecting facial profile and increasing chewing efficiency. In the robust australopithecines, this tucking-under character is particularly marked. The robusticity of the lower jaw (mandible) in hominins compared with the lower jaws in apes is particularly apparent in the robust species, reflecting their more powerful chewing action.

The extra muscle power necessary for this chewing action in the robust species has two anatomical consequences. First, one of the muscles that powers the lower jaw – the temporal muscle – is anchored to a raised bony crest that runs along the top of the cranium, front to back. This sagittal crest, which is also found in gorillas, is absent in gracile australopithecines. Second, the great size of the temporal muscle in robust australopithecines and the existence of a second chewing muscle, the masseter, causes the cheek bones (the zygomatic arch) to become exaggerated and flared forward. This feature and the strengthening of the central part of the face by pillars of bone give the robust australopithecine face a characteristic "dished" appearance.

In terms of function and overall size, the postcranial skeletons of gracile and robust australopithecines are very similar to one another, as far as can be deduced from the limited amount of fossil material available. The australopithecine pelvis of 2 million years ago was very much like that of Lucy, who lived a million years earlier; Lucy, as we noted in chapter 9, was adapted to upright walking rather than quadrupedalism. The thigh bone of australopithecines, however, diverges from the typical *Homo* pattern: the head of the femur is smaller than in *Homo* and is attached to a longer, more slender neck.

A recent analysis of an *A. africanus* partial skeleton discovered in 1987 produced some surprising results related to the species' locomotion. Carried out by Henry McHenry and Lee Berger, of the University of the Witwatersrand, and published in 1998, the study revealed that the joints of the arm bones of this specimen were more robust than in modern humans.[314] This finding implies that this species employed a more forelimb-dominated mode of locomotion than that used by modern humans, probably including a good deal of arboreality. Interestingly, the *africanus* forelimb seems to be more robust than that of *A. afarensis*, its presumed ancestor, and so *africanus* may have been the more arboreal. An ancestral progression of *afarensis* to *africanus* to *Homo* would therefore have involved a reversal of this trait. As evolutionary biologists believe

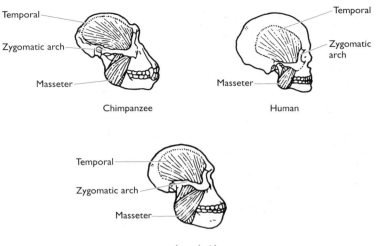

Chimpanzee

Human

*Australopithecus*

**FIGURE 10.16**

**Anatomy of chewing:** Two muscles are important in moving the lower jaw during chewing: the masseter, which is attached to the zygomatic arch (cheek bone), and the temporal, which passes through the arch. The larger the masseter and temporal muscles, the larger the arch. Chimpanzees have approximately three times as much chewing-muscle bulk as modern humans, and the australopithecines even more.

What were austalopithecine adaptations?

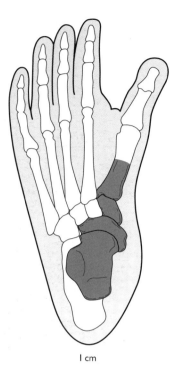

I cm

**FIGURE 10.17 Ancient foot:** The drawing shows the recently discovered four articulating foot bones (dark areas) of *Australopithecus africanus;* dated at 3.5 million years, this species is the oldest known hominin in south Africa. The angle of articulation of the bones implies that the great toe diverges from the other toes, as in apes, but to a lesser degree. This feature might have been an adaptation to a degree of arboreality. (Courtesy of R. J. Clarke.)

such reversals are unlikely, some observers have suggested that this new study throws doubt on the likelihood of this evolutionary sequence.

The recent discovery of four articulating foot bones from Sterkfontein also implies some arboreality in early australopithecines (Fig. 10.17). Reconstruction of part of the foot undertaken by Ronald Clarke and Phillip Tobias, of the University of the Witwatersrand, indicated that the great toe was somewhat splayed – a character seen in apes, but here to a lesser degree.[311] If this feature has functional significance, it may mean that the 3.5-million-year-old australopithecine was an adept climber. In their report, Clarke and Tobias drew the following conclusion: "It is becoming clear that *Australopithecus* was likely not an obligate terrestrial biped, but rather a facultative biped and climber."

Recent evidence from an unusual anatomical source – the inner ear – also implies that australopithecine locomotion was not identical to that of a fully committed biped. Three bony tubes arranged as arches at right angles to one another form an important organ of balance, known as the semicircular canals or vestibular system. Fred Spoor, an anatomist at University College London, measured the dimensions of these three arches (the anterior, posterior, and lateral semicircular canals) in living primates, including humans, and found an important difference between humans and apes. In humans, the anterior and posterior canals are larger than in apes, while the lateral canal is smaller. Spoor interprets the difference in humans as an adaptation to the demands of bipedal locomotion. He used computerized tomography to measure the dimensions of semicircular canals in a series of hominin fossils. In all australopithecines, the pattern was apelike; in contrast, it was humanlike in early *Homo*. He concluded that australopithecines did not move bipedally in the same way as modern humans or even early *Homo*.[315]

An analysis of the trunk of *Australopithecus* (as seen in Lucy) implies that, however well adapted this species was for bipedal walking, bipedal running was not part of its repertoire (Fig. 10.18). Peter Schmid concluded that Lucy's chest was funnel-shaped, not barrel-shaped as in modern humans. The shoulders, trunk, and waist are important elements in human running: the shoulders enable arm swinging and balance; the trunk is used for balance and breathing; and the waist permits flexibility and swinging of the hips. "What you see in *Australopithecus* is not what you'd want in an efficient bipedal running animal," noted Schmid:

The shoulders were high, and, combined with the funnel-shaped chest, would have made arm swinging improbable in the human sense. It wouldn't have been able to lift its thorax for the kind of deep breathing that we do when we run. The abdomen was potbellied, and there was no waist, so that would have restricted the flexibility that's essential to human running.

In other words, Lucy and other australopithecines may have been bipeds, but active, prolonged running was an adaptation that came only with *Homo*.

## Australopithecus, a tool maker?

The identity of the maker of the stone tools in the archeological record is a constant question, although many paleoanthropologists assign this role to *Homo*, not *Australopithecus*. Evidence on this issue is necessarily indirect, such as the anatomy of the hands. No hands of *A. africanus* have been discovered. The hand bones of *Australopithecus afarensis* (as known from the Hadar) were strikingly apelike, having curved phalanges, thin tips to the fingers, and a short thumb. By contrast, recent analysis of robust australopithecine hand bones from the Swartkrans site indicates that they were much more human-like. Randall Susman reports that the thumb is longer and more mobile and the fingertips much broader (the latter, as mentioned above, is a feature thought to be associated with the supply of blood vessels and nerve endings to the sensitive fingerpads).[316] According to Susman, the robust australopithecines' anatomy probably allowed sufficient manipulative skills to enable stonetool making, an ability that has usually been thought of as strictly within the domain of *Homo*.

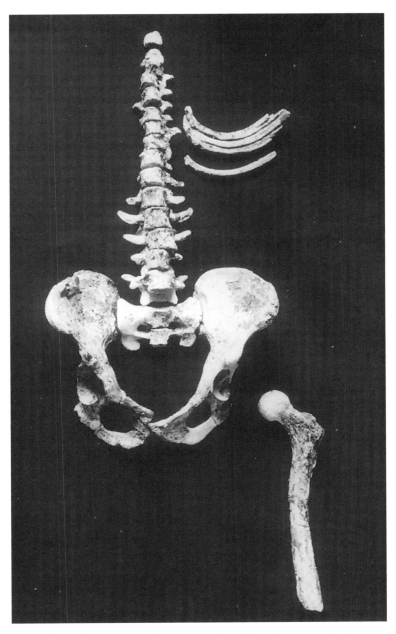

**FIGURE 10.18 Partial skeleton:** Found by Robert Broom and John Robinson in the late 1940s (and partially reconstructed by Robinson), these bones clearly show the bipedal anatomy of *A. africanus* (museum number, Sts 14). (Courtesy of Peter Kain and Richard Leakey.)

Differences of opinion have arisen over these conclusions, however. The recently discovered simple bone tools – digging sticks – may be taken as support for Susman's hypothesis.[274] Nevertheless, the tools could have been made by a species of *Homo*, whose fossils are also known at Swartkrans. Furthermore, some scholars question whether the fossil hand bones that Susman studied might have been those of *Homo* and not *A. robustus*, as he believes.

## Relationship between robust and gracile australopithecines

The gracile and robust australopithecines have often been viewed as basically the same animal, but built on different scales. Functionally speaking, this notion is accurate in many respects. The relationship may also be viewed in terms of evolutionary progression, however, with the gracile species being seen as ancestral to the robust species, in whom the australopithecine traits had become extremely exaggerated: specifically, the chewing apparatus became increasingly robust. If true, then the fossil record should have revealed a steady increase through time in dental, facial, and jaw robusticity.

The 1985 discovery of the *Australopithecus aethiopicus* cranium KNM-WT 17000) from the west side of Lake Turkana finally put to rest this simple relationship.[317] The cranium was as robust as any yet known, but was 2.5 million years old. Clearly, the huge molars, flared cheek bones, and dished face could not be the end product of an evolutionary line if it were present at the origin of that supposed line. How this discovery affects the shape of the hominin family tree remains under discussion.

*What is unique about robust australopithecines?*

This cranium, known colloquially as the "black skull," was surprising not only because of its great age but also because it contained an unexpected combination of anatomical characteristics. Although the face was distinctly like that of that most robust of robust australopithecines, *Australopithecus boisei*, the cranium – particularly the top and back – were not: they were similar to those of *Australopithecus afarensis*. Such an anatomical combination surprised most people, and reminded us that hominin biology of 3 million to 2 million years ago was more complicated than current hypotheses have allowed.

## Paleoenvironments of the early hominins

Analysis of the geology of the Aramis site and the fossils of other creatures found there indicates that this area was a closed woodland or forest setting at the time that these hominins lived there. For instance, 30% of the vertebrates were colobine monkeys, which are forest animals. The Allia Bay hominins apparently lived in or near gallery forest associated with a large river; at Kanapoi, the environment was more open, but probably close to gallery forest. The Hadar *afarensis* population lived in a woodland habitat, while Laetoli was much more open, possibly even grassland savannah. The *bahrelghazali* species apparently lived in a lakeside environment, incorporating rivers and streams and associated woodland. It is therefore apparent that the earliest hominins occupied a diversity of habitats, including closed forest and open terrain.[318] These reconstructions strongly influence the expected distributions of early hominins in relation to the changing climate (Fig. 10.19) and help to explain the trends of early hominin evolution (Fig. 10.20).

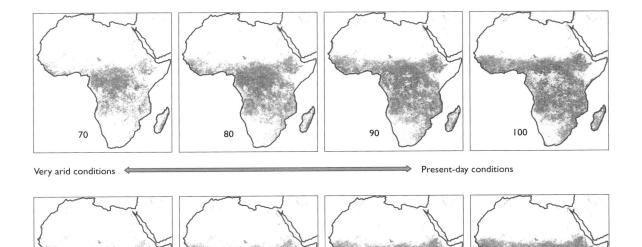

**FIGURE 10.19 Possible distribution of hominins in relation to African climatic change:** The distribution of early hominins would depend upon the available habitats and the tolerance of the hominins to them. Modeling of distributions can indicate when particular hominins may have been most successful. The unshaded areas on this series of maps shows the distribution of available habitat for a woodland/savannah-tolerant species as rainfall varies from 70% to 140% of contemporary levels. (Courtesy of Piers Gollop and Robert Foley.)

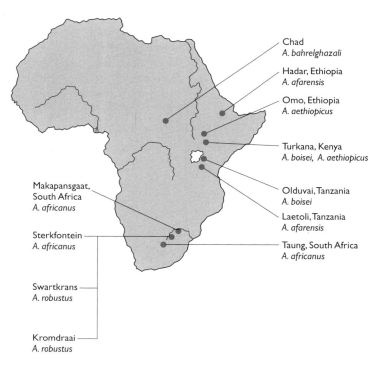

**FIGURE 10.20** Australopithecine sites mentioned in the text

# BEYOND THE FACTS
## OUT OF THE SEA?

*The issue: the idea that human evolution was triggered by an aquatic phase is widely supported and discussed outside the mainstream of paleoanthropology, but is dismissed by most scientists working in the field. How do we determine what models are reasonable and plausible, and which ones are worthy of serious scientific study?*

Paleoanthropology has a reputation for controversies and arguments, with major disagreements about who is who, and who is related to whom, among the fossil hominins. However, although there is considerable debate about the details, there is nonetheless remarkable consensus about the major aspects of human evolution – that our ancestors were derived from a population of African apes, adapting in increasingly open and savannah environments to the changing conditions. Most features, especially bipedalism, are seen as related to this change.

There is, though, a vociferous minority on the margins of the discipline who argue something completely different: that the hominin lineage went through an aquatic phase, and it was during this time that the key characteristics of humanity – bipedalism, hairlessness, language, tool making, etc. – all evolved. Rather than our features being adaptations to drier terrestrial environments, they are adaptations to living in water. The key figure in this model is Elaine Morgan, who has written a number of highly persuasive books making these claims, and who has a strong following on the web and in the more popular literature. She has, however, failed to make many if any converts among the mainstream of the discipline, so that to the uninitiated it might appear that there are two models of human evolution "out there," talking past each other.

Indeed, it is one of Elaine Morgan's complaints that her ideas have been ignored rather than criticized or dismissed, and that this is a case of "normal science," in the terms of philosopher of science Thomas Kuhn, ignoring the radical alternative paradigm rather than engaging with it. In fact there have been a number of serious examinations of the theory, most of which have failed to find support for it, but it is certainly the case that most textbooks on human evolution – this one included – simply ignore the aquatic ape model.

This model is one among many "alternative theories" of human origins, and indeed in that light is one of the most cogent and best argued. Others posit visitors from outer space, or strange racial theories of events deep under ice (for a survey of these theories see *Strange Creations* by Donna Kossy [Feral House, 2001]).

The existence of such models does raise the question of what it is that distinguishes a plausible model from an implausible one. What is it that makes it reasonable to discuss one model and to dismiss another out of hand? Is the aquatic ape hypothesis a reasonable explanation for many unique features of humanity, and ignored because it is a challenge to scientific orthodoxy, or is it a crackpot theory? If it is the latter, then should the scientific community spend time and resources refuting it? If it is the former, how can it become accepted as a good model?

# CHAPTER 11

# ORIGINS OF *HOMO*

In the previous chapter, we saw that a radiation of robust australopithecines occurred a little earlier than 2.5 million years ago. At about the
same time, a second radiation took place in a very different evolutionary
direction. Instead of an exaggeration of jaw and tooth size, as happened
with the robusts, this second radiation involved a reduction in jaw and
tooth size, particularly in the molar teeth. In addition, brain size increased
significantly, from approximately 500 $cm^3$ to more than 640 $cm^3$. A slight
increase in body size occurred as well – albeit not enough to account for
the larger brain size. For the first time, simple stone tools are found in the
record, and diet may have shifted to include more meat, procured either by
scavenging, by simple hunting, or by a combination of both. The increase
in brain size and decrease in jaw and tooth size are associated with the first
appearance of the genus *Homo*; the archeological evidence of a shift in subsistence patterns is often assumed to be associated with behaviors unique
to *Homo*, although this point remains to be definitively demonstrated. The
taxonomic interpretation of early *Homo* fossils was considered contentious
when they were first found, and in many ways it remains so today.

This chapter will describe some of the most important fossil discoveries
and the early evolution of the genus *Homo*.

## THE GENUS *HOMO*

### The first discoveries

The first discoveries of early *Homo* fossils
were made at Olduvai Gorge, not long
after Mary Leakey had found Zinj (now
*Australopithecus boisei*) and Louis Leakey pro-

**KEY QUESTION** What is the pattern of
diversity among early members of the genus
*Homo*?

nounced it to be the maker of the gorge's stone tools. Between 1960 and 1963, a series of fossils was uncovered close to the Zinj site, including hand and foot bones, a lower jaw, and parts of the top of a cranium. The fossils, which were judged to be slightly older than Zinj (therefore older than 1.75 million years), were less robust than Zinj; in addition, the teeth were smaller and the brain was calculated to be significantly larger, with a volume estimated at 640 cm$^3$. Displaying uncharacteristic caution, Louis Leakey initially referred to the fossils simply as "pre-Zinj," but he quickly became convinced that this creature, not Zinj, was the tool maker, and the ancestor of modern humans.

Much of the analysis of these fossils was carried out by John Napier, of the University of London, and Phillip Tobias, of the University of the Witwatersrand.[275,319] Leakey believed that the pre-Zinj specimen should be named as a new species of *Homo* (Fig. 11.1), making it the earliest known. His colleagues – particularly Tobias – were reluctant to follow this course. Leakey eventually prevailed, and in April 1964 the three workers published a paper in the journal *Nature* announcing *Homo habilis* (handy man), a name that had been suggested to them by Raymond Dart. The publication provoked near outrage among anthropologists, and it created a deep rift between Leakey and Sir Wilfrid Le Gros Clark, a particularly severe critic who had previously enjoyed a close association with Leakey. The reason for the outrage was twofold: (1) the naming of *habilis* as *Homo* required a redefinition of the genus, and (2) many argued that insufficient "morphological space" divided *Australopithecus africanus* (the presumed ancestor of *habilis*) and *Homo erectus* (the presumed descendant).

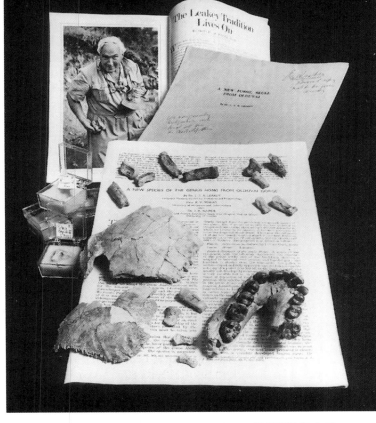

**FIGURE 11.1 Type specimen of *Homo habilis*:** The establishment of the species *Homo habilis* in 1964 involved a redefinition of the genus *Homo*. This development, among other things, provoked a strong reaction to its validity. (Courtesy of John Reader.)

## Critique of *Homo habilis*

The definition of the genus *Homo* has always been somewhat contentious, not least because it is tied – consciously or unconsciously – to the state of "being human." A series of anatomical characters is to be found uniquely in *Homo* – for example, an increase in cranial vault height and thickness, reduced lower facial prognathism, and reduction in the size of premolars and molars and the length of the molar row – but what has always stood out prominently in scholars' definitions is the size of the brain. To be *Homo* is to be a large-brained hominin, one presumably more technologically accomplished than the australopithecines. The question is: what brain size qualifies for admission to the genus *Homo*?

The average modern human brain is 1350 cm$^3$ in capacity, with a range of 1000 cm$^3$ to approximately 2000 cm$^3$. How much smaller than the lower limit can a hominin's brain be, and still be counted as *Homo*? Before 1964 several estimates had been made of this "cerebral Rubicon," ranging from 700 cm$^3$ to 800 cm$^3$. In the late 1940s, the British anthropologist Sir Arthur Keith proposed a figure of 750 cm$^3$ – a size midway between the largest known gorilla brain and the smallest human brain. Keith's proposal had been widely accepted by 1964, when Leakey, Tobias, and Napier advanced their new definition of the genus *Homo,* which included a cerebral Rubicon of only 600 cm$^3$. The reason for the reduction in this new definition of *Homo* was obvious: the fossil cranium that was part of the new species had a capacity of only 640 cm$^3$, a figure that would have failed under Keith's standard.

As part of the description of *habilis* in the *Nature* paper, and implied as part of the new definition of *Homo*, was the fact that the species was a tool maker, which Leakey described as a unique human trait. Anatomy, not behavior, is typically the basis of formal descriptions of new species, so this criterion was seen as both unusual and inappropriate. In any case, no method of determining whether *habilis* or some other hominin was the tool maker existed.

The objection relating to inadequate morphological space flowed from the prevailing ethos of "lumping" rather than "splitting." As mentioned earlier, the discovery of individual australopithecine specimens in south Africa was often accompanied by the proposal of a new species. The tendency to name new species on the basis of small anatomical differences between specimens is known as splitting. By the 1960s, anthropologists recognized what they should already have known – namely, that considerable anatomical variation appears within populations. The tendency to designate significant anatomical variation between specimens as intraspecific rather than interspecific variation is known as lumping. Splitters see many species in the record; lumpers see few. (The splitting tendency had indeed gone too far, making anthropological interpretations biologically unrealistic; it is increasingly realized, however, that lumpers also went too far, creating a different kind of biological unreality.)

As mentioned earlier, many anthropologists assumed an ancestor/descendant relationship between *Australopithecus africanus* and *Homo erectus*. For *Homo habilis* to be a valid species, it would have to be intermediate between the two, because it was of intermediate age. Lumpers expected considerable anatomical variation in both *africanus* and *erectus*, which left little or no room for an equally variable intermediate. The putative *Homo habilis* fossils therefore had to be either *Australopithecus africanus* or *Homo erectus*. Unfortunately, the critics of *habilis* could not decide to which species it belonged; some said that it was a large *africanus*, while others argued that it was a small *erectus*.

Eventually, *Homo habilis* was accepted by most anthropologists as a valid species, partly through the discovery of other, similar specimens, and partly because of a recognition of the excessive lumping tendency. Nevertheless, the species' history in the science has been rocky, principally because of the large degree of anatomical variation found among specimens that are intermediate between *africanus* and *erectus*, which are therefore putative members of *habilis*. Ironically, a current resolution of this dilemma that is gaining much favor involves a recognition of two species of *Homo* at this early time (close to 2 million years ago) – not just one, the point to which earlier workers objected so stridently.

## KNM-ER 1470 and diversity in early *Homo*

In 1972, three years after he initiated the expedition at Lake Turkana, Richard Leakey announced the discovery of a fossil that was to make him world-famous and subject the early history of the *Homo* clade to further scrutiny. That fossil, KNM-ER 1470, was the larger part of a cranium pieced together from hundreds fragments by Alan Walker and Meave Leakey.[276] The face was large and flat, the palate was blunt and wide, and, judging by their roots, the absent teeth would have been large (Fig. 11.2). These features are reminiscent of australopithecines. Nevertheless, the cranium was large, estimated at 750 cm³, which betokened *Homo*.

Minimal bone remained in the specimen, which prevented making an unequivocal attachment of the face to the brain case. As a result, uncertainty arose over the appropriate angle of the face: a small angle made the cranium look like *Homo*, while a larger angle was australopithecine-like. Leakey and his colleagues also debated the proper species attribution for the fossil; Leakey favored *Homo*, for example, while Walker preferred *Australopithecus*. Eventually, the fossil was described in a *Nature* publication as *Homo*, but with its species undetermined.

The reluctance to name 1470 as *Homo habilis*, the obvious choice, reflected two problems: (1) its brain was considerably larger than the 640 cm³ of the Olduvai *habilis*, and (2) 1470 was thought to be much older. The Olduvai *habilis* less than 2 million years old, while state-of-the-art radiometric dating gave an age of 2.6 million years for 1470. It later transpired that the dating had been erroneous, and that the true date was 1.9 million years, making the specimen a virtual contemporary of the Olduvai hominin. Despite the uncertainties surrounding 1470, its

Early *Homo* – one or two species?

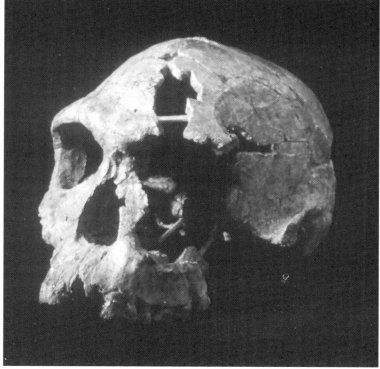

**FIGURE 11.2 Two skulls from Koobi Fora, Kenya:** The cranium KNM-ER 1470 (top) was found in 1972 and recognized as belonging to the genus *Homo*, although no species attribution was made initially. A second, smaller cranium, KNM-ER 1813 (bottom), was found a year later and was thought by some to be *Homo* and by others *Australopithecus*. It is now attributed to *Homo* by most observers. (Courtesy of Richard Leakey and Peter Kain.)

discovery helped convince anthropologists of the validity of *Homo habilis* as a species, because it demonstrated the existence of a creature that was definitely different from both *A. africanus* and *H. erectus.*

A year after the announcement of 1470's discovery, a second cranium was found at Lake Turkana, which was to play an important role in the resolution of early *Homo*. Known as KNM-ER 1813,[276] its face and palate are similar to those of *Homo habilis* from Olduvai and different from those of 1470; the brain is small, however – not much more than 500 cm³. Despite this disparity, 1813 has been described by some as a female *Homo habilis*, though Leakey himself has not made this claim.

## OH 62: further diversity

In 1986, Donald Johanson, Tim White, and a large team of colleagues discovered an extremely fragmented hominin skeleton at Olduvai Gorge, comprising part of the upper jaw, some cranial fragments, most of the right arm, and parts of both legs. The following year they published details of the fossils, numbered OH 62, which they attributed to *Homo habilis*.[320] An influential reason why they designated the specimen as *Homo habilis* was the resemblance of the palate to that of a skull found at Sterkfontein a decade earlier, numbered Stw 53, which was assigned to *habilis*.[321] Cranial remains were insufficient to estimate a brain size. The limb proportions, however, were both interesting and surprising.

OH 62 was a small, mature female, comparable to Lucy in being approximately 1 meter tall. As with Lucy, the arms were long and the legs short, compared with later *Homo*. The unexpected aspect, as shown by Robert Martin, of the University of Chicago, and Sigrid Hartwig-Scherer, was that OH 62's arms were even longer than those possessed by Lucy, and its legs shorter (Fig. 11.3). Thus, the specimen was even more apelike than *afarensis*, its presumed ancestor. The ratio of humerus length to femur length (the intermembral index) was 95% as compared with 70% in modern humans, 97 to 100% for chimpanzees, and 85% for Lucy. As Martin and Hartwig-Scherer titled their paper on their analysis, "Was 'Lucy' more human than her 'child'?"[322]

OH 62 was found in the lowest layers of Olduvai Gorge, and was dated at between 1.85 million and 1.75 million years old. The year before its discovery, the *Homo erectus* (*ergaster*) youth had been unearthed on the west side of Lake Turkana. This specimen was tall (almost 2 meters) and had very humanlike limb proportions, but lived only 200,000 years later than OH 62. If *Homo habilis* is ancestral to *Homo erectus* (*ergaster*), then evolution from an apelike to a humanlike condition must have occurred relatively rapidly. It would also require an evolutionary reversal, from moderately apelike limb proportions in *afarensis*, to more apelike proportions in *Homo habilis*, to humanlike proportions in *Homo erectus* (*ergaster*) – that is, if OH 62 was indeed a member of *Homo habilis*.

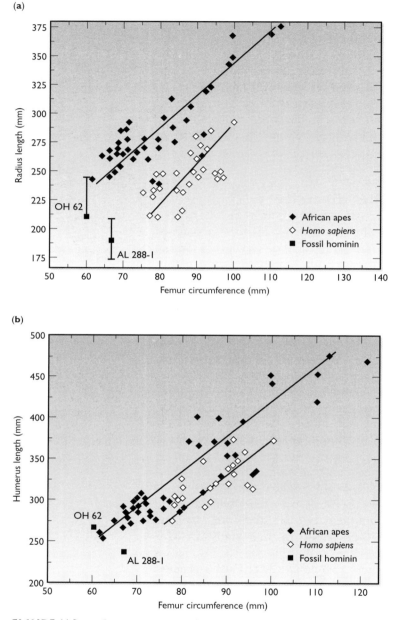

**FIGURE 11.3 Body proportions of Lucy and "Lucy's child":** Comparisons of radius length against femur circumference (a) and humerus length against femur circumference (b) indicate that "Lucy's child" (OH 62) is more apelike than its presumed ancestor, Lucy (AL 288-1). (Courtesy of S. Hartwig-Scherer and R. D. Martin.)

By this time (the mid-1980s), *Homo habilis* had become something of a grab-bag of specimens, different from its presumed ancestor and its presumed descendant. The question, "Do you accept *Homo habilis* as a valid species?", would likely draw the response, "Well, it depends on which specimens you want to include." As a result, some scholars began to contemplate splitting "*Homo habilis*" into more than one species.

## Kenyanthropus and yet more diversity

As discussed above, when *Homo habilis* was first announced, Le Gros Clark among others had thought there was not sufficient evolutionary space between the australopithecines and later *Homo* to allow for a further species. Ironically, as more and more fossils have been discovered it has increasingly appeared that there is too much variability in *Homo habilis* for it to be a single species. Now, more recent discoveries have cast even greater doubt on the extent to which the early hominins fall neatly into australopithecines and *Homo*, and led to problems about their relationships.

In 1999 Meave Leakey, of the National Museums of Kenya, and her colleagues discovered hominin fossils at the site of Lomekwi on the western shores of Lake Turkana in Kenya (Fig. 11.4). The specimens

(KNM-WT 40000) are dated to around 3.5 million years ago, making them contemporary with *Australopithecus afarensis*. The fossils consist of an almost complete cranium, plus other isolated elements. The discoverers of this specimen have argued that this is sufficiently different from all other known early hominins to be placed in a new genus (*Kenyanthropus*). Its most distinctive feature is its flat face (hence the species name *platyops*). There are other differences to be found, in the ear bones and the shape of the frontal. If there is any other fossil which it resembles, the best candidate is 1470, which, as we have discussed above, is placed in the genus *Homo* as part of *Homo habilis* in the broadest sense. This suggests that there may be a link between this new genus and the origins of the genus *Homo*.

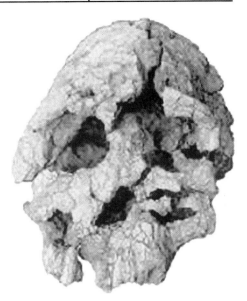

**FIGURE 11.4**
*Kenyanthropus platyops:* This recent find from west of Lake Turkana, dated to over 3 million years, has thrown early hominin taxonomy into turmoil. Although placed in a new and separate genus, it has been compared to specimens of *Homo rudolfensis*. (Courtesy of M. Leakey.)

## The earliest known *Homo*

Finally, in addition to these more substantial remains which throw light on the transition from the earlier hominins such as *Australopithecus* and *Homo*, there are also some fragmentary fossils which have been claimed as evidence for the genus *Homo* at earlier than 2 million years. These include a collection of isolated teeth from the Shungura formation in the lower Omo Valley and for a partial cranium from Sterkfontein (Sts 19); both collections are equivocal at best. Other claims come from the reassessment of a cranial fragment from Kenya and a recently discovered mandible from the site of Uraha in Malawi, which lies between east Africa and south Africa.[323]

In 1967, the temporal bone (side of the head) of a hominin was discovered in the Chemeron formation near Lake Baringo, in central Kenya. The structure around the ear – specifically the mandibular fossa, or jaw joint – is diagnostic of *Homo*; the age of the specimen was unknown. In February 1992, Andrew Hill (of Yale University), Steven Ward (of Northeastern Ohio Universities), and several colleagues announced a secure radiometric age for the fossil of 2.4 million years. This age is close to that of the oldest-known stone tools, from Kenya and Ethiopia, and indicated that this find represented the oldest-known specimen of *Homo*. No species attribution was made.

In October 1993, an international team, led by Friedemann Schrenk, of Darmstadt Museum, and Timothy Bromage, of Hunter College, New York, published a description of a partial hominin mandible that they had discovered near Lake Malawi. The mandible is less robust than that in australopithecines and the cheek teeth smaller, indicating its association with *Homo* (Fig. 11.5); the specimen has been dated by faunal

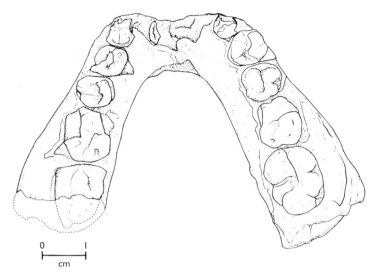

**FIGURE 11.5 Mandible
of *Homo rudolfensis*
from Malawi:**
(Courtesy of F. Schrenk
and T. G. Bromage.)

correlation at between 2.5 million
and 2.3 million years old, an age
comparable to that of the Chemeron
hominin. The authors assigned the
Malawi specimen to *Homo rudolfensis*,
a contemporary of *Homo habilis*
that is also found at Lake Turkana
(as described later in this chapter).
The Malawi hominin, together with
other fauna that are characteristic
of east Africa, indicates significant
faunal movement between the two
regions (Fig. 11.6).

# Anatomy and
biology of early *Homo*

As previously noted, the brain capacity of early *Homo* is larger than that
of the australopithecines, a change that produces several associated characteristics. For instance, the temple areas in australopithecines narrow
markedly (best seen from top view), forming what is known as the
postorbital constriction. In early *Homo*, this constriction is much reduced
because of the expanded brain.
In addition, the face of an
australopithecine is large relative to the size of its cranial vault,
a ratio that is reduced in the
larger-brained *Homo* species.
The cranial bone itself is thicker
in *Homo* than in *Australopithecus*
(Fig. 11.7).

The tooth rows in early
*Homo* are tucked under the face
as in other early hominins,
a feature that becomes even
more exaggerated in later
species of *Homo*. The jaw and
dentition of *Homo*, however,
are less massive than in the
australopithecines. Although
the teeth are capped with a
thick layer of enamel, their
overall appearance gives less
of an impression of a grinding

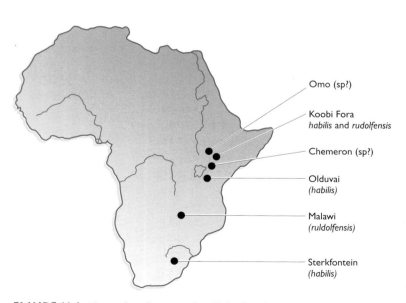

Omo (sp?)

Koobi Fora
*habilis* and *rudolfensis*

Chemeron (sp?)

Olduvai
*(habilis)*

Malawi
*(ruldolfensis)*

Sterkfontein
*(habilis)*

**FIGURE 11.6 Sites of early *Homo* fossil finds:** This shows sites in
Africa with specimens that have been attributed to genera other than
*Australopithecus* and which are Pliocene or very early Pleistocene.

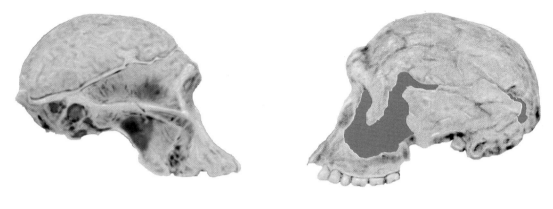

A. *africanus* (Sts5)          H. *habilis* (KNM-ER 1813)

**FIGURE 11.7** **Comparison of** *Australopithecus africanus* **and** *Homo habilis*

machine than appears in the small-brained hominins: the cheek teeth are smaller and the front teeth larger than in australopithecines, and the premolars are narrower. The patterns of wear on early *Homo* teeth are, however, indistinguishable from those of the australopithecines: the pattern is that of a generalized fruit-eater. Only with the evolution of *Homo ergaster* (*erectus*) about 1.9 million years ago does the toothwear pattern make a shift, perhaps indicating the inclusion of a significant amount of meat in the diet.

The original set of *Homo habilis* fossils from Olduvai Gorge included a relatively complete hand; its structure was compatible with an ability to make and use tools, concluded John Napier. The evolution of technological skills associated with stone-tool making has always appeared to be a satisfactory explanation for the expansion of brain capacity in the *Homo* lineage. If australopithecines were equally skillful, then this explanation fails. Presumably, some selection pressure on mental skills must have separated the *Homo* and australopithecine lineages. Whether this separation was associated with the development of more complex subsistence activities or lay in the realm of more complex social interaction is difficult to determine.

What are the adaptations of early *Homo*?

In an analysis of the body proportions of the early hominins, Leslie Aiello of University College, London, found a distinctly human form – that of small body bulk for stature – as well as an apelike form – that of high body bulk for stature. All australopithecines are characterized by the apelike form; *Homo erectus* is humanlike, as are certain specimens attributed to *Homo habilis*. OH 62, however, would fit best in the apelike group. The shift from apelike body proportions to humanlike proportions is seen only in *Homo*, and is assumed to be associated with an adaptive shift that includes greater routine activity.

# Taxonomic turmoil

As noted earlier, the OH 62 partial skeleton, with its primitive post-cranium, was influential in spurring a revision of the *Homo habilis* taxon. At the time of its discovery, the taxon included dozens of specimens (from Olduvai, Lake Turkana, and Sterkfontein) that displayed an uncomfortably wide range of anatomical variation. Several workers, including Chris Stringer (of the Natural History Museum, London)[324] and Bernard Wood (of George Washington University Washington DC),[325] had already expressed the opinion that the fossils belonged to two species, not one. Although alternatives to this terminology have been proposed since then, Wood's analysis is the most generally accepted at present.

The two species proposed by Wood are *Homo habilis* and *Homo rudolfensis*. They are distinguished as follows: *Homo rudolfensis* has a "flatter, broader face and broader postcanine teeth with more complex crowns and roots and thicker enamel." *H. rudolfensis* also has a larger cranium (Fig. 11.8). Wood includes all non-australopithecine specimens at Olduvai in *Homo habilis*, whereas the Lake Turkana fossils are divided between *H. habilis* and *H. rudolfensis*. The small, enigmatic cranium 1813 is included in *H. habilis*, as is a partial skeleton, KNM-ER 3735, which has primitive limb proportions like those of OH 62. The famous 1470 skull is designated as *Homo rudolfensis*, together with a collection of other specimens that includes examples of modern-looking leg bones. The Malawi hominin is designated as *H. rudolfensis*. The Chemeron hominin does not possess characteristics that are diagnostic of either species.

Other workers, such as Christopher Stringer and Richard Leakey, agree that two species existed. They suggest, however, that the Olduvai specimens should be split into two species: *Homo habilis*, as originally designated by the type material, and a smaller, more archaic form represented by OH 13 and OH 62. Other suggestions have been put forth as well.

Whatever form a consensus might eventually take, there is now general agreement that two species of *Homo* coexisted 2 million years ago. Although Wood's taxonomic distinction is based principally on certain cranial and dental characters, it is useful to think of *Homo habilis* as a smaller-brained creature with archaic postcranium, and *H. rudolfensis* as larger-brained with a more modern postcranium. Which of the two (if either) gave rise to later *Homo* is still debated. *H. rudolfensis* appears to have a good claim based on brain size and modern postcranium, but some insist that its facial and dental anatomy disqualify it from this role; *H. habilis* has a better claim in this latter respect, but its smaller brain and archaic postcranium militate against it.

To this turmoil must now be added the question of whether the newly discovered *Kenyanthropus* is allied to the genus *Homo* in particular, or whether this find casts doubt on the ability to make a clear distinction between *Homo* and other genera of the hominins. These issues will be

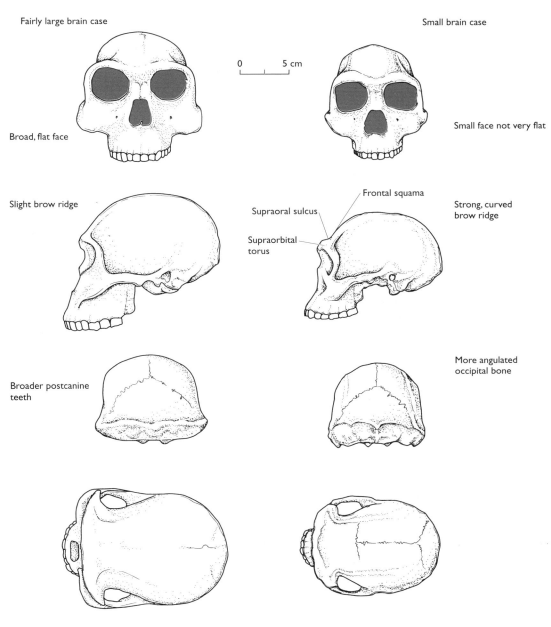

Fairly large brain case

Small brain case

Broad, flat face

Small face not very flat

Slight brow ridge

Frontal squama

Supraoral sulcus

Supraorbital torus

Strong, curved brow ridge

Broader postcanine teeth

More angulated occipital bone

KNM-ER 1470

KNM-ER 1813

**FIGURE 11.8 Comparisons of *Homo rudolfensis* and *Homo habilis* showing key differences in anatomy:** (left) KNM-ER 1470 has been attributed to *H. rudolfensis*; (right) KNM-ER 1813 is *H. habilis*.

considered later. What is clear from this and the preceding chapter is that the Pliocene is a period of remarkable hominin diversity, and there has been considerable progress in unraveling the hominin tree since the first tentative finds were made in Africa during the middle part of the last century.

## HOMININ RELATIONSHIPS

**KEY QUESTION** What are the phylogenetic relationships among the earlier homins, and what are the problems in reconstructing them?

This section will explore recent developments and current thinking about how early hominins were evolutionarily related to one another. This subject – phylogeny – has always attracted the attention of anthropologists, often overshadowing the more basic questions of hominin biology, such as subsistence strategies and behavior.

It is worth stating that evolutionary trees – phylogenies – can be constructed only with those species that are recognized; this limitation is important because the known fossil record (and its interpretation) is likely to underrepresent (and underestimate) the diversity of species that existed in the past. The discovery in the 1990s of seven more species of early hominin is testimony to that incompleteness. Moreover, support is increasing for the notion that the fossils once described as one species, *Homo habilis*, actually include two species, *H. habilis* and *H. rudolfensis*. In addition, debate continues as to whether *Australopithecus afarensis* is one species or more. The probable inadequacy of the fossil record and its interpretation should therefore be borne in mind whenever a diagram is offered as a hypothesis of "the" hominin phylogeny. It is considered legitimate to include putative, but as yet unknown, intermediate species that are inferred from the patterns of known species.

## How many hominin species should there be?

Should early hominins be diverse?

Assuming that primate species on average have a longevity of approximately 1 million years, Robert Martin calculated that roughly 6000 primate species have existed throughout the entire history of primate evolution.[154] The 200 living primate species represent just 3% of this total. More particularly, although 84 species of hominoids are estimated to have existed during the past 35 million years, less than half are known from the fossil record. Attempts have also been made to calculate the number of hominin species that probably lived in the period from 5 million years ago (the origin of the clade) to 1 million years ago (the dispersal of *Homo*). One model used both average mammalian speciation and longevity rates, and taxonomic specific rates of diversity in geographical areas to calculate species numbers, and suggested between 12 and 14[176] (Fig. 11.9).

Before the discoveries of the new taxa of the 1990s – *Ardipithecus ramidus, Australopithecus anamensis, Australopithecus garhi, A. bahrelghazali, Orrorin tugenensis, Sahelanthropus tchadensis,* and *Kenyanthropus platyops* – and the splitting of *Homo habilis* into two species, less than half the predicted number of hominin species had been identified. The number of known species of early hominin has now increased, but may still be an underestimate. Thus, the family trees that are constructed today are probably missing at least a quarter of their branches, and possibly more.

## The problems of "lumping"

As noted earlier, some of those missing branches may be occupied when new fossil species are recovered from the ground. But, as Ian Tattersall of the American Museum of Natural History has recently argued, other factors may apply:

> Over the past several years increasing attention has been paid to the search for patterns in the human fossil record. The reliability of any attempt to recognize pattern, however, is constrained by the accuracy with which we are able to recognize species in that record. . . . [It] is hard to avoid the conclusion that under current taxonomic practice there is a distinct tendency to under-estimate the abundance of species in the primate, and notably the hominin, fossil record.[177]

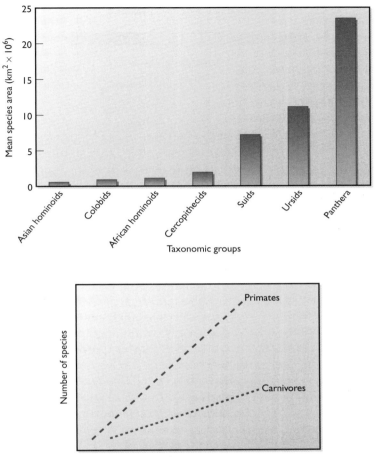

**FIGURE 11.9 Species area and taxonomic diversity:** Among extant species, the larger the geographical area, the greater the number of species is likely to be. However, the rate at which more species occur differs according to the type of animal, with primates being far more speciose than carnivores. Using these models, it is possible to predict the number of species likely to have occurred in different hominin groups according to their known geographical range, and compare this with the observed number.

The taxonomic practice to which Tattersall refers relates to the interpretation of anatomical variability. During the first half of this century,

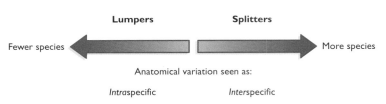

**FIGURE 11.10**
**Lumpers and splitters:**
Different philosophical
and methodological
approaches yield
different views of the
species richness of the
fossil record. In its early
years, anthropology was
dominated by splitters,
which yielded a plethora
of species. Sentiment
then switched to
lumping, which
underestimated species
richness. Recently, a
swing away from
lumping has occurred,
but not a return to the
previous excesses.

scholars commonly assigned a new species name to virtually each new fossil unearthed. In this "splitting" paradigm, each variant in anatomical structure was taken as indicating a separate species. The result was a plethora of names in the hominoid record. In 1965, Elwyn Simons and David Pilbeam, both then at Yale University, rationalized this paleontological mess and reduced the number of genera and species to a mere handful (the "lumping" paradigm)[242] (Fig. 11.10).

Lumping became the guiding ethic of anthropology. Taken to its extreme, it led to the "single-species hypothesis," which became popular during the 1960s and early 1970s. Promulgated by Loring Brace and Milford Wolpoff of the University of Michigan, this notion explained all anatomical differences among hominin fossils at any point in time as within-species variation. In other words, only one hominin species existed at any one time, making the hominin family tree a progression of species through time, with each following on from the one before.

Although the single-species hypothesis is no longer considered valid, Tattersall notes the persistence of a tendency to interpret anatomical differences as within-species variation rather than among-species variation. One reason for this trend is that, because of the nature of the system, no practical guide has been developed to explain how much anatomical difference between two fossils signals the existence of separate species. "The reason for this is, of course, that there is no direct relationship, indeed no consistent relationship at all, between speciation and morphological change," says Tattersall.

In other words, a daughter species might sometimes diverge from the parental species but develop very little obvious anatomical difference, while considerable differences might arise in other cases. Unless the living animals are available so that you can observe their behavior, it is often impossible to know whether the individuals belong to one species or two. Even with living species it is often difficult to be sure whether they are true species in the biological sense (Fig. 11.11). As a result, it is obviously easier to subsume anatomical differences under within-species

**FIGURE 11.11 Splitters and lumpers:** Louis Leakey (seated) was a keen splitter, reflecting the philosophy of his time; his son Richard Leakey was more cautious, reflecting changing times. (Courtesy of the L. S. B. Leakey Archives.)

variation rather than to argue for separate species. This tendency has certainly become a tradition in anthropology. The result, argues Tattersall, "is simply to blind oneself to the complex realities of phylogeny." In other words, the true hominin family tree – the one that actually happened in evolutionary history – almost certainly is more bushy than the ones currently drawn by anthropologists.

Obviously it is important to determine the number of species of hominin for a variety of reasons, including the better reconstruction of phylogenetic relationships. However, it is also important to remember that the way species are defined, and the number that emerge from analyses, influence answers to questions about the evolutionary process itself.[64,65] In Chapter 3 we looked at how different models of evolutionary change produced somewhat different patterns – punctuated equilibrium, Red Queen, evolutionary geography, etc. – and the pattern of hominin diversity can be fed back into these debates.

## Which data are the most reliable phylogenetic indicators?

Paleontologists reconstruct phylogenies from comparisons of anatomical similarities present in fossil specimens. As discussed in chapter 5, only those similarities that result from a shared evolutionary history (homologies) can reliably lead to accurate phylogenies. Similarities that result from independent, parallel evolution (homoplasies) may lead to erroneous phylogenies. Most anthropologists now accept that homoplasy has been common in hominin evolution,[127,129,326] but, as we will see later in this chapter, less agreement has been reached regarding which traits are homoplasies between certain lineages and which are not. Again, cladistic analysis should, in principle, help resolve this issue.

A further obstacle to accurate phylogenetic reconstruction arises from the way in which different traits are treated. In anthropology, phylogenetic reconstruction is based almost exclusively on cranial traits, for the very good reason that postcranial fossils are much rarer. In one of the more complete cladistic analyses of hominin phylogenetics, Randall Skelton (of the University of Montana) and Henry McHenry (of the University of California, Davis) employed 77 such traits. In their 1992 paper, Skelton and McHenry addressed the issue of the values assigned to these traits, identifying two problems: the independence of the traits and sample bias.[326]

Is it possible to reconstruct a reliable hominin phylogeny?

If all 77 traits were independent, then they would provide information on 77 evolutionary transformations, forming a powerful body of evidence. Anatomical traits are not independent, however, but form parts of trait complexes (Fig. 11.12). For instance, an important trend in early hominin evolution was toward heavy chewing in order to process tough plant foods. This development is seen, for example, in an increase in the size of

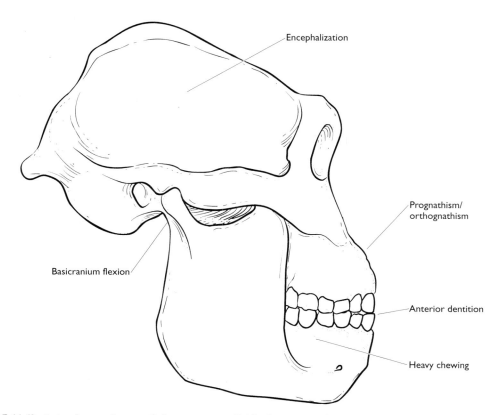

**FIGURE 11.12 Interdependence of characters:** Individual anatomical traits are typically parts of functional complexes and are not evolutionarily independent. These five functional complexes are associated with the hominin cranium.

molar teeth and in the thickness and depth of the mandible. Bigger teeth and more powerful chewing also require a more robust mandible, changes in face structure, and possibly alterations in the mechanics of muscles that move the jaws. Changes in the size of molar teeth and the robusticity of the mandible are therefore linked as part of an evolutionary package and are not independent of one another.

Thus, phylogenetic analyses should logically group traits into functional packages, rather than treat them as independent. In their analysis of hominin phylogeny, Skelton and McHenry identified five such functional complexes among the 77 traits: heavy chewing (34 traits), anterior dentition (11 traits), basicranium flexion (11 traits), prognathism/orthognathism (8 traits), and encephalization (3 traits). Grouped in this way, the 77 traits give phylogenetic information on just five evolutionary transformations – not 77 (Fig. 11.13). Even these five functional complexes are not completely independent, however, because the masticatory system involves many parts of the cranium. For instance, the evolution of traits associated with anterior dentition is linked in part to the evolution of heavy chewing, as is the shape of the face and certain cranial traits, such as the possession of a sagittal crest.

Biased sampling, which Skelton and McHenry identified as the second problem, is evident from the traits listed above – namely, some aspects of anatomy are more widely represented than others in the fossil record. Traits associated with heavy chewing are obviously the most common, because teeth and jaws are the most resilient parts of the cranium and consequently become part of the fossil record much more frequently. For this reason, anthropologists have concentrated much of their work on teeth and jaws, including basing phylogenetic reconstruction on them. Teeth and jaws, however, are particularly susceptible to homoplasy: species with similar diets will develop similar dentition through natural selection. Teeth and jaws, and their interpretation, may therefore receive more attention than their phylogenetic reliability justifies.

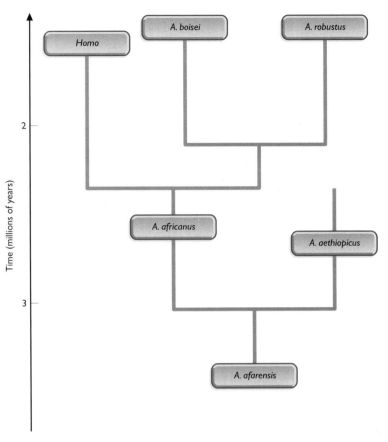

**FIGURE 11.13**
**Hominin phylogeny according to Skelton and McHenry**

## Key questions in hominin phylogeny

Three key questions have been the focus of most of the phylogenetic reconstructions of early hominins (Fig. 11.14):

♦ the relationship of *Australopithecus afarensis* to early and later hominins;
♦ the relationships among the robust australopithecines (*A. aethiopicus, robustus, and boisei*); and
♦ the origin of the genus *Homo*.

As we saw earlier, two decades after the first specimens of *A. afarensis* were discovered no consensus had been reached on whether they represent one extremely sexually dimorphic species or two less variable species (one large and one small). Until recently, the majority view held that just one species was present between 3.9 million and 2.9 million years ago, and that this species was ancestral to all later hominins. The recent

Relationship of A. afarensis
to early and later hominins

Relationships among
the robust australopithecines

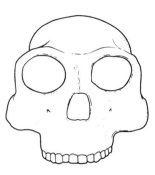

Origin of the genus Homo

**FIGURE 11.14 Key issues in early hominin phylogeny**

discovery of more *A. afarensis* fossils from Ethiopia did not resolve this difference of opinion.

The recent discoveries of new taxa for the period prior to 3 million years proves false the often implicit assumption that *A. afarensis* was the founding species of the hominin clade. The likelihood that *ramidus*, *platyops*, and *anamensis* were part of a bushy phylogeny prior to *afarensis*, rather than being stages in a single, transforming lineage, impacts on the status of *afarensis*. (The notion of a single, transforming lineage does have its supporters, however.) It is unlikely that a phylogenetically bushy clade would be reduced to a single species, which then gives rise to further bushiness. Unlikely – but not impossible. Further fossil finds in the period 5 million to 3 million years ago will be necessary to resolve this issue.

The question of robust australopithecine relationships affects the placement of *A. aethiopicus* in the evolutionary tree: is it ancestral to the two later robust australopithecines, or is it separate from them? The issue of the origin of the genus *Homo* concerns the identity of its direct ancestor: is it *A. afarensis*, *A. africanus*, or some as yet unknown third species? These two questions will be considered through Skelton and McHenry's cladistic analysis, not because it is universally accepted (it is not, though it is widely respected), but because it offers a strategy for addressing some key problems, particularly that of homoplasy.

## The Skelton/McHenry analysis

As stated above, Skelton and McHenry performed a cladistic analysis of the 77 cranial traits in several ways: they treated the traits as if they were independent; they compared the five functional complexes discerned; and they grouped the traits by anatomical region (face, anterior dentition, posterior dentition, mandible, palate, basicranium, and cranial vault), which is another way of overcoming linkage between traits. They then compared the results from these various analyses. Their study was performed prior to the discovery of *Ardipithecus ramidus* and *Australopithecus anamensis*, and it took the conservative position that *Australopithecus afarensis* is indeed a single species. The analysis of the later hominins is unaffected by these recent discoveries. One of the most important, and controversial, conclusions of their work was that traits associated with heavy chewing in hominins are subject to homoplasy.

Mentioned earlier was the trend in early hominin evolution toward ever-heavier chewing. Traits associated with heavy chewing are least developed in *A. afarensis* and most strongly developed in *A. boisei*. The "black skull," *A. aethiopicus* (see Fig. 10.14 above), also possesses large cheek teeth and a robust mandible, which many anthropologists interpret as indicating an ancestral relationship to *A. boisei* and the South African robust australopithecine, *A. robustus*. The anterior dentition of *A. aethiopicus*, however, is more similar to that of *A. afarensis* than to that of the other robust australopithecines. The degree of prognathism in *A. aethiopicus* resembles that in *A. afarensis*, while the other robust australopithecines are much less prognathic and more similar to *Homo*. The most parsimonious tree from a phylogenetic analysis using only traits related to the functional complex of heavy chewing gives a cladogram that links all three robust australopithecines as a clade. Analyses using posterior dentition, an anatomical region associated with heavy chewing, produce the same phylogenies.

By contrast, most other types of analysis (taking the 77 traits independently, and assessing the other functional and regional complexes, either independently or grouped) yield a different series of possible cladograms, with one being most common. In this tree, *A. aethiopicus* is not ancestral to the other australopithecines, but rather a megadont form of *A. afarensis* that became extinct with no descendants. The persistence of this particular cladogram is evidence of its strength, say Skelton and McHenry, which implies that the traits associated with heavy chewing shared by *A. aethiopicus* and the other two robust australopithecines are homoplasies – not the result of common ancestry. A second aspect of Skelton and McHenry's phylogeny that differs from phylogenies constructed by other workers is its proposal of a close link between the other robust australopithecines (*A. boisei* and *robustus*) and earliest *Homo* (discussed below). The proposed phylogeny requires three hypothetical ancestors – species that are as yet unknown, but are implied by the evolutionary transitions in the phylogeny.

Skelton and McHenry's phylogeny is as follows. *A. afarensis* is the most primitive early hominin after *Ardipithecus ramidus* and *Australopithecus anamensis*, from which it probably derived. They propose that *afarensis* gave rise to an as yet unknown species that was *aethiopicus*-like in some ways (in traits not related to heavy chewing); this species was the common ancestor of *aethiopicus* on one hand, and gave rise to *A. africanus*, early *Homo*, and the later robust australopithecines on the other. *A. aethiopicus* is therefore viewed as a side branch that became extinct, while *A. afarensis*

Are the robust australopithecines monophyletic?

was ancestral to all later hominins (but was not their common ancestor). *A. africanus* is derived from the *aethiopicus*-like ancestor, and in its turn gave rise to another proposed *africanus*-like species; this species was the common ancestor of earliest *Homo* on one hand and the robust australopithecines (via a proposed *robustus*-like common ancestor) on the other. Many anthropologists agree that *robustus*-like anatomy is likely to be ancestral to *boisei*. The close relationship between *Homo* and *A. robustus* and *A. boisei* (they have a common ancestor, to the exclusion of other hominins) is reflected in a more flexed cranial base, a deeper jaw joint, less prognathism, and greater encephalization than *A. africanus*.

This phylogenetic scheme, like other proposed alternatives, implies considerable homoplasy in hominin evolution, particularly in the heavy chewing complex. In contrast to Skelton and McHenry's proposal, other schemes have proposed that *A. aethiopicus* was ancestral to the other robust australopithecines, and that heavy chewing traits are homologous (not homoplasic). A recent cladistic analysis by David Strait and Frederick Grine, at the State University of New York, Stony Brook, strongly supports this view (the monophyly of the three robust australopithecines).[327] This phylogeny shifts the requirement for homoplasy to other traits – namely, anterior dentition, basicranial flexion, encephalization, and prognathism/orthognathism – that *A. aethiopicus* shares with other species.

A second area of homoplasy appears in the evolution of *Homo*. The shape of the face and small cheek teeth superficially resemble those of *A. afarensis*. Thus, these traits in *Homo* must have resulted from the retention of primitive traits present in *afarensis*, in which case *afarensis* would be the direct ancestor of *Homo*, or via a reversal of the hominin trend, in which case *africanus* would be the ancestor. A study of the ontogeny of facial development reveals that the formation of facial anatomy in *Homo* is unique, not a primitive retention. The well-documented reduction in the size of cheek teeth later in the *Homo* lineage also leads to the conclusion that this trend began with early *Homo*, and thus was not a primitive retention at this stage. If, as Skelton and McHenry point out in their analysis, the face and dentition of *Homo* are indeed uniquely derived, then these traits provide no useful information about the large-toothed australopithecine (known or yet to be discovered) from which it evolved; other, shared traits, such as basicranial flexion and orthognathism, are necessary to link *Homo* to *A. africanus*.

Skelton and McHenry's preferred phylogeny is one of several that can be seen in the anthropological literature; its strength, however, lies in its cladistic methodology and thoughtful treatment of potential biases. Many other schemes derive more than one lineage from *A. afarensis*, for instance, and designate *A. aethiopicus* as the ancestor of the other robust australopithecines. The most controversial aspect of the Skelton/McHenry phylogeny is its suggestion that the robust australopithecines are not monophyletic.

What are the phylogenetic origins of *Homo*?

# Phylogeny and increasing hominin diversity

When the pioneers of paleoanthropology were reconstructing human ancestry it was a relatively straightforward process, albeit one that engendered enormous controversy. The simplicity came from the fact that there were relatively few fossils, few taxa, and a largely qualitative method. The resulting phylogenies differed more in style of presentation than in substance. As discussed above, cladistics and the increased fossil record have changed this situation.

It is probably fair to say that there is no firm consensus on the phylogeny of the early hominin at the moment (Fig. 11.15). This is not to say that there is complete anarchy in the field. There is a broad acceptance that the early australopithecines such as *anamensis*, *africanus*, and *afarensis* give rise to the later and more robust ones, and that these forms also make possible ancestors of *Homo*. Where there is disagreement is over details of the particular branching sequence. The position of *Sahelanthropus*, *Orrorin*, *Ardipithecus*, and *Kenyanthropus* are all uncertain, as they have yet to be studied in detail, while other finds such as *Australopithecus garhi* do not fit neatly into any particular scheme.

If this state of affairs is a result of increased fossil evidence, it might also reflect the actual complexity of the evolutionary process. In finishing, it is worth looking briefly at what comes out of the details of any cladistic analysis. As discussed above, most analyses are based on parsimony, which means that they look for the shortest and simplest evolutionary tree that can account for all the evidence. The best tree is the shortest tree, and in theory that tree is also perfect in having no reversals or homoplasies (products of convergent evolution). However, as we saw from the Skelton and McHenry analyses, that is seldom the outcome. In practice, for most hominin phylogenetic analyses the shortest tree is generally one that may have up to 40% homoplasies.[325,328] This is not that unusual compared with

**FIGURE 11.15**
**A forest of hominin evolutionary trees:** Numerous phylogenetic interpretations of hominin history have been proposed. Hypothesis four is based on Skelton and McHenry's analysis. Hypothesis three shows the three robust australopithecines as being monophyletic; some scholars place these species in the genus *Paranthropus*.

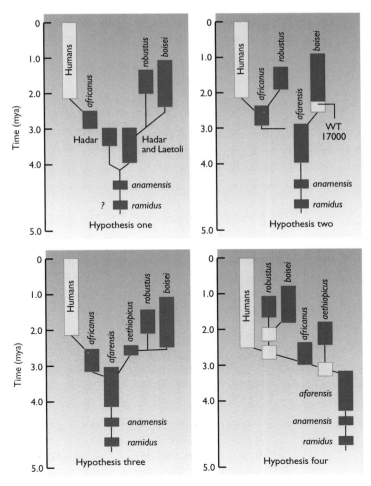

mammals as a whole. In other words, homoplasy is common in evolution, and the hominins in this as in many other ways prove to be unexceptional.

This should not surprise us. It is possible to see that evolutionary change is a response to environmental conditions, and these are local, so where different populations experience the same selective pressures, they will tend to evolve the same characteristics, especially if the populations are phylogenetically similar anyway. The difficulty of successfully reconstructing the evolutionary tree of the early hominins may be disappointing, but the consolation is perhaps that the underlying reasons are informative about the pattern and process of evolution.

## BEYOND THE FACTS
# THE PROBLEM OF PROGRESS

*The issue: when evolutionary theory was first widely accepted in the nineteenth century, it was considered to be broadly synonymous with the idea of progress. It took most of the twentieth century to shake off this association, and to many evolutionary biologists the term "progress" is anathema. However, it remains the case that evolution is often concerned with directional change, and the development of new features that are in some senses an "advance" on the ancestral form. Is there therefore a case for reconsidering the value of the concept of progress in evolution?*

If asked, most people would equate the word "evolution" with progress. Certainly during the early days of the Darwinian revolution there was a strong tendency to link the new ideas of biology – focusing on change over time – with the widely held ideas of the eighteenth-century Enlightenment, that human history was one of gradual progression from an age of darkness and ignorance, toward a more liberal and tolerant (and capitalist) society. Evolution provided a scientific underpinning to the political ethos of the time.

In contrast, if asked, most evolutionary biologists today would deny strongly that there is any relationship between the idea of progressive change and the principles of evolution. Evolution, they would say, is not about directional change, and certainly has no predetermined goal, in the way that social theorists might recognize. Rather, evolutionary change is about the local, the immediate, the demands of adaptive needs at a particular time, and the way that each lineage changes according to its own environmental demands. Progress, therefore, has no place in evolutionary biology.

Anthropologists feel this even more strongly. Early anthropologists equated different races and cultures with stages of evolution, with a progressive trend from the primitive (hunter-gatherers, savages, etc.) to the civilized (and largely European). One of the great achievements of anthropology in the last century has been to overthrow these simplistic notions, in the same way as evolutionary biology has replaced the idea of progress with that of adaptation.

We might therefore conclude that terms such as "progress" should be consigned to the dustbin of scientific history. But it is worth reflecting on what it was, beyond their prejudices, that made earlier generations interested in the idea of progress. Two observations were of importance.

The first was that, across geological time, and across human prehistory, entirely novel forms do appear and develop. There was a time when there were no vertebrates, no mammals, no humans, and so the explanation of entirely novel forms is a major problem, regardless of the persistence of forms that had been established for a long period of time. The world of the Pre-Cambrian, for example, is clearly less diverse in terms of levels of organization than is the present day. Progressive trends – that is, directional patterns in evolution – do occur and require explanation.

The second observation is that the early anthropologists could observe the way in which one society could come to replace another; progress was one way of explaining that change. In ecological contexts, we can also see this in the way in which one species can often displace another that has been present for enormously long spans of time – for example, the way eucalyptus trees are gradually spreading through the forests of the world and replacing indigenous forms.

While no one would argue for a return to Victorian notions of the primitive and advanced, nonetheless the biological and human world is made up of events where novel forms have emerged and directional change has occurred. Progress remains a concept to be grappled with in human evolution, and it is necessary to ask what "progressive" trends do occur, and how they can be accounted for.

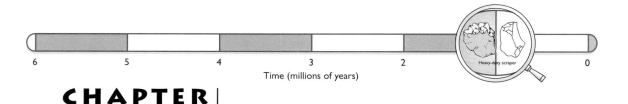

# CHAPTER 12

# BEHAVIOR AND EVOLUTION OF EARLY HOMININS

## EARLY TOOL TECHNOLOGIES

**KEY QUESTION** What additional light does the archeological record throw on the early evolution of the hominins?

The British anthropologist Kenneth Oakley once defined humans with the phrase "Man the Toolmaker." In fact, humans are not the only animals that use tools; some birds, sea otters, and chimpanzees regularly employ such instruments.[329] Humans, however, whatever their subsistence, are the only animals that have become dependent on the fruits of technology. For this reason, and because the archeological record provides good evidence for the evolution of technology, it has become important to consider how stone-tool technology relates to early hominin evolution, and at what point the hominins can be said to have become technologically dependent. Of course, as is so often the case in paleoanthropology, what was once a simple question has become far more difficult. This is partly because of the primatological evidence showing that chimpanzees can at least use stone tools, and partly because there are now so many different hominin species that it is difficult to determine which one may have been responsible for technological innovation and use. Here we will briefly consider the earliest evidence for stone tools, while in later chapters we will consider broader issues relating to the development of technology and its implications.

Many forms of material may be used as tools, such as stone, wood, bone and antler, bark and leaves. Because of the vagaries of preservation, only implements constructed from stone are likely to enter the earlier reaches of the archeological record. Archeological bias aside, stone artifacts may be extremely useful in simple subsistence strategies, both for the multifarious functions they may perform (such as slicing, scraping, and hammering) and in making other tools (such as digging sticks or spears). A record of stone-tool technologies in the earliest human groups therefore provides an important, though incomplete, insight into subsistence activities. In any case, stone artifacts offer the sole evidence of the technological life of all but our most recent ancestors.

## The archeology of stone tools

Stone artifacts have been collected by amateurs and professionals alike for centuries and studied as evidence of earlier societies. The mode of study, however, often focused on the implements as phenomena in themselves, with a great emphasis on classification of types. Today a strong interest has developed in studying artifacts within the subsistence context of early hominins. In addition to attempting to understand the functions of individual artifact types, archeologists use these relics to answer the following kinds of questions: how broad was the diet? Specifically, to what extent was hunting an important subsistence activity? Did the social context of subsistence activity include a "home base," such as occurs in modern foraging people? How did the hominins exploit their range, and how large was it? Thus, experimental archeology, once practiced by a small group of experts in a limited way, has emerged as an important research technique, allowing researchers to use stone implements with the aim of understanding early tool technologies.

How do stone tools vary?

Stone-tool assemblages have been classified into five categories, or modes, that are defined by characteristic artifacts in them[329] (Fig. 12.1); these categories appear sequentially through time, but may overlap when earlier modes persist after the appearance of later modes. Mode 1 technology, the earliest, is based on simple chopping tools that are made by knocking a few flakes off a small cobble. Mode 2 is characterized by tools that require more extensive conceptualization and preparation, and in particular the bifacially flaked handaxes. In Mode 3, large cores are preshaped by the removal of large flakes and then used as a source of more standardized flakes that are retouched to produce a large range of artifacts. Mode 4 technology is characterized by narrow stone blades struck from a prepared core. Mode 5 consists of microlith technology, which, as implied, constitutes the production of small, delicate artifacts.

This classification system, which was developed by the late Grahame Clark, of the University

**FIGURE 12.1**
**Technological modes:** Grahame Clark suggested that there were five basic modes of producing stone tools, and these can be said to cover the fundamental technological variation of the Pleistocene, although there is considerable variation within them (see text).

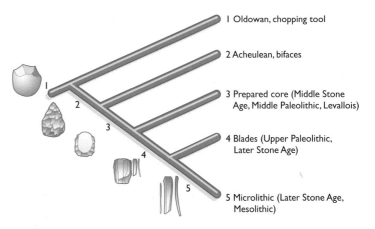

1 Oldowan, chopping tool

2 Acheulean, bifaces

3 Prepared core (Middle Stone Age, Middle Paleolithic, Levallois)

4 Blades (Upper Paleolithic, Later Stone Age)

5 Microlithic (Later Stone Age, Mesolithic)

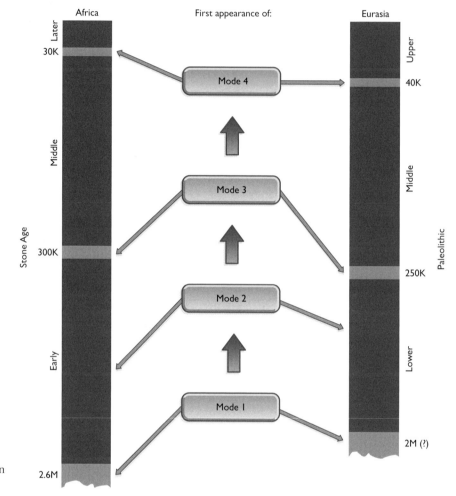

**FIGURE 12.2**
**Archeological units and periods:** For reasons related to the history of the science of archeology and the impact of new discoveries, the classification of the different periods and stages of cultural development in sub-Saharan Africa and Eurasia represents an uneven mixture of technological innovation and chronology.

of Cambridge, permits a description of the characteristics of archeological assemblages, not of archeological time periods.[330] For instance, Mode 1 technology appeared in Africa some 2.6 million years ago and persisted (as an opportunistic practice) until historical times. Moreover, the first appearance of a particular mode often differs in Africa and Eurasia. For instance, Mode 4 blade tools were produced in Africa intermittently over the last 100,000 years, but did not enter the European record until 40,000 years ago. Such differences almost certainly reflect the dynamics of the adaptation and migration among hominins.

For reasons related to the development of the science of archeology, a different terminology is used to describe archeological time periods in sub-Saharan Africa and those in Eurasia (Fig. 12.2). In Africa, the time before the appearance of agriculture and iron is known as the Stone Age. It is divided into three parts: the Early Stone Age (ESA), the Middle Stone Age (MSA), and the Late Stone Age (LSA). In northern Africa and Eurasia, stone-tool cultures prior to the Neolithic are termed the Paleolithic and are divided into three stages that are roughly equivalent to those in the Stone Age: the Lower Paleolithic, the Middle Paleolithic, and the Upper Paleolithic. These stages have been defined according to cultural evolution – a somewhat confusing system given that, while the boundaries between the stages are relatively clear in Eurasia, Africa has been associated with a more continuous flow of development. This difference in character may reflect *in situ* cultural change in Africa and evidence of population incursions in Eurasia. As mentioned earlier, the timing of first appearance of characteristic cultural artifacts (such as blades) often varies between the two geographic regions.

Bearing in mind the elasticity of stage boundaries, technology development unfolded as follows. The beginning of the ESA corresponds with the first appearance of Mode 1 tools, 2.6 million years ago; the entire ESA includes the first appearance of Mode 2, approximately 1.5 million years ago, and ends with the first appearance of prepared cores (Mode 3), which also marks the beginning of the MSA, 300,000 years ago. Traditionally, Mode 4 is most strongly associated with the Upper Paleolithic and the LSA, although some elements can be found intermittently in the MSA.

In Eurasia, the Lower Paleolithic begins when humans moved beyond Africa, perhaps close to 2 million years ago, and ends with the first appearance of prepared cores (Mode 3), some 200,000 years ago. The Middle Paleolithic begins with the first appearance of Mode 3 and ends with the first appearance of Mode 4 tools, 40,000 years ago. The Upper Paleolithic, and its extension the Mesolithic, begins with the first appearance of Mode 4, encompasses the appearance of Mode 5, and ends with the agricultural revolution. It must be stressed that all these dates are approximate, and it is important to remember that changes in technology are not synchronous over large areas. The divisions of the archeological record are not strict markers of time in the way that geological divisions, such as the Pliocene and Pleistocene, are.

This section will focus on the first part of the African ESA. The next will describe the technologies in the remainder of the ESA and the MSA, and the Lower and Middle Paleolithic of Eurasia. The archeology associated with the origin of modern humans (the MSA and LSA of Africa and the Upper Paleolithic of Eurasia) is the subject of chapter 16.

## The earliest known tools

The oldest stone tools in the archeological record are dated to approximately 2.6 million years ago, and are known from sites in the Lower Omo Valley, the Hadar region, and the Gona region of Ethiopia, and the western shore of Lake Turkana, Kenya.[331] The artifacts from the Lower Omo Valley are atypical, in that they are small quartz pebbles that were shattered to yield sharp-edged implements. Most tools dating from the period

What is the earliest evidence for stone tools?

**FIGURE 12.3**
**Experimental archeology:** These artifacts were made by Nicholas Toth as a way of understanding the principles of manufacturing the Oldowan assemblage. (top row) Hammerstone, unifacial chopper, bifacial chopper, polyhedron, core scraper, bifacial discoid. (bottom row) Flake scraper, six flakes. An actual tool kit would comprise mainly flakes. (Courtesy of Nicholas Toth.)

2.6 million to 1.5 million years ago were made from lava cobbles, and constitute a range of so-called core tools and small, sharp flakes. Generically, the technology is known as Oldowan, after Olduvai Gorge, Tanzania. (The gorge was once called Oldoway Gorge; hence the derivation of the tool technology's name.)

The technology, which is Mode 1, was defined on the basis of the artifact assemblages found in bed I and lower bed II at Olduvai Gorge (1.9 to 1.6 million years old) through the long and meticulous work of Mary Leakey; the results of her research were published as a monograph in 1971.[111] The artifacts fall into four categories (Fig. 12.3):

♦ tools, which include types such as scrapers, choppers, discoids, and polyhedrons;

♦ utilized pieces, such as large flakes produced in the manufacture of tools, having sharp edges useful for cutting;

♦ waste, or small pieces produced in the manufacture or retouching of tools and utilized pieces; and

♦ manuports, which are pieces of rock carried to a site but not modified.

The half-dozen or so tool types named in the typical Oldowan assemblage were not tightly restricted categories such as would be produced by a stone knapper with distinct mental templates for specific implements. The different forms tended to flow into one another typologically, and they carry an air of opportunistic production. This process contrasts with later finds in the archeological record, which exhibit evidence of tighter control over the production of specific tool types.

Frequently the labels applied to the various core forms implied function, such as scrapers and choppers. The small flakes removed from the cores were initially assumed to be waste, but may sometimes have proved useful as cutting tools. In the early 1980s, however, a series of experimental studies by Indiana University archeologist Nicholas Toth led to the conclusion that the real tools in the Oldowan assemblages were the flakes, and that the core forms represented the by-products of flake production[332] (Fig. 12.4). Toth discovered that undirected flaking of cobbles of different shapes led automatically to specific core forms, depending on the shape of the cobble used. For instance, a wedge-shaped cobble becomes a unifacial chopper with the removal of a few flakes; further flaking gives a bifacial chopper; and still further flaking yields a polyhedron. A hemispherical

**FIGURE 12.4 Cores compared:** A simple chopper from an archeological site (light color) compared with the same tool made recently. (Courtesy of Nicholas Toth.)

cobble is more varied in its potential transformations and can yield a core scraper in addition to unifacial and bifacial choppers and discoids (Fig. 12.5).

In a survey of the various archeological sites on the east side of Lake Turkana, in northern Kenya, Toth found that locations near the ancient lake edge provided relatively small lava cobbles as raw material, whereas sites nearer the margin of the lake basin, which is closer to the source of the volcanoes, offered the choice of larger cobbles. Toth's experimental stone knapping indicated that the number of core tool types produced during undirected knapping rose with an increase in the size of the starting material. Therefore, concluded Toth, "Much of the observed range of variation in core forms could be, to a large extent, due to variation in the size of the available raw material."

Toth did not suggest that the core forms were never used as tools; rather he concluded that they were not manufactured specifically for use as scrapers, choppers, or similar tools. In experimental butchering, Toth found that the most effective implement for slicing through hide was a small

**FIGURE 12.5 Tool profile:** A chopper and the flakes produced during its manufacture. (Courtesy of Nicholas Toth.)

flake (Fig. 12.6); a similar finding applied to dismembering and defleshing. For chopping residual dried meat from a scavenged carcass, however, a heavier implement was best, such as a large flake or a sharp-edged core (for example, a chopper). A heavy core or unmodified cobble was effective for breaking bone to gain access to marrow or brain. The manufacture of digging sticks was achieved with a range of implements: a sharp-edged chopper was useful for cutting a suitable limb from a tree, a flake or a flake scraper for fashioning the point, and a rough stone surface for honing the point. Flakes and scrapers offered an effective method for removing fat from hide. Nuts could be cracked easily with an unmodified stone hammer and anvil. Toth also discovered that antelope horns and broken bones from large animals could function as effective digging implements.

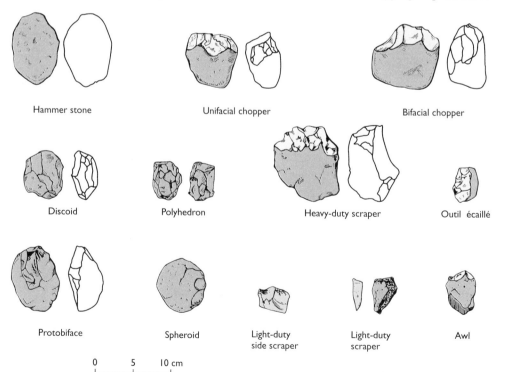

Hammer stone

Unifacial chopper

Bifacial chopper

Discoid

Polyhedron

Heavy-duty scraper

Outil écaillé

Protobiface

Spheroid

Light-duty side scraper

Light-duty scraper

Awl

0　　5　　10 cm

**FIGURE 12.6 Oldowan artifacts:** The manufacture of these simple pebble tools requires considerable skill.

Direct evidence of the application of an ancient tool is difficult to obtain, not least because the coarse nature of lava flakes does not sustain clear signals of the material with which it has been in contact. Nevertheless, Lawrence Keeley, of the University of Illinois, examined 54 flakes from a 1.5-million-year-old site from Koobi Fora, on the eastern side of Lake Turkana, and found evidence of use-wear on nine of them. Four had been used in butchering, three were applied to wood, and two were associated with soft vegetation.[117]

These and other studies give a sense of the variety of subsistence activities that became possible with the adoption of simple stone-tool technology. The small, sharp flake is, however, probably the most important implement and represents a technological and economic revolution. It allowed hominins to slice through hide and gain access to meat, with the stone flake literally opening up a new world of resources: potentially significant quantities of meat. The use of digging sticks permitted more efficient access to underground food sources, such as tubers. By broadening the diet in this way, hominins enriched and introduced a potential stability into their source of energy, which was important in the further expansion of the brain.[235]

What does the Oldowan tell us about early hominin behavior and abilities?

## Skillful Oldowan tool makers

The hominins' skill at producing flakes represented a technological revolution. Although the Oldowan industry is technically rather crude, the regular production of flakes is not a matter of chance. Three conditions must be met by a stone knapper who wishes to produce flakes routinely by percussion. First, the core must have an acute edge, one less than 90 degrees, near which the hammer can strike. Second, the core must be struck with a glancing blow about 1 centimeter from the acute edge. Third, the blow must be directed through an area of high mass, such as a ridge or a bulge. By examining the composition of cores and flakes at archeological sites, Toth could infer that the tool makers of 2.6 million to 1.5 million years ago had indeed mastered the percussion stone-knapping skill.

Similar comparative studies have shown that the ancient tool makers used the percussion technique exclusively to produce flakes. Toth demonstrated that of the three possible techniques for producing flakes – percussion, anvil (striking the core on a stationary anvil), and bipolar (striking the core with a hammerstone while it rests on an anvil) – percussion was the most efficient. Again, the ancient tool makers showed their skill, as they also did in avoiding flawed cobbles, which flake in unpredictable ways.

A debate over how much skill is required to carry out this simplest of stone knapping has recently been addressed in a most interesting fashion: by asking a bonobo (pygmy chimpanzee) to make Oldowan tools. This debate was initiated by Thomas Wynn, an archeologist, and William

McGrew, a primatologist. In 1989, the two researchers published a paper called "An ape's view of the Oldowan,"[333] in which they asked the following question: "When in human evolution did our ancestors cease behaving like apes?" In other words, given the opportunity and motivation, could an ape make Oldowan tools?

Wynn and McGrew analyzed the cognitive demands they deemed to be required in producing core tools and flakes, and concluded the following: "The spatial concepts required for Oldowan tools are primitive. The maker need not have paid any attention to the overall shape of the tool; instead, his focus appears to have been exclusively on the configuration of edges." They also concluded that these skills were no more demanding than those demonstrated by chimpanzees in their manufacture of certain tools from twigs. "All the spatial concepts for Oldowan tools can be found in the minds of apes," they noted. "Indeed, the spatial competence described above is probably true of all great apes and does not make Oldowan toolmakers unique."

Toth had an opportunity to test this assertion experimentally, when he collaborated with Sue Savage-Rumbaugh. Savage-Rumbaugh had spent 10 years working with a male bonobo, Kanzi, who had learned to use a large vocabulary of words displayed on a computerized keyboard and understood complex spoken English sentences. Toth encouraged Kanzi to make sharp stone flakes in order to gain access to favored food items enclosed in a box that was secured with string. Kanzi was an enthusiastic participant in the experiment over a period of several years. Despite being shown the percussion knapping technique, however, he never used it. Sometimes Kanzi produced flakes by knocking cobbles together, but without the precision inherent in the Oldowan technique; often he would simply smash the cobble by throwing it at another hard object, including the floor. Kanzi knew what he needed (sharp flakes) and figured out ways to obtain them (banging or throwing rocks), but he was not an Oldowan tool maker[334] (Fig. 12.7).

*Can apes make stone tools?*

Toth, Savage-Rumbaugh, and their colleagues described the first 18 months of observation as follows:

> So far Kanzi has exhibited a relatively low degree of technological finesse . . . compared to that seen in the Early Stone Age record. The amount of force he uses in hard-hammer percussion is normally less than ideal for fracturing these rocks. His flake angles when using hard-hammer percussion tend to be steep (approaching 90°), while Oldowan flakes were generally detached from more acute-edged cores (flake angles typically 75–80°). As yet, Kanzi's cores retain a very high proportion of their original cortex and are steep-edged and rather battered. The flakes he produces tend to be relatively small (generally less than 4 cm long) and often stepped or hinged, and his cores generally exhibit marginal (non-invasive) flake scars.

Thus, a clear difference separates the stone-knapping skills of Kanzi and the Oldowan tool makers, which appears to imply that these early humans

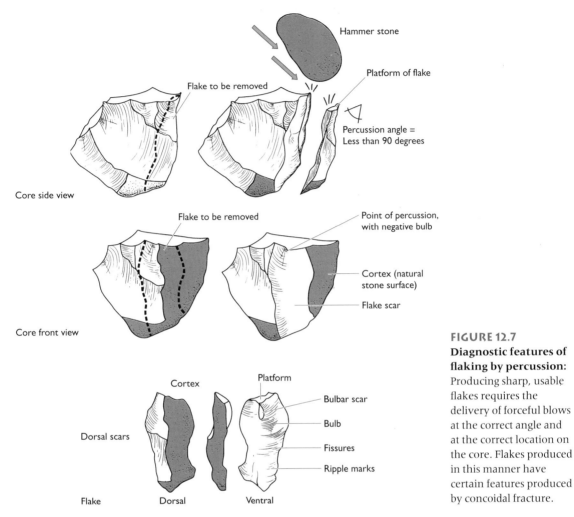

**FIGURE 12.7**
**Diagnostic features of flaking by percussion:** Producing sharp, usable flakes requires the delivery of forceful blows at the correct angle and at the correct location on the core. Flakes produced in this manner have certain features produced by concoidal fracture.

had indeed ceased to be apes. It remains uncertain, however, whether Kanzi's poorer performance reflects a cognitive or an anatomical limitation. The structures of bonobos' arms, wrists, and hands are different from those of humans, and may constrain the ability to deliver a sharp blow by snapping the wrist (a movement that is important in effective tool making). Because Kanzi produced sharp flakes that suited his purpose, it can be argued that he had no motivation to learn the percussion technique. Whatever the case, the Kanzi experiment gives some insight into possible stone-tool making prior to 2.6 million years ago. Archeologists recognize stone tools in the prehistoric record, on the basis of criteria that are appropriate for the Oldowan assemblage. If Kanzi can manufacture sharp flakes by non-percussive methods and use them to obtain food, perhaps hominins earlier than 2.6 million years ago did as well. But how could this process be recognized?

**FIGURE 12.8**
**Relationship between hominins and artifacts:** It is impossible to be sure about the association between particular taxa and the presence of artifacts, but in general terms there do appear to be some associations. (Courtesy of R. Foley and M. Lahr.)

Kanzi's tool-making sessions leave behind rocks that are battered or broken in a way that occurs through non-human forces in nature. Although enigmatic battered rocks older than 2.6 million years have been noted as possible artifacts, it has proved very difficult to distinguish between natural damage and human-caused damage. One conclusion of the Kanzi experiment, therefore, is that hominins prior to 2.6 million years ago might have made sharp stone flakes and used them in their subsistence activities, but this process would be very hard to demonstrate in the archeological record.

## Who made the tools?

In the period 2.6 million to 1.5 million years ago, several hominin species (*Homo* and *Australopithecus*) lived as contemporaries. How, then, is the identity of the tool maker to be discerned? Was it *Homo, Australopithecus*, or both? After some 1 million years ago, when only *Homo* existed, tool-making technology certainly continued – some of it very Oldowan-like. The argument from parsimony, therefore, would be that the earliest technology was also the product of *Homo*. In addition, the earliest evidence of stone-tool making coincides with the first appearance of *Homo*, approximately 2.5 million years ago (Fig. 12.8).

Randall Susman, of the State University of New York, Stony Brook, argues that robust australopithecines also had the manipulative potential to make tools. He bases his contention on the anatomy of the hand bones, and particularly the thumb, gathered from deposits in the cave of Swartkrans, South Africa.[316] The deposits, which are thought to date to roughly 1.8 million years ago, also contain stone tools and putative digging sticks. The breadth of

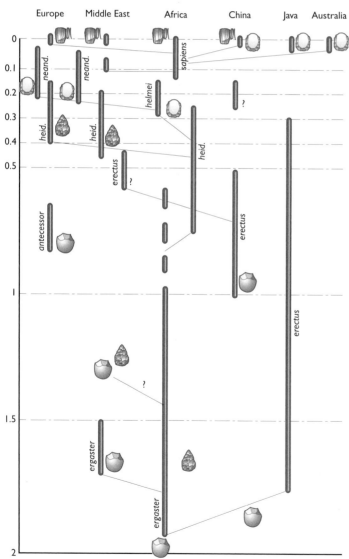

the thumb and the fingertips in the Swartkrans fossils indicates a degree of vascularity and innervation consistent with increased manipulative skill. Recent detailed studies of the thumb have shown that it was capable of forming a power grip, which is important in percussion stone knapping. The fingertips of modern apes and of *Australopithecus afarensis* are narrow; those of modern humans are broad. Susman concludes that, although early australopithecines were unable to make tools, later species, including early *Homo*, may have possessed this capacity.

Complicating the putative attribution of the finger bones to the *Australopithecus* species at Swartkrans is the fact that the same sedimentary layers have yielded fragments of *Homo*. As Susman points out, 95% of the hominin cranial bones found are those of *Australopithecus*, suggesting "an overwhelming probability" that the hand bones are indeed remnants of this species. He also notes apparent differences in the morphology of the thumb in the Swartkrans material and thumb morphology in a known *Homo erectus* specimen. Some observers contend that this evidence is too tenuous for definitive conclusions to be drawn.

## Absence of innovation in early technologies

About 1.6 million years ago, a new form of Oldowan industry emerged; this technology was named the Developed Oldowan by Mary Leakey.[111] It is characterized by a smaller percentage of choppers (less than 28% of all core tools), and a greater abundance of spheroids, subspheroids, and small scrapers. Bifacial tools appear for the first time in this new industry, including prototype handaxes and cleavers, which later come to characterize the Acheulean industry.

Because the Oldowan was defined on the basis of the assemblages found at Olduvai Gorge between 1.9 million and 1.6 million years ago, artifact assemblages taken from other sites are sometimes called by other names, even though they are very similar overall. More generic terms for this kind of artifact assembly are pebble-tool culture and chopping-tool culture. Such cultures persisted in Africa until at last half a million years ago and in eastern Asia until as late as 200,000 years ago. One very strong impression of the earliest technologies produced by hominins, therefore, is their tremendous stability and the lack of innovation carried through a huge tract of time. Such a lack is unthinkable to the modern human mind.

Why do early technologies stay the same for so long?

Aside from the redating of the Javan fossil sites, the oldest archeological sites outside of Africa are generally dated at less than 1 million years. Two claims, however, have been made for earlier occurrences apart from Asia: one in France and the other in Pakistan. Both finds include artifacts of the chopping-tool type, and both are said to be 2 million years old. Considerable doubt surrounds the validity of these dates and the reality of the artifacts at both sites.

# THE PATTERN OF EARLY HOMININ EVOLUTION

**KEY QUESTION** What is the overall pattern of evolutionary change among early hominins, and how does this relate to more general understandings of the evolutionary process?

We have now examined the evidence for the emergence and early evolution of the hominins. As has been shown, the evidence shows quite clearly that there is a good fit between the expectations drawn from molecular genetics and comparative anatomy of the living apes, and the fossil record – that is, the emergence of the hominin clade in sub-Saharan Africa, at some time toward the end of the Miocene. The exact details of the earliest hominins remain obscure, and current discoveries are changing the picture to a considerable extent. However, what has been demonstrated is that this early phase of evolution does not indicate a linear path to modern humans. Rather, we have evidence of a phase of hominin evolution that requires its own explanation, independent of what may have happened later in the genus *Homo*. In this section we will examine this pattern, and link it to the points about the general processes of evolution discussed in the first chapters of this book.

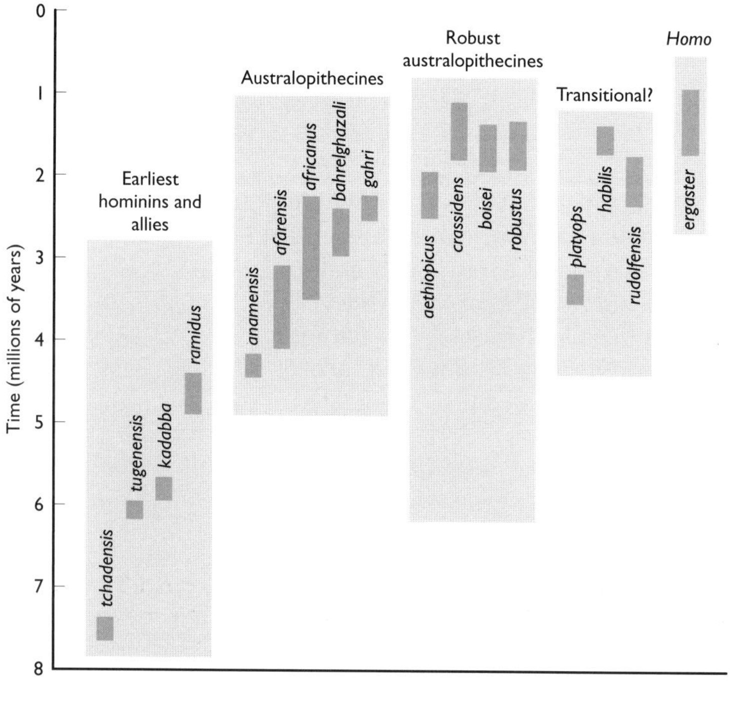

**FIGURE 12.9**
**Summary of hominin lineage distribution:** This shows current estimates of the chronological ranges of the early African hominins.

## Summary of early hominin evolution

Molecular data suggest that the hominin clade has its origins around 7–5 million years ago, and given the distribution of the African apes, to which humans are most closely related, in sub-Saharan Africa. *Sahelanthropus* or *Orrorin tugenensis*, although poorly known, may fit this prediction very well (Fig. 12.9). The slightly later *Ardipithecus ramidus*, which possesses a number of apelike features, however, indicates that this pattern is unlikely to be simple. It is likely that during the later Miocene there would have been a number of lineages, all closely related to each other, which could have been the common ancestor of chimps and humans, an ancestor of one or other, or an ancestor of neither. Because

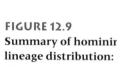

they were all at the base of a radiation, distinguishing between them is likely to be difficult.

These two genera represent a primitive group of hominins or hominin-like apes, whose adaptations are poorly known. By 4 million years ago the fossil evidence suggests that not only was the hominin lineage in existence in a cladistic sense, but the key adaptive traits that characterize the australopithecines were also present. *Australopithecus anamensis* quite clearly shows bipedalism, as is confirmed by the later hominins *A. afarensis, A. africanus*, and *A. garhi*. That this is again not entirely a simple linear development is indicated by the existence of another genus, *Kenyanthropus platyops*, at 3.5 million years ago in east Africa, and the probable existence of a further species in south Africa ("Little Foot": see chapter 10). It is also clear that this early phase involved an expansion of geographical range (*A. africanus* to the south, *A. bahrelghazali* to the north and west), and considerable adaptive diversity (the differences, for example, between the persisting arboreal adaptation of *A. afarensis* and the more terrestrial specializations of *A. africanus* and *A. garhi*) (Fig. 12.10).

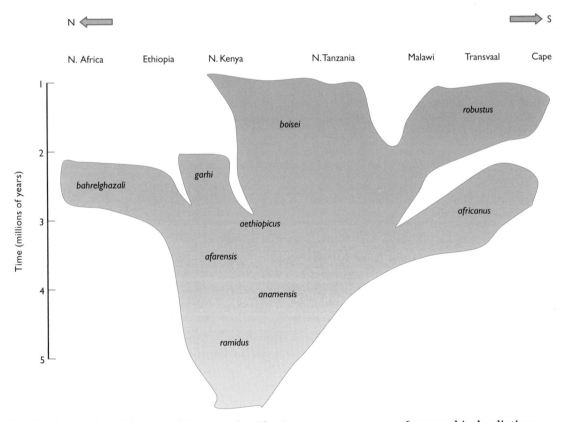

**FIGURE 12.10 Diversification of the australopithecines seen as a process of geographical radiation:** Most theories of species diversity suggest that lineages differentiate from each other as a result of becoming geographically isolated and dispersed. The australopithecines of Africa can be seen in this context. (Courtesy of Robert Foley.[82])

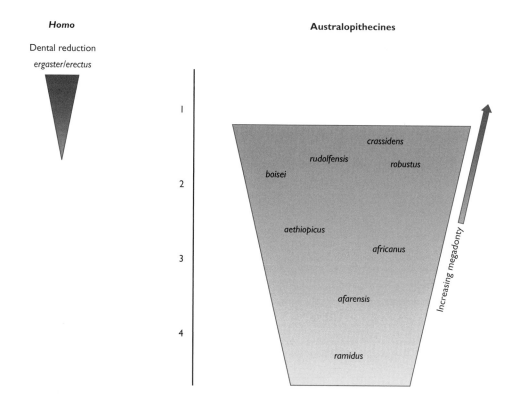

**FIGURE 12.11 Trends in early hominin evolution:** Once bipedalism is established, the dominant trend among early hominins is the increase of posterior tooth size – that is, megadonty. *Homo* represents a different trend.

From 2.5 million years ago or more the dominant trend of early hominin evolution is visible – that of megadonty (Fig. 12.11), seen in the enlargement of the molars and premolars, and the associated specialized musculature, of the later australopithecines. These forms become the typical hominins as the earlier and less specialized ones become less common and ultimately disappear. Robust forms are known in both the east (*A. aethiopicus, A. boisei*) and the south (*A. robustus*) of Africa. Whether they repesent a single evolutionary event, and hence deserve their own genus (*Paranthropus*), or are the result of independent evolutionary trends in different populations, remains to be determined.

Perhaps the most obscure event at the moment is the emergence of the genus *Homo*. While there are elements of *Homo* ancestry to be found in a number of species – *A. anamensis, A. africanus, K. platyops,* and *A. garhi* in particular – none of them is unambiguous, and there is a curious mosaicism about the pattern. This is further confounded by the difficulty of relating the earliest specimens of the genus *Homo*, namely *habilis* and *rudolfensis*, to either their putative ancestors or their later descendants (*H. ergaster* and *H. erectus*: see part 3 below). All that can be said with any certainty at the moment is that between 2.5 million and 2 million years ago there is a trend toward larger brain size, and either the proliferation of

What are the evolutionary trends of early hominins?

a surviving smaller-toothed primitive hominin, or the reversal in dental trends back to smaller and less specialized dentitions. These changes may be associated with more unambiguous stone-tool manufacture, a greater presence in environments for which there is less evidence for significant tree cover, and possibly an increase in body size.

## Adaptive radiations of early hominins

A major theme of this book is that human evolution must be studied in the context of evolutionary theory more generally, and so we can usefully ask how the pattern of early hominin evolution relates to the evolutionary processes discussed in the first part of this book. It will be remembered that in chapter 3 the normal pattern of evolution was described as an adaptive radiation. This occurs when a group of descendant species diversifies from a single ancestral stock or lineage to occupy different ecological niches. It is this process that produces the characteristic bushy-shaped diagram that Darwin himself identified, which is used to illustrate evolutionary patterns. This contrasts with the model of a ladder, where there may be a progressive linear trend in evolutionary direction. The bushy pattern results in multiple species and cladogenesis; the linear patter is dominated by anagenesis. The fact that early hominin evolution appears to be like that of any other group is indicative that the processes underlying this particular evolutionary event are no different from that of any other group. The fossil evidence for early hominin evolution suggests that there may have been several phases of radiation.

The first of these we might tentatively, characterize as the radiation of the African ape lineage, perhaps as it dispersed across Africa in the Late Miocene following its immigration there from Asia. In this scenario the earliest and primitive hominins are the sister clades of the other apes, coming to exploit the environments of the equatorial tropics. This radiation would include, in addition to the ancestors of the living great apes, *Orrorin* and *Ardipithecus*. The second radiation is probably that of the early australopithecines, originating in eastern Africa, and diversifying and spreading across southern and eastern parts of the continent. This radiation would include *A. anamensis*, *A. afarensis*, *A. africanus*, *A. bahrelghazali*, and *A. garhi*, and may also be considered to include *K. platyops*. The third radiation is derived from that of the early australopithecines, and comprises the robust australopithecines, characterized by their megadontic specializations. This would take in *A. aethiopicus*, *A. boisei*, and *A. robustus*. Finally, there is perhaps a radiation of more large-brained hominins, represented by *H. rudolfensis* and *H. habilis*, and linked to the later species of *H. ergaster*. It should be stressed that these radiations may not be entirely discrete events, and that there is clearly continuity between the earlier and later australopithecines, the differences in dental size being ones of degree, not kind.

If the pattern of early hominin evolution is that of an adaptive radiation, or a series of adaptive radiations, we can ask what underlies it. Adaptive radiations occur when a new way of exploiting an environment arises, and when species disperse into new environments in response to the opportunities offered by them. What were the special traits of the early hominin radiations?

The adaptive basis for the first, African basal radiation is extremely hard to pin down, as the evidence is so scarce, and any interpretation is dependent upon which evolutionary origin model is accepted. As has been discussed in chapter 8, if the African clade is itself a later Miocene immigrant into Africa, then to some extent this radiation is the result of dispersals of a new fauna spreading into all the available habitat, either at the expense of the last surviving Proconsulids, etc., if there were any, or into niches for which there were no indigenous primate candidates. What may have given them their advantage are the traits we associate with the apes more generally: high sociality, intelligence, flexibility in social organization, and perhaps basic meat-eating and tool use.

*Is early hominin evolution a series of adaptive changes?*

Although it may be the case that some of these early African apes were bipedal, it is with the second radiation, that of the australopithecines, that we can perhaps put forward a case for bipedalism being the fundamental basis for their success. This is not to say that they were fully bipedal in the sense that modern humans are, but they do show a more significant level of bipedal locomotion than any living ape, even if they may have retained some arboreal capabilities. It can be argued that it was this bipedal adaptation that led both to the more widespread African distribution, as they dispersed to exploit the expanding open savannah environments of the Pliocene, and to the diverse forms of early australopithecine.

Once the bipedal australopithecines had become established across a wide range of biomes within Africa, then the basis for their subsequent specialization was established. That specialization was megadonty. At the time of the appearance of the robust australopithecines there was a decrease in global temperatures, and an increased aridity in eastern Africa. This meant more open and more seasonal environments (Fig. 12.12). The large molars of the robust australopithecines are most probably adaptations to eating coarse, fibrous, and gritty plant materials – in other words, the vegetation of the more seasonal and drier environments of Africa. In order for these australopithecines to survive the dry seasons of this time, it was essential for them to be able to process large quantities of such food. That they did so is evidenced by both the microwear on the teeth and the overall high rate of wear. By the time an *A. boisei* was reaching maturity – that is, its third molar was erupting – the first and second molars were already worn down to the dentine.

The robust australopithecines, then, appear to be bipedal apes specialized to survive on coarse plant foods in highly seasonal and dry environments. This does not preclude them from also being scavengers and hunters, and eating other foods when the opportunity arose, but it was

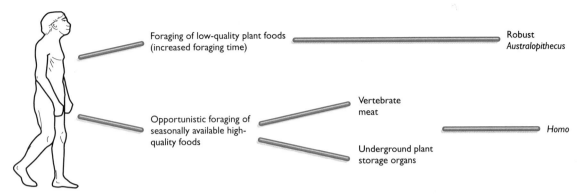

**FIGURE 12.12 Seasonality and the adaptive diversification of the hominins:** The different evolutionary trends among early hominins probably reflect adaptive differences between them. These may be the result of selection favoring different ways of coping with dry seasons – either specializing in coarse plant foods (robust *Australopithecus* with megadontic adaptation) or increasing the quantity of meat in the diet (*Homo*). (Courtesy of Robert Foley.[30])

this particular resource base that actually shaped their selective environment. That it was a successful one is attested by the diversity of the robust australopithecines, their widespread distribution, the fact that they are the most common Pliocene hominin fossil, and their persistence for over a million years.

The adaptive basis of the final radiation, that of *Homo*, will be dealt with in more detail in the next section. The Pliocene elements of it, characterized by the early, transitional *Homo* forms such as *H. rudolfensis* and *H. habilis*, are not a major part of the fossil record, and perhaps indicate not so much a radiation as the beginnings of a new trend in hominin evolution.

## Evolutionary trends and grade shifts

The problem of early *Homo* is perhaps the major area of uncertainty in early hominin evolution. The difficulty arises partly because the fossil record is both patchy and conflicting. The starting point for considering the problem is to recognize that the primary trend of hominin evolution from 5 million years to close to 1.5 million years is toward megadonty. We can see this just by plotting tooth size against time over this period. In that sense, the evolution of bipedalism and the development of a specialist, ground-dwelling ape, with ever larger molars, is a straightforward evolutionary trend. It can furthermore be accounted for by the increasing aridity of the environments across this time.

*Homo*, on the other hand, represents a reversal of that trend. The dentition is reduced and the face and cranium lack muscular specializations. In addition, where the brain size of the australopithecines remains relatively stable throughout their existence, in *Homo* there is an increase in brain size.

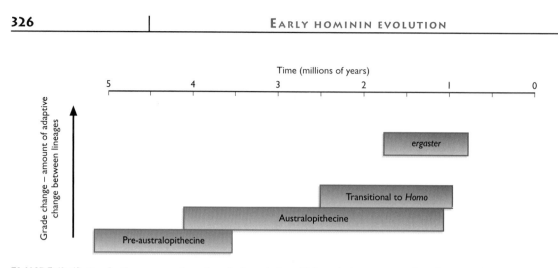

**FIGURE 12.13 Evolutionary grades of early hominins:** Although the concept of grades in evolution has been criticized, nonetheless there are marked changes in the course of hominin evolution, and many commentators consider that between the australopithecines/early *Homo* and *Homo ergaster/erectus* there is a major adaptive change.

Quite when, how, and why this reversal occurred is still disputed. Traditionally the trend toward *Homo*, and hence the recognition of the clade to which later members of *Homo* belong, were placed close to 2.5 million years ago, and associated with either *Homo habilis* or *H. rudolfensis* or both. The basis for this is the shared larger cranial capacity and smaller dentition. However, in some recent analyses, Bernard Wood and Mark Collard (an anthropologist at Washington State University) have suggested that when body size is taken into account, and a rigorous cladistic analysis is employed, early *Homo* tends to fall in adaptively with the australopithecines, rather than later *Homo*. If correct, this would imply that although the lineage that might give rise to later *Homo* was in existence by soon after 2.5 million years ago, the adaptive shift that led to its evolutionary success occurs only with *Homo ergaster*, and it is at that point that there is a grade shift (Fig. 12.13). Grades are defined as groups of organism, usually closely related, which share a common adaptation or type of biological organization. They are hard to define in practice, because of the continuity of evolution, but are useful ways of trying to identify when major changes may have occurred – in this case, the shift from a basically apelike creature to one which is more akin to modern humans. When this grade shift occurred has been a matter of considerable dispute.

What, in the end, can we make of the early hominins? They can be said to provide ample evidence for the evolution of a lineage closer to humans than to the African apes, and are thus a central part of the human story. They are also clearly a successful and diverse group, persisting for several million years. However, over recent years there has also been an increasing recognition that they are probably, in behavioral and ecological terms, closer to the apes than to humans. They are highly sexually dimorphic, have largely apelike guts and brains, and have a pattern of growth and life history that does not differ much from that of the living African apes.

Are there grade shifts in early hominin evolution?

In this sense, if there is a grade shift between apes and humans, then the australopithecines and their evolutionary allies are on the ape side, not the human one. Furthermore, their patterns of evolution, with a high rate of speciation and extinction, and a geographical restriction to Africa, are similar to those of other apes. That is why many researchers have taken to referring to these early hominins as "bipedal apes" rather than "proto-hominins." Both terms are probably caricatures that do not capture the complexity of these intermediate forms, but it is probably better to err on the side of the former than the latter on the basis of present evidence.

## Evolutionary geography, climate, and early hominin evolution

In chapter 3 we considered the factors that drive evolutionary change, and issues ranging from the role of the physical environment and competition to geographical patterns that shape evolutionary change. The early hominins, because of their diversity and complexity, should be a product of these processes. Certainly we have been able to see that there are trends in the adaptive characteristics of the hominins that are understandable in terms of environmental change (megadonty and greater aridity and seasonality); and the adaptive radiations fit a model of geographical dispersals in response to changing climatic conditions.

These dispersals can also be used to consider the phylogenetic issues that have been discussed (Fig. 12.14). Phylogenetic reconstruction is based on similarities and dissimilarities in morphology, which reflect evolutionary relationships. However, these relationships should also reflect spatial proximity and distance, given the allopatric nature of speciation and divergence. It is thus possible to explore the pattern of hominin relationships, as they change over the Pliocene, in the light of geographical patterns.[82]

◆ Around 5 million to 6 million years ago the ancestors of later hominins diverged from the ancestors of *Pan*. At some stage this probably involved an eastward dispersal of apes into the more seasonal and open environments, probably with some greater level of bipedalism than occurs in chimpanzees. *Orrorin tugenensis* and *Ardipithecus ramidus* are the current best contenders for some of the taxa involved in this dispersal.

*Does early hominin evolution reflect climatic and geographical change?*

◆ These early forms probably survived unchanged for a considerable period of time, but some gave rise to the australopithecine clade, as indicated by *Australopithecus anamensis*. Between 4 million and 3.5 million years ago, these "bipedal apes" in eastern Africa underwent range expansion. The ultimate evolutionary consequence of this was the southern African clade represented by *Australopithecus africanus*. More local dispersals within eastern Africa probably also occurred, as did isolation, leading to the occurrence in eastern Africa of several species – at least *Kenyanthropus platyops*, *Australopithecus anamensis*, and *Australopithecus afarensis*.

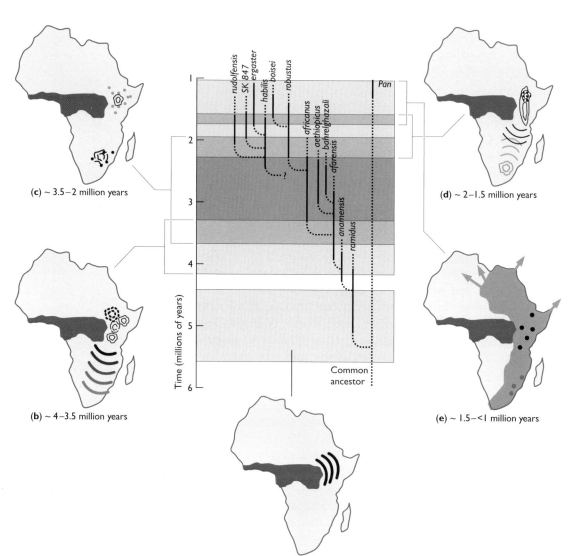

**FIGURE 12.14 Evolutionary geography and early hominin evolution:** Ultimately, the phylogeny of hominins should reflect both geographical and temporal patterns. (See text for details.) (Courtesy of Robert Foley.[82])

♦ The period 3.5–>2 million years ago must have seen major range contractions and ultimately extinctions of the more "primitive" hominins (represented by isolated "dots" of population). Within the overall range there were also new evolutionary events – in particular a probable trend toward megadonty in both eastern and southern Africa. One can speculate that *Homo*, which was probably around by the later part of this period, was a much more isolated and insignificant trend.

♦ By 2 million years ago, the fossil record suggests that all the "primitive" hominins were extinct, although it is possible that small "refuge" populations continued to survive. The primary events of this period appear to be the range expansion southward of at least one *Homo* taxon, and the expansion of robust australopithecines. This is represented here as occurring from south to north, but alternative interpretations are possible.

♦ After 1.5 million years ago, there was a range contraction of the robust australopithecines into refuge areas, with their extinction in place by 1 million years ago. Real or pseudo extinctions of *Homo* populations would have occurred in this period, as would the dispersal out of its core area (and ultimately out of Africa) of *H. ergaster*.

These patterns certainly reflect the changing climate of the Pliocene and early Pleistocene, and the ways in which hominins responded by either dispersing or contracting. As discussed in chapter 3, there has been considerable debate as to whether the climatic changes are the primary and only cause of hominin evolutionary change. Perhaps one of the strongest conclusions to come out of the material discussed in this chapter and the last one is that the hominin fossil record is sufficiently complex, and the patterns of change sufficiently spread across the entire time, to make a simple punctuated model of change relatively unlikely.

## BEYOND THE FACTS
# TOOLS – FORM OR FUNCTION?

*The issue: one of the great advantages paleoanthropologists have over other paleontologists is the presence in the archeological record of stone tools, which give access to behavior. However, determining how to interpret the evidence of stone tools is not a simple task.*

Stone tools are the most prolific source of evidence about human evolution. While bones only rarely fossilize, stones, needless to say, take a lot of destruction. It is the acquisition of stone-tool manufacture that so greatly enriches the history of the human evolutionary record compared to that of other species.

It is a truism to say that the stone tools provide the evidence for behavior. The information comes from the size and shape of the tools, the materials they are made from, where they come from, and the complexity of the process of manufacture. The more difficult question is determining what exactly they provide evidence for. The most obvious answer is that as the stone tools were used as part of the adaptive strategies of the hominins – to acquire and process food, for example – they provide evidence about function, the actual adaptations of which technology is a part. Thus, if one species or population makes simple chopping tools and another makes blades, this reflects the different adaptations of the populations and the demands of the environment. Differences between stone-tool assemblages reflect function, according to this view. This is perhaps the standard archeological model.

However, another view is that technology reflects not the demands of the environment, or at least not only the demands of the environment, but the constraints of the mind that made the technology. If one population made simple flakes and another complex tools, then this reflects perhaps that one is much more cognitively limited than the other. This might be consistent with the fact that

*Homo habilis*, with a brain size of around 600 cm³, makes simpler tools than does *Homo sapiens*, with a brain size of over 1300 cm³. In this case technology is providing information less about function (the simple and the complex tools might actually be doing the same thing) than about the hominins' cognition.

Yet another view might be based on the observation that technology is socially transmitted and acquired, and that for the most part hominins make the same stone tools as other members of their population, in the same way as members of the same population share the same features of anatomy. Here technology is seen as part of their phylogenetic heritage, and similarity reflects history rather than function. This is like the idea in evolutionary biology more generally that any feature reflects patterns of descent, and on that basis phylogenies can be reconstructed.

The issue here is that while stone tools are such a rich source of information, they are not simple to interpret, and in their shape and size can be said to reside evidence for function, for cognition, and for social tradition and history. The question is how to determine which is the most important.

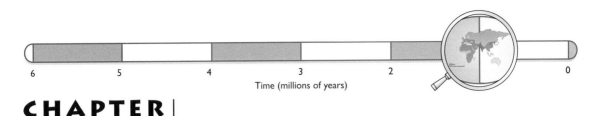
# CHAPTER 13

# AFRICA AND BEYOND: THE EVOLUTION OF *HOMO*

This chapter deals with the species of *Homo* that are assumed to be inter-mediate between early *Homo* (*habilis*/*rudolfensis*) and modern-day humans, *Homo sapiens*. Until recently, the story would have been portrayed as rel-atively straightforward: early *Homo* gave rise to a larger-bodied, larger-brained species, *Homo erectus*, approximately 2 million years ago, in Africa. Roughly 1 million years ago, *Homo erectus* expanded its range beyond Africa, first into Asia and then into Europe, developing geographically variable populations. *Homo erectus* then became the direct ancestor of *Homo sapiens*, either by a speciation event in a single population in Africa, which then spread throughout the Old World and replaced established populations of *Homo erectus* (the "out-of-Africa" or single-origin model), or by a gradual, worldwide (excluding the Americas and Australia) evolutionary transforma-tion of all populations of *Homo erectus* (the multiregional evolution model).

Much that was assumed to have been settled about the earlier events in this scenario has been overturned in recent years, through the discovery of new fossils and the redating and reinterpretation of known fossils. Before recounting the major fossil discoveries and their interpretations, it would be helpful to give a snapshot of evolutionary events as currently viewed by most anthropologists (Fig. 13.1).

Early *Homo* gave rise to a large-bodied, large-brained species in Africa approximately 2 million years ago, but this species is now called *Homo ergaster* by many anthropologists. *Homo ergaster* expanded its range beyond Africa and into Asia soon after its origin and at least by 1.8 million years ago; it then gave rise to *Homo erectus* in those areas. *Homo erectus* expanded its range throughout Asia, back into Africa, and presumably into Europe, although few unequivocal fossils have been found (most evidence takes the form of the stone-tool technology often associated with the species). In Africa and possibly Europe as well this lineage (*Homo erectus*/*ergaster*) evolved into a larger-brained form known as *Homo heidelbergensis*. Appro-ximately 150,000 years ago, a speciation event in Africa gave rise to *Homo*

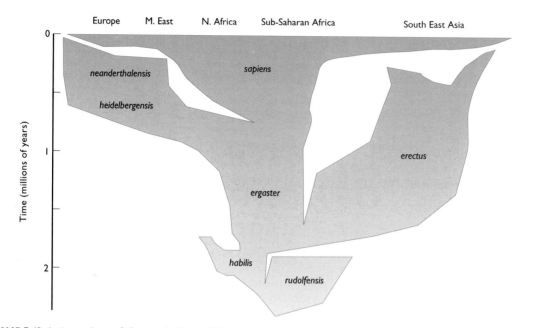

**FIGURE 13.1  Overview of the evolution of _Homo_:** The evolution of _Homo_ is seen here in terms of an expanding geographical distribution and the divergence of lineages. (Courtesy of Robert Foley.[82])

_sapiens_ (probably from _Homo heidelbergensis_ or a later derivative), which then spread into the rest of the Old World, and subsequently into Australia and the Americas. This broad outline receives widespread support, but it is in the details and the evolutionary mechanisms involved that there has been considerable debate.

In this chapter we will look at the broad patterns of Lower and Middle Pleistocene hominin evolution, from the fossil finds to the way these have influenced interpretations. We will then look at the archeological evidence that provides insights into hominin behavior over this period. The core taxon around which this chapter will be built is _Homo erectus_, one of the earliest of the hominin discoveries, but as we will see, exactly what is "meant" by _Homo erectus_ is highly controversial. In the first part of this chapter we will use the term _Homo erectus_ to refer to a wide range of material, from Africa and Asia. Later we will refine the use of the term.

## EVOLUTIONARY PATTERNS

### Early fossil finds

**KEY QUESTION** What pattern of genus _Homo_ evolution is shown in the fossil homins of the Pleistocene?

The discovery of the first _Homo erectus_ fossils is a storybook tale in anthropology.[15,16] In 1887, Eugene Dubois, a Dutch medical doctor, set off for Indonesia (then known as the

Dutch East Indies) in search of "the missing link." He had been inspired by Ernst Haeckel's description of a hypothetical human ancestor, which he called *Pithecanthropus*, or ape man (see chapter 3). Darwin and others had predicted that human forebears would have lived in the tropics. Consequently, taking advantage of Indonesia's status as a Dutch colony, Dubois obtained a post as a medical officer with the Dutch East Indies Army, with the intention of finding Haeckel's *Pithecanthropus*. Within four years, Dubois had achieved his goal, although he did not immediately recognize his success.

After two years of unsuccessful fossil prospecting on the island of Sumatra, Dubois shifted his efforts to the neighboring island of Java. Aided by gangs of convicts, he excavated fossil-bearing deposits along the banks of the Solo River, near the village of Trinil (Fig. 13.2). In 1891, one of his workmen found the top of a hominin cranium. Because this specimen had very low, prominent brow ridges and the bone was extremely thick, Dubois concluded that it must be a giant manlike ape. The following year, a complete femur was discovered that was unquestionably humanlike, despite being considerably more robust than the same bone in modern humans. The specimen had obviously belonged to a muscular creature that walked erect. Because the cranium and femur came from the same sedimentary layer (but roughly 10 meters apart horizontally), Dubois reasoned that they came from the same creature, which he named *Anthropithecus erectus*, or erect manlike ape. Only after more detailed study, particularly of the size of the cranium (850 cm$^3$), did Dubois realize that he had found the "missing link"; he then changed the name of this species to *Pithecanthropus erectus* (Fig. 13.3).

Great controversy greeted Dubois's announcement, and no agreement could be reached as to whether *Pithecanthropus* was human, ape, or something in between. Dubois suffered so much scorn at the hands of professional anthropologists

**FIGURE 13.2  Trinil:** The site on the Solo River where Eugene Dubois found the first *Pithecanthropus* specimens in the early 1890s.

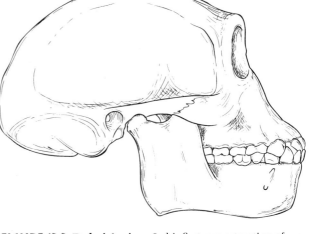

**FIGURE 13.3  Dubois's view:** In his first reconstruction of *Pithecanthropus* (1896), Dubois reflected his ambivalence over the human nature of the fossil, and chose to emphasize an apelike nature, seen in the prognathism and large canines.

**FIGURE 13.4 Zhoukoudian:** Excavation at the Zhoukoudian cave site, in 1936. (Courtesy of the Institute of Paleontology, Beijing.)

**FIGURE 13.5**
*Sinanthropus* **fossils:**
Some of the earliest finds are seen here being dried shortly after excavation in 1929. (Courtesy of the Institute of Paleontology, Beijing.)

that he eventually withdrew from the scientific limelight, taking his fossils with him. Only in the early 1920s did he once again allow the fossils to be shown in scientific circles, encouraged by the American anthropologists Henry Fairfield Osborn and Aleš Hrdlička. The Trinil fossils eventually became accepted as those of a human ancestor, even though Dubois's doubts about his find persisted (he believed the creature to be a giant gibbon, on the basis of calculations of brain size, a subject that fascinated him). It now seems likely that the femur is younger than the skull, and it may well belong to *Homo sapiens*.

The rehabilitation of *Pithecanthropus erectus* as an important discovery in human evolution coincided with discoveries in China, at the Choukoutien (now Zhoukoudian) site near Peking (now Beijing) (Fig. 13.4). In 1927, Davidson Black, the Canadian-born director of the Peking Medical College, recognized the human affinities of a tooth that had been found at the site. He named it *Sinanthropus pekinensis*, or Chinese man from Peking. An immense effort was mounted toward uncovering more fossils. Within a decade a rich haul had accumulated, including 14 partial or fragmentary crania, 14 mandibles, more than 100 teeth, and many other fragments (Fig. 13.5). Black concluded that *Sinanthropus* and *Pithecanthropus* were similar creatures, having a long, low, thick-boned skull, with a brain size intermediate between that of a human and an ape. Black died prematurely of a heart attack in 1934, and his work was continued by the German anatomist Franz Weidenreich.[335] As well as conducting detailed anatomical study of the fossils, Weidenreich directed the production of exquisite casts of them. This latter step proved providential, as virtually the entire collection was lost when the chaos of World War II descended on the region.

Meanwhile, fossil prospecting was continuing in Java, under the eye of the German anatomist G. H. Ralph von Koenigswald. Many *Pithecanthropus* teeth, and jaw and cranial fragments, were recovered, including the almost complete cranium of a child from the Modjokerto site. One problem with fossil collecting in Java was that it was often performed by local farmers, who came across specimens in their work or developed a talent for finding them. Von Koenigswald paid for each specimen turned over to him, a practice that encouraged some fossil finders to break large pieces into many fragments, thus securing a greater income. Equally serious was the issue of provenance of the fossil, or its exact location in the sediments from which it was recovered. This factor is essential if the fossil is to be dated accurately. Amateur collectors were not always as assiduous in marking sites of discovery as they were in collecting their rewards. Even where amateur collectors were conscientious, the fact that inaccurate information was sometimes disseminated throws doubt on all but the most secure finds. This caveat applies particularly to the Modjokerto skull, found in 1936, as we will see below.

In 1939, von Koenigswald took his *Pithecanthropus* collection to Beijing, so that he and Weidenreich could compare them with *Sinanthropus*. Like Black before them, von Koenigswald and Weidenreich concluded that the two fossil collections represented very similar creatures, but kept their distinct species names (Fig. 13.6). Increasingly, however, *Sinanthropus* came to be viewed as a Chinese variant of *Pithecanthropus*. In 1951, *Sinanthropus* and *Pithecanthropus* were subsumed under a single nomen that reflected their greater affinity to humans (belonging to the same genus), *Homo erectus*, which was recognized as a widespread species that exhibited significant geographical variation. *Homo erectus* became the "standard" for a Middle Pleistocene hominin.

## More recent finds

Since the 1950s, discoveries of *Homo erectus* fossils have been made sporadically, principally in Africa, but also in Asia (Fig. 13.7). The first of these discoveries took place at Ternifine (now Tighenif), in Algeria, where three jaws, a cranial bone, and some teeth of *Homo erectus*, dated at between 600,000 and 700,000 years old, were discovered in the mid-1950s. Later finds in northern Africa were made at Sidi Abderrahman (a jaw) in Morocco, soon after the first Ternifine find, and at Salé (cranial fragments), also in Morocco, in 1971.[336] Meanwhile, several specimens attributed to *Homo erectus* were found at Olduvai Gorge, in east Africa, including a rather robustly built, large-brained cranium, OH 9, initially dated at 1.2 million years (although it is probably younger). The South African cave site of Swartkrans also yielded *Homo erectus* fossils, which were originally

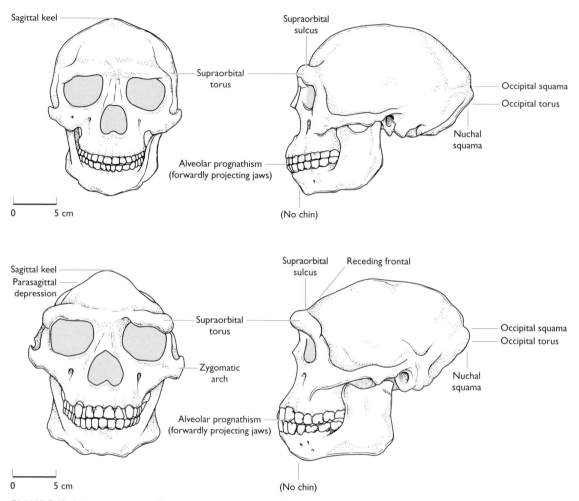

**FIGURE 13.6** *Homo erectus*: These two reconstructions by Weidenreich of Zhoukoudian *Homo erectus* (top) and Indonesian *Homo erectus* (bottom) show some of the anatomical variations present in Asian *Homo erectus*.

classified as *Telanthropus capensis*. Fossil prospecting in Java contributed an important cranium (Sangiran 17) in 1969 and a face and cranium (Sangiran 27 and 31) in the late 1970s, in the Sangiran dome region of the island. (These latter fossils have been referred to *Meganthropus*.)[337]

The richest source of fossils, however, has been the Lake Turkana region of northern Kenya, on both the east side (Koobi Fora) and the west side. These sites have yielded both the oldest known and the most complete specimens. In 1975, an almost complete cranium was recovered from Koobi Fora (KNM-ER 3733), with an age of 1.8 million years, and a brain size of 850 cm³.[276] A decade later, in the paleontological discovery of the century, the virtually complete skeleton of a *Homo erectus* boy was

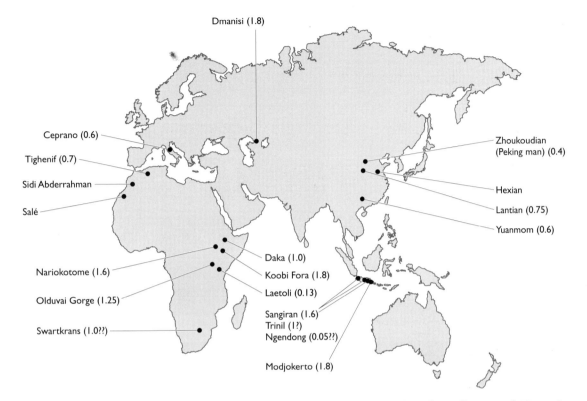

**FIGURE 13.7 Distribution of sites where *Homo erectus* and similar forms have been discovered:** Figures in brackets indicate the estimated age in million of years (where known). Until recently, no fossil specimen outside of Africa was dated as being older than approximately 1 million years. The recent redating of fossils from Java, if correct, suggests that this species expanded its range beyond Africa soon after it evolved.

unearthed at Nariokotome, on the west side of the lake (KNM-WT 15000)[183] (Fig. 13.8).

This skeleton, which came from 1.6-million-year-old deposits, is important not only because it represents the first discovery of many postcranial elements of the species, but also because it enables an assessment of overall body proportions and relationships. The skeleton is "the first [early fossil hominin] in which brain and body size can be measured accurately on the same individual," noted Richard Leakey and his colleagues, when they published their remarkable find. The boy stood more than 1.53 meters tall when he died, and would have exceeded 1.84 meters had he lived to maturity. His cranial capacity was 880 cm³. And his body stature and proportions – tall, thin, long arms and legs – are typical of humans adapted to open, tropical environments (see chapter 6). The anatomical detail available in this specimen has permitted paleoanthropologists to draw interesting inferences about many aspects of the species' mode of life, including life-history factors, subsistence, and language capacity.

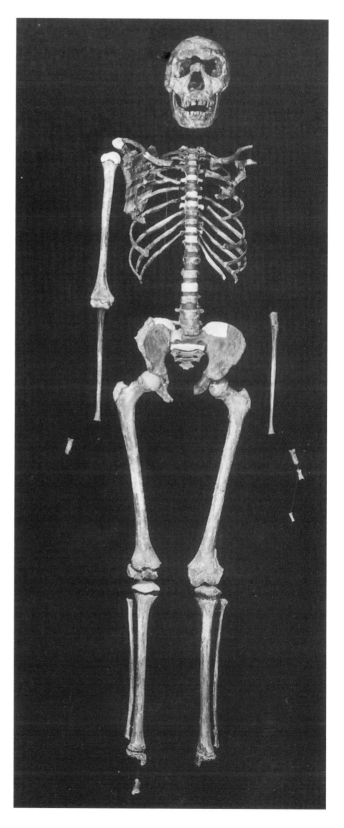

**FIGURE 13.8 The Nariokotome (Turkana) boy:** Discovered in 1984 on the west side of Lake Turkana, Kenya, this virtually complete specimen includes many skeletal elements not previously known. (Courtesy of Alan Walker/National Museums of Kenya.)

# Changing views: dates and evolutionary pattern

As the finds of putative *Homo erectus* fossils accumulated, two conclusions seemed to emerge. First, anatomical variations, which were seen initially in Asia, appeared to have proliferated elsewhere. Second, the species appeared to have originated in Africa close to 2 million years ago, and first set foot outside of Africa not much earlier than 1 million years ago. In recent years, both of these assumptions have been challenged.

Few of the Asian *Homo erectus* fossils have secure radiometric dates, with faunal correlation and paleomagnetic dating often used to approximate their age instead. Even where the presence of volcanic tuffs makes radiometric dating possible, as in Java, uncertainty has arisen over the reliability of such dates because of questions about provenance, as explained earlier. Consequently, the consensus was that no *Homo erectus* specimen outside of Africa was older than approximately 1 million years. The Beijing fossils were estimated to be roughly 300,000 years old (but have recently been shown to be at least 400,000 years old), though another Chinese site, Lantian, may be more than 700,000 years old. The oldest non-African *Homo erectus* sites were held to be in Java, with estimates of a little more than 1 million years for the Modjokerto child and something close to 750,000 years for Sangiran 27/31.[337]

What is the pattern of the dispersal of *Homo* out of Africa?

Until recently, the oldest specimen attributed to *Homo erectus* in Africa was KNM-ER 3733, from Koobi Fora, which was radiometrically dated to 1.8 million years. It was therefore assumed that *Homo erectus* originated in Africa and then, after a delay of almost 1 million years, spread into Asia. This apparent delay constituted a major puzzle to be explained in the overall history of *Homo erectus*. Some suggested that early *erectus* populations lacked a sufficiently sophisticated technology for moving beyond the traditional hominin geographic range. This technology, the Acheulean industry (Mode 2), is first seen in the archeological record some 1.4 million years ago (a date that still left an apparent, albeit smaller, delay). A new fossil find in 1992 and the redating of certain Javan fossils in 1994 implied one of two things: either no delay occurred, and *Homo erectus* expanded its range beyond Africa as soon as it evolved there, or *Homo erectus* evolved in Asia, not Africa, close to 2 million years ago.

In 1992, two German researchers announced the discovery of a *Homo erectus* mandible at Dmanisi, in Georgia, eastern Asia.[338] Its age – inferred from faunal correlation – was said to be 1.6 million to 1.8 million years. Nevertheless, the Dmanisi mandible prompted anthropologists to question the traditional interpretation of *Homo erectus* history, which held that *Homo erectus* did not move out of Africa until about 1 million years ago. More recently, new finds have been announced from Dmanisi which have bolstered this position[339] (Fig. 13.9). These comprise several crania. Although they are broadly similar to what we have been describing here as *Homo erectus*, they also display considerable variation, including

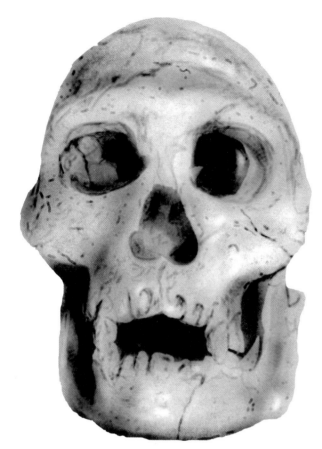

**FIGURE 13.9 Fossils from Dmanisi in Georgia:** These fossils, which now comprise several specimens, show that *Homo ergaster* had reached this part of Eurasia by over 1.5 million years ago. The most recent finds show that the population or populations from which they are drawn were variable.

specimens which are reminiscent of what in Java has been called *Meganthropus,* as well as very small and small-brained individuals.

Then, in early 1994, Carl Swisher and Garniss Curtis, of the Geochronology Center, Berkeley, announced new dates (based on single-crystal laser fusion) for the Modjokerto and Sangiran fossils: 1.8 million and 1.6 million years, respectively.[340] If this were correct, the Modjokerto skull would be equivalent in age to KNM-ER 3733, from Koobi Fora.

Many anthropologists are reluctant to accept the new dates, however. Their hesitation stems not from the radiometric dating of the volcanic material sampled, but from lingering uncertainties about the provenance of the Modjokerto find. Swisher and Curtis remain confident of the skull's provenance, and in any case point out the lack of uncertainty attached to the Sangiran fossil. If they are right, the new work changes the question anthropologists must answer about *Homo erectus:* there is now no delay to be explained, but the pattern of the species' origin is less clear. Although KNM-ER 3733 and the Modjokerto skull are of equivalent age, a sufficient margin of error exists in the dates to permit a gap in age of at least 100,000 years. A quick calculation shows that, even at the glacial pace of population expansion of 16 km per generation, *Homo erectus* could move from east Africa to east Asia in a mere 25,000 years. An African origin followed by population expansion into Asia is therefore consistent with the dates as currently known. It also reminds us that the time periods we are measuring are vast, and that there is room over these eons for many more events than we currently recognize.

Some anthropologists argue that an alternative pattern is equally plausible, with *Homo erectus* originating in Asia and then moving into Africa. The absence of unequivocal pre-*erectus* fossils outside of Africa militated against this hypothesis. At the end of 1995, however, an international team of researchers offered such a precursor, when they published their interpretation of a mandibular fragment (containing a premolar and a molar tooth) found in south-central China in 1988.[341] This fossil, which is dated at close to 2 million years, is interpreted as a form of *Homo* more primitive than *erectus*. If that is so, it would represent the first pre-*erectus* fossil known outside of Africa. Considerable uncertainty continues to plague the proper taxonomic status of the fossil, with some observers suggesting that it is not hominin at all, but rather an ape. Certainly it is not much on which to base a complete revision of the essentially "out of Africa" model for the origin and dispersal of *Homo erectus*.

At present, the most conservative position holds that the descendant of *Homo habilis/rudolfensis* evolved in Africa nearly 2 million years ago, and expanded its range into Asia with no significant delay. This scenario implies that the species was a new kind of creature in terms of behavior, which enabled it to expand beyond the traditional hominin geographical range.

Although there has been general agreement among most anthropologists that *Homo erectus* in the sense used here was not known in Europe, a recent find has raised doubts about that situation. From the site of Ceprano in Italy, Giorgio Manzi, of the University of Rome, and his colleagues have described an almost complete skull[342,343] (Fig. 13.10). Dated to around 600,000 years old, this specimen has been claimed to show some affinities to the *H. erectus* material from eastern Asia, although recent work has also noted unique differences. Whatever the taxonomic status of the Ceprano material, it does offer glimpses into the possibility of links between Europe and Asia during the Middle Pleistocene.

Finally, in 2001 Tim White and colleagues announced the discovery of a well-preserved cranium from the Middle Awash in Ethiopia, known as 'Daka', which yet again shows the same basic mid-Pleistocene morphology.[280] Along with a find from Buia in Eritrea, this extends the range of *Homo erectus*-like fossils in Africa.[281]

From a few fragmentary specimens in Java at the end of the nineteenth century, to the variable and scattered fossils we now know, it has been a considerable journey for a taxon that even its originator came to doubt. The exact status of *Homo erectus* is the next issue.

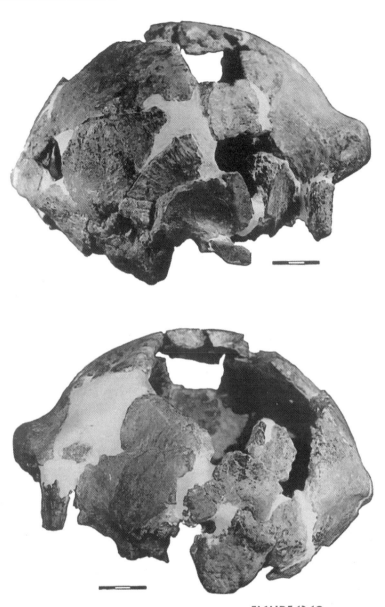

**FIGURE 13.10**

**A possible *Homo erectus* from Ceprano, Italy:** This specimen, recently discovered in Italy, may suggest that *Homo erectus* or a closely related form was present in Europe prior to half a million years ago. (From Ascenzi et al./*Journal of Human Evolution*.) (With permission from Elsevier.)

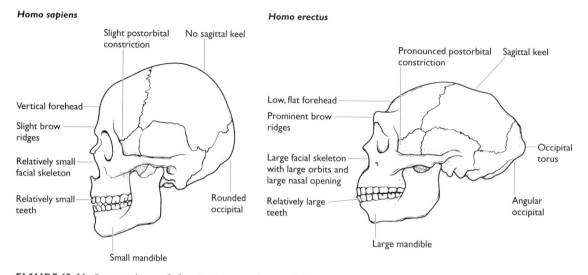

**FIGURE 13.11** Comparison of classic *Homo sapiens* and *Homo erectus*

## Changing views: anatomy and evolutionary pattern

As mentioned earlier, many anthropologists have recently concluded that the anatomical variations seen between different geographical populations of *Homo erectus* reflect the existence of more than one species, a view that is supported by cladistic analysis. The early African specimens, such as KNM-ER 3733 and the slightly younger 3883, have been assigned to a new species, *Homo ergaster*,[344] while the Asian specimens remain as classic *Homo erectus*. The two species are viewed as having an ancestor/descendant relationship, with *ergaster* originating in Africa close to 2 million years ago and then quickly expanding its range into Asia, where it probably gave rise to *erectus*. In this hypothesis, the later presence of *erectus* in Africa (such as the robust OH 9 from Olduvai Gorge) is interpreted as an Asia-to-Africa population expansion. Alternatively, *ergaster* might have given rise to *erectus* in Africa.[345,346] Whatever its precise origins, the key point here is that *H. erectus* is, according to this model, a specifically Asian species, and furthermore, that hominin evolution during the middle parts of the Pleistocene is not a single, general trend, but is structured around independent lineages in different parts of the world (Fig. 13.11).

What are the adaptive traits of *Homo ergaster/erectus*?

Many aspects of *ergaster* and *erectus* anatomy are, of course, similar, with the principal differences being a higher cranial vault, thinner cranial bone, absence of a sagittal keel, and certain cranial base characteristics in *ergaster*. One distinguishing feature between early *Homo* and *ergaster/erectus* involves increased brain size (ranging between 850 and 1100 cm³, with an increase over time), although the concomitant increase in body size (described later) means that the encephalization quotient, or relative brain size, may have increased but little. Other distinguishing features include a long, low

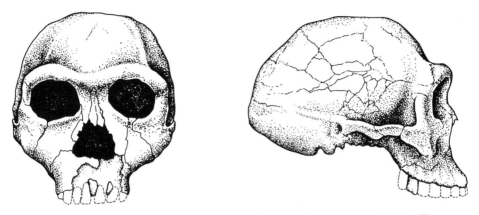

**FIGURE 13.12** *Homo ergaster:* This cranium, KNM-ER 3733 from Koobi Fora, Kenya, is 1.8 million years old. It has many similarities to Asian *Homo erectus* (more particularly with the Chinese specimens), but is judged by some anthropologists to be a different species, *Homo ergaster*.

cranium (particularly in *erectus*), thick cranial bone (particularly in *erectus*), the presence of brow ridges, a shortened face, and a projecting nasal aperture, suggesting the first appearance of the typical human external nose with the nostrils facing downward (Fig. 13.12). The structure of the nose, according to Erik Trinkaus of Washington University in St Louis, would permit the condensation of moisture from exhaled air, which would have proved beneficial in a species that pursued an active subsistence strategy in warm, arid habitats, such as those occupied by early *Homo ergaster*.

The body size of *ergaster/erectus* also represents an increase relative to that of early *Homo*, and reached nearly 1.8 meters and 63 kilograms in males and about 1.55 meters and 52 kilograms in females; this size compares with 52 and 32 kilograms, respectively, for male and female *habilis*. The larger body size is consistent with a more wide-ranging subsistence strategy.[30] Equally significant is the fact that the difference in body size between males and females is far less than that observed in all earlier hominins. Sexual dimorphism in earlier hominins was large, with males being almost twice as bulky as females. This situation has several possible behavioral implications – for instance, it might imply significant competition between males for access to females. With *ergaster/erectus*, this ratio declined considerably, with males being only 20 to 30% larger than females, perhaps implying a significant reduction in competition between males. Did the greater complexity of *ergaster/erectus* lifestyles include a degree of male–male cooperation?

Until the discovery of the Nariokotome (Turkana) boy skeleton, the postcranial anatomy was known from only a few elements, such as the femur and pelvis. The wealth of information provided by the boy's skeleton indicates that the postcranium is similar to that of modern humans, but more robust and heavily muscled; this structure implies routine heavy physical exertion. The thigh bone is unusual, in that the femoral neck is relatively long but the femoral head – part of the ball-and-socket joint with the pelvis – is large. This combination represents something of a mix between modern human and australopithecine anatomy: modern humans have a short femoral neck attached to a large head, while australopithecines possessed a long neck and a small head.

In the cervical and thoracic vertebrae, the hole through which the spinal cord runs is significantly smaller than in modern humans, which presumably indicates a smaller demand for nerve signal traffic. Furthermore, the absence of an expanded neural canal in the lower thoracic vertebrae of *Homo erectus* suggests the absence of an ability for spoken language.[347] In addition, the spines on all vertebrae are longer and do not point as far back as in modern humans; the significance of this feature remains an enigma.

## Changing patterns of behavior

A number of important "firsts" were recorded in human prehistory during the existence of *ergaster/erectus*:
- the first appearance of hominins outside Africa;
- the first appearance of systematic hunting;
- the first appearance of anything like "home bases";
- the first systematic tool making;
- the first use of fire; and
- the first indication of extended childhood.

Thus, these species were apparently capable of a life more complex and varied than had previously been possible.

The anatomy of the Nariokotome (Turkana) boy's pelvis provides an important insight into the changing patterns of behavior brought by this new species. The birth canal was smaller than in modern humans, but its absolute size suggests that humanlike infant development appeared for the first time. If humans resembled apes in terms of the way that the brain grows, then a human neonate could expect to double its brain size by the time it was mature. Given that the average adult human brain is approximately 1350 cm$^3$, a baby's brain at birth would be 725 cm$^3$ if it followed the typical primate pattern. In fact, the average brain size at birth is 385 cm$^3$, and even this size strains the mechanics of the system often enough to make birth much more hazardous in humans than it is in apes. As a result, human babies' brains continue to grow rapidly for roughly a year after birth, giving an effective gestation period of 21 months. Ultimately, the human brain more than triples its size between birth and maturity.

Because human babies are forced to come into the world with a relatively unformed brain, they are much more helpless than ape babies. This factor alone effectively lengthens childhood and demands a greater devotion to care taking in the social milieu. The necessity for social learning then extends childhood even further.

What about *ergaster/erectus*? Alan Walker, who directed much of the study of the Nariokotome (Turkana) boy's skeleton, calculated from the birth canal size in the boy's pelvis that the brain size in *ergaster* neonates would have been 275 cm$^3$. An apelike pattern of development (a brain size doubling from birth to maturity) would lead to an adult brain of less than 600 cm$^3$, which is significantly smaller than that of adult *H. erectus*. Continued brain growth at a high rate for a time after birth would be necessary to achieve the observed adult brain capacity of at least 850 cm$^3$ – the pattern seen in *Homo sapiens*. Infant helplessness and prolonged childhood would therefore have already begun in *Homo ergaster*, thus giving an opportunity for more cultural learning.

In an analysis of tooth development as an indicator of life-history patterns, Holly Smith,[231] of the University of Michigan, has also produced evidence for a shift to a life history pattern similar to that seen in modern humans. In apes, first molar eruption occurs at a little over 3 years, and lifespan is about 40 years; in humans, the corresponding figures are 5.9 years and 66 years, respectively. In other words, human life-history patterns have slowed relative to those of the great apes, including factors such as age at weaning, age at sexual maturity, and effective gestation length. While late *Homo erectus* fits the modern human pattern, as do Neanderthals and other archaic *sapiens*, *Homo ergaster* was somewhat intermediate between humans and apes; its first molar eruption occurred at 4.6 years, and its lifespan averaged 52 years.

One hallmark of *ergaster*/*erectus* is a particular type of stone tool, the teardrop-shaped handaxe (Mode 2 technology). These implements, which are usually called Acheulean handaxes after the site in northern France where they were first discovered, appear in 1.2-million-year-old deposits at Olduvai Gorge, where they are contemporaneous with the appearance of putative *Homo erectus* specimens. The oldest known specimens have been found at the site of Konso-Gardula, in southern Ethiopia, and date to 1.4 million years.[348] The time gap between the origin of *Homo ergaster* and the first appearance of handaxes is puzzling, if indeed this tool is an invention of this species and the known record reflects reality.

*Is Homo ergaster/erectus a new "grade" of hominin?*

Handaxes – sometimes crudely made, sometimes beautifully fashioned, and sometimes showing indications of individual or local styles – are found from these early dates through to 200,000 years ago in Europe and Africa. These artifacts require greater cognitive insight and more manipulative skill in their manufacture than do Oldowan artifacts. Curiously, and importantly, Acheulean handaxes were not an important feature of stone-tool technologies in east Asia.

The accumulations of bones and stones that appear in the archeological record coincidentally with the origin of the genus *Homo* become more frequent through *ergaster* and *erectus* times, giving an increasingly clear putative signal of some hunting activity. Some investigators speculate that a more broadly based diet, which included a greater proportion of meat than was eaten by earlier hominin species, was a factor in the population expansion out of Africa.[30,349] Whatever the niceties of taxonomy, the evolution of *ergaster*/*erectus* signals the appearance of a new grade of hominin evolution.

From half a million years ago onward, fossil hominins throughout much of the Old World began to display what may be described as anatomical features intermediate between *Homo erectus* and *Homo sapiens*. The interpretation of these specimens is important in hypotheses about the origin of modern humans.

## NEW TECHNOLOGIES

**KEY QUESTION** What role did the development of bifacial handaxes play in human evolution?

As we saw above, the evolution of *Homo ergaster* and subsequent appearance of *Homo erectus* brought many changes in the biology of our direct ancestors. Variations in life-history factors, in social structure, and in subsistence patterns combined to make the species a great deal "more human" than earlier species of *Homo* or the contemporary species of *Australopithecus*. In particular, the further development of meat as a significant component of diet must have been very important, both in increasing the stability and richness of energy resources and in allowing new habitats to be exploited. *Homo ergaster/erectus* was the first hominin to move beyond the bounds of the African continent. It might be expected that these developments would be accompanied by significant enhancement of stone-tool technologies.

## The Acheulean assemblage

As noted earlier, a significant innovation is seen in the archeological record with the appearance of the Acheulean assemblage (Mode 2 technology). The earliest known example of this assemblage comes from Konso-Gardula, Ethiopia, and is 1.4 million years old.[348] The name derives from the site of St Acheul, in northern France, where many examples of handaxes were discovered in the last century. The innovation consisted of the introduction of larger tools – known as handaxes, picks, and cleavers – than appear in Oldowan assemblages (Fig. 13.13). Although each of these tools is bifacially shaped, the teardrop-shaped handaxe is regarded as characterizing the new technology. Compared with Oldowan choppers, Acheulean handaxes required a higher level of cognitive ability in the conceptualization of the end product and its manufacture.

The earliest known *Homo ergaster* fossils appear in the record close to 2 million years ago, while the earliest known Acheulean element occurs some half million years later (Fig. 13.14). Several interpretations of this temporal gap have been suggested. For instance, the innovation may have been cultural, with later *Homo ergaster* populations inventing the new tool technology after having employed the simpler Oldowan technique for half

(a)

**FIGURE 13.13 Representative examples of Acheulean tools:** (top row) Ovate handaxe; pointed handaxe; cleaver, pick. (bottom row) Spheroid (quartz); flake scraper; biface trimming flake; two simple flakes. (All artifacts, except the spheroid, are lava replicas made by Nicolas Toth.) (Courtesy of Nicholas Toth.)

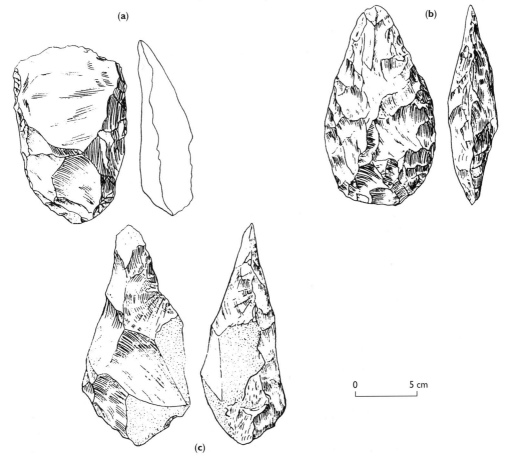

(b)

(c)

0      5 cm

**FIGURE 13.14 Early bifaces from Africa:** Drawings of (a) a pointed handaxe, (b) an ovate handaxe, and (c) a cleaver. The key innovation of the Acheulean industry was the ability to strike large flakes and retouch them on both sides to produce particular shapes.

a million years. Alternatively, Acheulean tools may have been a *Homo erectus* innovation. This latter explanation seems less likely, as archeological assemblages of the appropriate age in eastern Asia lack characteristic Acheulean artifacts (as discussed later in this chapter).

The precise path through which the Acheulean innovation emerged is not clear. The late Glynn Isaac argued, for instance, that it required the production of large ovoid flakes, more than 10 cm long, which were then trimmed by a few or many repeated blows along both edges.[350] Some large flakes were apparently functional without further trimming. The regular production of large flakes according to a preferred shape would have represented a major and sudden shift in technological expression upon which other bifacial implements could be built. The emergence of the handaxe may, however, have been more gradual. The Developed Oldowan included small bifaces, sometimes constructed from ovoid cobbles and sometimes derived from relatively large flakes. Acheulean bifaces may be envisioned as a further development of the technique that emerged earlier.

Once the large, bifacial handaxe appeared, it remained a characteristic of Acheulean assemblages for a very long time, in both Africa and Eurasia. Production became refined through the millennia, so that some late examples appear finely hewn compared with the crude earlier specimens. Nonetheless, while no early handaxe was the product of long, careful flaking to yield an aesthetically pleasing, perfectly symmetrical teardrop shape, many late examples appeared as crude as the earlier versions. Part of the development of the Acheulean included an increasing reliance on more detailed preparation of the piece upon which the handaxe was then made. This core preparation, one method of which is known as the Levallois technique (named after the site in France where the first examples of later prepared-core assemblages were found), became especially dominant in Middle Stone Age and Middle Paleolithic technologies (Mode 3 technologies). Overall, then, the Acheulean, like the Oldowan before it, was marked by a tremendous technological continuity maintained through a very long period of time.

Acheulean assemblages are known from many sites in Africa, some of which are spectacularly rich. At Olorgesailie (700,000 years old), discovered 80 km south of Nairobi, Kenya, by Louis and Mary Leakey and excavated by Glynn Isaac, hundreds of handaxes were strewn over the land surface.[351] This industry persisted until roughly 200,000 years ago, when it is superseded by Middle Stone Age (Middle Paleolithic) assemblages. Chopping-tool assemblages (like the Oldowan) never completely disappear during the 1.3 million years of the Acheulean period, for reasons that remain unclear. One interpretation is that this persistence simply reflects sites of different functional activities;[107] another that it reflects different populations or species of hominin in different parts of the world.[352]

The earliest Acheulean site outside Africa is Ubeidiya, west of the Sea of Galilee, in Israel. Dated at approximately 1 million years old, the site occurs along the natural migration route out of Africa into Asia. Migration

*What does the Acheulean tell us about the mind of early Homo?*

into Europe may have followed the same route, or moved across the narrow Straits of Gibraltar from northwestern Africa to Iberia, or involved island hopping across the Mediterranean; it may also have occurred via any combination of these. Dating early sites in Europe is difficult because of the lack of volcanic rocks suitable for radiometric dating. Early sites include Isernia in Italy (700,000 years) and Vértesszöllös in Hungary and Arago in France (both over 300,000 years). All three of these sites exhibit chopping-tool assemblages. Acheulean sites in Europe begin to appear soon after 500,000 years ago. The many famous later sites include Terra Amata (France), Torralba and Ambrona (Spain), and Swanscombe and Hoxne (both in England).[353] However, the most secure and important evidence for the earliest occupation of Europe comes from Atapuerca in Spain (800,000 years).

Many Acheulean industries in Africa, Europe, and Asia bear local names, implying local populational expressions of the same kind of technology. Overall, however, the continuity of form over a vast period of time and over a huge geographical area is more impressive than the local variation.

## Geographical distribution of the Acheulean

The earliest Acheulean assemblages are located in Africa, but later sites are found in western Asia,[354] southern Asia,[355] and Europe.[356] They remain absent in eastern Asia, however – a curious pattern that was first emphasized by Hallam Movius in the 1940s.[357] Stone-tool assemblages east of the so-called Movius line (Fig. 13.15) take on the chopping-tool form – that is, they remain Mode 1. Many hypotheses have been put forth in an attempt to explain this pattern. Movius, for example, considered the hominins in the east to be less evolutionarily developed than hominins elsewhere. In 1948, he claimed that the people of the east could not have "played a vital and dynamic role in early human evolution."

Some scholars suggest that the pattern is simply the result of an absence of suitable raw material for fashioning large bifaces east of the Movius line or that other material allowed the manufacture of tools that substituted for Acheulean handaxes. For instance, William Paterson University archeologist Geoffrey Pope suggests that bamboo may have been used extensively by the Lower Paleolithic people east of this line. He points out that the region is rich in bamboo, an extremely versatile raw material that is used in the modern world for applications ranging from furniture to scaffolding in the building of skyscrapers. Simple, effective knives can be made from bamboo, which may have obviated the need to fashion handaxes; the latter tools require more work and a less abundant raw material.

Others suggest that the pattern reflects a division of cultural tradition, and has no functional or technological significance. Recent ideas about the evolution of the genus *Homo* and the redating of fossils in Java offer a simpler alternative. If, as seems likely, *Homo ergaster* extended its range

**FIGURE 13.15  The Movius line: distribution of biface and non-biface industries:** Biface assemblages are confined principally to Africa, western Asia, and Europe; they are absent in eastern and southern Asia, where chopping-tool industries are found. The dividing line between the two regions is called the Movius line. (Courtesy of R. Foley and M. Lahr.)

Bifaces

Non-bifaces (chopping tools)

beyond Africa soon after it arose, then the first occupants of Asia would have long predated the first appearance of the Acheulean technology. Later incursions into eastern and southern Asia by Acheulean-bearing hominins might have been prevented if populations there were already well established.[352,358]

## The function of Acheulean handaxes

The function of Acheulean handaxes has long been a subject of speculation. A particularly unlikely explanation is that they were used as lethal projectiles, thrown like discuses as a means of killing prey. More prosaic suggestions hypothesize that they were used as axes or heavy-duty knives. In experimental studies, Indiana University archeologist Nicholas Toth found that handaxes (and cleavers) were highly effective at slicing tough hide, such as that of elephants.[359] The combination of weight and relatively sharp edges gives them greater efficacy than the ubiquitous small, sharp flakes. Microwear studies by Lawrence Keeley, of the University of Illinois, reveal that handaxes were used for many functions, and for materials ranging from meat and bone to wood and hide.[360] Thus, the Acheulean handaxe may have been the Swiss army knife of the Lower Paleolithic.

The end of the Acheulean industries, which occurred from 300,000 to 200,000 years ago throughout the Old World, marked the end of these stone-tool assemblies that had few artifact types and enjoyed enormous longevity. Both the Oldowan and the Acheulean lasted at least 1 million years, and both produced a dozen or fewer identifiable implements. The end of the Acheulean brought the Lower Paleolithic (Early Stone Age) to a close and marked the beginning of the industries of the Middle Paleolithic (Middle Stone Age). This period lasted only from 300,000 years ago to roughly 40,000 years ago, and included many more identifiable tool types. Real technical innovation had begun, although even this development was overshadowed by what followed in the Upper Paleolithic (Later Stone Age).

## Hunter or scavenger?

Some time between the beginning of the hominin lineage and the evolution of *Homo sapiens*, an essentially apelike behavioral adaptation was replaced by what we would recognize as human behavior – namely, the hunter-gatherer way of life. How and when this development occurred is central to paleoanthropological concerns. As we have seen, fossil evidence reveals the fundamental anatomical changes during this period, but it is to archeology that one turns for direct evidence of behavior.

**KEY QUESTION** Were early hominins acquiring meat from active hunting or from scavenging the carcasses of already dead animals?

The earliest stone artifacts recognized in the record are dated to approximately 2.6 million years ago. From their earliest appearance in the record, stone tools occur both as isolated scatters and, significantly, in association with concentrations of animal bones. What this association between bones and stones means in terms of early hominin behavior has become the subject of heated debate among archeologists.

Until recently, some archeologists argued, from analogy with modern hunter-gatherer societies, that the associations represented remains of ancient campsites, or fossil home bases, to which meat and plant food were brought to be shared and consumed amidst a complex social environment.[361] Others have countered by suggesting that these combinations merely indicate that hominins used the stones to scavenge for meat scraps and marrow bones at carnivores' kill sites; according to this hypothesis, the associations had no social implications.[120] Hence the debate, which has often been characterized as "hunting versus scavenging," is being fought over how "human" the behavior of hominins 2 million years ago was.

## Early hypotheses and recent developments

During the 1960s and early 1970s, paleoanthropologists considered hunting to be the primary human adaptation, a notion that has deep intellectual roots, reaching back as far as Darwin's *Descent of Man*. The apogee of the "hunting hypothesis" was marked by a Wenner-Gren Foundation conference in Chicago in 1966, titled "Man the Hunter." The conference not only stressed the idyllic nature of the hunter-gatherer existence – "the first affluent society," as one authority termed it – but also firmly identified the technical and organizational demands of hunting as the driving force of hominin evolution.[285]

A shift of paradigms occurred in the mid- to late 1970s, when Glynn Isaac proposed the "food-sharing hypothesis." Cooperation was what made us human, argued Isaac – specifically, cooperation in the sharing of meat and plant-food resources that routinely were brought back to a social focus, the home base. In this system, the males did the hunting while the females were responsible for gathering plant foods. As for "Man the Hunter," Isaac claimed that it was not possible to determine the importance of hunting relative to that of scavenging. "For the present it seems less reasonable to assume that protohumans, armed primitively if at all, would be particularly effective hunters," he concluded in 1978.[361]

Although the shift from the hunting hypothesis to the food-sharing hypothesis changed what was perceived to be the principal evolutionary force in early hominins, it nevertheless left them recognizably human. Specifically, the conclusion that the coexistence of bones and stones on Plio/Pleistocene landscapes implied a hominin home base immediately invoked a hunter-gatherer social package. Although the food-sharing hypothesis was often described by proponents as merely one of many possible candidates for explaining the evolution of human behavior, it proved very seductive. As Smithsonian Institution paleoanthropologist Richard Potts has observed: "The home base/food sharing hypothesis [was] a very attractive idea because it integrates many aspects of human behavior and social life [that] are important to anthropologists – reciprocity systems, exchange, kinship, subsistence, division of labor, technology, and language."[362]

To some extent the last 40 or 50 years have seen a full circle in hypotheses – from the "Man the Hunter" of the 1960s, through the rejection of meat being important in the 1970s, to a view of scavenging in the 1980s, and more recently to an emphasis on the nutritional importance of protein and the significance of hunting for primates (Fig. 13.16).

*Is there evidence for home-base/food-sharing in early* Homo?

## Testing assumptions

Realizing that several assumptions were implicit in these interpretations, in the late 1970s Isaac initiated a program of research that would test the

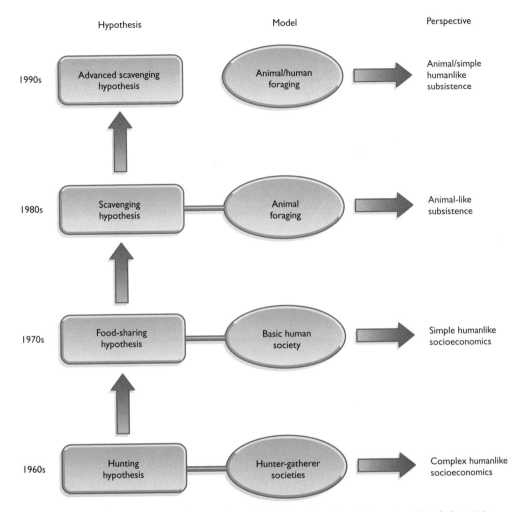

Hypothesis · Model · Perspective

1990s — Advanced scavenging hypothesis — Animal/human foraging — Animal/simple humanlike subsistence

1980s — Scavenging hypothesis — Animal foraging — Animal-like subsistence

1970s — Food-sharing hypothesis — Basic human society — Simple humanlike socioeconomics

1960s — Hunting hypothesis — Hunter-gatherer societies — Complex humanlike socioeconomics

**FIGURE 13.16  An evolution of hypotheses:** Since the 1960s, ideas about the nature of early hominin subsistence (social and economic) activities have passed through several important stages. In the 1960s, anthropologists envisioned hominin evolution in terms of the impact of cooperative hunting. In the 1970s, the image shifted, with the focus emphasizing social and economic cooperation through a mixture of hunting and gathering in a protohuman context. This view changed again in the 1980s, effectively taking any "humanity" out of the picture and attributing a marginal scavenging behavior to hominines. The current position is that scavenging was probably a very important route of meat acquisition, but not the exclusive one; this view is taken within the context of a human/animal model.

food-sharing and alternative hypotheses. Lewis Binford, of Southern Methodist University, independently embarked on a similar venture (Fig. 13.17). Both studies addressed several basic issues. First, what processes brought concentrations of stone artifacts and animal bones together in particular sites? Second, if the bones and stones are causally associated at these sites, what behavioral implications are possible? For Isaac and his associates, these questions were addressed by re-examining fossil bones from several

**FIGURE 13.17 Rival hypotheses:** Accumulations of stone artifacts and broken animal bones in the same location form an important element of the early archeological record. Traditionally interpreted as the remains of some kind of hominin home base (hypothesis Z), these accumulations are now subject to other interpretations. For instance, hypothesis Y suggests that the accumulation occurs at one location because hominines used trees there to escape competition from other carnivores while eating scavenged meat. Hypothesis X argues that hominines made caches of stones, to which they brought the more easily transported carcass fragments. Both cases produce the same result: an accumulation of bones and stones in one location. (Courtesy of Glynn Isaac.)

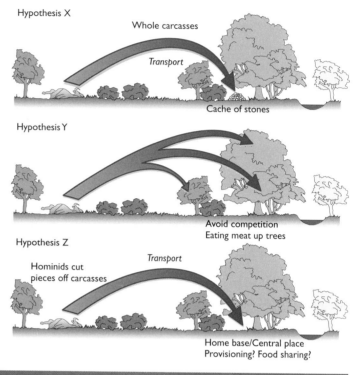

Hypothesis X

Whole carcasses

Transport

Cache of stones

Hypothesis Y

Avoid competition
Eating meat up trees

Hypothesis Z

Transport

Hominids cut
pieces off carcasses

Home base/Central place
Provisioning? Food sharing?

**FIGURE 13.18 Excavation in progress:** Site 50, on the eastern shore of Lake Turkana, Kenya, has yielded important information with which to test the hypothesis that the co-occurrence of bones and stones resulted from hominin activity.

already excavated, 1.8-million-year-old sites at Olduvai Gorge and a newly excavated, 1.5-million-year-old site at Koobi Fora, known as site 50 (Figs 13.18, 13.19). For Binford, the exercise entailed the scrutiny of published material on the Olduvai sites.[120]

In fact, bone fragments and stone artifacts might accumulate at the same site and yet be causally unrelated for several reasons (Fig. 13.20). For instance, they might be independently washed along by a stream and then

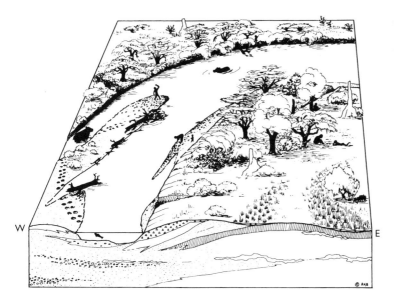

**FIGURE 13.19  A 1.5-million-year-old site:** Excavated on the floodplain east of Lake Turkana, site 50 has yielded 1405 stone fragments and 2100 pieces of animal bone. Nearly 1.5 million years ago, the site, which was located in the crook of a river course, was used for only a relatively short period of time. Stone fragments and debris struck during their manufacture could be reconstructed to form the original pebble used by the tool makers, and smashed animal bones could be conjoined to establish whole sections. Microscopic patterns on stone-tool edges indicate their use in cutting meat, soft plant material, and wood. This body of evidence invokes a picture of a rather humanlike subsistence behavior. (Courtesy of A. K. Behrensmeyer.)

deposited together – a hydraulic jumble, as it is known. Alternatively, carnivores might use a particular site for feeding on carcasses, while hominins might use the same site for stone knapping and whittling wood, having no interest in the bones whatsoever. The first possibility can be tested by the detailed stratigraphy of the site. The second hypothesis would require some indication that the stones were used on the bones in a particular way.

Of the six major early bone and artifact sites at Olduvai bed 1, the most famous site is the *Zinjanthropus* "living floor," which includes an accumulation of more than 40,000 bones and 2647 stones. Geological analysis indicates that hydraulic processes probably had little or no influence in the formation of most of the bed 1 sites. Binford's analysis of the sites compared the pattern of bone composition with that of modern carnivore sites, using the assumption that any difference could be attributed to hominin activity – residual analysis, it is called. His conclusion was forthright: "The only clear picture obtained is that of a hominin scavenging the kills and death sites of other predator-scavengers for abandoned anatomical parts of low food utility, primarily for purposes of extracting bone marrow. . . . [There] is no evidence of 'carrying food home.' "

For Binford, therefore, the Plio/Pleistocene bone accumulations of the

**FIGURE 13.20  Site dynamics:** Many factors influence the materials that might be brought to a locality and those that might be removed from it. Archeological excavations can recover only what remains at a site and what can be preserved (bones and stones, not plant and soft animal material). (Courtesy of Glynn Isaac.)

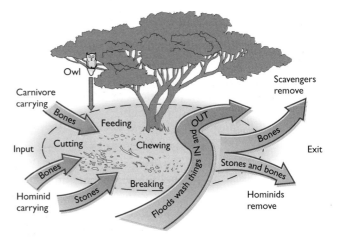

oldest archeological sites at Olduvai were principally the result of carnivore activity, with hominins playing the role of marginal scavengers. No humanlike social implications can be made for such species. "The famous Olduvai sites are not living floors," he concluded.

This last conclusion has also been reached by several of Isaac's associates, including Potts, Pat Shipman (of Pennsylvania State University), and Henry Bunn (of the University of Wisconsin). Their interpretations of the bone accumulations, however, differ widely. Specifically, none of the three agrees with Binford that the accumulations are primarily the result of carnivore activity. All see the collections as the work of hominins, with carnivores visiting these sites only occasionally. The assessments made by Potts, Bunn, and Shipman differ in terms of how much of the accumulations are attributed to hunting and how much to scavenging.[114,120,362–4]

Binford's analysis has been criticized on a number of grounds. For example, as Potts points out, this version of residual analysis makes the *a priori* assumption that hominins displayed no carnivore-like activity. If hominins hunted and consumed animals as other carnivores do, then the resulting bone fragment pattern would be subsumed under "carnivore activity," leaving no residual. Potts's own analysis of the Olduvai archeological sites indicates that the pattern of bone accumulation is more diverse than would be expected at exclusively carnivore sites. He concludes that the accumulations probably represent a mixture of scavenging and hunting, and argues that it is difficult – if not impossible – to distinguish between the bone-accumulation patterns that would result from hunting and the patterns from what he terms "early scavenging." Early scavenging could occur when, for example, a hominin located a dead animal that had not yet been partially eaten by a non-human carnivore.

Did early hominins acquire meat, and if so, how?

## Cutmarks and their significance

In 1979, Potts, Shipman, and Bunn simultaneously discovered cutmarks on fossil bones at Olduvai, which apparently had been inflicted by stone flakes used to deflesh or disarticulate the bones. Cutmarks stand as perhaps the most direct evidence possible that hominins used the bones at the archeological sites (Fig. 13.21). Percussion fractures on bones can also be informative (Fig. 13.22). Once again, however, the researchers' interpretations of this phenomenon differ somewhat.

Shipman, for instance, sees little or no indication that the Olduvai hominins were disarticulating bones and therefore concludes that the bone accumulations were principally the fruits of scavenging from other carnivore kills. Both Potts and Bunn observe what they interpret as evidence of disarticulation of bones, which could indicate hunting or early scavenging. Of the two, Bunn more strongly favors hunting as an important aspect of the Olduvai hominins' behavior. Potts points out, incidentally, that nature includes very few pure hunters and pure scavengers, with

**FIGURE 13.21 Cutmarks in closeup:** This fragment of bone, from a 1.5-million-year-old site in northern Kenya, bears characteristic marks that are left when a stone tool is used to deflesh a bone. The discovery of cutmarks provided an important method of testing the hypothesis that bones and stones at ancient sites were causally related.

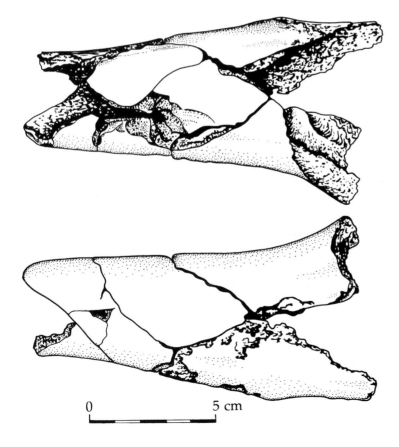

0                5 cm

**FIGURE 13.22 Bone jigsaw puzzle:** Fragments of bone found at site 50 were conjoined, producing these two ends of the humerus of a large, extinct antelope. The pattern of fracture indicates that it was the result of percussion by early hominins. Cutmarks were also present on the bone.

most carnivores participating in both activities to some extent. "To ask whether early hominids were hunters or scavengers is therefore probably not an appropriate question," he says.

Nevertheless, whether they were hunted or scavenged, the remains of animals at the Olduvai sites could, in principle, serve as an indication of hominin home bases. This explanation seems unlikely, however. Typical hunter-gatherer home bases are places of intense social activity and havens of safety that are occupied for periods of a few weeks and then abandoned. In contrast, the Olduvai sites apparently accumulated over periods of between 5 and 10 years, and they were obviously visited by carnivores. The carnivores left their signatures on the sites in the form of tooth marks on certain bones. Some tooth marks overlap cutmarks, which seems to imply that hominins got to the bones first. Other tooth marks are overlapped by cutmarks, which appears to confirm that the hominins occasionally scavenged from carnivore kills.

A recent analysis by Robert Blumenschine, of Rutgers University, carried out on the *Zinjanthropus* site bones suggests that, although the Olduvai hominins were not minimal scavengers of bones discarded by other carnivores (as Binford argues), they were principally scavengers and not significant hunters (as Bunn claims). This work was based on a comparison of tooth marks and percussion marks on fossil bones from Olduvai with marks produced experimentally.[365]

If the Olduvai sites were not typical home bases, what were they? Potts has suggested that they formed around stone caches[366] – places at which hominins accumulated raw material for making artifacts. Potts' computer simulations appeared to show that, on energetic grounds, forming stone caches and bringing carcasses to them would be an optimal strategy. In any case, the raw material for the artifacts at some sites apparently came from sources as far as 11 kilometers away. Some of this raw material was never processed, but was left as lumps called manuports.

Thus, the Olduvai sites appear to have been formed by hominins transporting stones to particular localities; they probably also brought meat-bearing bones to these sites, the result most likely of scavenging but possibly of some hunting. Instead of home bases, these sites appear to have been meat-processing and consumption places. Not all early sites are identical, however. For instance, some locations at Koobi Fora, including site 50, are clearly not stone caches because the stone artifact raw material is sourced on the spot. Moreover, several of the stone flakes at site 50 show signs of wood whittling and processing of soft plant material, which might imply a more leisurely use of the site than might otherwise have been envisaged. Whether this development represents a change through time – site 50 is approximately 300,000 years younger than the Olduvai sites – or differences in ecological context remains unknown.

Isaac's response to the findings was to suggest that the food-sharing hypothesis be replaced by the central-place foraging hypothesis.[367]

"Conscious motivation for 'sharing' need not have been involved," he wrote in 1982. "My guess now is that in various ways, the behavior system was less human than I originally envisaged, but that it did involve food transport and de facto, if not purposive, food sharing and provisioning."

Here Isaac was touching on a difficult methodological issue – trying to imagine the lives of humanlike creatures in unhumanlike terms. Modern hunter-gatherers operate with sophisticated organization and (relatively speaking) technology. Lacking weapons to kill at a distance, as humans did until late in prehistory, hunters could achieve only very limited goals and might not qualify as hunters in the commonly understood sense. Scavenging, on the other hand, would have been both technologically and ecologically feasible.

## Current developments

The debate of the 1970s and 1980s concerning the role of hunting and scavenging played an important part in questioning many assumptions about the nature of human evolution in general, and the role of meat-eating in particular. Its effect was to make archeologists and anthropologists adopt a more critical approach to the evidence, rather than assuming that it could be interpreted in a straightforward way. At one extreme it led to ideas of "Woman the Gatherer," rather than "Man the Hunter," a useful addition to models of human evolution.

However, as is often the case in paleoanthropology, in recent years the pendulum has swung again, not toward an extreme "Man the Hunter" model, but toward a recognition that meat-eating probably did play an important role in human evolution. The main sources of information for this swing have been as follows:

♦ There is increasing evidence for significant levels of hunting among chimpanzees, and for the role it can play in social interactions, as has been most cogently argued by Craig Stanford of the University of Southern California.[368]

♦ Most of the debate about the role of meat-eating has focused on the means by which it was acquired – hunting or scavenging – rather than its quantity. It is probably the case that the scale of meat-eating only has to be in the realm of 10% to 20% of the diet to be significant behaviorally and nutritionally.[349]

♦ The evidence of cutmarks on bones associated with *Australopithecus garhi* at 2.5 million years extends back the date for the first signs of meat-eating in hominin evolution.[279]

♦ The apes tend to have large guts, a feature associated with an essentially herbivorous diet. From the rib-cage reconstructions of *Australopithecus afarensis* it seems that the early australopithecines also had large guts. However, with *Homo ergaster* there is a change in gut size, a reduction that could be correlated with an increasingly carnivorous diet.[369]

♦ Meat is a highly nutritious food, with high protein content. The additional costs of development associated with the evolution of the human brain would have required a secure source of protein, and meat would, in the environments in which hominins evolved, have been an important source.[370]

♦ Once *Homo ergaster* is present there does appear to be a new scale to the archeological record, with a much larger quantity of stone tools, a clearer structure to archeological sites, and a strong association with mammalian bones. This suggests a change in foraging behavior, even if the precise nature of it cannot be pinpointed.

♦ The discovery of wooden spears in Shoeningen, Germany, dated at about 400,000 years old, published in February 1997, implies that systematic hunting had been well developed by that date.[371]

♦ Detailed analyses of the faunal remains of many sites during the Lower and Middle Pleistocene have suggested that there is an association between early hominins and the very large mammals that were abundant at that time, although changes in the fossil record after 300,000 years, as shown by Mary Stiner of the University of Arizona, indicate that however they may have hunted, early hominins were nonetheless different from modern humans.[372]

♦ The dispersal of *Homo* around the Old World, with relatively little taxonomic diversification, and hence a lower level of taxonomic diversity than that of the australopithecines, is perhaps a sign of a more carnivorous species, for this is the pattern found among other carnivores.[176]

Overall, therefore, meat-eating does appear to have played a role in the evolution of humans, although current models are a far cry from the extreme, bloodthirsty carnivory of Raymond Dart's early formulation. However, some authorities, such as James O'Connell and Kristen Hawkes of the University of Utah, would still argue that meat-eating's importance is relatively low among the causal factors in human evolution.[373,374]

Meat-eating, however it was acquired, is clearly an important part of human evolution. From the point of view of evolutionary theory, it is apparent that one of the ways in which major changes occur in evolution is through changes in diet, and so it is important to establish what aspects of the diet changed, and when, and with what consequences. Discussions about meat-eating and hunting are likely to remain lively in paleoanthropology.

# BEYOND THE FACTS
## ARE GRADES BAD FOR YOU?

*The issue: with the growth of cladistics and stricter rules of phylogenetic analysis, it has become clear that while clades or branches are real biological phenomena, grades, an established concept in biology, are far more problematic. Should the notion of grades be abandoned, or do they still have a use?*

To American college students struggling with this text there might be a simple and straightforward answer to this question; grades, unless of course they are As, are the curse of education. In evolutionary biology also, grades have been problematic.

An important distinction to make is between the words "clade" and "grade." "Clade" (derived from Greek) simply means a branch, and refers in evolution to the branches on a phylogenetic tree. Evolutionary history can be represented as the "tree of life," with every group representing the limbs, branches, and twigs as life forms diversify. While there is often considerable dispute as to what the actual tree should look like, there is general acceptance that there is such a tree that could be reconstructed and would represent what actually happened in evolution. Clades, therefore, might be elusive, but they are real and are here to stay in evolutionary biology.

Grades, on the other hand, are more difficult as a concept. For a start, they are harder to define. Basically, a grade is a group of organisms that are considered to share a similar level of biological organization, or share a particular system of adaptation. Mammals, for example, could be considered to represent a grade on the grounds that they share the fundamental adaptation of vivipary – giving birth to live young. Grades are not specific to any particular taxonomic level – for example, among the mammals all the ruminants could be said to represent a grade, on the basis of their gut morphology and function.

For the most part, grades and clades coincide. The mammals can be said to be both a grade and a clade, and the evolution of particular branches has coincided with the evolution of new systems of biological adaptation. At this level the notion of a biological grade is relatively unproblematic. However, this is not always the case, and then it becomes questionable whether "grade" is a useful term. Examples are easy enough to find among the primates; monkeys can be said to represent a grade relative to both apes and prosimians, but the groups that constitute them – the platyrrhines and catarrhines – are not a monophyletic clade. The tarsier is to all intents and purposes a prosimian grade of animal, yet in terms of its clade, it belongs to the anthropoids as a group. As we have seen, hominins are clearly a grade under most definitions, but where to draw the line for that grade is a difficult question.

This might seem to be reason enough to abandon the notion of a grade, which is a somewhat old-fashioned concept harking back to the *scala naturae* or Great Chain of Being discussed at the beginning of this book. As many taxonomic purists have correctly argued, clades are real and testable, grades merely a product of a biologist's prejudices. Certainly the term "grade" has declined in popularity, in direct proportion to the increased use of clades and cladistics as a technique.

However, evolutionary biology is more than just the search for that Holy Grail, the evolutionary tree, important as that may be. It is also about the study of adaptive patterns. We have seen that there are relationships between brain size and body size that are independent of phylogeny, and these can emerge from considering grades. To throw grades out might be to throw the adaptive baby out with the bathwater, for it is still useful to be able to treat the prosimians as a group, and compare them with the anthropoids. Biology, in the end, might be more than phylogeny.

# PART 3

# 3

# LATER HOMININ EVOLUTION

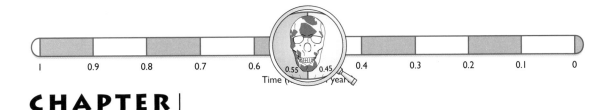
# CHAPTER
# 14

# THE ORIGIN OF MODERN HUMANS: BACKGROUND AND FOSSIL EVIDENCE

Anthropologists agree on the overall anatomical and behavioral shifts that accompanied the evolutionary transformation from *Homo erectus* to *Homo sapiens*. Anatomically, it involved a decrease of skeletal and dental robusticity, modifications of certain functional – particularly locomotor – anatomy, and an increase in cranial volume. Behaviorally, the transition brought more finely crafted and diverse tool technologies, more efficient foraging strategies, more complex social organization, the full development of spoken language, and artistic (symbolic) expression. Less agreement, however, has been reached over the mode of this evolutionary transformation – in particular, the origin of what are generally termed anatomically modern humans.

Since the 1980s, the question of the origin of modern humans has been among the most hotly debated issues in paleoanthropology, with very divergent opinions being vigorously expressed. At one extreme, the hypothesis is that the transformation occurred as a gradual change within all populations of *Homo erectus* wherever they existed, leading to the near-simultaneous appearance of multiple populations of modern humans in Africa and Eurasia. In this view, the genetic roots of modern geographical populations of *Homo sapiens* are deep, reaching back to the earliest populations of *Homo erectus* as they became established throughout much of the Old World (almost 2 million years in some cases). At the other extreme, an alternative hypothesis views modern humans as having a single origin (in Africa), followed by population expansion into the rest of the Old World that replaced established non-modern populations. In this scenario, the genetic roots of modern geographical populations of *Homo sapiens* are very shallow, going back perhaps 100,000 years. As we will see, other possibilities exist as intermediate positions between these two extremes.

Three lines of evidence are relevant here: anatomical, genetic, and archeological (behavioral). Inasmuch as they address the same evolutionary event, these independent lines of evidence, interpreted correctly, should be congruent. In fact, considerable fuzziness afflicts the overall picture, perhaps reflecting the complexity of the evolutionary process. Nevertheless, the weight of evidence increasingly supports some form of recent, single-origin model. This chapter will explore the anatomical evidence, while the subsequent two chapters will examine the genetic and archeological evidence.

## BACKGROUND FOR THE EVOLUTION OF MODERN HUMANS

**KEY QUESTION** When looking at the origin of modern humans, what is the context in terms of the evolution of *Homo* as a whole?

In this part of the book, the evidence for the evolutionary history of later *Homo* is described and discussed. At one level, this is a relatively straightforward story. When discussing the early hominins in the previous part, the emphasis was on complexity and taxonomic confusion, over a very long period, with a lineage (the hominins) confined to Africa. The contrast with what happens in the Pleistocene to *Homo* is quite striking.

First, where the early hominin events are purely African, this history includes Europe and Asia as well (but not the Americas and Australasia). Second, we are dealing with a single genus, compared to the four or more that characterize the early hominins. The scale of diversity is thus considerably less. This should, however, be put into context to some extent, as the period we are dealing with now is not the broad sweep from more than 6 million years ago down to less than 2 million, but the period from a little more than 2 million years ago onwards. Third, although it is clear that there is some taxonomic diversity within *Homo*, it is nonetheless relatively limited. While some authors, principally Milford Wolpoff of the University of Michigan,[123] have suggested that all Pleistocene *Homo* material be placed within a wide-ranging *Homo sapiens*, most authorities recognize there is some diversity, but less than that seen in the australopithecines. And fourth, there are marked changes in the biology and behavior of the hominins with the appearance of the genus *Homo* – changes in life history, body size and proportions, brain size, technology, and foraging behavior.

*What is the general pattern of evolution in Homo?*

In other words, there is something very much like a grade shift between the australopithecines and *Homo ergaster*. This contrast is between the essentially bipedal apes that characterize early hominin evolution, and the humanlike creatures of Middle Pleistocene *Homo*.

That pattern is fairly clear. Where there is controversy and disagreement, it largely relates to two issues. The first of these is the actual level of taxonomic diversity, and in particular whether it is possible to distinguish at the species level populations such as the early African material (*H. ergaster*) from that of Middle Pleistocene eastern Asia (*H. erectus*), and if so, what is the significance of anomalous material such as that from Ceprano in Italy, or OH 9 at Olduvai Gorge. The second is the extent to which these early representatives of the genus *Homo* are different from or similar to modern humans, and the extent to which they may be said to grade from one into the other.

Both of these issues take on considerable significance for the topic of this part of the book, the evolution of modern humans. As we shall see, the extent to which there is continuous evolution over time, and a continuous population geographically, is the central issue in determining when, how, and why *Homo sapiens* evolved. The importance of the contextual background can be shown by briefly describing two scenarios for the evolution of *Homo*.

In one scenario, once *Homo* is established as a new grade of hominin in Africa nearly 2 million years ago, it successfully expands in population size, and disperses across the habitable parts of Africa, Asia, and Europe. Being a large and widespread population, it would be expected to have considerable local variation, with the result that small local differences would emerge, but these would be transitory and largely insignificant. Occasionally major changes would evolve – larger brains, for example – or new behaviors would develop – handaxes, perhaps – and these would spread through gene flow to adjacent populations, or else would be culturally transmitted, so that eventually at the level of major adaptations the population would be relatively homogenized. The key element of this model is that we are dealing with a single population, and hence there can be no speciation. Geographical differences can only be relatively minor, and any changes must be global in their impact, if not in their origin. In this model, the context for the origin and evolution of modern humans is a large, well-established, only slightly differentiated hominin population. We can think of this as the "*Homo erectus* as general ancestor" model (Fig. 14.1).

The second scenario has much the same starting point – that is, the evolution of a new grade in Africa, *H. ergaster* – and its dispersal out of Africa at some later point. However, in this case the difference is that the dispersing populations have different demographic parameters. Rather than being large, stable populations, they are relatively small and prone to fluctuations. Thus, although the hominins do achieve a widespread distribution across Africa and Eurasia, they do so under the demographic conditions of small, isolated populations. They are therefore more prone to local differentiation through either drift or selection, and also more prone to extinction. The result is a more

**FIGURE 14.1** *"Homo erectus* **as general ancestor" model:** One model for the evolution of *Homo* involves the evolution and dispersal across the world of a single species, in which there may be minor regional differences, but homogeneity is maintained through gene flow.

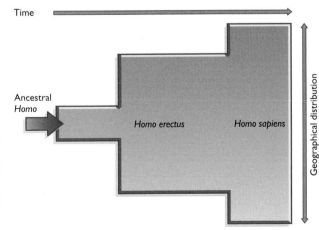

Time

Ancestral Homo

Homo erectus

Homo sapiens

Geographical distribution

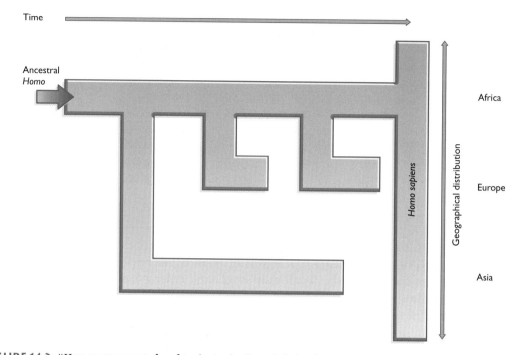

**FIGURE 14.2** *"Homo* **as separate local trajectories" model:** An alternative model for the evolution of *Homo* views hominin populations as more fragmented and fluctuating, resulting in the buildup of breaks in distribution and a proneness to species boundary formation.

differentiated hominin population, with breaks in distribution. This would lead to a greater proneness for species boundaries to form, and, most importantly from the point of view of later human evolution, for local evolutionary trajectories to become independently established. This model would thus recognize *H. erectus, H. ergaster,* and indeed any other species as the result of evolutionary divergence. If this model – the *"Homo* as separate local trajectories" model – is correct, then *Homo sapiens* can only come from one of these evolutionary trajectories (Fig. 14.2).

We can add a behavioral dimension to these two scenarios or models. In the first of these, *"Homo erectus* as general ancestor," major behavioral differences between populations would not develop to any great extent, as the effects of gene flow and cultural transmission would harmonize these. However, in the *"Homo* as separate local trajectories" model, such differences could become very marked – for example, as the presence of handaxes on one continent and an absence on the other – due to complete isolation. There could thus be several modes of hominin behavior present on the Earth at any one time. What happens to these over the long term, and when they perhaps meet up, is one of the big issues of later human evolution.

These two scenarios are extreme interpretations of the fossil evidence of the Pleistocene, and they provide the context in which the evolution of *Homo sapiens* evolved. Which of these is the correct one can be determined in two ways. One way is obviously to explore the fossil and archeological evidence and to look for the extent to which different trajectories are established and persist, and new research in this area is likely to provide the important answers. The other way is to consider, as we did in the first part of this book, which model of evolutionary process is more likely to have occurred, both in terms of general evolutionary theory, and in terms of the specific conditions of evolution of an intelligent, highly social, large-bodied primate and mammal. Both these elements will be considered.

## A note on species and names

Before doing so, however, we must consider briefly the issue of terminology in relation to later hominin evolution. This is necessary as there is no clear consensus about what species exist, and more importantly, the terms used are themselves dependent upon the interpretation being offered. For example, if one believes that there was interbreeding between modern humans and Neanderthals, then it follows that they must be the same species, and so the term *Homo sapiens* must be used to include both modern humans and Neanderthals, and any differences must be reduced to those of subspecies. Although it does not necessarily follow that if there is no admixture between Neanderthals and modern humans, or indeed between *erectus* and modern humans, they are separate species, it would at least be possible to use the species rank to distinguish between them without any major problem.

Earlier we saw how very difficult it is to identify species from fragmentary fossil remains. For instance, the identifications of many species of Old World monkey rely on the presence of soft parts of the body; if skeletons alone were available, far fewer species distinctions would be evident. Furthermore, different concepts of the species are available, from the biological species concept, which is dependent upon the potential for interbreeding, to the phylogenetic species concept, which is dependent upon morphological characters defined cladistically. Thus when it comes to the hominins of the last half million years or so, it is a matter of considerable controversy what terms are used.

If we accept the first scenario outlined above – the evolution of *Homo sapiens* from a generalized *H. erectus* ancestor – then the nomenclature is very simple: "*Homo erectus*" is used to describe all *Homo* post-*habilis*, regardless of any chronological or geographical patterns, and "*H. sapiens*" is used to describe all hominins that show certain developments, in particular larger brain size and loss of the particular cranial pattern of *Homo erectus*,

How should *Homo* species be classified?

namely the supraorbital torus as a single continuous bar, and the increasing verticality of the occipital bone. This nomenclature is consistent with a process of anagenesis and a generalized transition toward the modern human condition. Indeed, as discussed earlier, Milford Wolpoff has argued for a simplification of this nomenclature by removing *H. erectus* altogether and simply having *Homo sapiens* as a single species covering all of Pleistocene *Homo*.[123]

Without necessarily saying this is wrong, a disadvantage of this nomenclature is that it makes it very difficult to describe any pattern of later human evolution usefully. The origins of *Homo sapiens* in this system have nothing whatever to do with the evolution of the features, behavioral or anatomical, that we associate with modern humans. This view uses nomenclatural simplicity to remove the problems of human evolution from the realm of evolutionary theory.

Partly because of this difficulty, and partly because there is increasing acceptance that this nomenclature does not reflect evolutionary reality, there has been a move toward a greater amount of splitting within the genus *Homo*. Starting with *H. erectus*, as we saw in chapter 13, this can be broadly divided into *H. erectus* in the strict sense and *H. ergaster*, the former being primarily an Asian lineage, and the latter primarily African. This allows for the different evolutionary lineages and outcomes to be recognized in terms of the specifics of *H. sapiens* ancestry.

The finer divisions of *Homo sapiens* in the more general sense are far more problematic. Even under the older usage, which was all-embracing, there was a general recognition of "archaic" *H. sapiens* and "modern" *H. sapiens*. Following principally the lead of Philip Rightmire of the State University of New York at Binghamton, the term *H. heidelbergensis* has been increasingly used to distinguish between archaic and more modern members of the larger-brained group of the genus *Homo*.[375] The more derived forms have then been seen as *Homo sapiens*, to include only those specimens that are anatomically similar to living humans, and *H. neanderthalensis*, to cover the later European archaic forms that have a number of distinctive characteristics.

According to this nomenclature there would therefore be five species of post-habiline *Homo*, namely *ergaster, erectus, heidelbergensis, neanderthalensis*, and *sapiens*, compared to the one or two of the more traditional perspective. However, various authors have also proposed an even greater degree of subdivision of later *Homo*. The principal ones are:

♦ *Homo antecessor*, to describe material in the region of 800,000 years old from Atapuerca in Spain;[376]

♦ *Homo helmei*, to describe material from Africa and Europe which may represent an intermediate form between the more primitive of the archaic *sapiens* and the more derived, such as *H. neanderthalensis* and *H. sapiens*;[377]

♦ *Homo soloensis*, as a possible species of later archaic *Homo* found in Java, although most authorities view this as *H. erectus*; and

♦ *Homo mapaensis*, to describe later archaic material from China.

Splitting ←————————————————————————————→ Lumping

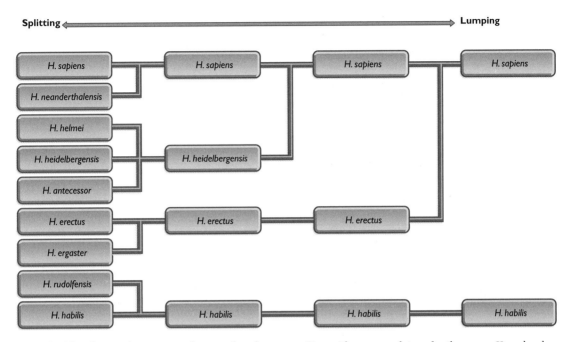

**FIGURE 14.3 Alternative nomenclatures for the genus *Homo*:** The nomenclature for the genus *Homo* has been a matter of considerable controversy, with "lumpers" and "splitters" strongly divided. This diagram provides a guide to how the various taxa may be lumped and split. For example, some authorities have suggested that all of *Homo* except for *H. habilis* and *H. rudolfensis* should be lumped into *Homo sapiens*.

It is important to remember that these distinctions are not necessarily true species in the full sense required by the biological species concept. Rather, they reflect hypotheses as to what the primary lineages of later human evolution, recognizable because they have an independent fossil manifestation, are. Ian Tattersall of the American Museum of Natural History has pointed out that the principal feature that unites hominins of the Middle and Late Pleistocene is a large brain, whereas many other features distinguish different groups within them and can be used for taxonomic purposes, and this also reflects the evolutionary process of speciation.[122] Clark Howell, of the University of California at Berekely, has argued that finer splitting is warranted, but that the lineages represented are "paleo-demes" rather than full-blown species.[378] The alternative nomenclatures are shown in Fig. 14.3.

*Homo – splitting or lumping?*

In the discussion which follows in subsequent chapters, a more fine-grained taxonomy will be used, as this allows the subtleties of the debate about modern human origins to be explored.

# COMPETING HYPOTHESES FOR MODERN HUMAN ORIGINS

**KEY QUESTION** How do the two hypotheses for the origins of modern humans – regional continuity and recent African – differ in their nature and expectations?

The major question in later human evolution relates to the origins of modern humans in particular, and within the framework of the more general two models outlined above, two extreme hypotheses have been developed.

The multiple-origins, or multiregional, hypothesis was the first comprehensive theory of the origin of modern humans. Its history stretches back to the 1940s, to the German anatomist Franz Weidenreich's formulation (see chapter 13). This hypothesis attempts to explain not only the origin of *Homo sapiens*, but also the existence of anatomical diversity in modern geographical populations. According to the multiregional hypothesis, this diversity resulted from the evolution of distinctive traits (through adaptation and genetic drift) in different geographical regions that became established in early populations of *Homo erectus* and persisted through to modern people. This persistence is known as regional continuity.[379]

In its original formulation, the multiregional hypothesis posited limited gene flow (mating) between different geographical populations, and was therefore dubbed the candelabra hypothesis from the shape of the diagram it gives rise to (see Fig. 14.4). It has since been modified, with gene flow between populations now viewed as an important component. This most recent formulation, developed principally by Alan Thorne (of the Australian National University, Canberra) and Milford Wolpoff, is now known as the multiregional evolution hypothesis. It views the *erectus*-to-*sapiens* transformation as a balance between the maintenance of distinctive regional traits in anatomy, through local regional adaptation, and the maintenance of a genetically coherent network of populations throughout the Old World, through significant gene flow.

The recent, single-origin hypothesis has a shorter history, dating back to Louis Leakey's ideas developed in the 1960s. Leakey considered the Early and Middle Pleistocene hominins of Africa to be better candidates for modern human ancestry than the *Homo erectus* fossils of Asia; the latter, he said, were an evolutionary dead end. Howell later dubbed the notion of a single origin the "Noah's ark" model. The most extreme form of this recent-African-origin (or "out of Africa") hypothesis, which assumes substantial replacement of archaic populations by invading modern humans, is most closely associated with Chris Stringer, of the Natural History Museum, London. The hypothesis views the establishment of regional anatomical traits in today's geographic populations as the result of adaptation and genetic drift in local populations during the last 100,000 years.[380]

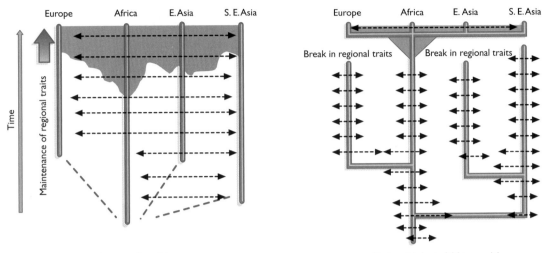

**FIGURE 14.4 Models for modern human origins:** In the single-recent-origin model, Africa serves as the source of modern humans, who then replaced established populations. At the other extreme lies the model of multiregional evolution, which posits a single evolving population, with distinct regional subdivisions. Variants between these two models invoke greater or lesser amounts of gene flow and admixture between populations, and specify when these might be occurring.

Between these two extremes it is possible to envision intermediate versions, depending on the primary mechanism involved (Fig. 14.4). In effect the single-origin model places most emphasis on local evolution and replacement, with little room left for gene flow; the multiregional model, on the other hand, places all, or most, emphasis on gene flow. However, the amount of gene flow is a quantitative matter, and one can up the level of gene flow from nothing (an extreme form of single-origin) to a little (as proposed by Gunter Bräuer[382]), to rather more (leading to actual assimilation of populations, as proposed by Fred Smith of Northern Illinois University[381]), to a major factor, as is integral to the original multi-regional model.

The multiregionalist view of the mode of evolution of modern humans has led to an interesting consequence for taxonomy. This theory indicates that the evolutionary transformation took place as continuous change within a genetically coherent lineage. If this scenario is correct, it rules out a clear break between *Homo erectus* and *Homo sapiens*, suggesting that they form an evolutionary continuum. As a result, there is no valid reason to distinguish between species, but instead the lineage should be regarded as a single species, *Homo sapiens*. Not surprisingly, many scholars question the biological reality of lumping together, for example, the robust *Homo erectus* people of Java and the gracile people of today's world into the same species.

# Predictions of different hypotheses

Competing hypotheses are tested, of course, by assessing how accurately their predictions are proved in the fossil record and how far they are consistent with genetic and archeological evidence. For the extreme hypotheses, the predictions are as follows. If the "out-of-Africa" model is correct, four principal predictions should hold:

♦ Anatomically modern humans should appear in one geographical region (Africa) significantly earlier than in others.

♦ Transitional fossils from archaic to early modern anatomy should be found only in Africa.

♦ Because traits that distinguish modern geographical populations are recently developed, they will show no necessary links with those of earlier populations in the same region (note that this caveat also applies to Africa, because 100,000 years of evolution in diverse populations in that continent will lead to a variety of local traits).

♦ Little or no evidence should suggest hybridization between archaic and early anatomically modern populations.

Predictions for the other single-origin models allow for more extensive evidence of hybridization and continuity.

In the multiregional evolution model, there are three expectations:

♦ Anatomically modern humans will appear throughout the Old World during a broadly similar period, although one area might see populations slightly earlier than the rest.

♦ Transitional fossils, from archaic to early modern anatomy, should be found in all parts of the Old World.

♦ In each region of the Old World, continuity of anatomy from ancient to modern populations should be apparent.

Hybridization between archaic and modern forms is not an issue with this hypothesis, as there are no separate "species" to hybridize.

The two fundamental questions in testing the hypotheses against the fossil record are the location of the earliest anatomically modern humans and the issue of regional continuity. Apart from the Neanderthals, the relevant fossil record is frustratingly sparse. No doubt this incompleteness contributes to the prevailing diversity of opinions. Other factors are important, however, because various authorities interpret the same body of evidence in very different ways. One issue here is the validity of particular traits for inferring phylogenetic histories: only derived traits are informative, whereas primitive traits have no such value in this context. Experts frequently disagree over the validity of such traits in the relevant specimens.

We will now look at the fossil record with these predictions in mind, beginning with the question of the location of the earliest modern forms, and then considering the issue of continuity by geographical region.

## CHRONOLOGICAL EVIDENCE

### Where were the earliest anatomically modern humans?

The first issue is the definition of "anatomically modern humans." As mentioned earlier, these individuals are characterized by a reduction in skeletal robusticity and the development of modern bipedal locomotion. Nevertheless, they were still more robust than modern-day humans. A general description of the skull would include a short, high, rounded cranium with a small face and the development of a chin.

**KEY QUESTION** What is the chronological pattern for the appearance of modern humans in the fossil record from around the world?

The identification of the earliest anatomically modern humans depends not only on an anatomy satisfying these criteria but also on a reliable age for the fossil specimens. Dating methods have improved greatly in recent years, particularly for the age range of 60,000 to 200,000 years that is relevant here. Nevertheless, some dates for specimens that have an anatomically modern appearance are less secure than anthropologists would prefer, often because of uncertainties over provenance.

Given these qualifications, specimens of anatomically modern humans from Africa and the Middle East stand out as significantly older than those seen elsewhere in the Old World. For instance, the Omo I (Kibish) brain case and postcranial material, found in southern Ethiopia in 1967, are strikingly modern; they are estimated to be around 130,000 years old (Fig. 14.5). (A second brain case, Omo II (Kibish), is slightly more primitive, but roughly the same age.)[383] Slightly younger specimens come from the Klasies River Mouth Cave in South Africa, dating between 70,000 and 120,000 years old. Border Cave, also in South Africa, has yielded modern-looking cranial and skeletal fragments that may be 70,000 years old. Provenance has been a concern in these cases, so that the true date may be substantially less than 70,000 years.[384] An important find of three remarkably complete crania with almost modern human anatomy from Herto in the Middle Awash region of Ethiopia was reported in mid-2003. Dated at 160,000 years old, these specimens are the best evidence of early *Homo sapiens* yet known. The discoverers, led by Tim White of the University of California, Berkeley, named a new subspecies, *Homo sapiens idaltu*.

In the Middle East, the Israeli cave sites of Skhūl and Qafzeh have yielded extensive fossil material, including partial skeletons. Most anthropologists judge these specimens to be essentially modern, even though they have some archaic features. Recent dating efforts (with electron spin resonance and thermoluminescence techniques) give these specimens' ages as close to 100,000 years.

If the earliest dates for modern humans in Africa and the immediately adjacent region of the Middle East are of the order of 100,000 years ago or

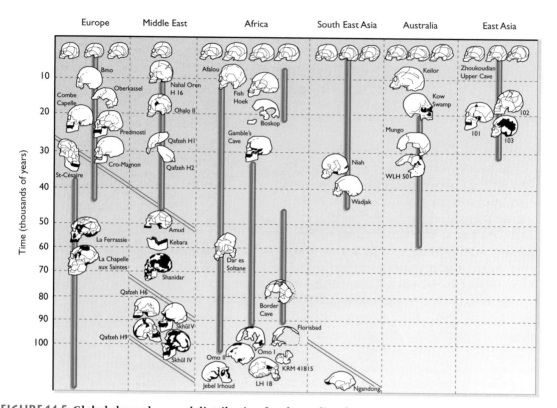

**FIGURE 14.5  Global chronology and distribution for the earliest known anatomically modern humans:** Although there are problems with dating, the fossil record shows a clear pattern of an earlier presence of modern humans in Africa and the Middle East than elsewhere. (Courtesy of M. Lahr and R. Foley.[399])

more, those in other parts of the world are considerably later. There is for the most part a chronological gap of tens of thousands of years between the earliest evidence in Africa and elsewhere. This conforms to the prediction of the single-origin/"out of Africa" model, and is not consistent with major gene flow across the world as a continuous part of the evolutionary process.

The earliest dates outside Africa are matters of considerable controversy. In China the earliest claimed fossil is from the cave site of Liujiang, in southern China, with a date of 67,000 years ago, but the association between the date and the fossil is not clear. In Australia there have been claims for a human presence at over 100,000 years, from the site of Jinmium, but these have largely been rejected. The current earliest claims are for 60,000 years ago for the southern Australian Lake Mungo burials, and between 50,000 and 60,000 years ago for sites in northern Australia. However, some authorities reject dates earlier than 40,000 years ago.[385]

In the Far East (Java), modern humans appear to be late arrivals, with *Homo erectus* coexisting with early moderns in Africa and the Middle East. Similarly, the earliest modern people in Europe are latecomers, appearing some 40,000 years ago.

What is the earliest evidence for anatomically modern humans?

What this means overall is that there is a much earlier presence of people in Africa than elsewhere. The pattern outside Africa may be an earlier presence in the southern parts of the Old World (around 60,000 years ago), and a later one (around 40,000 years ago) in the more northerly regions of Eurasia, although some would see a single event at the later date for both north and south.

The roughly similar dates of the African and Middle Eastern fossils have led some anthropologists to suggest a north African origin for modern humans, with the Middle East as part of the same ecological zone. Others leave open the possibility that the Middle East itself was the region of origin. The striking new finds from Herto in the Middle Awash, and the modern form of the Omo I (Kibish) brain case, dated at as much as 130,000 years old, provide sub-Saharan Africa's strongest claim to being the region of origin (Fig. 14.6).

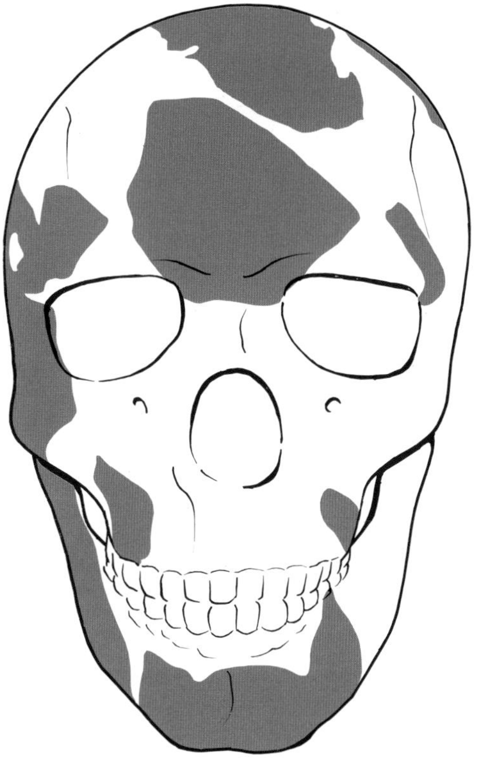

## THE QUESTION OF REGIONAL CONTINUITY

**FIGURE 14.6 Omo (Kibish) I:** This skull from Ethiopia is one of the oldest known anatomically modern humans, dated to around 130,000 years ago.

Regional continuity of anatomical traits from ancient to modern populations represents the cornerstone of the multiregional evolution hypothesis. The extreme form of the single-origin hypothesis denies such continuity, particularly through to the present day. The identification of such putative regional continuity in the Far East, in fact, led Weidenreich to formulate the multiregional hypothesis in the 1940s. Modern proponents of the hypothesis claim to find such continuity in Asia, Africa, the Middle East, and Europe, as well as the Far East. The issue of regional continuity remains the most contentious aspect of the current debate. We will present the evidence region by region, beginning with the Far East.

**KEY QUESTION** Does the hominin fossil record show a pattern of populational continuity between archaic and modern forms across some or all regions of the Old World?

## Southeast Asia: Australasia

Proponents of multiregionalism argue that a distinctive anatomy developed in the earliest settlers of Java and continued through to modern-day Australians. The claim is essentially based on three data points: the earliest inhabitants of Java, much more recent archaic forms in Java, and modern

Australians. The range of time involved here is considerable, which is pertinent to evaluating the putative continuity.

The earliest Javan inhabitants, *Homo erectus*, possessed especially thick skull bones, strong and continuous brow ridges, and a well-developed shelf of bones at the back of the skull. Their foreheads were flat and retreating, and the large, projecting faces sported massive cheek bones. Indeed, the teeth are the largest known in *Homo erectus*. As noted in chapter 13, these people may have lived in Java as long as 1.8 million years ago.

The next data point is taken from a dozen brain cases found in 1936 at Ngandong, in western Java. Colloquially known as Solo Man, these specimens have many *Homo erectus* features. Multiregionalists see them as descendants of the earlier Javan *Homo erectus* people, displaying many of the same anatomical features mentioned above, but with enlarged brain cases. The age of the Ngandong fossils is surprising. Until recently, they had been estimated at more than 100,000 years old, but dates newly obtained at the Berkeley Geochronology Center place them between 27,000 and 53,000 years old. If correct, this means that the archaic Ngandong population lived long after modern humans had appeared elsewhere in the Old World, and were contemporaries of the earliest *Homo sapiens* in the region.[386] This development is parallel to the situation in Europe, where Neanderthals and modern humans coexisted for a while. However, this possibility requires further confirmation of the actual date of the Ngandong specimens, which may still turn out to be considerably older.

Archeological evidence indicates that humans first reached Australia approximately 60,000 years ago, although most of the actual fossil evidence is considerably younger. According to multiregionalists, the earliest Australian fossils "show the Javan complex of features."[387] Moreover, these hypothesis proponents state that later Australian fossils "demonstrate that the same combination of features that distinguished those Indonesian people from their contemporaries distinguishes modern Australian Aborigines from other living people." Can the features cited as evidence of regional continuity truly be traced from ancient Javan *Homo erectus*, to the Ngandong specimens, to modern Australians? (The new dates for Ngandong make this chronological sequence implausible.) More particularly, are these features truly unique (that is, derived) to this region of the world?

A general anatomical similarity undoubtedly exists in these three populations, particularly in terms of their robusticity. Unfortunately, the comparison of facial and dental features cited as evidence of regional continuity cannot be tested with the Ngandong specimens because they comprise brain cases only. Two independent studies by Australian anthropologists Colin Groves and Phillip Habgood in the late 1980s, however, questioned the phylogenetic validity of several of these features, concluding that they are retained primitive traits common to *Homo erectus* and archaic *Homo sapiens*, not derived features unique to the region.[388,389]

Indeed, many of these features occur with greater frequency in other Asian populations. More recently, Marta Lahr, of Cambridge University, England, reached similar conclusions based on an examination of cranial features. Her conclusions were that there was little to connect modern Australians uniquely with any eastern Asian archaic populations of *H. erectus*.[390]

With regard to the single-origin hypothesis, Australasia offers its strongest challenge in the apparent existence of two populations in Australia between 10,000 and 30,000 years ago. One of these populations displayed significant cranial robusticity, the other much less so. Multiregionalists suggest that the robust population derived from Java whereas the less robust people came from China. Some anthropologists, however, suggest that the differences are merely within-population variation that developed in situ, with the entire population becoming more gracile within the last 10,000 years. In any case, if the single-African-origin hypothesis is correct, Australia would have been settled by African people. Can this scenario be proved? Not definitively, although Stringer and Bräuer point out that the Omo II (Kibish) brain case displays many of the anatomical features that are supposedly unique to Australasia. These researchers accept the possibility of some interbreeding between modern humans and, for instance, Ngandong-like people as they migrated through southeast Asia, but contend that these archaic features failed to pass through to modern populations. There is a further problem with the multireional interpretation of these morphologies. According to this model, the earlier forms should be the more robust, with the later ones developing more uniquely gracile proportions. However, most of the dating methods applied have shown that the most gracile specimens – those from Lake Mungo – are the earliest, and that the more heavily built forms known from Kow Swamp are all less than 20,000 years old. Furthermore, the most robust of all the Australian fossils, the Willandra Lakes 50 specimen, has been shown to be only 15,000 years old.[391]

The analyses by Lahr, Groves, and more recently Peter Brown, of the University of New England, Australia, all question the idea of continuity between ancient and archaic populations in Australia and southeast Asia, and thus undermine the basis for regional continuity. In another analysis Wolpoff and Hawks have compared the modern Australian Willandra Lake specimen and the early modern humans of the Middle East with Ngandong. The result, they claim, shows that the Australian Aborigines' closest relative is the Ngandong *Homo erectus*. This result has been challenged by Stringer and Brown.

Did Asian *H. erectus* contribute to the modern Australian ancestry?

## East Asia

Wolpoff and fellow proponents of the multiregional hypothesis view the east Asian hominin fossil record as showing a continuous sequence

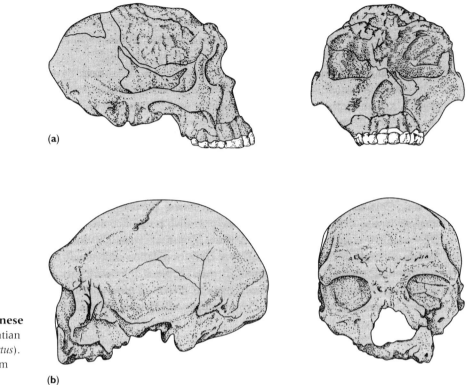

**FIGURE 14.7 Chinese fossils:** (a) The Lantian cranium (*Homo erectus*). (b) The Dali cranium (archaic).

of human fossil remains from almost 1 million years ago to the present day, and argue that this record shows a smooth transition to the living peoples of east Asia.[392] This view has not been supported by more recent work.

The earliest known human fossil material in the region is a cranium from Lantian, in northeastern China (Fig. 14.7); this *Homo erectus* specimen is dated at close to 1 million years. Among the later ones is the collection of cranial parts from the main cave of Zhoukoudian, which are also classic Asian *Homo erectus* (see chapter 13) and span a period from more than 500,000 to 200,000 years ago. *Homo erectus* remains of similar age to the youngest fossil at Zhoukoudian have also been found at Hexian. Fossils that display a mix of *erectus* and more derived traits have been discovered in China, including a partial skeleton from Jinniu Shan and a skull from Dali, both dated to approximately 200,000 years. These latter two specimens are generally known as archaic *sapiens*, and in the multiregionalists' scheme they represent forms transitional from *erectus* to modern *Homo sapiens*. They include the abundant fossil remains at the Upper Cave of Zhoukoudian, dated at 20,000 years.

This picture became more complicated in the early 1990s, after the discovery of two crushed crania at Yunxian, in east-central China.[393] Although their damaged condition makes anatomical analysis difficult, these specimens have been described as archaic *sapiens* and are dated to as long ago as 350,000 years. If their anatomical attribution and age are correct, they would be almost twice as old as some of the latest *Homo erectus* populations, undermining the pattern of regional transition as envisaged in the multiregional evolution hypothesis.

Multiregionalists argue that the fossils in eastern Asia differ from those in the southeast much in the same way as the modern populations vary. Eastern Asians, both ancient and modern, have smaller faces and teeth, flatter cheeks, and rounder foreheads; their noses are less prominent and are flattened on top. A feature that is particularly emphasized as reflecting regional continuity is the shovel-shaped upper incisors. Critics of multiregionalism point out that this supposed derived feature is also found in ancient populations elsewhere in the Old World, and therefore cannot be used to link Chinese *Homo erectus* to modern Chinese people. Lahr's recent analysis is also critical of this and other primitive traits that are used to infer regional continuity.[390] Moreover, a study by J. Kamminga and Richard Wright strongly indicates that the Upper Cave population individuals do not resemble modern Chinese people, as they should if they were part of a gradual, regional transition from *erectus* to archaic *sapiens* to modern *sapiens*; instead, this population is more closely allied with African morphology.[394] In other words, there is not even continuity within anatomically modern populations in east Asia, let alone between the moderns and the archaics.

Is there evidence for regional continuity in east Africa?

The sequence of evolutionary change in east Asia is poorly known and poorly dated, and it is far from well understood. However, whatever is the ultimate interpretation of the material, one of the least likely models is that of regional continuity over half a million years or more. Indeed, what is striking is the lack of continuity: the modern east Asians are different from the Late Pleistocene modern humans of the Upper Cave at Zhoukoudian, and these in turn are different from the so-called "archaic *sapiens*" or *H. heidelbergensis* specimens from Mapa and Dali, and indeed these two specimens are very different from each other; and none of these show any great continuity with the even more ancient *H. erectus*. Overall, the Chinese material is evidence of the complexity and discontinuity of the later hominin fossil record, even if much of it is still unclear.

## The Middle East

The fossil record of the Middle East is rich, including several partial skeletons. The first excavations began in the 1930s at the cave sites of

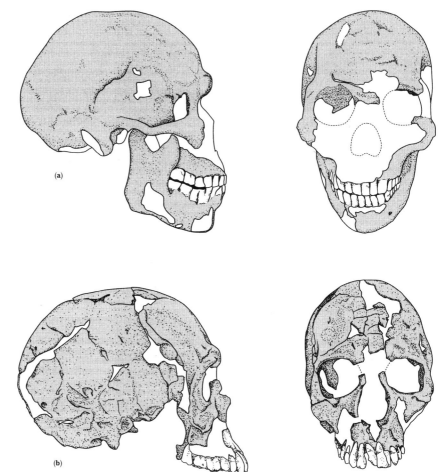

**FIGURE 14.8 Middle Eastern moderns:** Crania from (a) Skhūl and (b) Qafzeh. Both sites are dated at approximately 100,000 years, which means that modern humans moved out of Africa soon after they originated south of the Sahara; alternatively, it might indicate that north Africa or the Middle East was the site of origin.

Skhūl and Tabūn on Mount Carmel, Israel, and produced partial skeletons that probably resulted from deliberate burial. The Tabūn individuals are Neanderthals, while those from Skhūl are primitive-looking moderns. Excavations conducted over the next five decades yielded additional human fossils from these and three more sites (Kebara, also on Mount Carmel; Amud, near the Sea of Galilee; and Qafzeh, near Nazareth). Kebara and Amud yielded Neanderthals and Qafzeh moderns (Fig. 14.8).

Until a decade ago, the Neanderthals were thought to predate the modern population (with ages of 60,000 and 40,000 years old, respectively) and were assumed to be ancestral to them, in line with the multiregional hypothesis. Recent dating efforts, however, have revealed a more complicated picture that offers much less support for the multiregional hypothesis. Kebara and Amud appear to be nearly 60,000 years old, as believed earlier, but Tabūn has some much older layers (120,000 years)

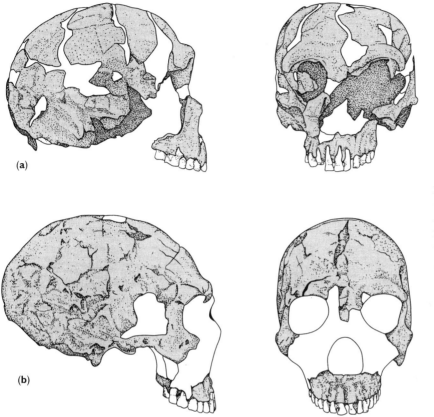

(a)

(b)

**FIGURE 14.9 Middle Eastern Neanderthals:** Crania from (a) Tabūn and (b) Amud. These Neanderthals were once thought to predate the modern population in the Middle East, suggesting an ancestor/descendant relationship. Although the Tabūn people predated the moderns, those from Amud did not (unlike those from Kebara), making such a relationship impossible.

(Fig. 14.9). A more significant redating affected the modern populations, with Skhūl and Qafzeh being placed at 100,000 years. Clearly, a simple ancestor/descendant relationship between Neanderthals and moderns is not possible, as the moderns are near contemporaries with the earliest Neanderthals of the region, and Neanderthals persist for at least 40,000 years after the first appearance of moderns.[100]

Some evidence suggests that the Neanderthals and moderns did not coexist in the region but rather displayed climate-driven fluctuating occupation, with Neanderthals inhabiting the region during colder times and moderns being present during warmer times. Nevertheless, the temporal overlap of the two populations is construed by proponents of a single-origin model as strong support for their hypothesis. The body proportions of the modern people more closely resemble those of warm-adapted Africans than those of cold-adapted Neanderthals, which provides additional support for the single-origin hypothesis.

Proponents of multiregionalism counter these conclusions by arguing that the region was occupied by one highly variable premodern group, not separate Neanderthals and moderns. (The fact that the two populations

Did Neanderthals and modern humans coexist in the Middle East?

used similar stone-tool technologies, the Mousterian, is sometimes adduced in support of this notion.) Most observers find the claim for a single, variable population unconvincing, noting that it would require a range of variation unknown in any other hominin population. Moreover, aspects of the Neanderthal postcranial anatomy show retention of certain primitive features (in the femur and pelvis) that the moderns lack.

Some of the Neanderthals are more generalized in their cranial anatomy than others, a disparity that has been inferred by some observers as evidence of hybridization between Neanderthals and anatomically modern people. The distinctive postcranial anatomy of the two groups does not prove the existence of hybridization, however.

Overall, the Middle East offers more support for the single-origin hypothesis than for the multiregional hypothesis, and may even refute the latter.

## Europe

The Middle to Late Pleistocene hominin fossil record of Europe is dominated by the Neanderthals. For this discussion, the pertinent question involves the identity and fate of their ancestors. According to the multiregional evolution hypothesis, the Neanderthals were part of a gradually evolving lineage that eventually yielded anatomically modern humans in Europe. In contrast, the single-origin hypothesis proposes that they represent a locally evolved species that became extinct approximately 30,000 years ago and that contributed nothing to modern European populations.

As noted earlier, the fossil of Ceprano might be the only evidence to prove the presence of *Homo erectus* in Europe. Older than Ceprano are, however, some recent spectacular finds at Gran Dolina, Atapuerca, Spain. These remains may be 780,000 years old. In May 1997, the discoverers of the fossils elected to name the fossils as a new species, *Homo antecessor*.[376] The relationship between Ceprano and the Gran Dolina fossils remains unclear.

Despite the paucity of early remains, many examples of so-called archaic *sapiens* or *Homo heidelbergensis* have been discovered in Europe. According to proponents of the single-origin hypothesis, most of these fossils are the same as their African contemporaries, and would have been ancestral to Neanderthals in Europe and to *Homo sapiens* in Africa. Multiregionalists view this group as evidence of continuity.

The Mauer mandible, found in 1907 and dated at roughly 500,000 years old, combines primitive features (robusticity) with modern features (molar size) (Fig. 14.10). It was given the species name *Homo heidelbergensis* in 1908. Other fossils with a similar mix of ancient and modern were found in the mid-1930s, such as a cranium at Steinheim, Germany, and skull fragments at Swanscombe, England. Both of these items date to

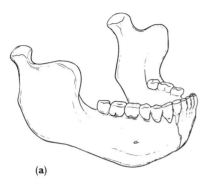

(a)

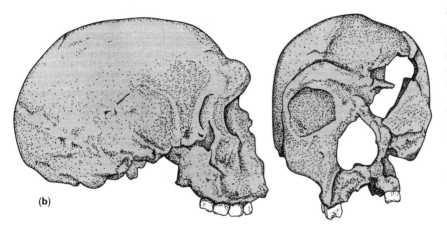

(b)

**FIGURE 14.10 Two German fossils:** (a) The Mauer mandible, found in 1907, combines archaic features (robusticity) with modern features (molar size). It is dated at 500,000 years and is the type specimen of *Homo heidelbergensis*. (b) The Steinheim cranium, found in 1933 and dated between 200,000 and 300,000 years, displays a mix of archaic features (heavy brow ridges) and modern features (large brain).

between 200,000 and 300,000 years old (although the Swanscombe specimen may now be considered to be earlier, perhaps over 400,000 years old). The Steinheim skull possesses heavy brow ridges and a low forehead that betray its primitive status – albeit one not equivalent to *Homo erectus*. Another German site, Bilzingsleben, has yielded cranial fragments that fit the Steinheim pattern and that may be as old as 350,000 years.[375]

In 1960, Greece joined in the panoply of European archaic human sites, with the discovery of a robust but large cranium in a cave at Petralona (Fig. 14.11). Dating this fossil has long posed a challenge. In the early 1970s, the face, forehead, and two jaws of an archaic form were found at the cave of Arago, near Tautavel, in southwest France. The face protrudes forward, the brow ridges are heavy, the forehead is slanting, and the brain is smaller than the modern average. Overall, the Arago fossil is more primitive than the Steinheim one, and perhaps 100,000 years older. In 1993, a massive tibia, or shin bone, was found at Boxgrove, England, together with some Acheulean tools. Its age has been estimated at 500,000 years, or similar to that of the Mauer mandible.

The most spectacular finds of recent times, however, are those at the fossil-rich Atapuerca Hills in northern Spain. In 1993, a team of Spanish

What is the evolutionary history of European hominin populations?

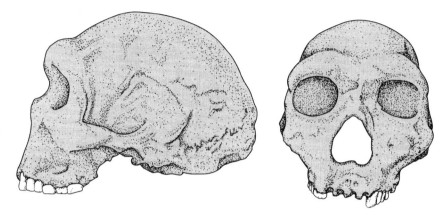

**FIGURE 14.11 The Petralona cranium:** Found in 1960, the cranium is robust but has a large brain case, thus combining archaic and modern features. It has recently been redated at 200,000 years.

researchers reported the discovery of 1300 human fossil remains (representing 30 individuals) from a single site (Sima de los Huesos), dated at 300,000 years old. It represents the largest single collection of early human fossils bones anywhere in the world. Like other human fossils of this age in Europe, the specimens display a mix of ancient and modern features, and may be assigned to *Homo heidelbergensis*.

In 1995, the same workers reported the discovery of more human fossils, representing four individuals, at the site of Gran Dolina, almost a mile away from the previous discovery. The Gran Dolina fossils are much more primitive than those from Sima de los Huesos, and are said to bear some resemblance to *Homo ergaster* from Africa. In their preliminary report, the researchers conclude that the fossils may constitute a primitive form of *Homo heidelbergensis*, or perhaps even a separate species.

These various specimens represent the potential ancestors of the Neanderthals. As we have already seen in the section on the Middle Eastern fossils, the Neanderthals are a Later Pleistocene group of large-brained, robust hominins with particular cranial and facial features, which has been claimed by some to have played a role in the evolution of the modern human species. The chronological and morphological evidence from the Middle East, which showed them to be both different from and later than modern humans in the region, contradicted this view.

However, the Neanderthals, discussed more fully in a section at the end of this chapter, are primarily a European population.[432] Forms transitional between *Homo heidelbergensis* and the Neanderthals have been claimed for as early as 400,000 years ago, notably the specimen of a fragmentary cranium from Swanscombe in England. This specimen possesses an iniac fossa, a Neanderthal characteristic. More convincing is a relatively complete skull from the Sima de los Huesos at Atapuerca, dated to around 300,000 years ago.[375,380] These and other specimens at the end of the Middle Pleistocene may indicate a long and local evolution of

Neanderthals in Europe, although some researchers have suggested that the Neanderthals themselves may have been a population that arrived in Europe somewhat later and either replaced or admixed with the European *H. heidelbergensis*.[86]

Whatever the details of their origin, there is little doubt that the Neanderthals were a widespread and successful population of hominins that flourished in Europe (and extended into the Middle East and Central Asia) between 70,000 years ago and 30,000 years ago.[353,432] They are characterized by a distinctive anatomy. They are robustly built, heavily muscled, and short in stature, with relatively short limbs. Their skulls have large brains, in the range of and exceeding the cranial capacity of modern humans. Compared to those of modern humans the skulls are long and low, with sloping frontals and often a protuberance in the occipital. Their most distinctive characteristic are their faces, which are large and prognathic. This prognathism is different from that seen in earlier hominins, in that it is associated with the mid-facial region; in other words they had large noses, pulling the whole face forward, producing distinctive, swept-back zygomatic arches. This is often regarded as an adaptation to the extreme cold of Europe during the last glaciation.[184] The Neanderthals possess large brow ridges, in common with other archaic hominins, but these are very curved and reduced laterally.

Many fossils display these traits. Key specimens include those from La Chapelle-aux-Saints, Le Moustier, and La Ferrassie in France, and from the Feldhofer cave in the Neander Valley in Germany (the type specimen). It is important to note that during the course of their existence in Europe they do not show a trend toward the modern human condition – rather, if anything they become "more Neanderthal." One of the latest specimens, that from Saint-Césaire, dated to less than 35,000 years old, is a very typical Neanderthal.

What of their fate? Beginning some 40,000 years ago, classic Neanderthal anatomy disappears in Europe, with an east-to-west progression that ends nearly 30,000 years ago. The latest evidence of Neanderthals is found at the site of Zafarraya, southern Spain. Fossil evidence indicating the presence of anatomically modern humans follows the same trajectory. For instance, modern jaw and tooth fragments from the cave of Bacho Kiro, Bulgaria, are dated at 43,000 years. A frontal bone with a high forehead and small brow ridges has been found at Velíka Pečina in Croatia and dated at 34,000 years. A similar specimen, but with a more robust frontal bone, from Hahnöfersand, Germany, has been dated at 33,000 years. A large collection of somewhat robust modern human remains was found at Mladeč, Czechoslovakia. The age of the famous Cro-Magnon fossils, from France, is placed at approximately 30,000 years (Fig. 14.12).

What is more, at the time of these modern humans in Europe, the surviving Neanderthals, such as those found in Spain, do not show any progressive trend toward a modern human condition, but do appear to have a more and more marginal ecological and geographical distribution,

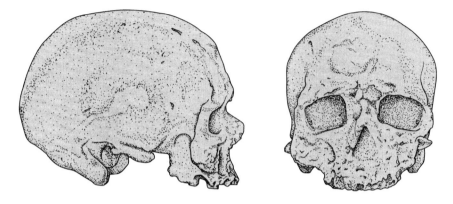

**FIGURE 14.12**
**Cro-Magnon:** The famous cranium from Les Ezyies, France, dated at 30,000 years, provides an example of early modern people in western Europe.

as if they are under considerable competitive pressure. Does this progressive disappearance of Neanderthal morphology and appearance of modern morphology signal population replacement by modern humans as they moved in a westerly direction across the continent, from the Middle East and ultimately from Africa, as the single-origin hypothesis holds? Or is it a morphological transformation of Neanderthals to modern, Upper Paleolithic people, as the multiregional evolution hypothesis contends?

When Bräuer and his colleague K. W. Rimbach compared the crania of the early moderns of Europe, the early moderns of Africa, and the Neanderthals, they found a close morphological similarity between the first two but saw no link between early European moderns and Neanderthals. Similarly, Cro-Magnon skeletons exhibit a warm-adapted body stature, not the cold-adapted formula seen in Neanderthals. This character may be taken as strong evidence of the replacement of Neanderthals and supports the single-African-origin hypothesis.

Although some proponents of the multiregional hypothesis accept that Neanderthals were replaced, at least in the west, most argue for continuity. As evidence, they adduce the size of the nose in Neanderthals and later Europeans, some details of the back of the skull, and, most particularly, the shape of the mandibular nerve canal (Fig. 14.13). This opening is grooved in most living people, but it is surrounded by a bony ridge in 53% of Neanderthals. The incidence in later, modern Europeans is just 6%. According to multiregionalists, this incidence is 44% in early moderns in Europe, indicating continuity. Stringer and Bräuer have recently criticized this claim, saying that while it might indicate gene flow between Neanderthals and early moderns, it is just as likely to be a statistical fluke. The sample used by multiregionalists comprises just four individuals, including one from Vindija, Croatia, that many consider to be Neanderthal. Given the small sample size of just three individuals, the inclusion of just one with an infrequent feature would produce an erroneously high

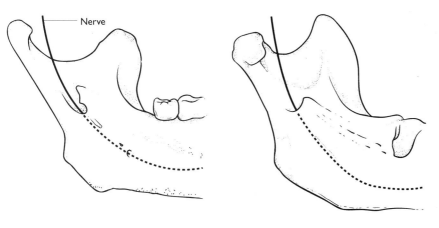

Nerve

**FIGURE 14.13**
**Mandibular nerve canal:** In most living and fossil people, the rim around the nerve canal is grooved (left); in roughly half of all Neanderthals, it is surrounded by a bony ridge. Multiregionalists argue that early modern people in Europe also had a high incidence of the bony ridge, indicating important morphological continuity.

incidence. Of course, the chances of this type of occurrence in a population with low incidence is not great.

For the most part, the fossil evidence of Europe seems to show that there is a replacement event of Neanderthals by modern humans, with little or more likely no interbreeding of populations. While this view has gained more and more support from both chronological and morphological evidence, it was challenged in the late 1990s by the discovery of a child's skeleton from the site of Lagar Vella in Portugal.[395] Dated to about 25,000 years ago, this specimen falls well within the time range of modern humans, and after the supposed extinction of the Neanderthals. According to its discoverer, Joao Zilhao, and Eric Trinkaus, of Washington University in St Louis, however, the child is evidence of hybridization between modern humans and Neanderthals, on a scale sufficient to be recognizable after thousands of years. The evidence comes from the limb proportions. While the skull, jaw, and dental morphology of the specimen are characteristically modern, the child had limb proportions that were similar to those of Neanderthals. These proportions are of course related to cold adaptation, so while Trinkaus has used this as evidence for hybridization, to others it is evidence of cold adaptation over a period of thousands of years. Certainly it would be unusual to find evidence for hybridization in only the postcranial elements of the skeleton.

# Africa

The Middle and Late Pleistocene human fossil record of Africa is not extensive, but a sufficient number of specimens has been found to prove a transition from primitive to modern humans. The first find was made in 1921 at a cave site at Kabwe (formerly Broken Hill), Zambia. The specimen,

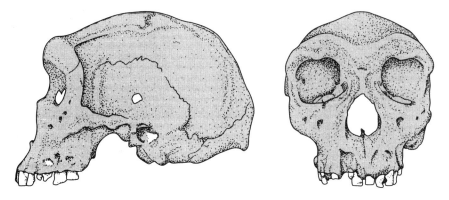

**FIGURE 14.14 The Kabwe cranium:** Estimated to be at least 200,000 years old, this cranium was the first early human fossil found in Africa.

a cranium, was originally called Rhodesian Man, but is now more generally referred to as the Kabwe cranium (Fig. 14.14). The cranium is large, having a capacity of 1280 cm$^3$, and possesses a sloping forehead and prominent brow ridges reminiscent of Neanderthals. Associated limb bones are, however, straighter and more slender than those of Neanderthals. The specimen's age is estimated to be at least 200,000 years.

Similar archaic forms have been found at Elandsfontein, South Africa (dated at 300,000 years); Bodo, Ethiopia (of similar age); and near Lake Ndutu, Tanzania (perhaps 100,000 years older than the other two finds). The cranial shape of most of these African archaics is long, as in *Homo erectus*, but more elevated; from the rear, it appears to be wider at the top than at the base, unlike the structure in *Homo erectus*. The Ndutu cranium is shorter and less flattened. In northern Africa, archaic forms of Middle Pleistocene age have been found at Salé and the Thomas Quarries, in Morocco.

Are transitional archaic–modern hominins found in Africa?

Some proponents of the single-origin hypothesis group these specimens in *Homo heidelbergensis*, which they claim evolved in Africa and then moved into other regions of the Old World. The species is held to be ancestral to modern humans, through a form represented by several specimens that are generally modern, but not yet fully modern (Fig. 14.15). These remnants include the following: cranial fragments from Florisbad, South Africa; a cranium and lower face from Ngaloba, Tanzania; a skull (KNM-ER 3884) from Koobi Fora, Kenya; the Omo II (Kibish) brain case from Ethiopia; and various cranial and postcranial fossils from Jebel Irhoud, Morocco. The recent discovery of a modern-looking specimen from Herto in the Middle Awash, Ethiopia, is an important addition to this material.

The dates of some of these specimens remain somewhat uncertain, but they are generally later than the above group of *Homo heidelbergensis* specimens. Recent dating of the Florisbad cranium indicates that it may be as old as 300,000 years. Whether this group can be contained within *Homo heidelbergensis* or should be assigned to a separate species (*Homo helmei*) is a matter of debate.[86,396] In any case, these individuals could

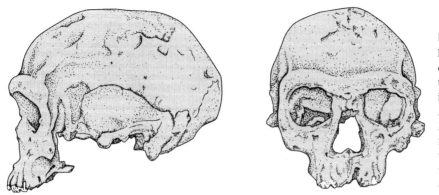

**FIGURE 14.15 Jebel Irhoud:** One of a group of later human fossils that were perhaps intermediate between *Homo heidelbergensis* and *Homo sapiens*. The species name of *Homo helmei* would be appropriate for this specimen, if a specific designation is justified.

represent a form transitional to modern humans, such as those found at Omo I (Kibish), Klasies River Mouth, and Border Cave.

This pattern of transitional forms from archaic to modern fits both the single-origin and multiregional evolution hypotheses, of course. The fact that it occurred earlier in Africa than elsewhere provides support for the former concept. In addition, the anatomical similarities between some of these African archaic forms and archaics elsewhere in the Old World supports the single-African-origin hypothesis rather than the multiregional evolution hypothesis.

## The fossil evidence as a whole

The hominin fossil record for the later Pleistocene is, by comparison with other periods, relatively rich (Fig. 14.16), and this has both helped and hindered the debate. The large quantity of material shows that it is not a simple problem, while the gaps can mean that there is much that remains obscure. However, overall the fossil record seems to show a pattern that is closer to the out-of-Africa than the multiregional one. If this is correct, then the phylogeny of *Homo* will be closer to the "*Homo* as separate local trajectories" model than the "*Homo erectus* as general ancestor" one (see beginning of chapter) (Fig. 14.17).

Because of the tenor of the interchange between the protagonists in the current debate, analyses by workers not directly involved in the issue are particularly useful. Aiello's overall conclusion, for instance, is that none of the current hypotheses completely explains the observed fossil evidence, but that some form of single-origin model is most strongly supported.[397] More recently, F. Clark Howell, of the University of California, Berkeley, reached a similar conclusion, stating that the anatomical data "afford no support to a continuity (so-called Multiregional Evolution or Regional Continuity) model of [hominin] evolution, but they are consistent with a variety of recent (and African) replacement models." He describes the

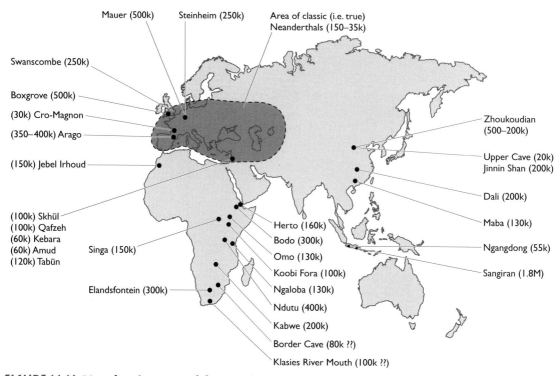

**FIGURE 14.16 Map showing some of the most important sites:** Dates are in thousands (k) and millions (M) of years.

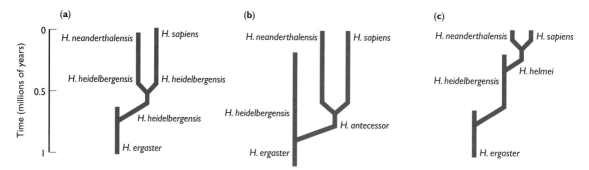

**FIGURE 14.17 Possible phylogenies of later *Homo*, within the framework of an out-of-Africa model:** There is no consensus on the phylogeny of later *Homo*. The three here are among the possibilities currently under discussion. (a) is favored by Stringer and Rightmire and many others; (b) is favored by Arsuaga and others in the light of recent discoveries at Atapuerca; and (c) is favored by Lahr and Foley. Those supporting the multiregional model would reject any species-based phylogenies for later *Homo*.

multiregional evolution hypothesis as "seriously flawed" and as being "a fragile, frayed, and unfructuous conceptual framework."[378]

Two other analyses are important here. The already-mentioned work of Lahr analyzed cranial features. She set out to test whether the key expectations of the multiregional hypothesis were based on sound premises – namely that the regional traits were indeed regional in modern human populations. What she found was that, when analysed statistically, they were not regional (Fig. 14.18). Many of the traits that have been adduced as unique to southeast Asia are actually correlates of cranial robusticity seen anywhere in the world where robust ancestors were present, not just

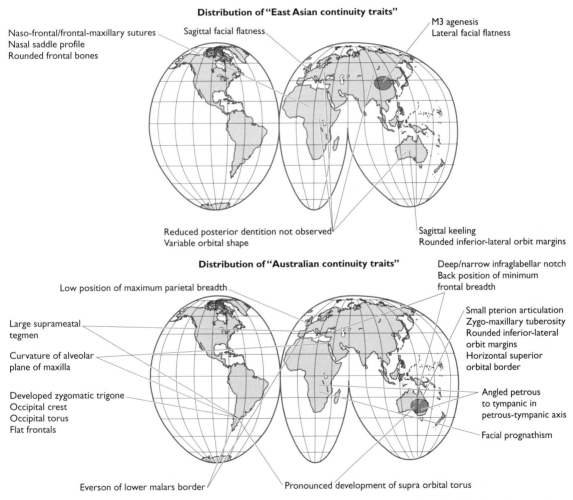

**FIGURE 14.18 Distribution of multiregional traits:** The multiregional model is based on the idea that certain cranial traits are specific to particular regions of the world, and that they represent long-term local populations. Marta Lahr looked at these in a sample from around the world today, and found them to be very variable in their regional expression. (Courtesy of M. Lahr and R. Foley.)

in southeast Asia. Other regional traits were either found not to occur in higher frequencies in those regions, or to be more global in their distribution. "The only conclusion to be drawn is that the morphological evidence does not support a multiregional model of modern human origins," she notes.[390,398]

The second study, by Diane Waddle, of the State University of New York, Stony Brook, is a matrix correlation test of cranial features. This analysis effectively compares measures of cranial features among all populations in the Old World and then assesses their consistency with hypothetical matrices that build in the assumptions of different models. For instance, the continuity model would show consistency of measures within geographical regions, while the single-origin model yields no such geographical groups. This approach provides the first quantitative test of the fossil evidence. Waddle concludes that her analysis results "support a single origin for modern humans as opposed to continuous long-term evolution within regions." Nevertheless, the analysis "cannot resolve the issue of the site of origin."

The origin of modern humans was undoubtedly complex, involving much population movement at different times, and local population expansions and extinctions. As Marta Lahr has shown, the pattern is too complex to be explained in terms of a single point of origin and immediate dispersal, but involves multiple dispersals.[399] The full pattern of these, however, requires consideration of the genetic and archeological evidence as well as the fossil.

## THE PLACE OF NEANDERTHALS IN HUMAN EVOLUTION

The Neanderthals have made an appearance in various parts of this book, and will continue to do so, as there are themes associated with them that have been pursued extensively in studies on human evolution – their body shape and size in relation to climate, their cognitive and linguistic abilities, and their role in the evolution of modern humans. It is worth, however, examining them in a little more detail, as they throw light on several important issues. The literature on them is vast, with many books devoted to them; of particular importance are Pat Shipman and Erik Trinkaus's *The Neanderthals*, Ian Tattersall's *The Last Neanderthal* and Chris Stringer and Clive Gamble's *In Search of the Neanderthals*.

Neanderthals are the proverbial "cavemen" of fact and fiction, largely because they were the first non-modern fossils to be discovered.[15,16,18] In the Feldhofer cave in the Neander Valley, near Dusseldorf in Germany, fragmentary remains of a cranium and some leg and arm bones were discovered in 1856. Although this was before the publication of Darwin's *Origin of Species* (1859), they were recognized by Hermann Schaafhausen as "primitive" – "one of the wild races of northern Europe." Thus began the shifting interpretations of the Neanderthals, from early European precursor, to pathological individuals, to cold-adapted European specialists, to a variant of modern humans.

The first important role they played was in establishing the reality of a fossil record for human, and in effect substantiating the implications of Darwin's theory of evolution. T. H. Huxley, in his book published in 1863,[2] was the first to make this point, showing that although Neanderthals shared many characteristics with modern humans, they nonetheless represented a different and in his view more primitive form of human, and thus evidence that humans had undergone evolutionary change. In the context of the turmoil following the establishment of evolutionary ideas, this was a major contribution. (It is worth noting that although the Neander Valley specimen was the first to be recognized as such, it was preceded by more than a decade by a discovery of a Neanderthal fossil in Gibraltar, but in the pre-Darwinian world, it was not given the significance it would now deserve.)

What is the role of Neanderthals in human evolution?

The following century produced many more discoveries of Neanderthal fossils, mostly from the archeologically rich caves of southwest France. These were important in establishing the context of the Neanderthals, both archeologically in terms of their association with Middle Paleolithic technology (the Mousterian), and also stratigraphically by showing that they preceded occurrences of modern humans. Although subsequent discoveries have shown that Neanderthals were preceded by many other types of hominin, and thus were a long way from being classical missing links (indeed, they should be seen more as contemporaries of early modern humans than as their antecedents), their role in establishing the study of human evolution should not be forgotten.

The second major role that Neanderthals have played in paleoanthropology has been the extent to which they have been the test-bed for many ideas about the human evolutionary past. The probable reason for this lies in the fact that not only are they relatively abundant, but that many of the specimens are relatively very complete, allowing more inferences about their adaptations than is often possible. What is striking is how wildly these inferences have fluctuated.

Early interpretations of Neanderthals focused on the possibility that they represented pathological humans rather than a particular type of human morphology.[2] This idea is most closely associated with the German anatomist Rudolf Virchow, who argued that their limb structure was a product of rickets. However, as more fossils were discovered showing the same features, this became increasingly untenable.

The next landmark in the study of Neanderthals was the work of the French anthropologist Marcellin Boule. He carried out a detailed study of the Neanderthal from La Chapelle-aux-Saints. His reconstruction suggested that Neanderthals were not fully bipedal, but had bent knees and bent hips, and were thus incapable of the full stride of modern humans. Boule's work was extremely influential, and led to the view that there was a gulf in ability and evolutionary status between Neanderthals and modern humans. It was this reconstruction that gave rise to many of the popular caricatures of Neanderthals, as shambling, brutish idiots. Within the discipline it also led to the view that Neanderthals were distant from the evolution of modern humans.

Subsequent research overturned Boule's views. The Chapelle-aux-Saints specimen is that of an old man, suffering from extreme arthritis, and many of the odd features of his gait were due to this. This work led to a rehabilitation of Neanderthals, and the establishment of the current view: that they are a robustly built but fully bipedal population of hominins, living in the cold conditions of Europe during the last glaciation. The facial prognathism, the body proportions, and the short stature are all traits that indicate adaptation to cooler conditions, much as is found in Inuit today. To this recognition were added ideas, drawn from the high level of bone breakage found, that Neanderthals led rough and arduous lives as hunters.

If morphological studies led to a more human interpretation of Neanderthals, archeological and behavioral studies went in a different direction. Partly under the influence of work such as Trinkaus's on bone breakage, but partly due to the impact of Lewis Binford's archeological analyses, it was increasingly argued that Neanderthals were behaviorally very different from modern humans, and that the contrast between the Middle and Upper Paleolithic represented a watershed from non-human to human behavior – the so-called "cultural revolution."[121,384,439,443,475] Evidence in support of this took many forms. Trinkaus, for example, used differences in pelvic size to argue for a different parenting strategy (a view he has now abandoned); Paul Mellars has argued that there are very substantial differences across a whole suite of archeological evidence between the Neanderthals and modern humans; Stephen Mithen has argued that Neanderthals were neurobiologically different, unable to have cross-module thought; Phillip Liebermann suggested that they lacked language; and Binford, as well as others, inferred an inability to hunt efficiently. The absence of beads and art was also seen as a significant difference. Taken together, these behavioral lines of evidence created a situation where the contrast between Neanderthals and modern humans was seen as very marked indeed. These ideas also fitted in well with the replacement elements of the out-of-Africa model.

Many of these inferences have remained reasonably well supported, although the high tide of Neanderthal behavioral difference has probably passed in current thinking. That early modern humans in Africa and the Middle East made essentially the same artifacts as Neanderthals is one factor suggesting a greater similarity. So too is the increasing evidence for language among Neanderthals (see chapter 18). Given their large brains, there is little doubt that Neanderthals are intelligent, flexible hominins, even if there may have been some differences between them and modern humans. It is probably the case that they shared more things with modern humans than with, for example, archaic hominins such as *Homo erectus*. The current view taken by many of those who support the out-of-Africa model is that Neanderthals belonged to a highly successful, cold-adapted population, capable of many sophisticated behaviors and strongly cultural, but that this did not prevent them from becoming extinct in the Later Pleistocene, either as a response to modern human intrusion and competition, or else because of changes in climate.

The changing perspective on the behavioral and adaptive characteristics of the Neanderthals has also influenced the way in which their phylogenetic relationship to modern humans has been perceived. While T. H. Huxley viewed them as very close to modern humans, indeed a minor variant, others placed them far farther away, and as an evolutionary dead end. This view predominated at the time that Boule's reconstruction of the Chapelle-aux-Saints skeleton was prevalent. This model is often referred to as the "pre-*sapiens* hypothesis," as it was believed that Neanderthals diverged from modern humans before the latter had "achieved" *sapiens* status. Once the Boule reconstruction had been overturned, there was a tendency to reduce the differences between the two lineages (the so-called "pre-Neanderthal hypothesis"). However, the major break in the way Neanderthals were viewed phylogenetically came in 1964, when Loring Brace of the University of Michigan wrote a highly polemical paper arguing that Neanderthals represented a direct ancestral phase in human evolution (the "Neanderthal phase hypothesis").

The current situation holds elements of all these views. The accumulating fossil and genetic evidence in favor of the out-of-Africa hypothesis, and the absence of any strong evidence for regional continuity between Neanderthals and modern humans, have effectively restored the principal ideas of the pre-Neanderthal and pre-*sapiens* hypotheses, namely that modern humans and Neanderthals diverged at some point, probably within the last half million years, and had parallel evolutionary histories in Africa and Europe respectively. Under this model there is a general recognition that Neanderthals and modern humans belong to separate species: *Homo neanderthalensis* and *Homo sapiens*. The Neanderthal phase hypothesis – the idea that humans are descended from a Neanderthal population, as opposed to sharing an ancestor with them – is held by few if any people now, although the view that Neanderthals contributed to the gene pool of anatomically modern peoples in Europe is a part of at least some versions of the multiregional model. The genetic evidence discussed in the next chapter has contributed to the abandonment of the idea that Neanderthals are ancestors of modern humans.

However, if the interpretation of the phylogenetic position of Neanderthals is close to that of Boule's, the behavioral and adaptive interpretation is not. The Neanderthals are now seen as a particularly successful population or species that evolved in Europe in the later Middle Pleistocene, dispersed into Central Asia and the Middle East in the Upper Pleistocene, and showed many characteristics which are synapomorphic with modern humans, including possibly language. Their extinction, at the hands of modern humans, climate, or some other factor, remains one of the fascinating questions of human evolutionary studies. With their extinction, hominins, possibly for the first time since early in their evolution, were represented by a single species.

## BEYOND THE FACTS
# TOO MANY SPECIES – OR NOT ENOUGH?

*The issue: over the course of its history, paleoanthropology has swung from a period of extreme splitting, where many species of hominin were recognized, to one of extreme lumping, and back to a high level of splitting. Is this just a matter of intellectual fashion, or is it possible to set criteria by which the number of species expected and observed can be scientifically established?*

A reading of much in paleoanthropology could lead to a serious sense of disillusion about the state of the discipline. There is apparently not even consensus on what are the species in the fossil record. How can one resolve the serious problems of reconstructing human evolution if the basic building blocks of which it must be made are not agreed? Those building blocks are the species that make up evolutionary history. As we have seen with regard to the early members of the genus *Homo*, there is debate as to whether there is one species or more; with the robust australopithecines, whether they are one species spread across Africa, or three or four local species; and whether *Australopithecus bahrelghazali* is another species of early australopithecine,

or just a local variant on *A. afarensis*. Each of the taxa of hominin evolution can be either lumped together or split.

As we have also seen, one view is that we have greatly overestimated the number due to the tendency of paleoanthropologists to read "new species" into the most trivial anatomical trait, and so not to recognize that each species will contain within it considerable variation. The usual model to determine how much variation there is within a species is that of the living apes and monkeys. After measuring their range of size and shape differences, you can then compare two or more fossils with that range. However, it could be argued that any sample drawn from a species today samples only a part of the variation, as all the specimens come from the same generations, whereas two fossils may be separated by hundreds of thousands of years and thousands of generations, and so much more variation would be expected in a single-species sample of fossils than of living beings. This would support the lumpers, by implying that there should be fewer species, and so a simpler phylogeny.

On the other hand, if we take another line we can just as well argue that among the primates, species are marked not by much in the way of measurable aspects of the skeleton, but by different coloration or different vocalizations. Small, seemingly trivial differences are the signals of species.

These are things that would not show up in the fossil record, and so bones alone would tend to underestimate the number of species found there.

So if paleoanthropology is to progress, it has to come up with an answer to the question of how many hominin species there are – but this in turn perhaps depends upon how many hominin species there ought to be. This latter is dependent upon understanding evolutionary theory and processes.

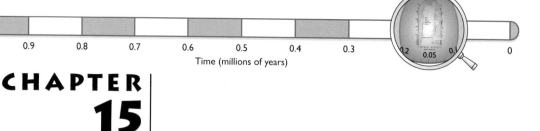

# CHAPTER
# 15

# THE ORIGIN OF MODERN HUMANS: GENETIC EVIDENCE

## THE IMPACT OF MOLECULAR EVOLUTIONARY GENETICS

> **KEY QUESTION** How has the development of molecular evolutionary genetics changed our understanding of human evolution?

Evidence from genetic studies has been applied to the question of the origin of modern humans to a greater extent than to any other topic in paleoanthropology, including the origin of the hominin clade. One consequence of this research is the growing realization of just how complex the event probably was compared with the earlier, rather simple hypotheses to describe it that were developed from fossil evidence. Another consequence is the recognition that interpreting genetic evidence is a far more difficult task than was imagined just a few years ago. Not surprisingly, a great deal of controversy has swirled around this topic in the past and continues to surround it today. This chapter will explore some of the strengths and uncertainties associated with this genetic evidence.

The first application of genetic data to the question of the origin of modern humans took place in the early 1980s, but not until 1987 did it become highly visible in this realm. The initial work, conducted first in Douglas Wallace's laboratory at Emory University and later in the University of California, Berkeley, laboratory of Allan Wilson, focused on mitochondrial DNA. It inspired the so-called mitochondrial Eve hypothesis, which posited that the mitochondrial DNA in all living people could be traced back to a single female who lived in Africa approximately 200,000 years ago (hence the inclusion of the term "Eve").[400] This female was a member of a population of an estimated 10,000 individuals, all of whom were related to the founding population of modern humans; descendants of this population spread into the rest of the Old World, and replaced existing populations of various species of archaic *sapiens* and *Homo erectus*. Thus, the mitochondrial Eve hypothesis was consistent with the recent-single-origin (out-of-Africa) model and gave no support for the multiregional evolution model.

The confidence with which proponents of the mitochondrial DNA work asserted that the question of the origin of modern humans had been decisively solved began to erode in the early 1990s. It was recognized that the analysis of the voluminous data had been inadequate, and that the conclusion of an African origin was not as clear cut as had been originally thought. Since then, genetic work has followed two pathways. First, mitochondrial DNA analysis has been extended and data from other genes added, including that from the nucleus, with the common objective of determining where and (more particularly) when modern humans evolved. Second, mitochondrial DNA data have been used to infer aspects of population dynamics of early modern humans. Both lines of enquiry remain consistent with the recent-single-origin theory of modern humans, and offer little or no support for a multiregional origin.

## The mitochondrial Eve story: briefly told

Most of the DNA in our cells is packaged within the 23 pairs of chromosomes in the nucleus, which in total measures about 3 billion base pairs in length; this structure is known as the nuclear genome. The cell also contains a second, much smaller genome that consists of a circular molecule of DNA, 16,569 base pairs long. Many copies of this second genome, called the mitochondrial genome, are found within each cell (Fig. 15.1). Mitochondria are the organelles responsible for the cell's energy metabolism, and each cell contains several hundred of these structures.

Mitochondrial DNA is considered useful for tracking relatively recent evolutionary events for two reasons. First, this DNA, which codes for 37 genes, accumulates mutations on average 10 times faster than occurs in nuclear DNA. Even in short periods of time, therefore, mitochondrial DNA will accumulate mutations that can be counted. In contrast, slowly evolving nuclear DNA may accumulate few or even no mutations over the same time. As mutations represent the equivalent of information,

**FIGURE 15.1 Nuclear and mitochondrial DNA:** Nuclear DNA, which controls the development of the phenotype, is found in the nucleus of the cell. The other type of DNA we all possess is mitochondrial DNA, which occurs in many copies outside the nucleus.

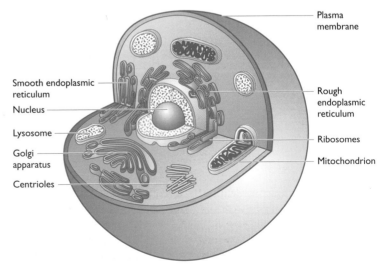

- Plasma membrane
- Rough endoplasmic reticulum
- Ribosomes
- Mitochondrion
- Smooth endoplasmic reticulum
- Nucleus
- Lysosome
- Golgi apparatus
- Centrioles

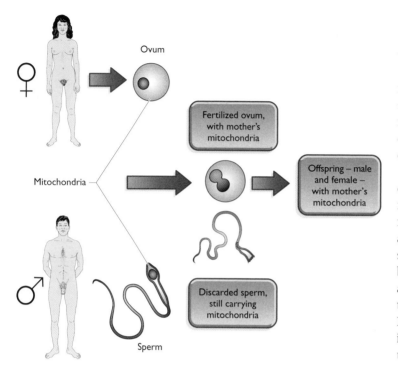

Ovum

Mitochondria

Sperm

Fertilized ovum, with mother's mitochondria

Offspring – male and female – with mother's mitochondria

Discarded sperm, still carrying mitochondria

**FIGURE 15.2 Patterns of inheritance:** Unlike nuclear DNA, which we inherit half from our mother and half from our father, mitochondrial DNA is passed on only by females. When the sperm fertilizes the egg, it leaves behind all of its mitochondria; the developing fetus therefore inherits mitochondria only from the mother's egg.

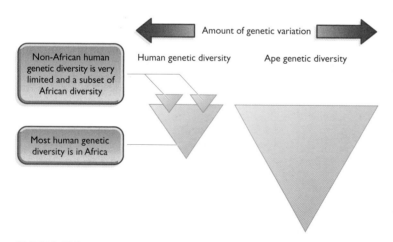

Amount of genetic variation

Human genetic diversity          Ape genetic diversity

Non-African human genetic diversity is very limited and a subset of African diversity

Most human genetic diversity is in Africa

**FIGURE 15.3 Human mitochondrial DNA variation:** The human population has very little genetic variation in its mitochondria when compared to the apes, and much of that variation is in Africa.

mitochondrial DNA provides more information over the short term than does nuclear DNA. Second, unlike an individual's nuclear genome, which consists of a combination of genes from both parents, the mitochondrial genome is inherited only from the mother (Fig. 15.2) (except under unusual circumstances). Because of this maternal mode of inheritance, no recombination of maternal and paternal genes occurs; such a mixture may sometimes blur the history of the genome as read by geneticists. Potentially, therefore, mitochondrial DNA offers a powerful way of inferring population history, unhindered by the genetic fog of recombination.

One of the first significant observations to emerge from this work was that the amount of variation of mitochondrial DNA types in the modern human population throughout the world is surprisingly low – just one-tenth of that known among chimpanzees, for instance (Fig. 15.3). One explanation is that modern humans evolved very recently, a view that Wallace and Wilson independently supported. A calculation based on the rate of accumulation of mutations of mitochondrial DNA gave a time of origin of 140,000 to 280,000 years ago. An alternative explanation holds that modern humans passed through a population bottleneck recently, which reduced genetic variation. These explanations are not mutually exclusive: modern humans may have evolved recently and experienced a population bottleneck. Another scenario would involve the evolution of modern humans in ancient times, followed by a recent population bottleneck. These possibilities are discussed below.

A second finding from the early work was that African populations display the greatest degree of variation in their mitochondrial DNA (Fig. 15.4). This discovery was taken to indicate that this population was oldest, and therefore represented the population of origin of modern humans. An alternative explanation, however, is that the early African population was larger than other populations, and its greater size promoted the accumulation of more extensive genetic variation.

Wilson's initial work sampled mitochondrial DNA data from 147 individuals representing Africa, Asia, Australia, Europe, and New Guinea. Using the same format as that followed by Wallace's team, the technique used to measure genetic variation was employed to produce a kind of map of the mitochondrial genome in each individual. The DNA was cut with a small number of so-called restriction enzymes, each of which cleaved a discrete sequence of nucleotide bases. If everyone's mitochondrial genomes were identical in sequence, the fragments produced by this procedure would give the same pattern in all cases. Any variation in DNA sequence that altered one of the restriction enzyme cleavage sites, however, would result in a different fragment pattern for different individuals. The technique, known as restriction fragment length polymorphism, effectively samples about 9% of the genome, giving a picture of the different mitochondrial genomes in different populations.

Wilson's initial analysis revealed 133 different types, which were organized into a genealogical tree by parsimony analysis (a technique, as discussed earlier, that searches for the smallest number of changes that will link the different patterns and eventually trace back to a common ancestor). The now-famous horseshoe diagram produced by this analysis contained two groups: one including members from all geographical locations, and one including only Africans. The latter group had the deeper root, which was interpreted as further support for an African origin of modern humans.

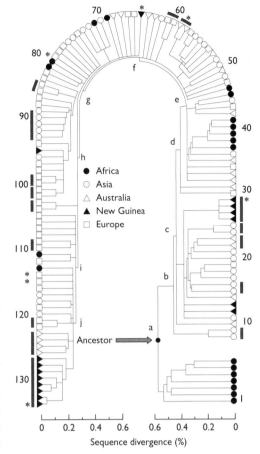

**FIGURE 15.4 Patterns of relatedness:**
The classic "Christmas tree" genealogy produced by Wilson and his colleagues in 1987 shows the genetic divergence among 147 individuals from different geographic populations, whose mitochondrial DNA was tested. The tree shows a split between African and non-African populations. The African population is the longest established, indicating the origin of modern humans in that continent. The different degrees of sequence divergence among the non-African populations give some indication of when different parts of the Old World were colonized. Recent analysis has shown that this tree, one of many possible, may not be the optimum interpretation. (Courtesy of Rebecca L. Cann et al./*Nature*.)

Colorful though it is, the term "Eve" in the hypothesis title is misleading, and it originally led to widespread misunderstanding of the implication of the analysis's results. The mitochondrial DNA types in today's human population can be traced back to a single female, not because she was the only woman living at the time, but because of the dynamics of loss of the DNA. This process is best explained by analogy. Imagine a population of 5000 mating pairs, each with a different family name. As time passes, the population remains stable (each couple produces only two offspring). In each generation, on average, one-fourth of the couples will have two boys, one-half will have a boy and a girl, and one-fourth will have two girls. In the first generation, therefore, one-fourth of the family names will be lost (assuming that family names are passed only through the male line). With each succeeding generation, more losses will occur, albeit at a slower rate. After approximately 10,000 generations (twice the number of original females), only one family name will remain (Fig. 15.5). The same pattern holds for the loss of mitochondrial DNA types, except that the transmission flows through the female line.

Wilson and his colleague's conclusion in their 1987 paper, therefore, merely pinpointed the region of origin of modern humans (Africa) and estimated the time at which they arose (roughly 200,000 years ago). Nevertheless, it was widely assumed that a severe population bottleneck (not necessarily a single pair, but perhaps just a few hundred individuals) was a necessary element of the Eve hypothesis. When data from a set

**FIGURE 15.5 Life of a lucky mother:** This illustrates the concept that all maternal lineages in a population trace back to a single lineage in an ancestral population. At each generation one-fourth of the mothers will have two male offspring, one-fourth will have two female offspring, and one-half will have one female and one male offspring. The mitochondrial lineages of mothers bearing only male offspring will come to an end, leading eventually to one lineage dominating the entire population. (Courtesy of Allan Wilson.)

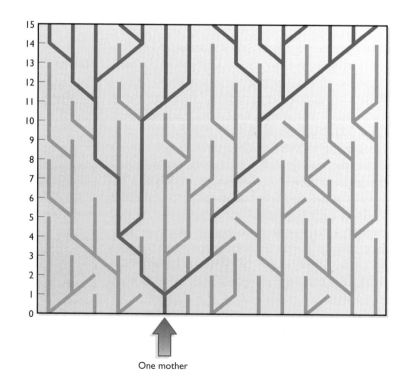

One mother

of genes associated with the immune system indicated that no severe bottleneck had occurred in recent human history, many observers declared that the hypothesis had been disproved. This assumption is not necessarily vindicated. Eve may be invalid, but not because recent human history experienced a population bottleneck at the level of a few hundred individuals.

Wilson's 1987 study was criticized on three grounds.[401,402] First, the technique culled only a small amount of information from the mitochondrial genome, and the methodology used was clearly inferior to sequencing sections of it. Second, the Africans in the sample were African Americans and not native Africans, which could have led to erroneous results. Third, the human data were not compared with those of a close relative (such as the chimpanzee), which would have been a suitable outgroup. The internal comparison that was employed instead may have yielded an unreliable result, it was suggested.

The Berkeley group responded by gathering sequence data on a region of the mitochondrial genome in many more individuals, including native Africans. The new studies produced much the same results as the earlier one, and were also able to demonstrate statistical significance to the conclusions. By this time, several thousand samples of mitochondrial DNA had been analyzed in various laboratories, and all were "young" – that is, they displayed little genetic variation. The absence of "old" mitochondrial lineages (which demonstrate widely divergent genetic variation) was taken as further evidence that modern humans originated recently. In addition, this observation implies that little or no interbreeding had occurred between the early modern human populations and existing archaic *sapiens* populations, which were assumed to have been replaced. At the time of this book's first publication (1997), not a single example of an ancient mitochondrial DNA lineage had been found among the more than 5000 individuals sampled. At the time of current writing this is still the case for samples drawn from living populations, although some ancient DNA from an Australian sample may reflect an older history (but see below) (Fig. 15.6).

> What is the basis for the "African Eve" hypothesis?

The most serious challenge to the validity of the Eve hypothesis came in 1992, when it was realized that the Berkeley laboratory's parsimony analysis of the new data had been inadequate, as had that of the original 1987 analysis. The technique of parsimony seeks to find the tree that joins together all observed variants via the minimum number of mutations. When the sample includes more than 100 individuals, and a similar number of informative sites on the genome, the number of possible trees becomes enormous. Vast amounts of powerful computer time are needed in such a case to sort out those having the fewest mutations. For instance, the number of possible trees derivable from the data in the 1991 paper is an astronomical $8 \times 10{,}264$; even the number of shortest trees exceeds 1 billion. No amount of currently available computing power can sort through this fog of possibilities. Even using the best algorithms and the

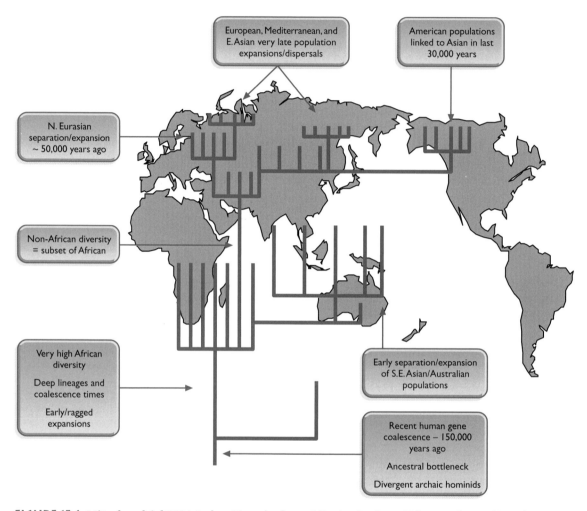

European, Mediterranean, and E. Asian very late population expansions/dispersals

American populations linked to Asian in last 30,000 years

N. Eurasian separation/expansion ~ 50,000 years ago

Non-African diversity = subset of African

Very high African diversity

Deep lineages and coalescence times

Early/ragged expansions

Early separation/expansion of S.E. Asian/Australian populations

Recent human gene coalescence – 150,000 years ago

Ancestral bottleneck

Divergent archaic hominids

**FIGURE 15.6  Mitochondrial DNA today:** Since the first publication by Cann, Wilson, and Stoneking, there have been many more studies, all of which broadly confirm the first. From them, it has been possible to reconstruct a tree of human mitochondrial DNA history. (Courtesy of R. Foley and M. Lahr.)

fastest computers available, the most exhaustive parsimony analyses have racked up only 50,000 trees.

Consequently, researchers must be selective about the trees they examine and must assume that their selection constitutes a representative sample. This assumption has proved not to be valid in all cases. For instance, 100 trees were examined in the 1991 paper. When two of the authors, Linda Vigilant and Mark Stoneking, of Pennsylvania State University, were prompted to check the validity of the conclusions, they produced 10,000 trees, finding just as many that suggested a non-African origin as an African one. Alan Templeton, of Washington University, and David Maddison and his colleagues, at Harvard University, obtained similar results.[403]

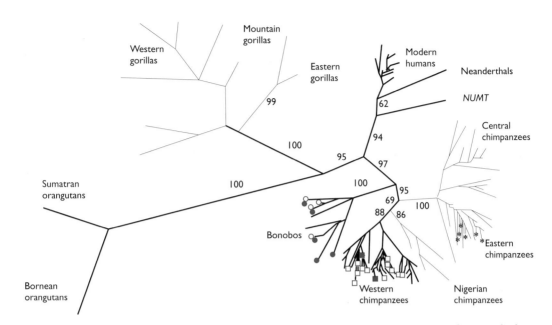

**FIGURE 15.7 Human and ape diversity:** The mitochondrial DNA of the great apes has now been studied, and shows a very different pattern to that of humans. Apes show much greater diversity. In this diagram, the relationships between the species in terms of mitochondrial DNA lineage are shown by the branching lines, and the length of the lines indicates the amount of difference between them. *NUMT* is a piece of mitochondria that acts as a marker for an ancestral mitochondrial genome. (Courtesy of Gagneux et al.[135])

At this point in the history, the mitochondrial Eve hypothesis, claiming a modern human origin rooted in Africa, could be supported from a statistical point of view – that is, was the most probable hypothesis – but was not sufficiently strong to reject completely a multiregional intepretation. This was taken by some as support for a multiregional hypothesis, and by others as calling for more evidence and better methods. Both of these last two have been forthcoming.

Since that time, further analyses have strongly supported the basis of the original mitochondrial evidence. First, a comparison of the human mitochondria compared to that of the great apes showed an enormous difference in the levels of variation among species. Pascal Gagneux (a geneticist at the University of California at San Diego) and colleagues showed that all the great apes show very large amounts of variation compared to that of humans (Fig. 15.7), suggesting very different evolutionary histories for humans and apes.[135] This again supports the idea of a recent, bottlenecked origin for the human lineage.

*What is the evidence for human diversity based on mitochondrial DNA?*

Second, Max Ingman, from Uppsala University, and Svante Pääbo, of the Max Planck Institute for Evolutionary Anthropology in Leipzig, and colleagues sequenced the entire mitochondrial genome, rather than just the fragments that were done in earlier analyses.[404] The results confirmed the picture obtained from the original work in all its substantial

**FIGURE 15.8**
**Dispersals and mitochondrial DNA:** Researchers have now used mitochondrial DNA to infer the history of migrations, by looking at the distribution of particular haplotypes and haplogroups (smaller and larger groupings of related genotypes) and their relatedness. The labels show mitochondrial lineages and range estimates for dispersal dates. (Courtesy of N. Maca-Meyer et al.)

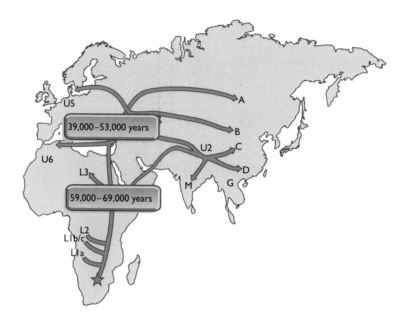

points, providing the statistical certainty that was previously absent. As Blair Hedges of Pennsylvania State University said in a commentary at the time: "The upshot is a robust tree rooted in Africa, which times the exodus from Africa to within the past 100,000 years (recent in evolutionary terms). With this result, the pendulum swings further towards the claim that modern humans, *Homo sapiens*, originated in Africa."

Third, two other analyses of global patterns, one by Peter Forster, at Cambridge University, and colleagues,[405] and one by a team from the Canary Islands led by Nicole Maca-Meyer,[406] both showed that the out-of-Africa/recent model was supported by the data, although they also showed that the pattern of dispersal and diversification was more complex and in line with the multiple-dispersals model[86,399] (see below) (Fig. 15.8).

The mitochondrial model showed that the genetics of living human populations could be a powerful tool in reconstructing evolutionary history, although the analytical difficulties could be formidable. The underlying reason for the difficulty is that what they provide is a gene history, and that is not the same as a population history. The mitochondria is just one (rather special) gene. From this emerged the need to explore other genetic systems to search for similar evidence, such as genes and other elements in the nuclear genome. The second conclusion was that the genes did not provide a pure picture of phylogeny, but rather one mediated through what had happened demographically to a population – whether it had grown or remained stable, etc. The bottleneck inference was one element of this. This led to a second strategy, developing models relating to population history. And third, the search began for ancient DNA to see whether the Neanderthals possessed genes related to Europeans (an expectation of the multiregional hypothesis) or outside the range of modern human variation (an expectation of the out-of-Africa model).

## RECENT DEVELOPMENTS

## Beyond the mitochondria: microsatellites, the Y chromosme, and the nuclear genome

No field of study in human evolution has expanded as greatly as that of evolutionary genetics, and it is not possible to summarize briefly all the results. We can, however, consider the main lines of new evidence and what they show.

> **KEY QUESTION** How has work on other genetic systems thrown light on recent human evolutionary history?

The most obvious line of approach is to look at genes other than mitochondrial DNA. New genetic data used in human origin analyses include two types that are particularly interesting: one is derived from microsatellite DNA and the other involves so-called Alu sequences. Although they may appear to represent arcane elements of modern molecular biology, these datasets offer important, practical tools for anthropologists. Both of these new, groundbreaking investigations appear to favor the recent-single-origin model.

Microsatellites, which are short stretches of DNA that contain many repeats of two-to-five-nucleotide segments, evolve very rapidly. Unlike the rates of mutation for most genetic elements, which often must be calculated by calibration against the fossil record, that of microsatellites can be determined by laboratory observation. This certainty adds some weight of confidence to the coalescence time calculated with this technique, which is 156,000 years according to work carried out by David Goldstein, of University College London, and others.[407–9]

Alu elements are sequences of DNA approximately 300 base pairs in length, which become inserted in large numbers over the nuclear genome. Once inserted, they are never removed (or at least not completely) and thus remain immune to the kinds of homoplastic changes that may obscure point mutations. A recent, multiauthored study on Alu elements in a large sample from around the world gave a coalescence time of 102,000 years.

Genes, microsatelites, and other genetic elements in the nuclear genome are spread across 23 chromosomes. Because of the process of recombination, it is not possible to treat these independently in the mitochondrial way that one can with DNA. There is one exception to this, and that is the Y chromosome. This is the sex-determining chromosome, and only males possess it. It is possible to treat the sections of the Y chromosome that do not recombine as a uniparental system, the male equivalent of mitochondrial DNA (although it is important to remember that although only females can pass on their mitochondrial DNA, males can be sampled for this system, whereas only males both possess and pass on their Y chromosome). In a way, mitochondrial DNA gives the history of women, the Y chromosome that of males. Hopefully they will tell much

Have recent developments in genetic research strengthened the out-of-Africa hypothesis?

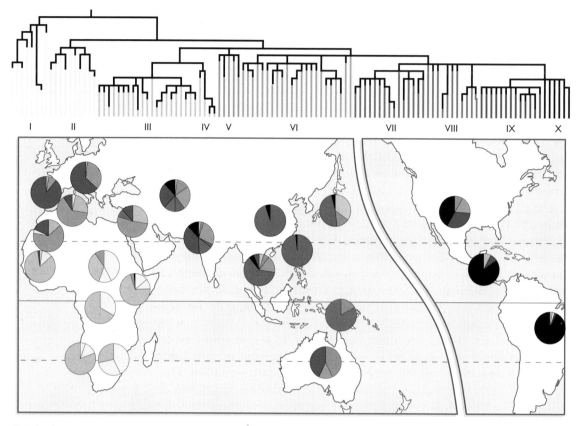

**FIGURE 15.9 Y chromosome diversity and relationships:** As with mitochondrial DNA, the Y chromosome shows that the greatest diversity is to be found in Africa. This tree represents the gene history of the Y chromosome, and hence the history of males. Roman numerals show the Y chromosome haplogroups that have been recognized and have a regional pattern; pie charts show their frequency in different parts of the world. (Courtesy of P. Underhill et al.[412])

the same evolutionary story, but there may also be differences which are of interest for understanding the process of evolution.

Work on the Y chromosome has been pioneered by Matt Hammer of the University of Arizona and a team led by Luca Cavalli-Sforza at Stanford University. Although there are some differences in their results, the basic picture is the same. The Y chromosome shows a recent coalescence – estimates vary, but between 100,000 and 150,000 years is the basic picture. Furthermore, the pattern again shows greater African diversity than that found in any other part of the world, with a tree rooted among African populations (Fig. 15.9). Again, this is consistent with the out-of-Africa model, and inconsistent with the multiregional one.[410–17]

Apart from looking at individual systems, it is also possible to consider several gene systems at once, and thus try to pinpoint whether the gene trees might reflect population history. The aim of this work, it is worth reiterating, is to test competing hypotheses. The multiregional hypothesis suggests that the roots of all modern human populations go back to *Homo erectus*, which originated in Africa almost 2 million years ago. By contrast, the single-origin hypothesis states that modern humans originated only 200,000 years ago, probably in Africa. Molecular evidence that indicates an African origin cannot differentiate between the hypotheses, because both claim an African origin. Instead, the time of origin distinguishes between them. The question is: how can the molecular data best be used to test the two hypotheses in terms of time of origin?

As we saw in chapters 5 and 8, many genes accumulate mutations at a rather regular rate, giving a potential molecular clock. With an extant population, the history of many different genetic variants of a gene, or alleles, can be traced by successive, inclusive steps, until a single ancestral type is reached. This ancestral type is known as the coalescent, and the time in history at which it is reached is called the coalescence time. If, when a new species is established, the population contains only a single allele of a particular gene, then the coalescence time for that gene may serve as a good indicator of the time of the speciation event. In other words, the gene tree is the same as the population (or species) tree. Frequently, however, the founding population of a new species will contain a subset of the existing genetic variation, so that the gene tree will show a more ancient divergence than the population tree, as we saw earlier. In this case, the coalescence time predates the time of the origin of the species. Under certain circumstances, the coalescence times may be substantially older than the time of origin of a species; in other (unusual) circumstances, the coalescence time may be younger.

For any particular species, a distribution of coalescence times of its various genes will exist. Some will coincide with the age of the species; many will be slightly older; some will be very much older; and a small number will be younger. Maryellen Ruvolo, of Harvard University, proposed that hypotheses of the time of modern human origins may be tested by examining the distribution of coalescence times of a range of genes in modern populations (Fig. 15.10).[418] If the multiregional model is correct, then those times should cluster around 1.8 million years ago (close to the time of origin of *Homo erectus*); if the recent-single-origin model is correct, those times will cluster around, for example, 200,000 years ago (the coalescence time of modern mitochondrial lineages). Ruvolo points out that, because only the distribution of coalescence times is informative, a single coalescence time cannot

**FIGURE 15.10**
**Method of testing a hypothesis:** When a population splits, it leads to a distribution of coalescence times (CTs) from many genes (denoted as G1, G2, and so on). Coalescence times can be expected to cluster around the time of population division, thereby indicating the time of origin of new species. No single coalescence time is a reliable indicator because some genes will have an older coalescence time than the population split, while others will be younger. (Courtesy of Maryellen Ruvolo.)

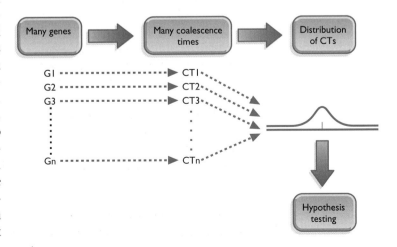

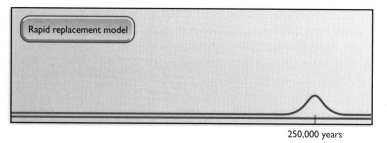

250,000 years

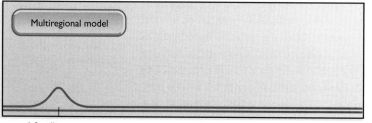

1.8 million years

**FIGURE 15.11 Coalescence times and the origin of modern humans:**
If modern humans originated close to 250,000 years ago, as implied by the
mitochondrial DNA hypothesis, then the distribution of coalescence times
would show a peak at that time (top). If the multiregional evolution model
is correct, then coalescence times would cluster around 1.8 million years
ago (bottom). (Courtesy of Maryellen Ruvolo.)

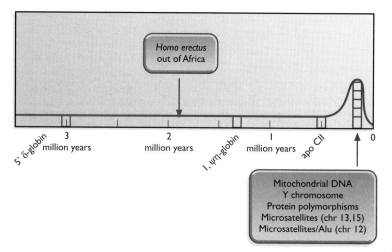

**FIGURE 15.12 Hypotheses tested:** The distribution of coalescence times
from mitochondrial and nuclear genes supports the recent-single-origin
model of modern humans. The globins and apo CII are proteins.
(Courtesy of Maryellen Ruvolo.)

prove or disprove either hypothesis. Even with a recent origin,
more ancient coalescence times
are expected; likewise, a certain
probability of recent coalescence times arises with an ancient origin as well (Fig. 15.11).

By 1997, 14 coalescence times
had been calculated for various
genetic loci, including 4 different measures in mitochondrial
DNA and 10 in different genes
in nuclear DNA. If the 4 mitochondrial results are counted
as a single data point (to reflect
their common inheritance),
then the remaining independent
coalescence times are as follows: 6 cluster around 200,000
years ago, while the rest are
scattered at 0.5, 1.2, 1.3, 3.0,
and 3.5 million years ago. (Two
independent studies on different regions of the Y chromosome, the male equivalent
of mitochondrial DNA, gave
coalescence times of 188,000
and 270,000 years.) Remember
that clustering of coalescence
times is the most important
criterion – not the position of
individual times. These results
clearly appear to favor the
recent-origin model (Fig. 15.12).

The idea that genes may
coalesce at different times is
an important element of new
developments, and underlies
the need to take a quantitative
and analytically sophisticated
approach to the problem of
modern human evolutionary
genetics. It is likely that even if
there was a harsh demographic
bottleneck around 150,000
years ago, nonetheless some

genetic systems will show older patterns of diversity. This is fairly obvious when we think that even in a very small population of several thousand, not all loci would be reduced to a single allele – if more than one allele gets through the bottleneck, then it will give an older signal. An example of this might be the beta-globin genes, which have older coalescence times (possibly around 700,000 years).[419]

Debate continues to swirl over the mutational dynamics of micro-satellite sequences and Alu elements, just as the interpretation of coalescence times has inspired controversy. In particular, population history may influence coalescence times in ways unrelated to the establishment of a species, usually leading to an erroneously young date. The fact that the inferences drawn from the mitochondrial DNA data are matched closely by a significant proportion of those from nuclear data, however, encourages the view that they are collectively providing insight into species events rather than identifying population events. For example, population crashes and explosions would affect mitochondrial DNA variation to a greater extent than nuclear DNA variation. While most observers accept the apparent implications of this body of work, a minority of critics remains unconvinced. As always, more data are required.

## Reconstructing population history from genetic evidence

Two factors play into the new line of investigation followed in population history analyses. The first stems from the difficulty that has been experienced in deriving an unequivocal phylogenetic tree from the mitochondrial DNA data. The low phylogenetic resolution in the data prompted certain researchers to seek other kinds of information that might be inferred from them, using a technique known as mismatch distribution. The insight gained with this technique can be applied to address the second factor – namely, the puzzle of the unusually low level of genetic diversity of mitochondrial DNA in modern populations.

The conclusion of this work is that, early in their history, the population of modern humans suffered a relatively severe bottleneck. As Henry Harpending, of Pennsylvania State University, has put it: "Our ancestors survived an episode where they were as endangered as pygmy chimpanzees or mountain gorillas are today."[420] Following that bottleneck, the population expanded explosively. These data imply that the multiregional evolution model cannot explain modern human origins. Rather, a modified form of the recent-single-origin model, known as the weak Garden of Eden hypothesis, is more likely to be correct.

Harpending and his colleague Alan Rogers, of the University of Utah, developed a hypothetical model of a population that expanded within a brief period of time. Genetic data culled from the modern descendants of

Did human populations experience bottlenecks in the past?

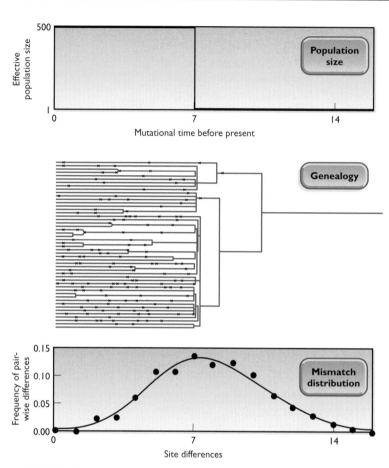

**FIGURE 15.13**
**Mismatch distribution:** This method uses genetic variation in modern populations to infer population events in the past. (See the text for details.) (Courtesy of Alan Rogers and Lynn Jorde.)

this population gave information about both the extent and timing of such an event (Fig. 15.13).

In their model, Harpending and Rogers assumed that mutations accumulate regularly in all lineages (mutations are shown as crosses on the horizontal lines in the middle panel of the figure). They then compared DNA sequences between all pairs of lineages in a sample of this population, and counted the number of mutational differences between each pair (a sample of 50 individuals gives 435 pairs for comparison). The time scale is measured in terms of mutational time, in which one unit represents the time needed for a single mutational difference to accumulate between two lineages; two units are sufficient for two mutational differences; and so on. The rate at which mutations accumulate is determined by both the rate of mutation at all sites in the DNA and the generation time. In this case, one mutational unit equates to 8333 years, given the known rate of mutation of certain mitochondrial sequences in humans.

Because the population underwent expansion at seven mutational units of time in the past (as Fig. 15.13 shows), a large proportion of lineages in the current population will include seven mutational differences between them. Some lineages split after the expansion event of course, and these lineages will differ by fewer than seven mutations. When all pairs of lineages have been compared and mutational differences counted, these numbers are then arrayed on a histogram, with the horizontal axis representing the mutational time, going from zero in the present to ever-increasing numbers as one moves back in time. The histogram shows a peak at seven mutational differences, with fewer points at older and younger times, forming a wave pattern (see the bottom panel of Fig. 15.13). Harpending and Rogers describe this pattern as "the signature of an ancient population expansion." The position of the crest of the wave indicates when population expansion occurred; the shape of the wave shows its magnitude (the sharper the peak, the more rapid was the expansion).[421,422]

When Harpending and Rogers applied the mismatch distribution analysis to real mitochondrial DNA data from modern human populations from around the globe, they found the same wave pattern. This discovery implies that the modern human population underwent a rapid expansion of numbers, the timing of which was centered around 60,000 years ago. Further analysis revealed that the expansion took place at different times for different geographical populations. The African population expanded first, followed later by expansions in the European and Asian populations. This conclusion came from a mismatch distribution analysis conducted within each geographical population, followed by a similar analysis performed between pairs of populations (this latter technique is termed "intermatch distribution").

Several possible scenarios exist to explain what happened here, the most persuasive of which is the weak Garden of Eden hypothesis (Fig. 15.14). Remember that the recent-single-origin hypothesis posits that modern humans arose as a small, isolated population, and that descendants of this

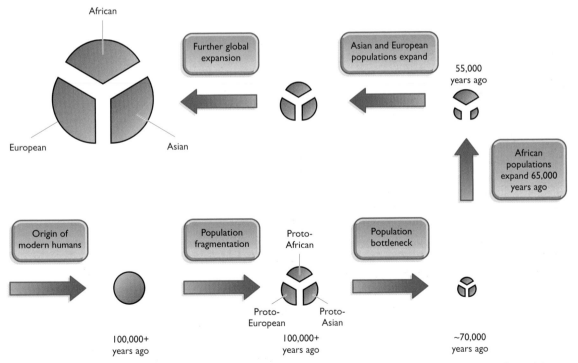

**FIGURE 15.14 The weak Garden-of-Eden hypothesis:** Developed from mismatch distribution analysis, this hypothesis represents a variant of the single-origin model. It posits the origin of modern humans in Africa, prior to 100,000 years ago. This population fragmented (within Africa), and the separate populations subsequently developed genetic distinctiveness. A population bottleneck reduced the size of populations and genetic variation within them. The African population was the first to expand, followed by the proto-Asian and proto-European populations, which migrated into these geographical regions. Population expansion then continued.

population spread throughout the Old Word, replacing existing populations of archaic *sapiens*. This concept is also called the Garden of Eden hypothesis. The intermatch distribution analysis implies a rather more complicated history. According to this hypothesis, once established (some 100,000 years ago), the founding population of modern humans fragmented into separate populations; these groups later spread out geographically to form the modern populations of Africa, Europe, and Asia. The genetic distinctiveness of these populations was therefore established prior to the expansion; the mismatch and intermatch distribution data indicate that these separate expansions took place at different times. Thus, replacement of archaic *sapiens* populations would still have occurred, but would not have involved the same dynamics as envisaged in the original Garden of Eden hypothesis.

According to this new line of investigation, the low level of mitochondrial DNA diversity reflects a population bottleneck after the establishment of the modern human population; this bottleneck was followed by sequential population expansions in different parts of the world. Several questions arise here. For example, what might have caused a population crash? Stanley Ambrose of the University of Illinois has argued that a gigantic eruption of Mount Toba, on the Indonesian island of Sumatra, 73,500 years ago might have been the culprit,[423] but the timing and location of this have not convinced other researchers.[424]

In addition, how large (or small) was the post-bottleneck population? The complicated calculation required to answer this question is based on the current genetic diversity of mitochondrial diversity in the world and on the mutation rate of these DNA sequences. The simplest answer indicates the existence of some 3500 breeding females, which would give a total population of approximately 10,000 individuals. (Similar numbers have been obtained from other data, including nuclear DNA data.) In fact, population genetics equations show that if this population was distributed in discrete geographical populations over the Old World, as required by the multiregional hypothesis, the number of females would have been smaller – close to 1500. This figure creates a fatal problem for the hypothesis because, as Harpending and Rogers note, "It is difficult to imagine that a population this small could have populated all of Europe, Africa, and Asia. . . . Knowledge that Eve lived recently would imply that the human population was . . . too small to have populated three continents." In other words, the numbers that flow from this analysis (if correct) make the multiregional hypothesis untenable. Some form of a recent-single-origin model would seem much more reasonable.

Since the work of Harpending and Rogers, there have been several developments in the field of ancient human demography. David Goldstein and David Reich used a multiple gene loci approach to calculate the age and size of the human bottleneck.[409] Their approach emphasized that such estimates came with broad confidence limits – between 49,000 and 640,000 years ago for the extreme levels of certainty – but that all the

What was ancient human population size?

evidence indicated that the expansion of human populations was significantly earlier in Africa than elsewhere. Their analysis of 30 loci suggested that there was a severe bottleneck – they tentatively suggest an effective population size of 6900 – most probably between about 150,000 and 287,000 years ago, a date consistent with the out-of-Africa model.

However, although there has been more and more evidence accumulating from genetics, most of which supports the out-of-Africa model, and none of which is inconsistent with it, nonetheless there is growing recognition of the complexity of the analyses involved. The developments have come from the use of wider and wider genetic systems, where before there were small segments sampled, and better and better mathematical models for analyzing them. The difficulties have arisen because of the computational complexity of simultaneously analyzing different loci, and taking into account the possibilities of (1) constant or expanding populations, and (2) selection playing a part even on apparently neutral parts of the genome. If selection is operating, then reading population history is much harder. There is certainly some evidence that selection is important, and good theoretical grounds for expecting that to be the case, but despite this, recent major new analyses by Santos Alonso and John Armour of the University of Nottingham show evidence for earlier population expansion in Africa.[425] Furthermore, supporters of the multiregional hypothesis have placed great emphasis on the level of uncertainty surrounding estimates for the chronology of past events based on genetics, leaving the door open for other interpretations than the recent-origin one.[426]

## Ancient DNA and Neanderthals

One of the most keenly awaited results was finding DNA in a Neanderthal. It is often forgotten that all that we know about the genetics of our ancestors comes not from sampling fossils, but from the living survivors of evolution – twenty-first-century *Homo sapiens*. It is based on a process of inference back into the past. One of the inferences to come from the out-of-Africa model was that Neanderthal DNA should lie outside the range of modern human variation, but to determine that it would be necessary to extract DNA from the fossils. This was not a trivial task, given the state of preservation and the age of the fossils concerned. However, in 1997, Matthias Krings and Svante Pääbo managed to sequence some mitochondrial DNA from the Neanderthal type specimen.[427] Only small fragments could be retrieved, but these were enough to obtain some results. The Feldhofer specimen showed no specifically close affinity to living Europeans, but lay outside the range of observed modern human variation – in other words, the last common ancestor of this Neanderthal and all modern humans was older than the last common ancestor of known living humans. The age put on that common ancestor was about 450,000 years ago, compared with the estimated coalescence for living

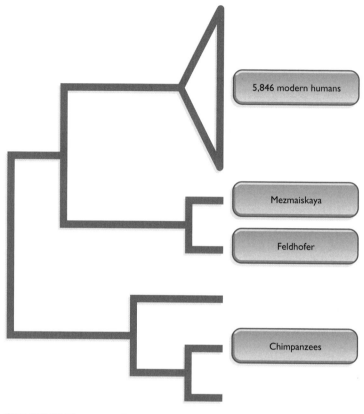

humans of around 150,000 years (Fig. 15.15). This seemed to be conclusive about the fate of the Neanderthals, and provided support for the out-of-Africa model.

Since the work on the Feldhofer material, two other Neanderthals have been sequenced, one from Mezmaiskaya Cave in the Caucasus, and another from Vindija in Croatia.[428] These showed the same divergence from modern humans. They also gave an insight into the amount of genetic diversity among the Neanderthals. On a phylogenetic gene tree, all the Neanderthals come out more closely related to each other than to any modern human. Like modern humans, they exhibited a lack of genetic variation compared to the great apes, implying that they too had experienced a population bottleneck and then population expansion.

These results do not rule out interbreeding between humans and Neanderthals, and say nothing about whether such interbreeding was possible or not. What they do show is that the Neanderthals had, from less than half a million years ago, a separate evolutionary trajectory from modern humans, and that if they did interbreed with *Homo sapiens* after 50,000 years or so ago, this did not have any impact on the modern human gene pool that can be detected.

Ancient DNA opens up many opportunities for insights into the past, but as yet still faces major technical difficulties. Survival of DNA seems to be very dependent upon thermal conditions, and only sustained cold climates lead to long-term survival. Very small quantities are preserved even then, so that only the more numerous mitochondria have been recovered systematically, and contamination has been a significant problem in most analyses; the highest standards, especially repeat results, have been required.

While Neanderthal genes have been discovered, a gap still remains with regard to the ancient DNA of the early modern human populations known from the fossil record. The only Pleistocene modern humans to have been sequenced are specimens from Australia. The Lake Mungo 3 burial, which may be as old as 60,000 years, was shown by Gregory Adcock and others

**FIGURE 15.15**
**Neanderthal DNA:**
DNA has been recovered from Neanderthals and this can be compared to that of modern humans alive today. Current evidence suggests that where all living humans coalesce to a common ancestor around 400,000 years ago, the coalescence with Neanderthals is closer to 650,000 years ago. While these dates should be treated with caution, the relative difference is important.

to have mitochondrial DNA that was more divergent than that seen in other modern humans.[429] On this basis, the researchers argued that the data showed Australian populations could have had contributions from non-modern populations, such as *Homo erectus*, as expected by the multiregional model. Other researchers, however, have questioned both the results and the interpretation, and it remains to be seen whether this will stand. Even if it does, it does not refute the out-of-Africa model, as modern humans 60,000 years ago, through the coalescence process described above, would be expected to possess diversity that has since been lost.

*How has ancient DNA contributed to our understanding of human origins?*

The 1990s onward have been a turbulent period for research in human evolutionary genetics. There is little doubt that genetics has transformed our understanding of modern human origins, and is likely to continue to do so. As a result of the application of molecular genetics to human evolution, we are beginning to gain an understanding of the long-term demographic history of humans[86] (Fig. 15.16). This is important when we remember, as discussed in chapters 2 and 3, the extent to which demography underlies evolutionary process. Broadly speaking, the evidence that has emerged has provided more and more support for the out-of-Africa model, and virtually none that uniquely supports the multiregional model

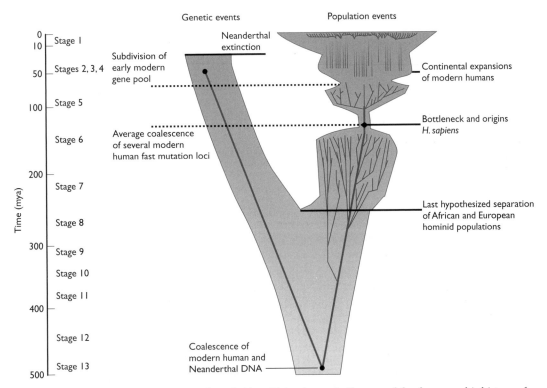

**FIGURE 15.16 Human demography and evolution:** This schematic diagram of the demographic history of *Homo sapiens* shows the proposed bottleneck and subsequent expansions and contractions. (Courtesy of Marta Lahr and Robert Foley/*American Yearbook of Physical Anthropology*.)

in the way it was originally formulated. However, it is also the case that the out-of-Africa model that is supported by the genetic evidence is not the simple, one-off, time-compressed event originally implied by the so-called Eve theory. What we can say is that living humans are derived from an ancestral population in Africa within the last quarter of a million years, that they spread around the world through a process dominated by replacement, that the founding population was very small, and therefore that there has been massive population expansion. To fill that story out, and to determine what it might mean in terms of behavior and adaptation rather than a few genes, we need to consider the archeological evidence.

## BEYOND THE FACTS
# IS THE FUTURE GENETIC?

*The issue: molecular genetics has had a massive impact on the study of human evolution, and opened up major new areas of research and new interpretations. Molecular techniques are likely to increase in their power and applicability. Does this signal the death of classic plaeoanthropology, or is there still a role for fossils and the archeological record?*

The growth of evolutionary genetics is a phenomenon that few would have predicted on the basis of the first tentative steps made in the 1960s by Allan Wilson, Vincent Sarich, and Morris Goodman. At that time new fossils were appearing from Africa, and the wealth of material that we now know from the area was only just being uncovered. Conflicts between fossil dates and genetic dates led to severe mistrust between the two communities of scientists, and a "my data are better than your data" standoff.

The successes of the last fifteen years or so have overturned this situation. While there are still some paleoanthropologists who refuse to recognize the significance and importance of genetics, for the most part the evidence from the genes has become both an important framework for understanding human evolution, and a rich source of hypotheses about the past – for example, the hypothesis of a bottleneck event associated with the origins of modern humans in Africa. Genetics has also opened up discussion of aspects of human evolution that were thought to be invisible, such as the estimates of effective population size.

All this might lead to the view that future developments in human evolutionary studies will come from genetics, a view bolstered perhaps by the fact that the whole debate on the evolution of modern humans has taken place in the absence of any major fossil finds of great relevance (although there have been a number of major advances in dating which have played their part). Add to this the fact that evolutionary genetics brings a quantitative precision to the subject that is often absent from the musings of paleoanthropoliogsts – David Goldstein of University College London, for example, estimating the effective population size as 6900. Carrying this to its logical conclusion, one might in the end decide to close down the classic search for "fossil man." Better the certainty of the laboratory than the haphazard romance of the dig and the guesswork of the archeologist.

Such a move, others would argue, would be not only premature but mistaken. This is to do not so much with the weaknesses of genetics as an approach to evolution, but with its limitations. Fossils and archeology add important elements to the past, some absolutely critical. Archaeology, for example, can provide insights into behavior, which many would argue is what drives evolutionary change. The fossils of the hominins indicate their actual phenotype, and remind many researchers that there is more to being a hominin than its genes. Finally, evolutionary genetics, as shown above, is for the most part the genetics of the survivors, and so does not provide insights into the paths down which human evolution might have gone. A geneticist's perspective on human evolution might well be that there was the divergence from the chimpanzees, and then modern humans. Only the fossils and archeology can provide the rich details of adaptive radiations from the australopithecines onward.

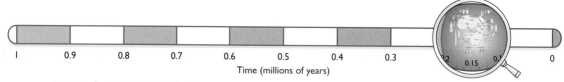
# CHAPTER 16

# THE ORIGIN OF MODERN HUMANS: ARCHEOLOGY, BEHAVIOR, AND EVOLUTIONARY PROCESS

Archeology, the third line of evidence in the study of the origin of modern humans, focuses on behavior rather than morphology or genetics. Interpreted correctly, however, it should be congruent with the other two lines of evidence if biology and behavior evolved in concert. As described in chapter 14, the weight of anatomical evidence is consistent with the evolution of modern humans first in Africa, with a subsequent expansion of range into Asia and Europe. The genetic evidence supports the same evolutionary pattern (see chapter 15). Thus, anatomical and genetic evidence is consistent with the out-of-Africa model of modern human origins and provides little support to the multiregional evolution model. If biological and behavioral evolution are indeed linked, then archeological evidence should also reveal modern human behavior first in Africa and later in the rest of the Old World. Furthermore, the archeological evidence may give important insights into such questions as whether the biological and behavioral changes associated with the evolution of modern humans were synchronous or occurred serially, and whether the anatomy led to the behavior or vice versa.

Although the archeological evidence related to this issue is relatively good in Europe and western Asia, it is poor in east Asia and to some extent in Africa. For instance, while more than 100 sites dating between 250,000 and 40,000 years old have been carefully excavated in southwestern France (and many more are known in less detail), only about a dozen such sites have been studied in east Africa, a region almost 100 times larger.

This disparity has led inevitably to a distinctly Eurocentric interpretation of the archeological record, which gives the impression that the pertinent behavioral changes principally took place in Europe. Several important discoveries have been made in Africa in recent years, however, and their interpretation is leading some archeologists to favor a different view of our behavioral evolution.

# ARCHEOLOGICAL EVIDENCE

## The archeological background

Before examining the evidence from the principal geographical regions that bears on the origin of modern humans, we will survey some of the general technological characteristics of the relevant archeological stages. Specifically, we are concerned with the Middle Stone Age (MSA) and Late Stone Age (LSA) in Africa, dated at some 300,000 to 30,000 years and 30,000 to 10,000 years, respectively. The equivalent stages in Europe, Asia, and north Africa are known as the Middle Paleolithic and Upper Paleolithic (see Fig. 12.2 above). It is important to remember that these stages are based on characteristics of tool assemblages in the different geographical regions; consequently, their chronological equivalence is not exact, nor is there exact equivalence of artifact types. In addition, because fossil specimens of anatomically modern humans from Africa have been identified from at least 130,000 years ago (and somewhat later in the Middle East), the earlier stage holds the most interest in relation to this chapter's main question.

> **KEY QUESTION** What light can archeology throw on the pattern of modern human evolution and the role of behavior in these evolutionary events?

It is perhaps worth reiterating the basic framework for the stone-tool technology on which the Paleolithic is organized (and see chapter 12). Although the stone tools made by early hominins come in all sorts of shapes and sizes, which can be described as various tool *types* (scrapers, burins, awls, axes, etc.), the means by which they were made – their technological mode of production – is standardized for long periods of time. This was the basis for Grahame Clark's, of Cambridge University, technological modes[329]: Mode 1 was the simplest, producing basic flakes and core tools; Mode 2 was represented by the development of bifacial secondary flaking of larger flakes, principally producing handaxes; Mode 3 involves the preparation of the core before the striking off of the flakes, which allows for thinner flakes and more careful shaping of the final tools; Mode 4 is the production of blades rather than flakes, especially off cylindrical cores, while Mode 5 consists of microlithic blade and flake production. The Lower Paleolithic or Early Stone Age (ESA) is characterized

**FIGURE 16.1 Middle Paleolithic artifacts:** These typically retouched flakes of various types were made between 250,000 and 40,000 years ago. (top row, left to right) Mousterian point; Levallois point; Levallois flake (tortoise); Levallois core; disc core. (bottom row, left to right) Mousterian point; Mousterian scraper; Quina scraper; limace; denticulate. (Scale bar: 5 cm.) (Courtesy of Roger Lewin and Bruce Bradley.)

by Modes 1 and 2; the Middle Paleolithic and MSA by Mode 3 (Fig. 16.1), and the Upper Paleolithic and LSA by Modes 4 and 5 (Fig. 16.2). The archeology associated with the origins of modern humans relates to the development of Modes 3 and 4 technologies. These terms refer to stone technologies specifically, but there is also the question of how the technologies may be associated more broadly with other aspects of behavior, such as hunting strategies or spatial patterns.

What is the pattern of technological variation in the Later Pleistocene?

The end of the Lower Paleolithic (Modes 1 and 2), 250,000 years ago, saw the end of innovation-poor, long-lasting stone-tool industries. With the beginning of the Middle Paleolithic and MSA (Mode 3), the number of identifiable tool types quadrupled, reaching perhaps 40. With the Upper Paleolithic (Mode 4), beginning 40,000 years ago, the number of tools more than doubled again, to as many as 100 (Fig. 16.3). Moreover, whereas regional and stylistic variations in earlier industries were less

**FIGURE 16.2 Upper Paleolithic artifacts:** These artifacts are typically formed from retouched blades and are finer than Middle Paleolithic tools. (top row, left to right) Burin on a truncated blade; dihedral burin; gravette point; backed knife; backed bladelet; strangulated blade; blade core. (bottom row, left to right) End scraper; double end scraper; end scraper/dihedral burin; Solutrean laurel leaf blade; Solutrean shouldered point; prismatic blade core. (Scale bar: 5 cm.) (Courtesy of Roger Lewin and Bruce Bradley.)

impressive than the overall stability, variability through space and time became a dominant theme in Upper Paleolithic industries. For instance, European tool industries cascade through at least four identifiable traditions in less than 30,000 years – a pace of innovation and change unprecedented in the archeological record. In addition to new forms of tools, raw materials that were only infrequently used earlier, such as bone, ivory, and antler, became very important in the Upper Paleolithic industries. For this reason the Upper Paleolithic has been the focus of ideas about the association between the evolution of modern humans and the archeological record.

The Middle Paleolithic and MSA (Mode 3) technologies were characterized by the predominance of the prepared core technique, such as the Levallois technique, which appeared earlier. In this tool-making method,

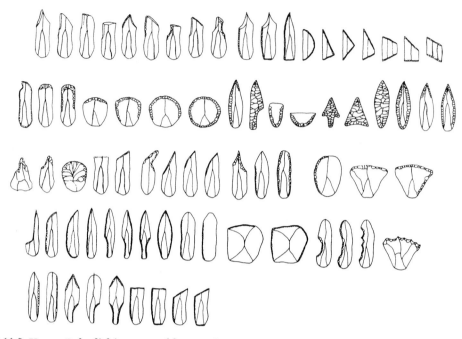

**FIGURE 16.3 Upper Paleolithic range of forms:** The French archeologist G. Laplace produced this typology of Upper Paleolithic tools in the late 1950s and early 1960s. It forms the basis of all Upper Paleolithic typologies. The intricacy as well as the variety of production can be seen.

a large core is prepared so that it has a flat upper surface and a convex lower surface. The considerable force required to detach the broad flakes is applied by bringing the striking platform of the core down sharply at an angle on an anvil. The relatively large, thin flakes conform to the shape of the outline of the prepared core. Smaller flakes may be removed using a hammerstone on the core. Many flakes of similar form may be produced by repeatedly striking around the edge of the core until it is virtually all used up. The Levallois technique is much more economical of the raw material than earlier flaking techniques, producing many more centimeters of working edge for each kilogram of core.

Once produced, the flakes may then be further fashioned to give what some archeologists identify as approximately 40 different implements, each with its own putative cutting, scraping, or piercing function. Some of the identified types may actually represent stages in the manufacture of other artifact types or the products of repeated resharpening – that is, they may not be real tool types. Some variation exists in Middle Paleolithic assemblages throughout the Old World, which has encouraged the development of a plethora of local names. In Europe these are considered to be variants within what is known as the Mousterian, while in Africa and Asia there are many MSA regional styles, such as the Stillbay, the Aterian, Howiesons Poort, etc. (Fig. 16.4).

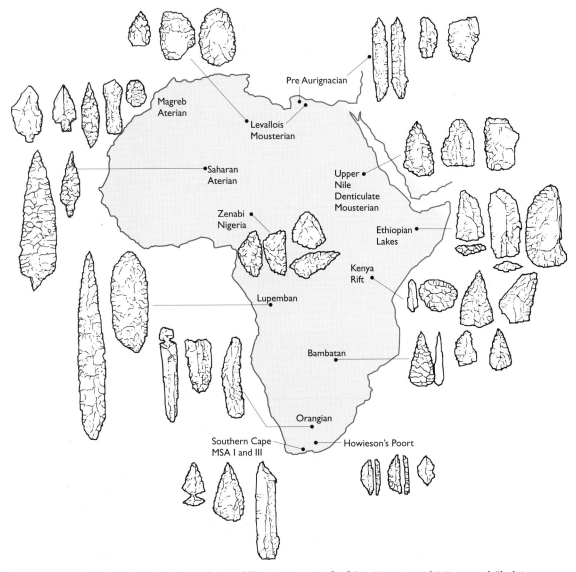

**FIGURE 16.4 Regional variation in the Middle Stone Age of Africa** (Courtesy of J. Desmond Clark.)

Just as flakes from prepared cores characterize Mousterian (and Mousterian-like) industries in the Middle Paleolithic, blades produced from prepared cores constitute something of a signature for the many industries in the European Upper Paleolithic (Mode 4). A more pertinent signature, however, is the much greater standardization in the form of tools made, indicating a clearer mental template and a greater manipulative skill applied to their manufacture (Fig. 16.5). In addition, strong evidence suggests the development of complex social behavior, including trading of material over long distances, larger settlements than hitherto

**FIGURE 16.5 Solutrean laurel leaf blade:** Some examples of these blades are so thin as to be translucent. They were possibly used in rituals rather than in practical affairs. (Scale bar: 5 cm.) (Courtesy of Roger Lewin and Bruce Bradley.)

known, and artistic expression in the form of body ornamentation and engraving and painting of objects (including cave walls).

Blades are defined as flakes that are at least twice as long as they are wide. The preparation of the cores used for their manufacture requires great skill and time. Many blades may then be detached sequentially using a pointed object, such as the end of an antler, hammered by a hammerstone. The blades, often small and delicate, may be functional without further preparation, or they may merely serve as the starting point for specifically shaped implements. In addition to the signature blade, Upper Paleolithic tool makers also made extensive use of bone, ivory, and antler as raw material for some of the most delicate implements (Fig. 16.6).

Thus, a strong sense of directed design and multi-functional use characterizes Upper Paleolithic tool assemblages. Some tools even appear to celebrate the tool maker's skill, such as the Solutrean blades, laurel-leaf-shaped objects that may be 20 centimeters long and so thin as to be translucent in places. Clearly of little use in normal subsistence activities, the blades may have been important in ritual ceremonies.

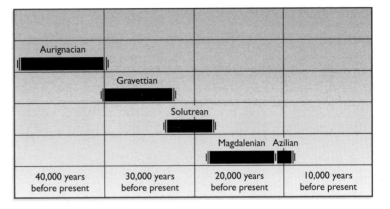

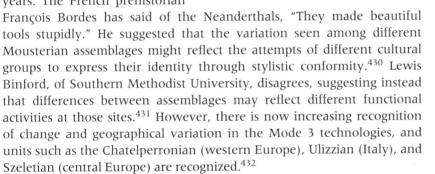

In contrast to the innovation of Upper Paleolithic tool makers, the Mousterians continued to produce their much more limited, though impressive, tool for almost 200,000 years. The French prehistorian François Bordes has said of the Neanderthals, "They made beautiful tools stupidly." He suggested that the variation seen among different Mousterian assemblages might reflect the attempts of different cultural groups to express their identity through stylistic conformity.[430] Lewis Binford, of Southern Methodist University, disagrees, suggesting instead that differences between assemblages may reflect different functional activities at those sites.[431] However, there is now increasing recognition of change and geographical variation in the Mode 3 technologies, and units such as the Chatelperronian (western Europe), Ulizzian (Italy), and Szeletian (central Europe) are recognized.[432]

This description and comparison of the Middle Paleolithic (and MSA) and the Upper Paleolithic (and LSA) tool-making processes mirrors the perspective that prevailed in the archeological community until recently. For many practitioners, this view still appears valid – namely, that the transition between the earlier and later stages was rapid, reflecting the first emergence of modern human behavior approximately 40,000 years ago (Fig. 16.7). As Stanford University archeologist Richard Klein has observed, where the evidence is plentiful it reveals "the most dramatic behavioral shift that archeologists will ever detect." For this reason, the transition has been regarded as revolutionary, not gradual. If true, this would imply that the evolution of modern morphology (which appeared more than 130,000 years ago) occurred separately from the evolution of modern behavior (40,000 years ago).[384,433] Recent discoveries in Africa may raise questions about this interpretation, however.[396,434]

This perspective can be characterized as the Mode 4 hypothesis, the link being between the evolution of modern humans and the development of what can be thought of as the Upper Paleolithic or Mode 4 technologies and associated behaviors. In recent years there has also been a greater interest in the role that Mode 3 technologies play in the evolutionary pattern – the Mode 3 hypothesis.[358,377] Part of the reason for the development of the Mode 3 hypothesis lies in the difficulty of associating the Mode 4 industries with modern humans, except in northern Eurasia

**FIGURE 16.6 Tool industries of the European Upper Paleolithic:** The pace of change of tool technologies becomes almost hectic from 40,000 years onward. In addition, the tool industries themselves take on a complexity and refinement unmatched in earlier periods. A distinct sense of fashion and geographic variation is also well developed.

Is the development of Mode 4/5 technologies significant in the evolution of modern humans?

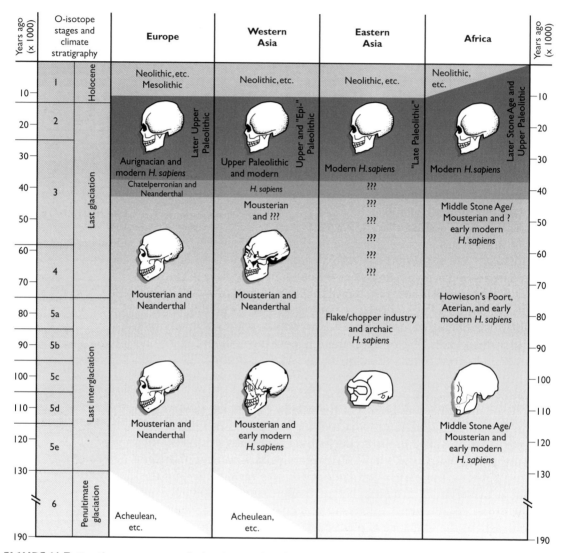

**FIGURE 16.7 Continents compared:** The picture of modern human origins derived from archeological evidence is at best incomplete. In Europe, where the evidence is most plentiful, the picture shows a sharp transition approximately 40,000 years ago that reflects the inward migration of anatomically modern humans carrying modern cultural behavior. In Asia, the picture is less clear. In Africa, new evidence suggests that modern human behavior begins to appear in the Middle Stone Age, congruent with the early appearance of anatomically modern humans in that continent. (Courtesy of Richard Klein/*Evolutionary Anthropology.*)

What does Mode 3 tell us about the origins of modern humans?

(Fig. 16.8). The Mode 3 model, developed by Robert Foley and Marta Lahr, proposes that there was a dispersal of archaic hominins in both Africa and parts of Eurasia, giving rise to both modern human and Neanderthal lineages, with a more recent common ancestor for them than in the Mode 4 model, and modern humans evolving very much in the context of Mode 3 technologies (Fig. 16.9).

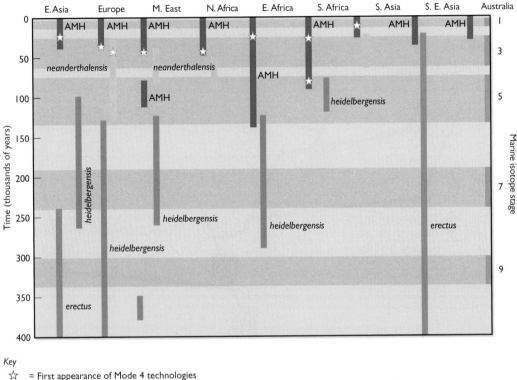

Key

☆ = First appearance of Mode 4 technologies

AMH = Anatomically modern humans

**FIGURE 16.8 Relationship between the presence of Mode 4 industries (blade technology) and anatomically modern humans:** Although the appearance of blades has often been associated with the appearance of modern humans, the relationship is in fact much more complex. The columns indicate the different hominin lineages. (Courtesy of Robert Foley and Marta Lahr/*Cambridge Archaeological Journal*.)

We will now examine the archeological evidence bearing on modern human behavior region by region, beginning with Europe, then Asia, and finally Africa, and then consider the way this integrates with the overall evolutionary history of modern humans and their antecedents.

## REGIONAL PATTERNS IN THE ARCHEOLOGY

### European evidence

As has been stated, the European archeological evidence for the stages in question is extensive, and it does appear to give a clear signal of a revolutionary change some 40,000 years ago. For this reason, the transition

**KEY QUESTION** What is the geographical and chronological pattern of Middle and Upper Paleolithic assemblages and how do they fit into the fossil evidence?

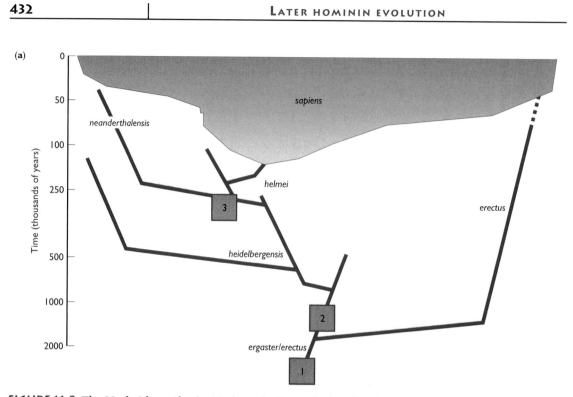

**(a)**

**FIGURE 16.9  The Mode 3 hypothesis:** (a) The early dispersals of modern humans are associated with Mode 3 technologies, whereas (b) the Mode 4 dispersals are confined to the Eurasian regions and occur much later. (Courtesy of Robert Foley and Marta Lahr/*Cambridge Archaeological Journal.*)

in Europe has been dubbed the Upper Paleolithic revolution. It coincides with the first appearance of modern humans in the region, carrying the cultural tradition known as the Aurignacian. Until recently, archeologists thought that this movement spread from east to west, beginning in the east 40,000 years ago and arriving in the west 5000 years later. New radio-carbon dates at two Upper Paleolithic sites in northern Spain, l'Arbreda and El Castillo, give dates of close to 40,000 years ago, however, which indicates that the influx of the modern populations through Europe may have been virtually (in archeological time) instantaneous.

Aurignacian sites throughout Europe show the typical blade-based technology and use of bone, ivory, and antler, not only to make points but also to create beads as body ornamentation. The sites are also associated with other characteristics of the Upper Paleolithic: they are larger than those of the Middle Paleolithic; open-air (as opposed to rock-shelter or cave) sites are more distinctive and organized; artifacts indicate the exist-ence of long-distance contact and even trade (shells and exotic stone that must have come from afar); and musical instruments, specifically simple flutes made from bone, are present.

As the Upper Paleolithic progressed, substantial temporal and spatial variability of style developed in artifact assemblages; the sense of cultural

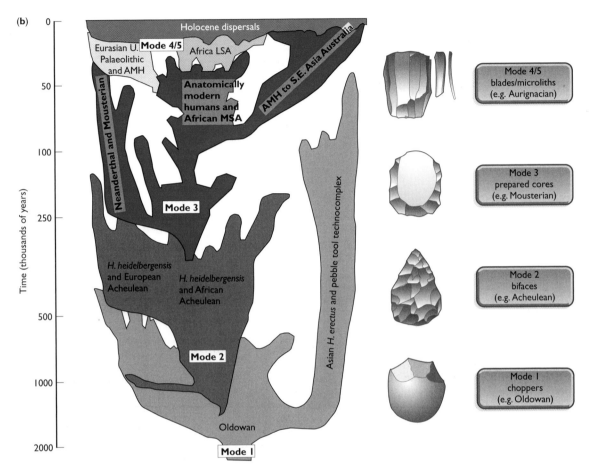

**FIGURE 16.9** (cont'd)

traditions in the way we would mean today was strongly present for the first time. Although sculpting and engraving appeared from the Aurignacian onward, evidence of cave painting did not become strong until the Gravettian, some 30,000 years ago.

The contrast between the Middle Paleolithic in Europe (specifically, the Mousterian) and the Upper Paleolithic is striking. Although not every aspect of Upper Paleolithic culture, especially technological advances and artistic traditions, was present from the beginning, overall it surely offers evidence of a revolutionary change. Opinion on this latter point – revolution or not – is divided, however, in terms of both its dynamics and its explanation.

For instance, Klein considers the magnitude of the collective change and the speed with which it occurs to be revolutionary, reflecting the influx of modern human populations who replaced local Neanderthals.[384,433,435] New York University archeologist Randall White independently came to the same conclusion.[436] This interpretation supports one aspect of the out-of-Africa hypothesis – namely, that the rapidity of the transition from the Middle to Upper Paleolithic in Europe suggests that modern humans replaced certain archaic *sapiens* populations such as the Neanderthals.

By contrast, Geoffrey Clark and John Lindly, of Arizona State University, suggest that evidence of modern behavior – such as the occasional production of blades, artistic expression, and ritual burial – is underplayed in European Neanderthals; they believe that the modern human behavior of Upper Paleolithic people developed more slowly than some scholars argue. As a result, Clark and Lindly suggest that the transition was more gradual than revolutionary, which speaks of the continuity of populations argued for in the multiregional evolution hypothesis, not replacement.[437]

The match between archeological and fossil evidence in Europe is quite good. For instance, wherever hominin remains have been found with Mousterian assemblages, they have been Neanderthal. Virtually all hominin fossils associated with Upper Paleolithic assemblages have been modern humans. Two exceptions to the latter generalization have been identified, at the French sites of Arcy-sur-Cure and Saint-Césaire. Although the fossil evidence at Arcy-sur-Cure is fragmentary, a classic Neanderthal partial skeleton has been found at Saint-Césaire.[438] These sites are interesting because the tool assemblages, although fundamentally Mode 3 (that is, Middle Paleolithic), contain elements that are similar to those of the Upper Paleolithic industries. This industry is called the Chatelperronian.

Some scholars have argued that the intermediate nature of the Chatelperronian technology indicates the presence of a population in biological transition – that is, changing from Neanderthal to modern humans. The anatomy of the Saint-Césaire individual shows no such characteristics, however. The age of the skeleton, recently dated at 36,000 years, leaves little or no time for an evolutionary transition to local modern human populations. In any case, the site postdates the earliest Aurignacian sites, which have no local precursors. One possible explanation of the Chatelperronian is that it was developed by late Neanderthal populations that had cultural contact with incoming modern human populations, although this view has been challenged.[439,440]

The situation concerning the terminal mode 3 technologies and the earliest Mode 4 – in other words, the end of the Middle Paleolithic and the beginnings of the Upper Paleolithic – has become more complex in the last few years. Scattered evidence for decoration in the Chatelperronian has led to the view that the differences between Neanderthals are more quantitative than qualitative. Francesco D'Errico, of the University of Bordeaux, and Joao Zilhao in particular have argued that the level of change within the terminal Middle Paleolithic is too great to be the result of the influence of modern humans, but shows indigenous evolution. Furthermore, work by Steve Kuhn of the University of Arizona in Turkey,

Is there continuity in Europe between Mode 3 and Mode 4 technologies?

and interpretations of other material in the more southeastern parts of Euorpe, have questioned the homogeneity of what has been called the Aurignacian.[441] These new observations do not necessarily cast doubt on the overall picture of the relationship between Neanderthals and modern humans, but they do suggest that it is not a single, simple event between two populations. A strong case can be made for its support of revolutionary change as a result of population replacement. It does not, however, address the issue of the origin of modern humans, for it is too late relative to what happened in Africa.

## Asian evidence

The archeological evidence in Asia is open to even more diverse interpretation than in Europe, partly because the data are fewer and partly because some apparent paradoxes exist. Great differences are also noted between western Asia and eastern Asia, where the evidence is sparsest of all. Western Asia, which includes the Middle East, is closely allied to Africa geographically and provides a natural migration route out of Africa. Between 200,000 and 50,000 years ago, this region was variously occupied by Neanderthal and early modern humans, while the east was inhabited by populations that were neither Neanderthal nor modern.

The archeological transition from Mode 3 to Mode 4 in the Middle East is typologically very similar to the Mousterian to Upper Paleolithic transition in Europe, and apparently occurs about the same time (40,000 years ago). If the transition tracks the migration of modern humans out of Africa, through the Middle East, and finally into western Europe, then the evidence for it in the Middle East might be expected to predate the evidence gleaned further west. Tentative confirmation of this movement might come from the site of Boker Tachtit in Israel, which dates to between 47,000 and 38,000 years ago. Evidence of Upper Paleolithic human remains in the Middle East is scarce, but is essentially that of modern humans.[441,442]

What are the implications of a shared Mode 3 technology between Neanderthals and modern humans in the Middle East?

Where western Asia differs from Europe is in the occurrence of anatomically modern humans with classic Mode 3 assemblages, at the Israeli sites of Skhūl and Qafzeh, which have been dated to approximately 100,000 years. These fossil remains are either equal in age to or predate Neanderthals of the region, and thus would seem to preclude an evolutionary transformation of Neanderthals into modern humans. Nevertheless, the occurrence of modern human anatomy with Mousterian assemblages some 60,000 years before Upper Paleolithic assemblages appear in the region represents a puzzle. It implies either that modern human anatomy evolved long before modern behavior or that the modernity of the Skhūl and Qafzeh remains has been overstated. (Recent analyses have implied that the two populations used different hunting strategies, with modern humans being more efficient.)[443]

Klein points out that the Skhūl/Qafzeh specimens are extremely variable anatomically and that they possess some archaic features, such as prominent brow ridges and large teeth. "Both cranially and postcranially, they clearly make far better ancestors for later modern humans than the Neanderthals do," he says. "However, it seems reasonable to suppose that they were not yet fully modern biologically – perhaps, above all, neurologically." Once again, Clark and Lindly's reading of the evidence differs from Klein's interpretation, arguing for continuity between the archaic and the modern species, in both the fossils and the archeology.

The interpretation of eastern Asian evidence poses a challenge because of the scarcity of sites and uncertain dating. There does appear to be a continuity of chopping-tool assemblages from *Homo erectus* times through approximately 10,000 years ago, with no dramatic shift equivalent to that seen in the European Upper Paleolithic. One site in Sri Lanka, Batadomba Iena cave, contains a modern tool assemblage and may be as old as 28,500 years. In addition, sites in Siberia, dated between 35,000 and 20,000 years old, contain Upper Paleolithic-like artifacts and art objects, suggesting a more European-like pattern. The migration from southeast Asia to Australia between 60,000 and 45,000 years ago possibly implies the evolution of modern behavior, and links in to ideas of multiple dispersals from Africa, of which the earliest are along the Indian Ocean Rim.[399]

The Asian evidence is therefore equivocal at best, but offers little to suggest the appearance of modern human behavior early in the record. Certainly it should not be treated as homogeneous, and there is a difference between the eastern and western parts of the continent, and probably between north and south, that persists for very long periods of time, and which mirrors the biological evidence. In this context it is interesting to note that southern Asia – the Indian subcontinent – tends to follow the pattern of the west rather than the east.

# African evidence

The African evidence is obviously key to understanding modern human origins from an archeological perspective. In Europe and Asia there is, broadly speaking a relationship between the appearance of modern humans after 40,000 years ago and the appearance of "modern" behavior, as evidenced by the appearance of Mode 4 industries and their associated material culture. This is understandable in terms of dispersals – that is, what these associations reflect is not the in situ evolution of behavior, but the bringing in of these behaviors with the first modern humans. The sub-Saharan African evidence should be different, because it is here that there should be evidence for the actual evolution of modern behavior. This leads to two expectations: that whatever the evidence for modern behavior might be, it should appear earlier than in other parts of the world; and that, assuming that the shift was not a single revolutionary step, there should be evidence for the transitional antecedents of modern human behavior.

To some extent, although the situation is far from clear, both expectations are fulfilled. The key elements of the record are the MSA and LSA – Modes 3 and 4/5 respectively, although as we shall see they are more complicated than that. The naturally expected association would be for the LSA, the technological equivalent of the Upper Paleolithic, to be associated with modern humans, and the MSA to be associated with archaic hominins, in the same way as the Mousterian is associated with Neanderthals. However, as we have already seen for western Asia, such simple associations do not hold. The MSA, which makes its appearance more than a quarter of a million years ago, is associated with both later "archaic" *Homo* and modern humans, while the LSA is much later, usually from around 30,000 years ago – more than 100,000 years after the appearance of modern humans.[444]

Clearly the situation is therefore more complex than a simple correlation between fossil taxa and archeology. One approach is to look for more derived or modern features and their antecedents in the MSA. Certainly there is considerable evidence for these:

*Is there evidence for a gradual transition to modern behavior in Africa?*

♦ the presence of blade technology within the MSA, possibly extending back as far as 240,000 years ago (findings by Sally McBrearty, of the University of Connecticut, in the Kapturin Beds in Kenya);[396]

♦ evidence for symbolic behavior in Zambia (findings by Lawrence Barham of the University of Bristol);[445]

♦ evidence for microliths in the Howiesons Poort of South Africa, dated to close to 100,000 years ago;[446]

♦ evidence for systematic exploitation of aquatic resources in the MSA of southern Africa;[447]

♦ the presence of red ocher, systematically engraved, at Blombos, in South Africa, dated to over 70,000 years ago;[448] and

♦ the discovery by Alison Brooks (of George Washington University, Washington, DC) of a barbed bone harpoon dated around 100,000 years ago, from Katanda in the Democratic Republic of the Congo[449] (Fig. 16.10).

Recently, Brooks and McBrearty have summarized the behavioral evidence for the Middle Stone Age, showing that the MSA of Africa (Fig. 16.11) – the context in which the earliest modern humans are found – is more derived in relation to the behavior of modern humans than is the case for the Middle Paleolithic of Europe. This argument was made previously by Hilary Deacon of Stellenbosch University on the basis of his work at Klasies River Mouth, perhaps the key site in early modern human archeology in Africa.[446] It should also be remembered that Africa to the north of the Sahara has its own record, one which interestingly shows links both to the south and to the Mediterranean and Europe. Perhaps the key site in the region is that of the Haua Fteah, excavated in the 1950s by Charles McBurney, which showed the ephemeral but early presence of Mode 4 technologies.[450]

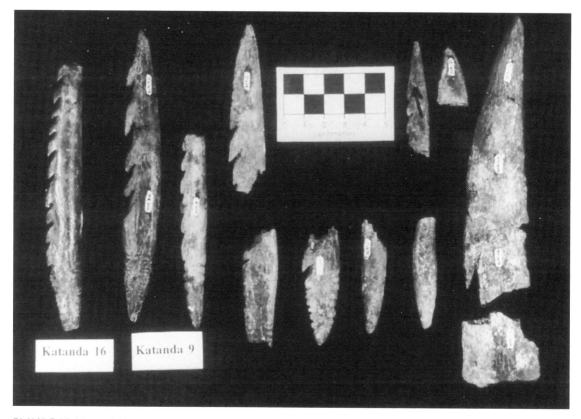

**FIGURE 16.10 Middle Stone Age bone tools:** Discovered recently in the Democratic Republic of the Congo, these harpoon-like bone points are the earliest known examples of worked bone, dated at between 90,000 and 160,000 years old. (Courtesy of Alison Brooks and John Yellen.)

Despite these discoveries, nothing discovered in Africa has matched the artistic expression for which the Upper Paleolithic of western Europe is so famous. The oldest reliably dated rock painting in Africa appears in the Apollo 11 cave, Namibia, dated at 27,000 years, which is equivalent to the oldest examples of art in Europe. In contrast, pigments and grinding stone for processing pigments have been found in many regions of Africa, dating from at least 80,000 years ago. If such pigments were used for body decoration, for example, rather than treating hide, it would be significant in the context of the current question. It is impossible to prove which of these possibilities is correct, however. Evidence of personal adornment, such as ostrich-eggshell beads, appears in the record relatively late, about 60,000 years ago. Are these artifacts to be taken as evidence of an absence of the early symbolic behavior that is so often considered as reflecting the modern human mind at work? Not necessarily so, argues Brooks, given the very unfavorable conditions of preservation in the African environment and the paucity of sites investigated.

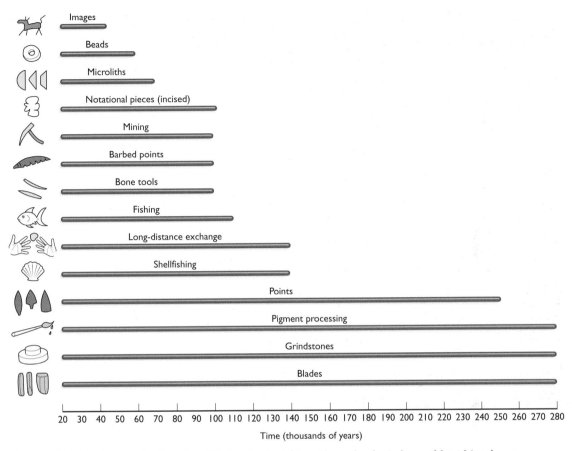

**FIGURE 16.11 Evidence for "modern" behavior in Africa:** The archeological record for Africa shows considerable time depth, as well as a mosaic pattern of trait development that has been associated with modern behavior. (Courtesy of McBrearty and Brooks/*Journal of Human Evolution*.)

Further indication of modern behavior in the MSA takes the form of long-distance transport of valued material. One of the most striking examples, found by Michael Mehlman, of the University of California, Santa Cruz, is the transport of obsidian (a kind of glass) from the central highlands of Kenya to the Mumba site, Tanzania – a distance of more than 300 kilometers. The Mumba site is dated at approximately 130,000 years old.[451] Such apparent depth of planning is not normally associated with what has been taken to represent the limited MSA mind.

For an increasing number of archeologists, these separate lines of evidence tell us something about a gradual emergence of modern human behavior. Once it passed a certain threshold, that behavior appears to have exploded, producing the rich fabric of social complexity associated with the Upper Paleolithic and LSA. That explosion was a cultural change, however, not a biological one. By contrast, Klein and others have argued that only with a critical biological change – such as facilitation of linguistic ability or symbolic thought – did modern human behavior become possible; they define modern human behavior as including the ability to produce the entire range of activities, not just one of them at different times and different places.

What is clear is that the African archeological record shows that the context in which modern humans evolved is the MSA or Mode 3 technologies. These industries and their associated behaviors persisted for tens of thousands of years after the appearance of anatomically modern humans, such that the development of the Upper Paleolithic or LSA cannot be seen as chronologically associated with the modern human emergence biologically. This chronological gap between behavior and biology is of great interest for the process of becoming modern, and the historical events associated with populations dispersing around the world.

## Hypotheses tested

As a test of competing hypotheses – out-of-Africa and multiregional evolution – the archeological evidence is not unequivocal, and certainly not as strong as the anatomical and genetic evidence. Nevertheless, it can be argued that a signal of modernity appears first in Africa, representing a chronological precursor of what later appears in Eurasia. The appearance of modern cultural activities in Europe seems to coincide with the first appearance of anatomically modern humans there – a culture brought by migrants, not developed locally. Thus, the out-of-Africa model is more strongly supported than the multiregional evolution model.

## TOWARD AN INTEGRATED MODEL OF MODERN HUMAN ORIGINS

**KEY QUESTION** What are the future avenues of research into modern human origins and how do we go beyond the classic "two-model" approach?

These three chapters on the origins of modern humans have been set in the context of two models: multiregionality versus out-of-Africa. This has been the framework for the last fifteen years or more. As we can see from the evidence presented here, there is stronger support for the out-of-Africa model. Virtually no evidence is directly supportive of the multiregional model, although there are sufficient ambiguities within the data not to rule it out. Given this state of affairs, one can ask whether these models should still provide the framework for current work, and indeed why it is that the debate still appears to be so prominent.

One answer to this latter question is that over the time since the models were first proposed, they have both changed subtly (and indeed not so subtly), so that their supporters can still claim that they are empirically useful. Early formulations of the multiregional hypothesis placed great emphasis on the continuity of lineage evolution within each region, and argued that this occurred worldwide. Current formulations play down ideas of independent regional evolution, and instead focus on the "absence of speciation" as the key criterion, allowing population replacement to be a part of the model where before it was anathema. At the other extreme, the out-of-Africa model as proposed on the basis of the early mitochondrial DNA data (unlike the original out-of-Africa model, developed by Harvard anthropologist W. W. Howells on the basis of cranial evidence) saw modern human origin as a single, local event, occurring in Africa, with a single major and simultaneous dispersal. Anatomy, genes, behavior, and dispersals were all compressed into the one event, generally associated with speciation. However, as we have seen, this is no longer an essential part of out-of-Africa, and while the emphasis remains on a local and recent origin in a small population in Africa, this is not necessarily the point of behavioral change or of anatomical change, or indeed of dispersal. Multiple rather than single dispersals are now the preferred model[86,390,399] (Fig. 16.12). Furthermore, the emphasis is no longer on speciation and reproductive incompatibility, but on the fluctuating demography of paleodemes. The models, in other words, have themselves evolved, and have become more complex in the case of out-of-Africa, and rather less clear cut and testable in the case of multiregionality.

*Multiple dispersals – the basis for an integrated model of human origins?*

We can therefore ask whether the time has come for the formulation of a new model, and if so, what form it should take. Certainly it will be broadly out-of-Africa, but what other elements might it have? The following framework should be the context for taking the debate beyond multiregionalism and out-of-Africa:

♦ All living humans are descended from a relatively small population, most probably living in Africa, dated to around 150,000 years ago. Although there is good, consistent evidence to place that event at that time, both fossil and genetic, and it is also consistent with a period of marked aridity in Africa when populations are likely to have been fragmented and reduced, nonetheless the genetic evidence still requires further chronological refinement and reduction of confidence limits on its estimates of coalescence (Fig. 16.13).

♦ Hominins, both archaic and modern, outside that population have not contributed significantly in terms of either genotype or phenotype to the modern population, although the genetic evidence as it stands cannot preclude minor admixture that has subsequently been lost. It follows from this that the primary process of modern human colonization both within and beyond Africa was one of population replacement.

♦ The "origins" of modern humans are not a single event focused on that time horizon.[399] Rather, the origins of modern humans are spread over the last 200,000 years, and include the progressive trend toward

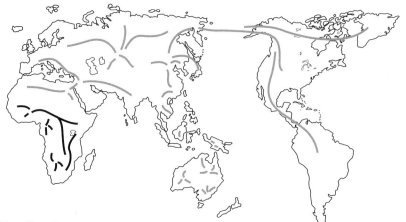

3 Late Upper Pleistocene: 15–0 thousand years

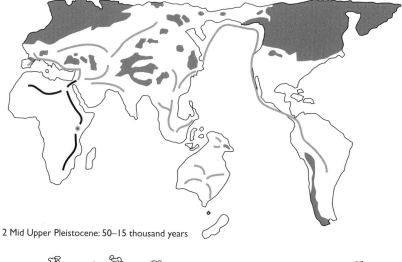

2 Mid Upper Pleistocene: 50–15 thousand years

**FIGURE 16.12**
**Multiple dispersals out of Africa:** The current view of the dispersals of modern humans assumes not a single event, but a cumulative one over time. Shaded areas show ice sheets. (Courtesy of M. Lahr and R. Foley.[399])

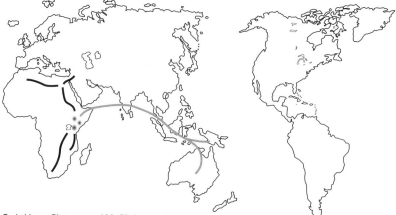

1 Early Upper Pleistocene: 100–50 thousand years

anatomical modernity in Africa; the development, albeit intermittent and geographically scattered, of traits associated with behavioral modernity; the early dispersals around Africa and into the Middle East during the last interglacial; subsequent population contractions in the early last glacial; and then the Eurasian dispersals after 40,000 years ago, associated with the Upper Paleolithic.

♦ Multiple dispersals, rather than a single one, characterize the spread of modern humans around the world, with the earliest being in Africa (~100,000), then along a southern route (~60,000), and then a northern Eurasian route (~40,000). Even after these dates, dispersals and replacements, as well as admixture and local adaptations, would have been the pattern of human demography.[399]

♦ Modern humans are not associated with the Upper Paleolithic, which develops some 100,000 years after the first appearance of anatomically modern humans. The context for the evolution of modern humans in archeological terms are the Mode 3 industries of Africa, and these are shared with Neanderthals. Furthermore, the shift in behavior cannot be associated with a biological or genetic change, as this would conflict with the genetic signature seen in living humans today.

♦ Not all archaic hominins are the same. The Neanderthals clearly have a different set of capabilities than other archaic hominins such as *erectus* and *heidelbergensis*. However, the similarity of their behavior to that of modern humans does not preclude replacement as a process of change (Fig. 16.14).

This multiple-event, multiple-dispersal model,[399] within the framework of an out-of-Africa consensus, opens up major new areas of research and questions. While it contains some elements of recent formulations of the multiregional model, especially for the later phases, it is not in that sense a compromise. The idea of a global evolution of modern humans, with distinct long-term regional trajectories, has no empirical support. The model also relates back to some of the more general principles of evolution outlined in the first part of this book. These include:

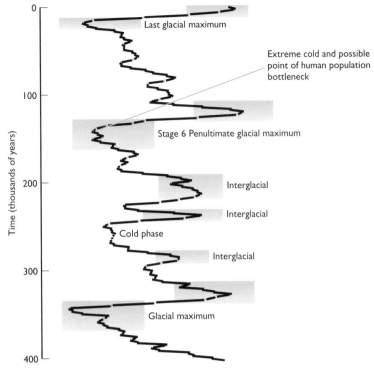

**FIGURE 16.13 Climatic change and modern human origins:** The proposed bottleneck in human populations may relate to the decreased temperatures and increased aridity of the penultimate glacial maximum (marine isotope stage 6).

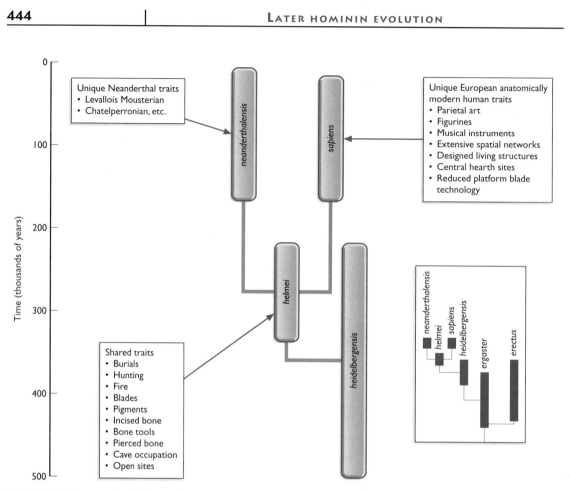

**FIGURE 16.14 Neanderthals and modern humans:** Neanderthals and modern humans share many characteristics that are most probably pleiomorphic and indicate their close relationship and relatively recent common ancestry compared to other archaic hominins. (Courtesy of M. Lahr and R. Foley.)

♦ the importance of geography in evolutionary processes, and the subtle shifts in demography (contractions and dispersals) as the driving force of microevolution (Fig. 16.15);

♦ the role of climate in shaping evolutionary patterns;

♦ the role of behavior in the evolutionary process;

♦ the importance of microevolutionary processes in examining evolutionary patterns, rather than all events being placed in a single macroevolutionary speciation event; and

♦ the importance of extinction as a mechanism for structuring evolutionary diversity.

This more complex approach to the problem of modern human origins and evolution should take what is certainly one of the most important issues in human evolution into a new era. There is little doubt that evolutionary genetics has radically transformed the way in which we look at

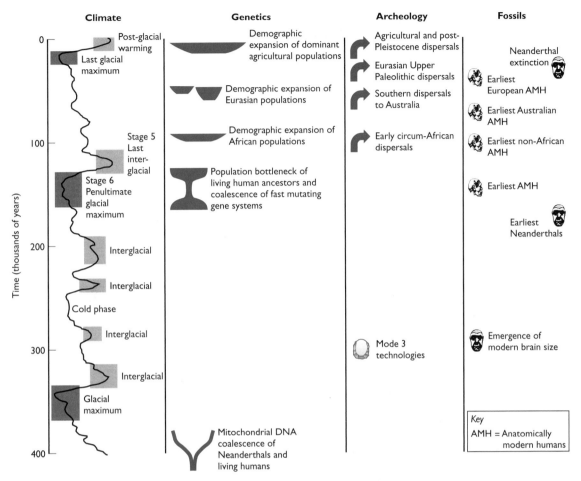

**FIGURE 16.15 Multiple events and modern human origins:** Taken as a whole, the genetic, archeological, and fossil evidence suggest that the evolution of modern humans is a complex set of events spread over the last quarter of a million years or so, perhaps influenced by the changing climates. (Courtesy of M. Lahr and R. Foley.)

evolutionary problems, and it has sparked new areas of controversy. Theoretically it has placed greater emphasis on demography and geography, and brought the question of population dispersals and movements back into the discipline. The next few years are likely to see this continue and expand, but this will place even greater emphasis on the need for paleoanthropology to provide the context in which genetic models can be assessed, and thus lead to a more integrated approach.

BEYOND THE FACTS
# ARE GOOD SCIENTIFIC THEORIES COMPLEX?

*The issue: a good axiom in science is that simple explanations are better than complex ones. This has been one of the strengths of Neo-Darwinism, which has remained at heart a very simple theory. The question is: is it too simple, and are more complex theories – such as those proposed by the macroevolutionists – needed to explain the complexity of life?*

One of the recurring critiques of Darwinian theory is that it is simplistic. This theme can be found in creationist tracts, which question whether it is possible that anything as complex as the eye could be produced by such a simple (and random) theory. It can also be found in the writings of social scientists, who would argue – probably quite rightly – that humans are too complex to be a result of "ordinary evolution." And it can also be found in much of the work underlying the punctuated equilibrium debate.

One of the lines of attack on the Neo-Darwinian synthesis is that because biological systems are hierarchically complex, they are unlikely to be the product of a unitary process. By "hierarchically complex," Stephen Jay Gould and others mean that biological systems are built from atomistic units such as DNA molecules, through cells, to organisms, to species, and indeed to entire ecosystems, each of which is greater than the sum of its constituent parts. In another dimension, biological systems are also organized hierarchically in terms of the groups to which they belong – populations, species, families, etc. It is argued that each of these cannot be reduced to the units below it, and evolutionary theory needs to be pluralistic and hierarchical to reflect the complexity of the biological world.

This might seem to be a powerful argument, for surely it is the case that more complex theories can explain more than simple ones. The complexity of evolution deserves something more than a unitary theory. But is this so? On the whole, and certainly in the physical sciences, the most powerful theories are considered to be those that can account for the most observations through the simplest models. This is what is often known as Occam's razor, after the proposition of the twelfth-century monk, William of Occam, that the simplest theories were the best. Certainly the principle underlies the great achievements of Newton and Einstein. Indeed, it is one of the reasons that Darwin and Wallace are among the greatest names of science, because the simplicity of their theory cut through swathes of misunderstood complexity.

There is thus a pull in evolutionary biology in two directions. On the one hand, there is the model of the other sciences and mathematics, where simplicity is considered a strength, not a weakness; and on the other hand, there is the pull toward the uniqueness of the biological sciences, which may have their own rules. In the end, it is a question that can only be solved by applying the principle of simplicity itself – that is, to see how simple an explanation one can get away with. However, many find the urge for complexity to be even more compelling. Perhaps this is why, in the words of a great evolutionary biologist, John Maynard Smith, biology is not a hard science, but it is a difficult one. Models of modern human origins are a further context in which the value of complexity versus simplicity needs to be assessed.

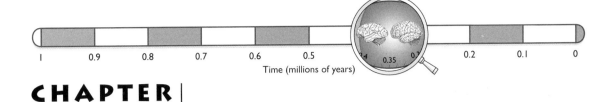

# CHAPTER 17

# EVOLUTION OF THE BRAIN, INTELLIGENCE, AND CULTURE

This chapter focuses on the context and consequences of human brain evolution. The context will expose some of the biological constraints under which this extraordinary evolutionary event occurred. The consequences should help to illuminate the selection pressures that fueled the event.

The question is: how did humans come to possess such extraordinary powers of creative intelligence, powers that surely outstrip what would have been necessary in the practical day-to-day life of technologically simple foraging? In exploring this conundrum we will cover three aspects of human mental evolution: the expansion of the brain and intelligence; consciousness; and language. Inevitably, these three qualities are tightly intertwined.

Although the hominin lineage stretches back approximately 5 million years, fossil evidence for brain size goes back only some 3 million years, to the specimens of *Australopithecus afarensis* found at the Hadar, Ethiopia. Given this somewhat limited view of hominin history, it is nevertheless apparent that brain expansion has been great and rapid: a threefold increase occurred in that 3 million years, with brain size expanding from 400 cm$^3$ to 1350 cm$^3$, the average of modern populations. Impressive in itself, this trend appears even more spectacular when, as pointed out by Harry Jerison of the University of California, Los Angeles, "there is no evidence of a change in any other mammals in [this same period]."[452] In other words, brain expansion among hominins was not merely part of a recent, general mammalian pattern.

To understand this expansion more fully, we will look first at some of the characteristics of the human brain in the context of primate biology. We will then turn to the fossil evidence of this expansion, and some of the ideas currently offered to explain the phenomenon. Finally we will explore whether encephalization reflects the evolution of something very new and different – culture.

## ENCEPHALIZATION

## The human brain in a primate context

**KEY QUESTION** When, how, and why did the hominin brain become so much larger than that of an ordinary primate?

The brain is a very expensive organ to maintain. In adult humans, for instance, even though it represents just 2% of the total body weight, the brain consumes some 18% of the energy budget. "One might therefore ask," says Robert Martin, of the University of Chicago, "how, in the course of human evolution, additional energy was made progressively available to meet the needs of an ever-increasing brain size."[453]

As we saw in chapter 6, life-history factors – gestation length, metabolic rate, precociality versus altriciality, and so on – have an important impact on the size of brain that a species can develop. In this context, two major ideas have been advanced in recent years that bear on the special problem faced by hominins in brain expansion.

The first, proposed by Martin, is that the mother's metabolic rate is the key to the size of brain a species can afford: the higher the metabolic rate, the bigger the relative brain size. The second, proposed by Mark Pagel of Reading University and Paul Harvey of Oxford University, is that gestation time and litter size represent the determining factors: long gestation, with a litter of one, is optimal for a large-brained species.[454] Although both hypotheses are said by their authors to have empirical support, debate continues as to which is the more germane. Whichever case proves to be correct, both pathways require the same kind of environmental context: a stable, high-energy food supply, with minimum predation pressure.

In being well endowed mentally (Fig. 17.1), humans and other primates are a part of a clear pattern among vertebrates as a whole. Depending somewhat on the measure used, mammals are approximately 10 times "brainier" than reptiles and amphibians. Underlying this stepwise progression, which takes into account successive major evolutionary innovations and radiations, is the building of more and more sophisticated "reality" in species' heads.

By being mammals, primates are better equipped mentally than any reptile. Two orders of mammal have significantly larger brains than the rest of mammalian life: Primates and Cetacea (toothed whales). And among primates, the anthropoids (monkeys and apes) are also very brainy. Only humans are outliers from the monkey/ape axis: the brain of *Homo sapiens* is three times bigger than that of an ape of the same body size.

The need to grow such a large brain has distorted several basic life-history characteristics seen in other primates. For instance, the adult ape brain is nearly 2.3 times bigger than the brain in the newborn (neonate); in humans, this difference is 3.5 times. More dramatic, however, is the size

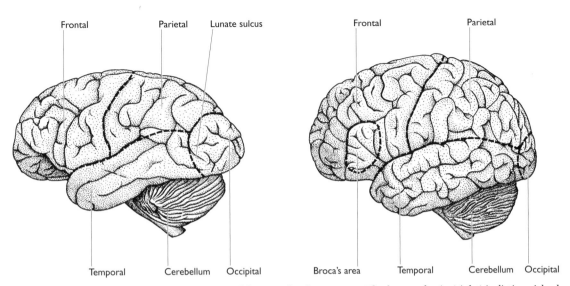

Frontal    Parietal    Lunate sulcus        Frontal        Parietal

Temporal    Cerebellum    Occipital    Broca's area    Temporal    Cerebellum    Occipital

**FIGURE 17.1 Diagram of the typical ape and human brain pattern:** The human brain (right) is distinguished from that of the chimpanzee (left) not only by size but by the relatively small human occipital lobe and large parietal lobe. (The brains are not shown to scale. The cerebellum is an enlarged part of the hindbrain, controlling movement among other things; see text on "Fossil evidence" below for other terms.) (Courtesy of Ralph Holloway/*Scientific American*, 1974, all rights reserved.)

of the human neonate compared with ape newborns. Even though humans are of similar body size to apes (57 kilograms for humans, compared with 30 to 100 kilograms for apes) and have a similar gestation period (270 days versus 245 to 270 days), human neonates are approximately twice as large and have brains twice as large as ape newborns. "From this it can be concluded that human mothers devote a relatively greater quantity of energy and other resources to fetal brain and body development over a standard time than do our closest relative, the great apes," notes Martin.

Another major difference is the pattern of growth. In mammals with precocial young, among which are primates, brain growth proceeds rapidly until birth, whereupon a slower phase ensues for roughly a year. In humans, the prenatal phase of rapid brain growth continues for a longer period after birth, a pattern that is seen in altricial species. Compared with other altricial species, however, the rapid postnatal phase (fetal rate) of brain growth continues for a relatively longer period in humans. This extension effectively gives humans the equivalent of a 21-month gestation period (9 months in the uterus, and 12 months outside).

This unique pattern of development has been called secondary altriciality. One important consequence is that human infants are far more helpless, for a much longer time, than the young of the great apes. This extended period of infant care and subsequent "schooling" must have had a major impact on the social life of hominins.

*What is special about the human brain?*

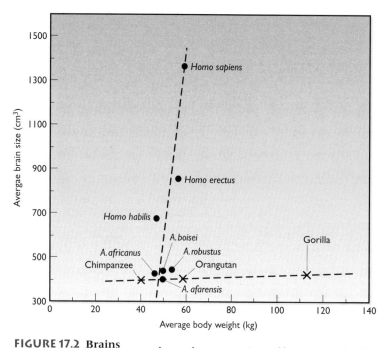

**FIGURE 17.2 Brains and bodies:** Even though a dramatic increase in body size did not occur in the *Homo* lineage, absolute (and therefore relative) brain size expanded significantly from *habilis* to *erectus* to *sapiens*. Brain size did not change significantly among the australopithecines or the modern apes, despite a large body size difference in the latter.

What is the pattern of hominin encephalization?

## Fossil evidence

Two types of fossil evidence are related to brain evolution: indications of absolute size, and information about the surface features – convolutions and fissures – of the brain. Both pieces of evidence can be obtained from either natural or artificial endocasts, which show the convolutions of the brain as they became impressed on the inner surface of the cranium.

Brain size is the first and most obvious piece of information to be gleaned, and it can often be gained even with partial crania. Thus, a fair number of data have been gathered about the expansion of brain size, beginning with *Australopithecus afarensis*, a little more than 3 million years ago (Fig. 17.2). Measured at a little less than 400 cm³, this early australopithecine brain is often said to be roughly the same size as modern gorilla and chimpanzee brains. This implication that the early hominin was no more expanded in terms of brain capacity than apes is misleading, however, for two reasons: (1) early australopithecines were smaller in body size than modern gorillas, and (2) modern ape brains may be larger than those of their 3 million-year-old ancestors. It is therefore safe to say that some brain expansion had already been established by the time *Australopithecus afarensis* appeared.

Put boldly, brain size for the australopithecines was close to 400 cm³, and it increased only a little throughout the tenure of this genus. More marked expansion is seen with the origin of the genus *Homo*, specifically *Homo habilis/rudolfensis*, which existed from 2.5 to 1.8 million years ago and had a range of brain size of 650 to 800 cm³. The size range for *Homo ergaster/erectus*, dated at 1.8 million to 300,000 years ago, is 850 to slightly more than 1000 cm³ (Fig. 17.3). The comparable measurements for archaic *Homo sapiens*, including Neanderthals, range from 1100 to more than 1400 cm³, or larger than in modern humans. Using the encephalization quotient (EQ), a measure of brain size in relation to body size, this progression can be discerned more objectively. The australopithecine species have EQs in the region of 2.5, compared with 2 for the common chimpanzee, 3.1 for early *Homo*, 3.3 for early *Homo ergaster/erectus*, and 5.8 for modern humans[455,456] (Fig. 17.4).

By looking at overall brain structure as revealed in endocasts, it is possible to differentiate between an apelike and a humanlike brain organization.

Each hemisphere contains four lobes: frontal, temporal, parietal, and occipital. Very briefly, a brain in which the parietal and temporal lobes predominate is considered humanlike, whereas apelike brains contain much smaller parietal and temporal lobes. In addition, human frontal lobes are considerably more convoluted than in apes.

Anthropologists find it very helpful to know when a human brain organization emerged in hominin history. Ralph Holloway, of Columbia University, examined in detail a wide range of hominin fossil endocasts, including *Australopithecus afarensis*, and concluded that brain organization was very humanlike. His analysis included the position of the lunate sulcus, a short groove that lies at the margin between the occipital and temporal lobes. In humans, the sulcus lies relatively further back than in apes. According to Holloway, in all fossil hominin endocasts in which the lunate sulcus could be discerned, this structure lies in the human position.[457]

In 1980, Dean Falk, of the State University of New York at Albany, challenged this view after a study of the hominin endocasts in south Africa.[458] The two researchers have since exchanged more than a dozen papers, each defending his position, but no resolution has been reached. Falk's position has recently received support independently from two researchers, Este Armstrong and Harry Jerison. If brain reorganization toward the human configuration began only with the origin of *Homo*, while the australopithecine brain remained essentially apelike, then it would be consistent with other events in human prehistory, including the shift toward a humanlike life history pattern with the origin of *Homo*, the evolution of humanlike body proportions, the reduction of body size dimorphism, and the first appearance of stone-tool technology. Falk has argued that an important anatomical feature in the expansion of the brain in *Homo* was the distributed structure of the blood vessels, which permits efficient cooling; this concept is known as the radiator hypothesis.

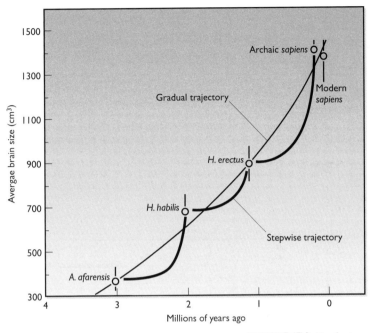

**FIGURE 17.3 Brains through time:** A threefold increase in absolute brain size occurred during the past 3 million years. Whether this increase took place gradually (as indicated by the smooth slope) or episodically (as indicated by the steps) is a matter that will be settled only with the discovery of additional, accurately dated fossils.

## Measures of intelligence

It is relatively easy to plot brain expansion through hominin history, but how are we to measure the rise of intelligence through time? The archeological

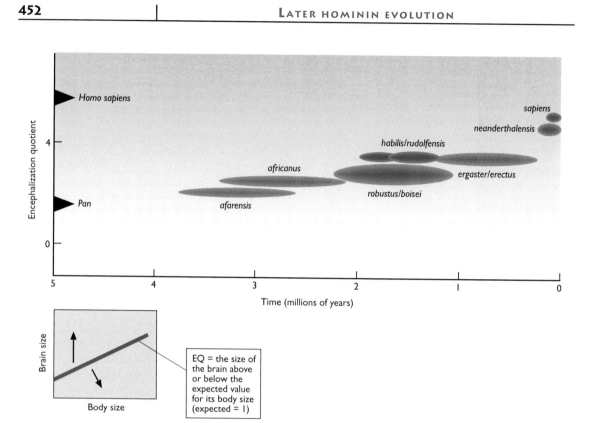

**FIGURE 17.4 EQ and time:** Brain size is strongly related to body size, and so it is necessary to look at how relative brain size (EQ) evolves over time.

record is notoriously lacking in tangible indications of the working of the mind. Tales told around a campfire, complex sand drawings, dances embodying social mythology, and symbolic body markings leave no trace, yet are the essence of humanity in a hunter-gatherer society. Although paintings and engravings betoken mental activities beyond basic subsistence – something we can identify as quintessentially human – they come very late in our history.

Thus, we are left with stone tools and other clues to economic activity as measures of intelligence. As we saw in earlier chapters, the imposition of standardization and expansion of complexity emerged very slowly in prehistoric stone-tool industries. Using the criteria of psychological theory developed by the Swiss psychologist Jean Piaget, archeologist Thomas Wynn of the University of Colorado has analyzed some of the early stone-tool industries, looking for signs of humanlike intelligence.[459] "The evolution of a uniquely hominid intelligence had not occurred by Oldowan times," he concludes, referring to the fossil and archeological remains at Olduvai Gorge, dating between 1.9 and 1.6 million years. This period was the time of *Homo habilis* but prior to *Homo erectus.* "This suggests that selection for a complex organizing intelligence was not part of the original hominid adaptation." Recent studies on the stone-tool manufacturing abilities of pygmy chimpanzees (bonobos) suggest, however, that the

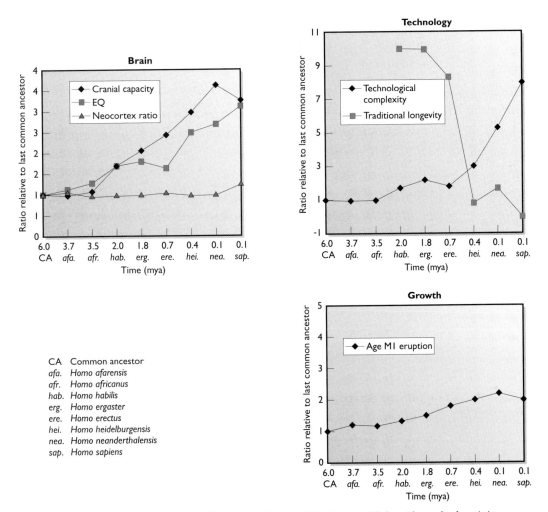

**FIGURE 17.5 Finding a measure of intelligence:** It is very difficult to establish evidence for hominin intelligence in the past. This diagram shows a measure of "intelligence" relative to a chimpanzee-like common ancestor, based on a series of different lines of evidence. Depending upon which one is used, the evidence can suggest different patterns of evolution. (Courtesy of Robert Foley.[460])

capacity for concoidal fracture that forms the basis of the Oldowan technology exceeds the capabilities of apes (see chapter 12). Apparently something changed in the brains of the earliest hominin tool makers to permit the development of this ability.

Finding a measure of intelligence in the fossil record is important, and it is not clear exactly what the best measure is (Fig. 17.5). Studies such as Wynn's look to technology as a manifestation of intelligence, while brain-size studies look at the actual level of encephalization. Changes in life history may also be indicative, for they may relate to the period available for learning in a maturing infant. However, these need not show the same pattern, which means that we do not necessarily know how the process of

What can we infer about the evolution of human intelligence?

encephalization may have related to the cognitive and behavioral abilities that were being selected.[460]

One other insight into how fossil evidence might show expanding brain size concerns the impact of brain expansion on social organization, specifically in infant care. Once hominins shifted from the basic primate pattern of brain growth, producing a much more helpless infant whose brain continued to grow at the fetal rate, then a great allocation of time and resources would be needed for rearing offspring. It is theoretically possible, as Robert Martin points out, that no change in infant care would be needed until after the adult *Homo ergaster/erectus* brain size exceeded 873 cm$^3$.[453] The argument is as follows.

Suppose that the size of the birth canal in early *Homo ergaster/erectus* was the same as in modern humans and therefore could accommodate a new-born with a 385 cm$^3$ brain. Suppose also that a *Homo ergaster/erectus* infant doubled its brain size to adulthood, as ape infants do. Would such a doubling produce a brain of the size we know *Homo ergaster/erectus* possessed? If the answer is "yes," then no lengthening of childhood would be necessary to allow for fast brain growth after birth. The real answer to this question is "no." If the human neonate brain size of 385 cm$^3$ is slightly more than doubled, it reaches approximately 800 cm$^3$, or smaller than the average size of approximately 900 cm$^3$ for adult early *Homo ergaster/erectus*. Brain growth would therefore have to continue at a high rate for a time after birth to achieve the extra brain capacity in the adult *Homo ergaster/ erectus*. Infant helplessness and prolongation of childhood would have already begun in early *Homo ergaster/erectus*.

In his calculations, Martin deliberately stacked the argument against *Homo ergaster/erectus* by assuming that its birth canal was as large as that seen in modern humans. If the birth canal was actually smaller, then the conclusion about extended childhood would be strengthened. The information on pelvic dimensions available from the Nariokotome (Turkana) boy's skeleton (see chapter 13) provided an opportunity to test this hypothesis. The width of the pelvic opening is 10 centimeters, as compared with 12.5 centimeters in modern humans. From this measurement one can calculate the size of the neonate's brain, which was roughly 275 cm$^3$, or not substantially greater than half the modern average. This figure implies that *ergaster/erectus* individuals had to triple the size of their brains between birth and maturity in order to give a 900 cm$^3$ brain in the adult. Tripling of brain size after birth, the human growth pattern, must have been associated with an extended period of infant care. Because of the absence of pelvic anatomy for *Homo rudolfensis/habilis*, nothing can be said about this pattern in the earliest species of *Homo*.

## Possible causes of brain expansion

From fossils we turn to theories about the selection pressure (or pressures) that powered hominin brain expansion. A long-popular notion was the hypothesis that the very obvious difference between hominins and apes – that humans made and used stone tools – was the most likely cause: the tripling of hominin brain size was seen as being accompanied by an ever-increasing complexity of tool technology. "Man the Tool Maker" was the encapsulation of this approach in the 1950s, followed a decade later by "Man the Hunter." In either case, the emphasis was placed on the mastering of practical affairs as the engine of hominin brain expansion.

New ideas have emerged more recently that might be described by the phrase "Man the Social Animal." The incentive for this shift of opinion has partly come from primate field studies, which are now reaching an important point of maturity. In addition, researchers have displayed a greater introspection about the human mind itself, particularly consciousness.

The new insight into "Man the Social Animal" begins with a paradox, similar in nature to the human paradox. Laboratory tests have demonstrated beyond doubt that monkeys and apes are extraordinarily intelligent, and yet field studies have revealed that the daily lives of these creatures are relatively undemanding, in the realm of subsistence at least. Why, then, did this high degree of intelligence develop?

The answer may lie in the realm of primate social life. Although, superficially, a primate's social environment does not appear to be more demanding than that of other mammals – the size and composition of social groups is matched among antelope species, for example – the interactions within the group are far more complex (Fig. 17.6). In other words, for a non-human primate in the field, learning the distribution and probable time of ripening of food sources in the environment is intellectual child's play compared with predicting – and manipulating – the behavior of other individuals in the group.[461] But why should social interactions be so complex – so Machiavellian – in primate societies?

When one observes other mammal species and sees instances of conflict between two individuals, it is usually easy to predict which animal will triumph: the larger one, or the one with bigger canines or bigger antlers (or whatever is the

**FIGURE 17.6**

**The social milieu:** Socializing has become an important part of primate life. Making alliances and exploiting knowledge of others' alliances are key to an individual's reproductive success. Biologists now believe that the intellectual demands of complex social interaction were an important force of natural selection in the expansion of primate – and, ultimately, human – brains.

appropriate weapon for combat). Not so in monkeys and apes. Individuals devote much time to establishing networks of "friendships" and observing the alliances of others. As a result, a physically inferior individual can triumph over a stronger one, provided the challenge is timed so that friends are at hand to help the challenger and that the victim's allies are absent.

"Alliances are far more complex social interactions than are two-animal contests," says Alexander Harcourt of the University of California at Davis.[462] "The information processing abilities required for success are far greater: complexity is geometrically, not arithmetically, increased with the addition of further participants in an interaction. . . . In sum, primates are consummate social tacticians."

What drives the evolution of increased brain size?

In a survey of much of the field data relevant to primate social intelligence, Dorothy Cheney, Robert Seyfarth (both of the University of Pennsylvania), and Barbara Smuts (of the University of Michigan) posed the following question: "Are [primates] capable of some of the higher cognitive processes that are central to human social interactions?" This question is important, because if anthropoid intellect, honed by complex social interaction, is merely sharper than that of the average mammal and more adept at solving psychologist's puzzles, then it does not qualify as creative intelligence.[463]

Cheney and her colleagues had no difficulty in finding many examples of primate behavior that appear to reflect humanlike social cognition. The researchers conclude that "primates can predict the consequences of their behavior for others and they understand enough about the motives of others to be able to be capable of deceit and other subtle forms of manipulation." Supporting this hypothesis, Robin Dunbar, of the University of Liverpool, has found that primate species with more complex social interaction have larger cerebral cortexes.[464]

If non-human primate intellect has truly been honed not in the realm of practical affairs, but in the hard school of social interaction, one is still left to explain why this situation has arisen. Why have primates found it advantageous to indulge in alliance building and manipulation? The answer, again gleaned from field studies, is that individuals that are adept at building and maintaining alliances are also reproductively more successful: making alliances opens up potential mating opportunities.

Once a lineage takes the evolutionary step of using social alliances to bolster reproductive success, it finds itself in what Nicholas Humphrey, a Tufts University psychologist, calls an evolutionary ratchet. "Once a society has reached a certain level of complexity, then new internal pressures must arise which act to increase its complexity still further," he explains. "For, in a society [of this kind], an animal's intellectual 'adversaries' are members of his own breeding community. And in these circumstances there can be no going back."[465]

Where does consciousness fit into this mix? Humphrey describes it as an "inner eye," with pun intended. Consciousness is a tool – the ultimate tool

– of the social animal. By being able to look into one's own mind and "see" its reactions to things and other individuals, one can more precisely predict how others will react to those same things and individuals. Consciousness builds a better reality – one that is attuned to the highly social world that humans inhabit.

## The ecology of intelligence

The most popular theory concerning the evolution of intelligence in hominins is undoubtedly the social hypothesis, and there is little doubt, following Robin Dunbar's work, that the size of the neocortex in primates correlates most strongly with group size[464] (Fig. 17.7). This also fits our understanding of both the importance of human social behavior in anthropological perspectives, and the role social factors play in primate behavior (Fig. 17.8). However, it is perhaps important not to lose sight of the ecological aspects of intelligence, and to remember that social life is not divorced from the environment. We saw, for example, in chapter 7, that the complexity of social organization among non-human primates was directly related to resource distribution. According to the classic ideas of socioecology, the way resources are available in the environment determines the distribution of females (whether they are solitary or gregarious, live in kin groups, etc.), which in turn determines the distribution of males. Thus if social complexity drives intelligence, it is ecology that drives social complexity, and so there is a resource-based element to the evolution of human intelligence – it is not just a matter of sociality and interpersonal relationships.

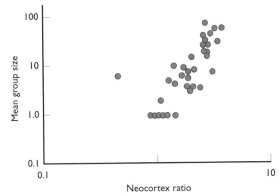

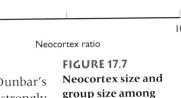

Neocortex ratio

**FIGURE 17.7 Neocortex size and group size among primates:** Among living primates, it appears that neocortex ratio (the size of the neocortex relative to the size of the brain as a whole) is related to the size of the group in which each primate species usually lives. (Courtesy of Robin Dunbar/*Journal of Human Evolution*.)

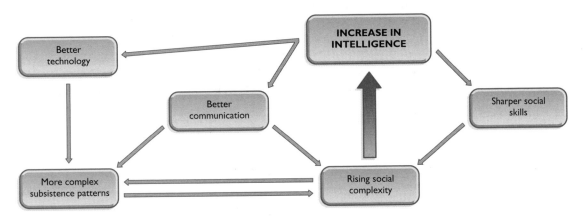

**FIGURE 17.8 Social complexity and increased intelligence:** The need to cope with rising social complexity – including increasingly demanding subsistence patterns but particularly a more ramified social structure and unpredictable social interactions – may have represented a key selection pressure for increased intelligence.

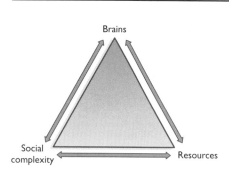

**FIGURE 17.9 Triangle of intelligence:** Understanding the evolution of intelligence is not just a question of the social context. Brain size is related to sociality in terms of the "reasons" why a species might need to be smart, but brain size is also influenced by ecology in terms of the nutrients needed to grow and maintain a large brain. There is thus an ecological context which can limit or enhance brain size. Furthermore, as the group size in which species live is determined by the distribution of resources, it is also the case that sociality too is determined by ecology. (Courtesy of Robert Foley.)

There is another way in which intelligence and ecology are linked, as we saw earlier in this chapter. Brains are metabolically expensive, and both the mothers of larger-brained infants and the infants themselves pay a high cost for their neural tissue. As we saw, that cost is paid in terms of life-history strategy. More directly, however, it is paid in terms of the actual energetic costs of growth and maintenance. It is probably this high energetic cost that inhibits the evolution of more large-brained lineages, regardless of the selective pressures imposed by a demand for greater social complexity. Basically, an animal may have as large a brain as it needs for social reasons, as Nicholas Humphrey has argued, but it also has as large a brain as it can afford energetically. Another way of looking at this, therefore, is to recognize that the enlargement of the brain in the course of human evolution must have rested upon certain ecological preconditions. Several possibilities can be put forward – for example, that the additional energetics were provided by provisioning of mothers by either males or female kin (such as grandmothers).[228] Alternatively it could have been the availability of new high-quality resources, such as meat or fish, or perhaps new processing techniques that would release nutrients, such as cooking. It might also have been linked to the development of better technologies, increasing the efficiency of food acquisition and therefore providing a better and more secure food base. In this way ecology is linked to brain size by providing the necessary conditions under which greater social complexity could evolve. There is thus a triangle of relationships – intelligence, ecology, and sociality – and while sociality may have been the ratchet (the cause in evolutionary terms: see chapter 4), ecology is the underlying environmental condition[235] (Fig. 17.9).

## CULTURAL EVOLUTION

**KEY QUESTION** Does brain expansion in human evolution involve the emergence of a new phenomenon – cultural evolution – and if so, how does it operate?

If one were to ask what the outcome of greater brain size and intelligence in humans was, the most obvious answer is that it provides the basis for culture. Indeed, human evolution has sometimes been thought of in terms of the transition from biological evolution, driven by classic Darwinian mechanisms, to cultural evolution, with a species freed from the constraints of natural selection. Certainly that is the basis for much of anthropology beyond the immediate field of human evolution.

Culture is a particularly difficult concept to place within an evolutionary framework. This is partly to do with the problems of defining it. To some it is simply what makes humans unique, and thus becomes a circular argument. To others it is a composite of all human behavior, from symbols and religion to technology and subsistence practices. It can be treated as both the outcome of a cognitive process and the process itself. Another way of looking at culture has been to associate it with the means by which information is acquired – learnt, not instinctive, and socially transmitted. This confusion has been added to by recent research into chimpanzees, which shows that they also have many of the basic traits associated with culture: variation in social traditions, tool use, communication skills, and elementary symbolic understanding.[329,466]

Despite these definitional difficulties, there is clearly a need for any comprehensive approach to human evolution to tackle the question of culture, however it is defined, to understand how it may have affected the pattern of human evolution. Indeed, in recent years cultural evolution has become an important branch of evolutionary theory, developing its own models.

*How does culture evolve?*

Several schools of thought on cultural evolution exist.[467] They all take as their starting point the idea that the culture represents a means of passing information from individual to individual that is different from the classic genetic one, and so has a different evolutionary process. In "normal" biological evolution, information is coded in the genes and transmitted only from parent to offspring, following strict rules of inheritance – those discovered by the botanist Gregor Mendel in the nineteenth century. This fact constrains the nature of evolution. If culture is taken to be another means of coding information – in this case, rules of thought and behavior – then we can recognize that its rules of transmission and inheritance are markedly different from those of genes. First, such information can be passed from one individual to another regardless of whether they are parent and offspring – this, after all is happening when anyone reads this book. As such, an idea or a technological innovation can spread rapidly through a population, much more rapidly than a gene can. Second, where the gene, the unit of inheritance in biology, mutates randomly and at a rate independent of selection, an idea can also change rapidly, and in direct response to selection. This means that cultural evolution can be expected to be a rapid process, as indeed we can observe in the present day, and that change can be directional. And third, cultural evolution can be independent of genetic evolution – in a sense, a parallel system running alongside biological systems.

These general characteristics have led to a number of different approaches that place greater or lesser emphasis on each aspect of this. For example, the dual-inheritance models – "gene–culture coevolution," as this paradigm is called – treat cultural inheritance as independent of the genetic system, and examine how it might evolve in relation to selection for particular characteristics. It has been developed most thoroughly by Marcus Feldman and Luca Cavalli-Sforza at Stanford University,[36] and Robert Boyd and Peter Richerson of the University of California.[37] A particular and important focus of their work is to determine the pattern of change in cultural units, and how these might be influenced by the environment.

Related to the coevolutionary models are those that employ the meme as the central concept. This was a term coined by Oxford biologist Richard Dawkins,[468] who saw it as the counterpart to the gene, and defined it as the minimal unit of an "idea" or a behavior. It is memes that are transmitted, and then either survive or become extinct. Memes, in other words, are under selection – cultural selection – and have differential reproductive and survival rates, just like the gene. Meme evolutionary studies have been especially concerned with determining what it is about a meme that gives it a greater or lesser chance of success, and how this relates to the environment in which memes find themselves. This approach has been particularly popular because of its elegance and analogy with genetic evolution, although it has proved to be difficult to apply to large-scale problems.[469,470]

These two approaches to cultural evolution share an interest in defining and understanding how a new mechanism of evolution can occur and what its consequences might be. Although often highly mathematical in approach, drawing ideas from population genetics, the meme and coevolutionary models are strongly anthropological in character in recognizing that something dramatically new may be occurring in the course of human evolution – the emergence of a new type of evolution. Archeologists and anthropologists have in turn tried to map the evidence of the fossil and archeological record on to such a process, to determine when it might have developed. Many have pointed to the apparent conservatism and lack of change in much of the stone-tool record of pre-modern hominins as evidence that this new mechanism, if it exists, did not become significant until relatively late in human evolution.[471,472]

Others, however, have questioned whether anything dramatically new is occurring in the evolution of human behavior, seeing it rooted more firmly in the basics of primate and animal behavior. In this approach, there is not so much a qualitative difference between human behavior and that of other animals as one of degree. This view is linked closely to primatology, and has drawn great strength from the increasing evidence for complex and variable behavior among chimpanzees. In a groundbreaking paper, Andrew Whiten (a psychologist at St Andrew's University) and colleagues mapped the diversity of social and behavioral traditions among chimpanzees.[466] They looked at attributes such as how nests were built, how termite fishing sticks were made, and whether or not hammerstones were used (Fig. 17.10). One of the criteria for "culture" is the presence of arbitrary categories, such as slightly different ways of building nests or extracting termites, ways shaped by social tradition and all equally good, rather than being the product of a particular need. Whiten and his colleagues were able to show that the variation observed went beyond simple environmentally determined patterns and seemed to reflect socially transmitted group norms.

Do chimpanzees have culture?

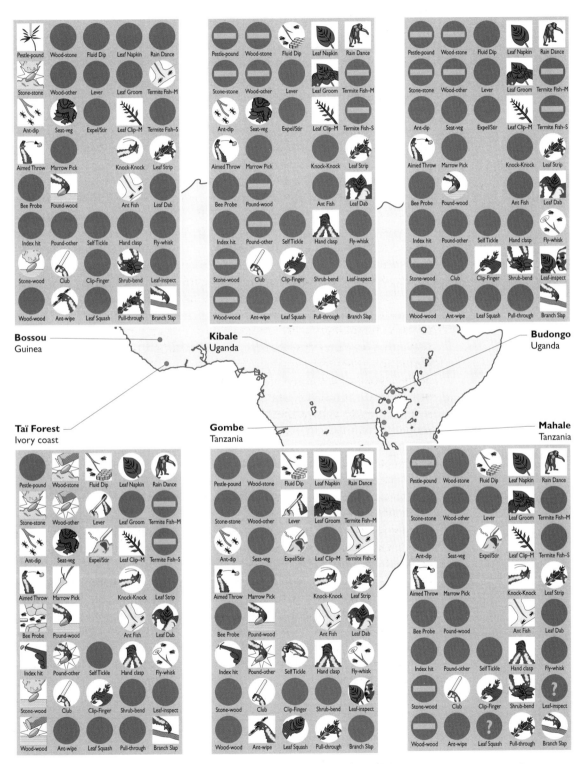

**FIGURE 17.10 Chimpanzee cultures:** Different communities of chimpanzee display their own particular patterns of behavior, giving rise to the idea that each has its unique cultural traditions. (Courtesy of Andrew Whiten et al./*Nature.*)

If some primatologists see the roots of culture in the behavior of extant primates, and therefore argue that the "emancipation from the genes" that is an integral part of culture occurs deep in the past, others prefer to see a stronger "hard-wiring" of behavior among living people. Particularly prominent here have been the evolutionary psychologists.[473,474] The fundamental premise of evolutionary psychology is that the human mind, which is ultimately the basis of all human culture, is not a blank slate on which culture may write what it wills, but rather a highly engineered system for succeeding in evolution. The mind, which evolutionary psychologists view as highly modularized, consists of ways of thinking that have been selected for, and hence that have been fixed by selection in the way people think. Evolutionary psychologists have used this approach to show that beneath the veneer of cultural variability lies a commonality of behaviors in terms of patterns of altruism, the way we choose mates, and so on. Stephen Mithen of Reading University has suggested that the evolution of modern humans involved a major change in the nature of the way in which the primate brain was modularized, and in particular that there was a breakdown of the level of modularization, leading to a greater fluidity in intelligence and ability to link, for example, social thinking with ecological thinking.[475]

We can see that culture and cultural evolution must be an important part of any attempt to explain human evolution. The approaches that have been developed all converge on the basic principle that behavior and cognition are as much a part of evolution as hard anatomy, and especially so for humans. Where they differ is in the extent to which the unique properties of human behavior have themselves been the product of a new evolutionary system, or whether they are merely additions to the primate repertoire, and whether they are universally under genetic control, or are flexible responses to environmental conditions. While there have been enormous advances in both the theory of cultural evolution and empirical studies on living humans and apes, the archeological record has not provided a sensitive test of these models. There would probably be general agreement among paleoanthropologsts that it is in the later parts of human evolution that cultural evolution becomes significant, but whether this is associated only with modern humans, or also with other species of hominin, remains hotly disputed. What would also probably be generally agreed is that language must have played an important part in this process.

## BEYOND THE FACTS
# EAT FIRST, THINK LATER

*The issue: it is generally accepted that human evolution involves three major changes – the evolution of upright walking, the evolution of reduced dentition, and the evolution of large brains (and the behaviors and abilities that derive from these). Is there any reason for thinking that these might all have evolved together, or in a particular order?*

If we think about human evolution in general terms, we can say that it consists of a major change in locomotion (the evolution of bipedalism), a major change in cognition (seen in the enlargement of the brain), and a change in dentition and digestion (reduction of the gut and of the teeth, albeit by a rather complex route). Reduced to these three simple facts, the glory of human evolution is less marked than might often be thought the case.

But one of the challenges of any evolutionary problem is to move from the particular to the general. There are two reasons for doing this. One is because a particular problem might well throw light on a general process, and the other is that in turn the general pattern may illuminate the particular case.

The generality in this case is the way evolutionary packages are put together over time. We can think of animals as systems which have to do a number of things (move around, feed, reproduce, and "think"). If we look across the animal kingdom as a whole, then we can ask whether there is any pattern to how these systems change. Perhaps it is the case that cognitive changes (how you think) come first, leading to the ability to seek out new resources (locomotor changes), leading to dietary changes (dental and digestive tract changes). Alternatively, perhaps it is the new diet that starts off the sequence of evolutionary change. Or perhaps it is a question of them all changing together as a package.

When we put the question in these general terms, it is clear that any lineage can be considered in this light. We can take cats or dolphins or horses and ask whether in each case there is a similar pattern to the way in which they evolved. Did they all go through the same sequence of change – for example, from locomotion to diet to cognition – or is it just a random process?

The answer to this question is obviously to be found in empiricism – in what, if anything, the fossil record shows. Can we discover a pattern, and if so, is it one that hominins fit? Or is it a case where the fact that humans do not fit the sequence can be used to buttress the argument that humans are different from other species, not just in their end product, but also in the way in which they arrived there?

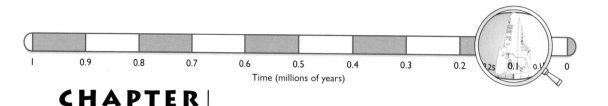

Time (millions of years)

# CHAPTER 18

# LANGUAGE AND SYMBOLISM

## THE EVOLUTION OF LANGUAGE

**KEY QUESTION** When does language evolve, and what factors led to it playing such a central role in human life?

One great frustration for anthropologists is that language – this species-specific quality with which we are so justifiably impressed – is virtually invisible in the archeological record. Not until permanent forms of writing emerged can one be certain that language existed. The first encryptions in clay tablets are found among relics of the Sumerian civilization of some 6000 years ago, but no one would argue that they mark the origin of the spoken word. Instead, other clues must be sought: in stone tools, among indications of social and economic organization, in the content and context of paintings and other forms of artistic expression, and in the fossil remains themselves.

One general question about the evolution of human language relates to the dynamics of its emergence. Was it a slow, gradual process, beginning early in hominin history and becoming fully modern only recently? Or was it a rapid process, beginning recently in hominin history? This chapter will examine several lines of evidence, taken from fossils and aspects of behavior identified in the archeological record.

### Fossil evidence

In recent years, researchers have pursued two kinds of evidence from fossil hominins. First, information is gleaned from endocasts, those crude maps of the surface features of the brain. Second, indications of the structure of the voice-producing structures in the neck (the larynx and pharynx) provide clues as to language ability.

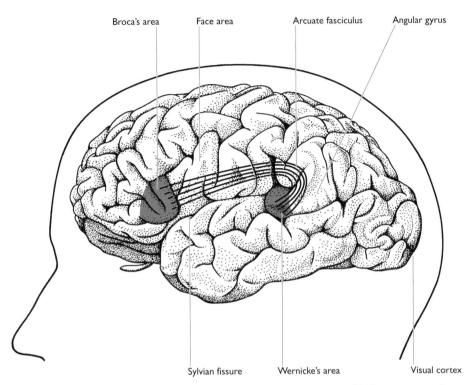

Broca's area     Face area     Arcuate fasciculus     Angular gyrus

Sylvian fissure     Wernicke's area     Visual cortex

**FIGURE 18.1 Language centers:** Wernicke's area, which appears to be responsible for content and comprehension of speech, is connected by a nerve bundle called the arcuate fasciculus to Broca's area, which influences the areas of the brain that control the muscles of the lips, jaw, tongue, soft palate, and vocal cords during speech. These language centers are usually located in the left cerebral hemisphere, even in many left-handers.

The major neural machinery for language functions is located in the left hemisphere in the great majority of modern humans, even in most left-handed people (Fig. 18.1). As with many complex mental functions, however, language capabilities cannot be pinpointed precisely to particular centers. Traditionally, Broca's area, visible as a small lump on the left side of the brain toward the front, has been associated with language, particularly with the production of sound. A second center, Wernicke's area, located somewhat behind Broca's area, is involved in the perception of sound. Recent brain scan studies, however, have shown that this concept oversimplifies the situation. Many aspects of language – for instance, the lexicon, or vocabulary with which we work – defy precise localization.

What are the anatomical traits associated with language?

Consequently, paleoneurologists can obtain few definite signs of language capacities from fossil endocasts. Signs of Broca's area have been found in *Homo rudolfensis* and later species of *Homo*, but not in australopithecines. For this reason, paleoneurologist Dean Falk believes that language capacity was already to some degree developed at the beginning of the *Homo* lineage. She disagrees with Ralph Holloway of Columbia University, however, who argues that language capacity began to develop

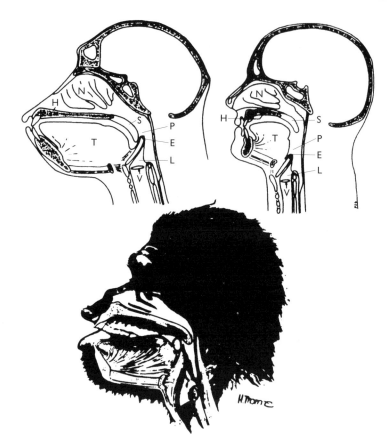

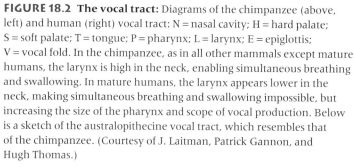

**FIGURE 18.2 The vocal tract:** Diagrams of the chimpanzee (above, left) and human (right) vocal tract: N = nasal cavity; H = hard palate; S = soft palate; T = tongue; P = pharynx; L = larynx; E = epiglottis; V = vocal fold. In the chimpanzee, as in all other mammals except mature humans, the larynx is high in the neck, enabling simultaneous breathing and swallowing. In mature humans, the larynx appears lower in the neck, making simultaneous breathing and swallowing impossible, but increasing the size of the pharynx and scope of vocal production. Below is a sketch of the australopithecine vocal tract, which resembles that of the chimpanzee. (Courtesy of J. Laitman, Patrick Gannon, and Hugh Thomas.)

earlier, among australopithecine species. His conclusion is based on the humanlike brain reorganization he detects in australopithecines. In contrast, Falk sees no reorganization in the human direction until *Homo* evolves.

If the fossil brains provide only tantalizing hints of verbal skills in our ancestors, what can we learn from the voice-producing apparatus? A number of researchers have pursued this question in recent years, in particular, Edmund Crelin, Philip Lieberman, and Jeffrey Laitman.[476,477] Perhaps not surprisingly, the human vocal tract is unique in the animal world (Fig. 18.2). In mammals, the position of the larynx in the neck assumes one of two basic patterns. One location is high up, which allows the animal simultaneously to swallow (food or liquid) and breathe. The second pattern places the larynx low in the neck, requiring temporary closing of the air passage during swallowing; otherwise solids or liquids will block it and cause choking. Adult humans have the second pattern, while all other mammals, and infant humans, possess the first.

The low position of the larynx greatly enlarges the space above it. "Consequently, sounds emitted from the larynx can be modified to a greater degree than is possible for newborns and any nonhuman mammal," explains Laitman. Non-human mammals are limited to modifying laryngeal sounds by altering the shape of the oral cavity and the lips. Human newborns maintain the basic mammalian pattern until about 1.5 to 2 years; the larynx then begins to migrate lower in the neck, achieving the adult configuration at approximately age 14 years.

Laitman and his colleagues discovered that the position of the larynx is reflected in the shape of the bottom of the skull, the basicranium. In adult humans, this structure is arched; in other mammals, and in human infants, it is much flatter. By looking at this feature in the fossil record, it should therefore be possible to discern something about the verbal skills of extinct hominin species.

What does the fossil record say? "In sum," says Laitman, "we find that the australopithecines probably had vocal tracts much like those of living monkeys or apes . . . The high position of their larynges would have made it impossible for them to produce some of the universal vowel sounds found in human speech." Unfortunately, the fossil record for *Homo rudolfensis/habilis* is poor as far as indications of the basicranium are concerned. Laitman and his colleagues have found that, in its putative evolutionary successor, *Homo ergaster/erectus*, "the larynx . . . may have begun to descend into the neck, increasing the area available to modify laryngeal sounds." The position of the larynx appears to be equivalent to that found in an 8-year-old human. Only with the origin of archaic *Homo sapiens*, some 300,000 years ago, does the fully modern pattern appear, indicating at least the mechanical potential for the full range of sounds produced by people today.

## The question of Neanderthals

A continuing controversy in the midst of this view of progressive evolution of language capacities concerns Neanderthals. Because they appeared 150,000 years after the fully arched basicranium evolved in archaic *sapiens*, implying fully developed speech potential in that species, Neanderthals might be expected to be similarly developed. Less basicranial flexion appears to characterize these beings than that observed in earlier archaic *sapiens*, as if the direction of evolution had been reversed, depriving Neanderthals of fully articulate speech. Laitman notes that the degree of basicranial flexion differs among different geographic specimens of Neanderthals, but suggests that their collective reduction in flexion may be related to their unusual upper respiratory tract anatomy, a possible adaptation to cold climes.

Did Neanderthals have language?

The notion that Neanderthals had poorly developed language abilities has become the majority position among anthropologists, who argue that it may have contributed to the extinction of the species. This conclusion has been challenged, however. In 1989, a team of researchers led by Baruch Arensburg, of Tel Aviv University, reported the discovery of a hyoid bone from a Neanderthal partial skeleton, at Kebara, Israel.[478] This small, U-shaped bone lies between the root of the tongue and the larynx, and is connected to the muscles of the jaw, larynx, and tongue. In size and shape, the Kebara hyoid is virtually identical to the modern bone. Arensburg and his colleagues claim this feature as proof that Neanderthals' language capacity resembled that of modern humans.

A second challenge to the accepted view comes from David Frayer, of the University of Kansas. He points to a new reconstruction of the famous La Chapelle-aux-Saints Neanderthal cranium, which, he says, indicates much more flexion in the basicranium than has been assumed. Frayer also argues that basicranial flexion in other Neanderthals falls within the range of other Upper Paleolithic and Mesolithic European populations.

Overall, fossil endocasts and laryngeal structure indicate a rather gradual acquisition of language capabilities through hominin history, possibly beginning with the origin of the genus *Homo*. Holloway would put language origins further back in time.

It should be remembered that higher primates are able to produce a wide range of sounds, which they use to subtle effect. For instance, when juvenile monkeys are threatened by an older opponent, they scream, which usually brings help. This scream differs subtly depending on the intensity of the threat and the dominance rank and kinship of the aggressor. Experiments with tape-recorded screams show that mothers' responses to the screams vary according to the indicated danger. In addition, some higher primates give different alarm calls for different predators (leopard, snake, and so on). Although the different calls are not "words," they do appear to be labels.

In thinking about the acquisition of spoken language by hominins, one must therefore imagine the buildup of an ever-greater range of primate sounds, and their eventual conjunction as words. Terrence Deacon, of Harvard University, suggests that neurological evidence supports such a scenario, and that language origins began with the genus *Homo* and developed gradually.[479] For some researchers, however, the structured use of words – syntax – that characterizes human speech differs so dramatically from primate vocalization that it is seen as disjunct. In other words, these researchers argue that human language is not part of a continuum with primate vocalization.

## New approaches to language origins

Finding anatomical evidence for language remains a difficult problem, and so anthropologists are always seeking new avenues by which to approach this issue. Two new ones have emerged recently. The first was a claim by Richard Kay and Matt Cartmill of Duke University concerning the hypoglossal canal.[480] This is a canal in the skull that takes nerves from the brainstem to the tongue – in other words, it innervates the tongue, the main organ of speech (Fig. 18.3). These researchers observed that the hypoglossal canal was larger in modern humans than in apes, which would fit with the idea that the movements of the tongue in humans are more finely coordinated for speech. Kay and Cartmill then looked at fossil hominins, and found that australopithecines and early *Homo* had ape-sized hypoglossal canals, whereas Neanderthals and other later archaics conformed to the modern human pattern. From this they inferred that while the

*Is there good fossil evidence for the evolution of language?*

australopithecines lacked speech, the later archaics did not – in other words, that speech was older than *Homo sapiens*. This work is still being debated, as others have claimed that in fact the pattern of variation is not related to language.

The second line of argument also relates to speech. Ann McLarnon of the Roehampton Institute in London looked at the size of the spinal cord in primates. She found that when body size was taken into account, the size of the canal reflected the amount of innervation to the limbs, upper and lower. A difference was found in humans, however, in that they showed higher levels of innervation, and hence a larger spinal cord volume, associated with the innervation of the diaphragm. This, McLarnon argued, reflected the control of the diaphragm during the fine breathing associated with speech. Here, perhaps, was another possible diagnostic tool for speech. The problem is, of course, that it requires high-quality fossil material, which is relatively rare. However, McLarnon was able to show that the Nariokotome (Turkana) boy (*Homo ergaster*), at 1.6 million years ago, lacked this diagnostic trait, and so by implication did not possess speech capabilities like those found in modern humans. In contrast, Neanderthals did show such a pattern. This observation not only supports the idea that Neanderthals possessed speech to at least some extent, contrary to the claims of some workers, but also that speech was most probably present in the common ancestor of modern humans and Neanderthals, thought to have lived about 400,000 to 500,000 years ago. This is further evidence that speech at least – and the relationship between language and speech is a complex one – was present long before the evidence for the explosion of symbolic culture at the end of the Pleistocene.[347]

Finally, geneticists are also beginning to work on the problem. Clearly the universal nature of human language abilities, and the extent to which language acquisition in infants follows a remarkably predictable pattern, suggest it has a genetic basis. Svante Pääbo, of the Max Planck Institute in Leipzig, and colleagues have looked at the genetics of individuals who have certain forms of language impairment, and then examined the

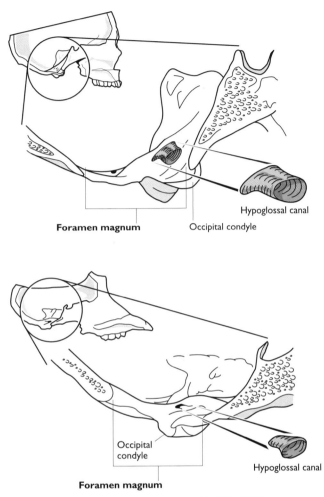

Hypoglossal canal

Foramen magnum          Occipital condyle

Occipital condyle

Foramen magnum          Hypoglossal canal

**FIGURE 18.3**
**Hypoglossal canal:**
The hypoglossal canal takes nerves to the tongue, which is more finely controlled in humans as a means of controlling sounds. Kay and Cartmill have suggested that the size of this may indicate the presence of speech in fossil hominins, although others have questioned their results.

genetic variation among them. The researchers suggest on the basis of this that language evolved relatively recently, although it must be admitted that errors on the chronological estimates are wide at the moment.[481]

# Archeological evidence: tools

Some anthropologists have argued that the pattern of tool manufacture and language production – essentially, a series of individual steps – implies a common cognitive basis. If that is true, then following the trajectory of the complexity of stone-tool technology through time should reveal something about the change in language capabilities.

Thomas Wynn, of the University of Colorado, has used psychological theory to examine the validity of this argument. "It is true," he says, "that language and tool making are sequential behaviors, but the relationship is more likely to be one of analogy rather than homology." In other words, only a superficial similarity connects the two, and their cognitive underpinnings remain quite separate. Thus, one cannot look at the complexity of a tool assemblage on one hand and learn anything directly about language abilities on the other.[459]

Glynn Isaac also searched for indications of language function in ancient tool technologies, albeit via a different approach. He has argued that the complexity of a tool assemblage might provide some information about social complexity, not cognitive complexity, relating to mechanical or verbal processes (Fig. 18.4). Beyond a certain degree of social complexity, there is an arbitrary imposition of standards and patterns. Discerning such a relationship is to some extent an abstract exercise, which would be impossible in the complete absence of language.

As we saw in previous chapters, the trajectory of technological change through hominin history falls into two phases: an almost incredibly slow phase leading from the earliest artifacts some 2.5 million years ago to approximately 250,000 years ago, followed by an ever-accelerating phase.

What lessons do we learn from this basic archeological evidence, in relation to origins of language? On the large scale, it seems reasonable to infer that a language complex enough to conjure the abstract elements of social rules, myths, and ritual is a rather late development in hominin history; that is, it began only with archaic *Homo sapiens*, and became fully expressed only with anatomically modern humans. If one adds the economic and social organization necessary in hunting and gathering activities, which ultimately would involve the need for efficient verbal communication, then the archeological record shows the same pattern. Only in the later stages of hominin history does this organization take on a degree of sophistication that would seem to demand language skills.

*Does technology imply language?*

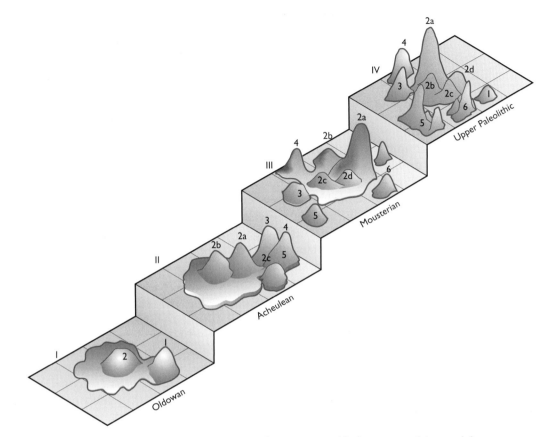

**FIGURE 18.4 Sharpening the mind, sharpening the tongue:** With the passage of time and the emergence of new species along the *Homo* lineage, stone-tool making became even more systematic and orderly. Numbered peaks in the diagram represent identifiable artifact modes, with tall, narrow peaks implying highly standardized products. The increased orderliness in stone-tool manufacture must, argued archeologist Glynn Isaac, reflect an increasingly ordered set of cognitive processes that eventually involved spoken language. (Courtesy of Glynn Isaac.)

## Archeological evidence: art

Australian scholars Iain Davidson and William Noble argue that spoken language is a very recent evolutionary development, closely tied to the cognitive processes of the development of imagery and art.[482]

Painting or engraving an image of, for example, a bison does not necessarily imply anything mystical about the motives in the artist's mind. Nevertheless, the creation of art represents an abstraction of the real world into a different form, a process that demands highly refined cognitive skills. But the art created in the Ice Age was not simply a series of simple abstractions of images to be seen in the real world; rather, it was a highly selective abstraction. Whether it represented hunting magic or an encapsulation of social structure, this art speaks of a world created by introspective consciousness and complex language. It was, in fact, a world like ours, just technologically more primitive.

If artistic expression can inform us about the possession of complex language, the question is: how far back in prehistory did it stretch? Not very far, it seems. Although some form of abstract artistic expression is claimed to date back to 300,000 years ago, it is not until a little more than 30,000 years ago that artistic expression really began to blossom (albeit slowly). One of the earliest sites to show this type of evidence is Vogelherd, Germany, from which were recovered several small, exquisitely carved animal figures, made from mammoth ivory. The most famous of the objects is the Vogelherd horse, dated at approximately 32,000 years.[483]

Is art the best evidence for symbolic thought and language?

Beyond Vogelherd, however, little art has been recovered. Two pendants – one made from reindeer bone, the other from a fox tooth – were discovered at the 35,000-year-old Neanderthal site of La Quina, France; an antelope shoulder blade etched with geometric pattern was found at another French site, La Ferrassie. Elsewhere in Europe, bones and elephant teeth with distinct zigzag markings have been discovered that were carved by Neanderthals at least 50,000 years ago. Recent finds from Blombos cave in South Africa may be evidence of symbolic behavior 70,000 years ago.

Bearing in mind the probable imperfections in the archeological record – in Europe, but especially in Africa – the inference to be drawn from artistic, abstract expression is that something important happened in the cultural milieu of hominins late in their history. The late British anthropologist Kenneth Oakley was one of the first to suggest, in 1951, that this "something important" was best explained by a quantum jump in the evolution of language. This development occurred, suggest Davidson and Noble, some 50,000 years ago.

Thus, the evidence from artistic expression suggests that the dynamic of language evolution was rapid and recent, although it is conceivable that extensive language abilities could have existed and not led to art.

## What caused the evolution of language?

The most obvious cause for the evolution of language was its development within the context in which it is so obviously proficient: communication. For a long time, this line of argument was pursued by a variety of anthropologists. The shift from the essentially individualistic subsistence activities of higher primates to the complex, cooperative venture of hunting and gathering surely demanded proficient communication. A popular hypothesis of language evolution included the notion that a first stage would have been a gesture language – gesturing, remember, is something humans do frequently, especially when lost for words.

In recent years, however, the explanatory emphasis has shifted, paralleling the shift in explanation for the evolution of intelligence. From the practical world of communication, explanation of language origins now turns to the inner mental world and social context (Fig. 18.5).

"The role of language in communication first evolved as a side effect of its basic role in the construction of reality," argues Harry Jerison of the University of California, Los Angeles. "We can think of language as being an expression of another neural contribution to the construction of mental imagery. . . . We need language more to tell stories than to direct actions." As we saw in chapter 7, anthropologists are beginning to recognize the importance of social interaction as the engine of the evolution of hominin intelligence. Consciousness and language go hand in hand with that view.

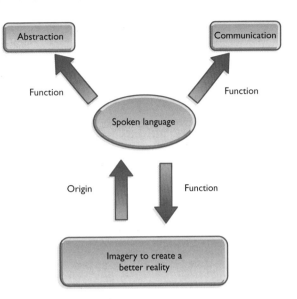

More recently, British primatologist Robin Dunbar has suggested that language may have evolved as a way of facilitating social interaction in human groups, the equivalent of grooming in non-human primates.[484] Beyond a certain group size, he argues, grooming becomes inefficient for maintaining social ties. Language is powerful because it can include individuals who are not present. These lines of investigation – the inner mental world and the social world – support an early, gradual dynamic of language evolution.

## Conclusion

**FIGURE 18.5 Origin and function of language:** Although communication is clearly an important function of spoken language, its origins (and continued functions) probably centered on creating a better image of our ancestors' social and material worlds.

We have seen that different lines of evidence, as currently interpreted, lead to different conclusions about the dynamic of language evolution (Fig. 18.6). Basically, in theoretical terms, there are two models under consideration. Under one model, the evolution of language is a long, gradual process, stretching back to the social grooming of non-human primates, and gradually developing as speech under the selective pressures of increased social complexity. The other model is that while there may have been a gradual buildup of intelligence, speech and language evolved rapidly in the most recent phase of human evolution, and brought about the explosive symbolic revolution represented most clearly by the Upper Paleolithic.

Language – recent and sudden, or old and gradual?

In recent years, especially given the growing evidence for the out-of-Africa model of modern human origins, archeologists have tended to see the origin of language as late and revolutionary, whereas biologists have probably been inclined to opt for at least some level of more gradual evolution, even if the rate was by no means constant. The current evidence would seem to suggest that the australopithecines show no evidence for

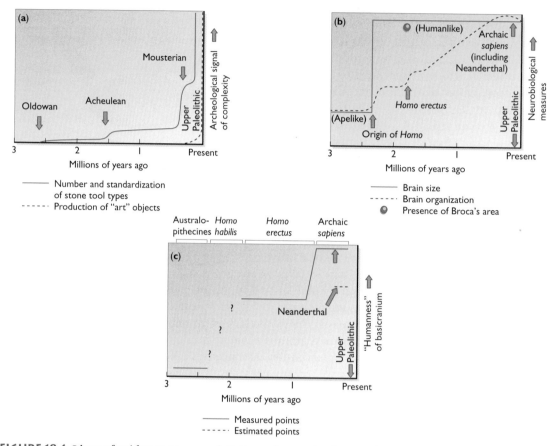

**FIGURE 18.6 Lines of evidence compared:** Evidence from archeology (a), brain size and brain organization (b), and indications of the structure of the larynx (c) is thought to be informative about the trajectory of the evolution of language. Archeological evidence indicates a recent, rapid evolution, whereas evidence about the brain and vocal tract implies an early, gradual evolution. (Courtesy of the Scientific American Library.)

language capabilities above those of the great apes; for early *Homo*, the evidence is also fairly equivocal. However, there seems to be little doubt that once hominins had brain sizes in excess of 1000 cm³, then linguistic capabilities existed, and thus language may have been present in Neanderthals at least.

# ART IN PREHISTORY

**KEY QUESTION** How does the appearance of art in the Upper Paleolithic throw light on human behavior, cognition, and social organization?

Prehistoric art is perhaps the most poignant of all the aspects of behavior that our ancestors left in the archeological record. Our feelings resonate with the painted and engraved images of an ancient world, regarding them as the products of minds – if not societies – like our own. Since their acceptance at the beginning of the twentieth century as being truly prehistoric, these ancient images have engaged archeologists in a constant and changing struggle for their interpretation. As we shall see, theories about their implications have changed through time, and in many ways prehistoric art remains as mysterious and beyond our understanding as ever. Because the art was produced by societies unlike our own, it might always remain so.

Traditionally, the study of prehistoric art meant the study of prehistoric art in Europe, specifically in southwest France and northern Spain, created during the period 35,000 to 10,000 years ago (the Upper Paleolithic). Artistic expression undoubtedly flowed elsewhere in the Old World at this time – in Africa and Australia – but accidents of history and preservation have endowed Europe with a rich record of painted, engraved, and carved images that, properly interpreted, might give some insight into the workings of the human mind at this point in our history.

Recent years have witnessed a number of important developments in the study of prehistoric art, including discoveries beyond Europe, such as an engraved antler from Longgu Cave in China, the first prehistoric art object to be found there. The most spectacular new finds, however, have occurred in France, with the discovery of Chauvet Cave in the Ardeche, southern France, and Cosquer Cave, on the southern coast near Marseilles. In some ways, these caves (particularly Chauvet) differ so dramatically from anything known previously that they upset important prevailing assumptions.

Other developments have occurred in the technical realm. Specifically, advances in our ability to analyze pigments used in paintings now sometimes yield precise dates for the work, something that was previously impossible. One dating study suggests that art at a site in northern Australia may date back some 60,000 years, which is almost twice as old as anything in Europe.

This section will examine the context of prehistoric art in Europe, and discuss the development of ideas for its interpretation.

## The context of Upper Paleolithic art

The beginnings of European prehistoric art apparently coincide closely with the arrival of anatomically modern humans (Fig. 18.7). At that time, the continent was held in the grip of the last great Pleistocene glaciation, which began 75,000 years ago and ended 10,000 years ago. After their arrival in western Europe, modern human populations would have eventually had to endure the glacial maximum, between 22,000 and 18,000 years ago, which just precedes the point that many scholars once considered to represent the high point of prehistoric art. The paintings in the famous cave of Lascaux, for instance, were done some 17,000 years

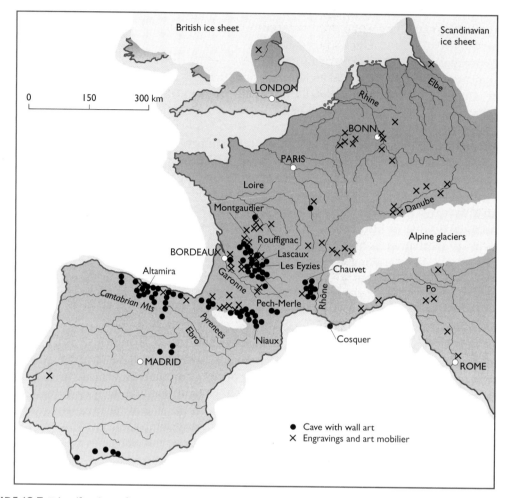

**FIGURE 18.7 Distribution of art sites in Europe:** The limestone caves of Ice Age Europe have preserved a rich legacy of Paleolithic art. Although a certain stylistic continuity characterizes cave painting, motifs in art mobilier (art on portable objects) display much more variability.

ago. Until the discovery of Chauvet cave at the end of 1994, Lascaux was the most spectacular of all known painted caves, and was known as "the Sistine Chapel of prehistory." Chauvet is not only more than equal to Lascaux in its artistic exuberance, it is claimed to be twice as old, with radiocarbon analysis dating it to a little over 32,400 years ago.

For much of the last glaciation, northern Europe lay buried beneath an ice sheet more than 1 kilometer thick in many places. To the south of the ice, the climate was colder and drier than in interglacial times, with open grassy plains replacing the dense woodland and forest that once mantled much of Europe. Regions with topographical relief experienced a greater variety of vegetation cover, with patches of woodland surviving in sheltered valleys. Herds of horses, bison, and aurochs (forerunners of today's cattle) roamed the plains of France and Spain, as did reindeer and ibex. Woolly mammoth and rhinoceros were to be seen as well, although they were more common further north and east.

During the glacial period, the climate fluctuated in severity, and the animal and plant communities rose and fell in tandem with it. Warmer climes brought a return of woodland, and woodland creatures, such as wild boar. Gone were the animals of the plains, temporarily at least. Sometimes the fluctuations extended over many millennia, sometimes within dramatically short periods. For instance, one region in southern France went from open grassland to oak forest and back to grassland, all in the space of a few hundred years. Thus, the artists of the Ice Age lived in changing times.

Caves may occasionally have been chosen as special sites for painting and engraving, perhaps being perceived as a sanctuary of some kind. But the cold climate of the Ice Age must have encouraged people to use caves for shelter, and people probably painted where they lived. This development, fortuitously, enabled a legacy of artistic expression to be preserved for us to view today. Although open-air rock faces were probably decorated as well, only engravings (not paint) had a chance of surviving. The discovery in 1994 of a series of more than 60 bold engravings of bison, deer, ibex, and other animals on rock faces along a stretch of the Côa River in northern Portugal attests to the importance of open-air sites.

## Features of Upper Paleolithic art

The discovery of Chauvet Cave has upset some of the generalities that could be adduced for Upper Paleolithic art. For instance, carved and engraved images were thought to have preceded painted images by at least 10,000 years. Chauvet, however, is as old as some of the oldest caved objects, such as the ivory animal figures from Vogelherd, Germany, that date to a little more than 30,000 years. Moreover, the painted wall art consists mainly of large mammals, such as bison, aurochs, deer, horses, mammoth, ibex, and so on; carnivores are rare and usually sequestered in the deepest recesses of caves. This latter fact was interpreted as signaling prehistoric people's fear and respect for a fellow predator. At Chauvet, however, carnivores are prominent among the painted images, and they include a hyena and a leopard, animals not previously seen in prehistoric art.

Birds, plants, and humans are only infrequently represented, and the last are often depicted quite schematically when they do appear. The painted images are often very good, naturalistic representations of single animals or small groups of individuals, but they convey little sense of natural scenes. Again, Chauvet has a scene of two rhinos fighting, a unique depiction of an aggressive scene. Hand stencils – produced by brushing or blowing pigment around the hand while placed on a rock surface – are relatively common, often revealing what appears to be missing fingers.

Some archeologists believe that, rather than representing mutilation, these stencils were produced by curling a finger under the palm, perhaps as a signature.

Painted images are usually scattered on rock surfaces in a seemingly random manner, often with one image superimposed partially or wholly on another. Sometimes interspersed among the animal images are simple geometric figures – some as simple as dots, others resembling grids and crescents.

Engraved or carved images, particularly on portable objects such as spear throwers, batons, pendants, and blade punches, often contain more detail in their execution (Fig. 18.8). Overall, they give a sense of a wider representation of nature, including the large mammals seen in wall art (although in different proportions). For instance, birds, fish, and plants are often depicted, sometimes in rich combination; again, this illustration seems to be the representation not of a scene so much as an idea, such as a season. Interestingly, carnivore teeth are present in high proportion in body ornamentation such as necklaces and pendants, in a striking contrast to the subjects of most wall art.

The human image occurs more frequently in carved and engraved images than in painting. Here again, these depictions are often schematic in nature, as in the famous "Venuses," female figures often with exaggerated features, such as breasts and buttocks, that are commonly thought of as related to fertility rituals. One site, however, contained a cache of more than 200 small engraved human faces, completely lifelike and individualistic – a portrait gallery from 20,000 years ago.

When the Ice Age finally came to a close, the art ended as well, at least in the generally naturalistic, representational style that had persisted for 25,000 years (Fig. 18.9). Geometric patterns became predominant, and people apparently no longer sought out deep caves in which to paint. It is quite possible, of course, that people painted just as much as before, but on open-air surfaces from which the images have disappeared.

The overriding sense with both wall and portable images is of art that is used, perhaps with portable objects being more personal than painted

**FIGURE 18.8** (*opposite*) **Examples of Paleolithic art:** (a) Fragment of reindeer antler from La Marche, France, approximately 12,000 years old. Apparently used as an implement for shaping flint tools, the antler fragment is engraved with a pregnant mare, which seems to have been symbolically killed by a series of engraved arrows. Above the horse is a set of notches that have been interpreted by Alexander Marshack of Harvard University as documenting the passing lunar cycles. (b) A drawing of the surface of the antler in (a), "unrolled." (c) An engraved antler baton from Montgaudier, France, dated at approximately 10,000 years old, and perhaps used in straightening the shafts of arrows or even spears. The baton's collection of engraved items suggests a representation of spring (see text). (d) A drawing of the antler baton in (c), "unrolled." (e) The Vogelherd horse, carved from mammoth ivory some 30,000 years ago and worn smooth by frequent handling over a long period of time. The horse, which is the oldest known animal carving, measures 5 centimeters. (f) The black outline of this horse was painted on the wall of a cave in Peche-Merle, France, approximately 15,000 years ago. Infrared analysis indicates that the mixture of black and red dots was added over a period of time. The black hand stencils are also later additions. Does the Peche-Merle horse, one of two in the cave, indicate the "use" of art? (Courtesy of Alexander Marshack.)

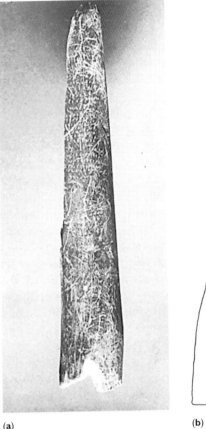

(a)

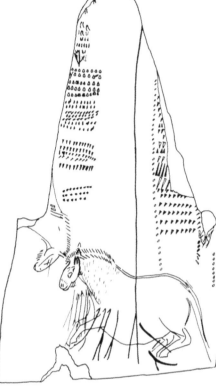

(b)

(c)

(d)

(e)

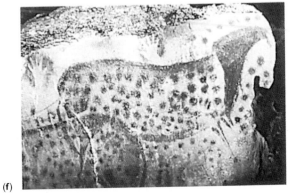

(f)

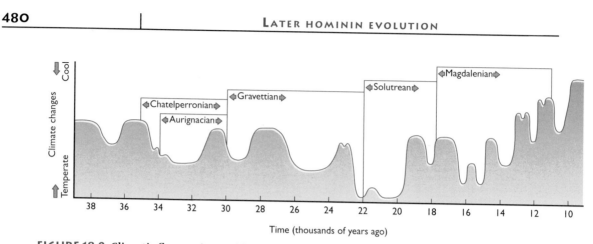

**FIGURE 18.9 Climatic fluctuations:** Although European prehistoric art was the product of the Ice Age, temperatures fluctuated somewhat throughout this period, driving dramatic shifts in ecological patterns. The most frigid period, from 22,000 to about 18,000 years ago, preceded the high point of prehistoric art, the Magdalenian. (Courtesy of the Randall White/American Museum of Natural History.)

or engraved images on walls. The task of the scholar, says Margaret Conkey of the University of California, Berkeley, is not to wonder what it all means but to ask: "What was the social context of the art that made it meaningful to the people who made and used these images?"

## Interpretations of prehistoric art

When late nineteenth- and early twentieth-century scientists accepted Upper Paleolithic art as being genuinely of great antiquity, it was interpreted merely as "art for art's sake" – an idea that has recently been revived. For instance, John Halverson, of the University of California at Santa Cruz, interprets the directness and simplicity of Paleolithic images as basic artistic expression "unmediated by cognitive reflection." What we see in the art, he suggests, is the product "not of 'primitive mind' but 'primal mind,' human consciousness in the process of growth." Most scholars, however, believe that more maturity and complexity characterize human consciousness and its products in the Upper Paleolithic.[485]

The first systematic study of Ice Age art was undertaken by the great French archeologist, the Abbé Henri Breuil (Fig. 18.10). Throughout the first half of the twentieth century, he carefully copied images from many sites and attempted a chronology based on artistic style. He, and later scholars, believed that the art would grow more sophisticated through time – hence the notion that Lascaux was the high point of prehistoric art, given its brilliance in color and incorporation of perspective. The discovery of Chauvet has upset this simple idea of progress in execution of images, because it is Lascaux's equal in these respects.

Guided by the late nineteenth-/early twentieth-century discovery that Australian Aboriginal art is sometimes related to this people's hunting activities, Breuil developed the hypothesis that prehistoric art was also

*What does cave art tell us about the mind and behavior of prehistoric people?*

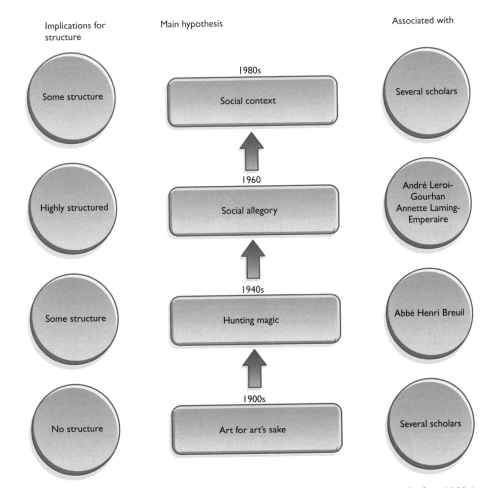

Implications for structure — Main hypothesis — Associated with

Some structure — 1980s Social context — Several scholars

Highly structured — 1960 Social allegory — André Leroi-Gourhan Annette Laming-Emperaire

Some structure — 1940s Hunting magic — Abbé Henri Breuil

No structure — 1900s Art for art's sake — Several scholars

**FIGURE 18.10 Changing theories:** After cave and portable art was finally accepted (in the late 1890s) as a genuine product of ancient people, scholars' interpretations of its meaning evolved through different stages. Shown here are the four major hypotheses, with dates and main proponents (where applicable). The different hypotheses offered different explanations of how the art was distributed – structured – within the caves.

"hunting magic" – that is, a way of ensuring fruitful hunts and propitiating the victims. Both the Aborigines and the Upper Paleolithic people were hunter-gatherer societies, and both produced paintings in which a few species were overrepresented by comparison with the prevailing environment. Breuil reasoned that Upper Paleolithic people may have made paintings to ensure the increase of totemic and prey animals. Supporting this idea is the presence among the images in many caves of animals apparently impaled on arrows or spears. Even the absence of such weapons does not militate against the idea, because an animal's image might be impaled symbolically during a ritual performance in front of it.

The hunting-magic hypothesis does have to face a problem in that the images painted in the caves often depicted animals not included in the painters' diet, as indicated by bones found at living sites. In many cases, these bones show that reindeer were important as food – yet reindeer images are few. The reverse was true for horses and bison. As the French philosopher Claude Lévi-Strauss once observed of art among San (southern African bushmen) and Australian Aborigines, certain animals are depicted frequently, not because they were "good to eat" but because they were "good to think."

Breuil's hunting-magic explanation persisted until his death in the 1960s, when it was replaced by the notion that the art somehow reflected the society that produced it. This thesis was developed independently by French archeologists André Leroi-Gourhan and Annette Laming-Emperaire. They noted that the inventory of animals depicted was comparable throughout Europe, and they described the presentation as remaining remarkably stable through time, an observation that contrasts with the much more locally idiosyncratic nature of portable art.

For Leroi-Gourhan and Laming-Emperaire, wall art reflected the duality of maleness and femaleness in society. Certain images were said to represent maleness, while others were female. The cave images were arranged so that female representations occurred at the center, with male representations located around the periphery, thereby reflecting a certain type of social structure. Although the two researchers did not fully agree on which images represented maleness and which femaleness, their work had the important effect of emphasizing social context in interpreting Paleolithic art.

Thus, where Breuil's explanation required no overall structure of the images within the caves, Leroi-Gourhan and Laming-Emperaire's clearly did. Both explanations, however, were essentially monolithic. In recent years, this concept has changed as well. "We are beginning to see a great deal more diversity and complexity in Upper Paleolithic art," explains Randall White of New York University. "And this affects the way we envisage what was going on during this important stage of human evolution."

The Upper Paleolithic is divided into different cultural periods, based upon the tool technologies of the time. Throughout these different cultures, different aspects of the art changed in various ways, as Breuil noted in his chronology. "It is important not to get the idea this pattern of change advanced on a broad front," cautions White. "In addition to differences through time, there are differences between regions, real geographic variations." These spatial and temporal variations in tool cultures are matched by similar variations in the art, although no precise correlation exists between a culture's technology and its art. Thus, a monolithic explanation of the meaning of the art is impossible.

Hunting magic may well explain some of the images. Ritual of other kinds almost certainly centered on the art as well. Something other than practicality drove Upper Paleolithic people to seek out and decorate deep caves, which appear to be otherwise unused. South African archeologists Davis Lewis-Williams and Thomas Dowson have suggested that the art is shamanistic[486] – that is, produced by shamans (priests who use magic) in or after a state of trance. The researchers base their conclusion on a study of San art, which is known to be shamanistic, and on a survey of psychological studies on the hallucinatory images produced during trance (Fig. 18.11).

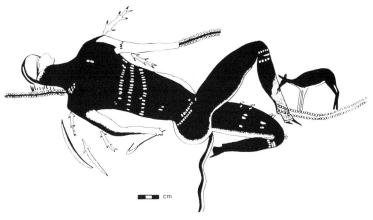

During trance-induced hallucination, the subject experiences a small set of so-called entoptic ("within the nervous system") images, such as grids, zigzags, dots, spirals, and curves. In deeper stages of trance, these images may be manipulated into recognizable objects, and subjects may eventually come to see therianthropes, or chimeras of animal and human forms. Images that reflect these trance experiences are common in shamanistic art, in south Africa and elsewhere; Lewis-Williams suggests that they may have been part of Upper Paleolithic art, too.

Still another possible role of the art is in reflecting society as a whole, as Leroi-Gourhan and Laming-Emperaire suggested. For instance, the famous cave of Altamira in northern Spain includes a circle of painted bison, surrounded by other animals, as the centerpiece of its ceiling. Archeological and other evidence appears to show that Altamira was the focus of seasonal aggregation of people from several different geographical regions, a phenomenon that occurs in modern hunter-gatherers. Conkey, who has studied the wall images and portable art at Altamira, has speculated that the circle of bison represents this process of aggregation. The recent discovery (based on radiocarbon analysis) that the paintings were created at three different times (some 13,600, 14,000, and 14,400 years ago) militates against this idea, however. This example illustrates the potential power of being able to date paintings directly.

The power of accurate dating in testing hypotheses was also demonstrated in 1992, when a team of French and Spanish scientists published radiocarbon dates on images taken from two caves in Spain and one cave in France. Remember that Breuil had suggested that chronology could be inferred from style, given that style was held to change and improve over time. The researchers derived dates for certain images from Altamira and El Castillo, in Spain. The images were stylistically similar; thus, under Breuil's scheme, they should have been the same age. In fact, they were separated by more than 1000 years. A third image, from the Niaux Cave in the French Pyrenees, differed stylistically from the Spanish images; under Breuil's scheme, it would be expected to have been made at a different time from those in the Spanish caves. In fact, it is almost identical in age to

**FIGURE 18.11**
**Shamanistic art:**
This shows a supine therianthrope (human/animal chimera) with fish. A small antelope, bleeding from the nose and therefore dying, stands on a double line of white dots. Such chimeras are a feature of shamanistic art. This image is from the site of Maclear, in the eastern Cape, south Africa. (Courtesy of David Lewis-Williams.)

the images found in El Castillo in the Pyrenees. Clearly, age and style do not always coincide.

Many portable art objects are decorated with geometric patterns. Some carry pictures of animals, fish, and plants; others include series of seemingly random notches. Alexander Marshack, an associate of the Peabody Museum at Harvard University, performed detailed studies of such objects. He suggested that some of the image combinations might represent seasons of the year: the images of a male and a female seal, a male salmon, two coiled snakes, and a flower in bloom, all engraved on a reindeer antler baton, are one such example (see Fig. 18.8 above).

In recent independent investigations, Denis Vialou, of the Musée de l'Homme in Paris, and Henri Delport, of the Musée des Antiquités Nationales, near Paris, conclude that less overall uniformity of structure connects the painted caves than originally envisaged by Leroi-Gourhan and Laming-Emperaire. The discovery of Chauvet reinforces this point. Vialou and Delport acknowledge that most of the caves follow some kind of structure, but caution that each cave should be viewed as a separate expression.

Diversity, then, begins to emerge as a more realistic interpretative lens through which to view the Upper Paleolithic – a diversity of people, a diversity of cultures, and a diversity of the art. Paleoanthropologists have now shifted from trying to understand what an individual image or set of images might mean to attempting to understand the social context in which those images were produced. Most of all, an attempt is being made to divest modern interpretations of the bias inherent in modern eyes and minds. As Conkey says, "Perhaps we have closed off certain lines of inquiry, simply by using the label 'art'."

## Precursors to Upper Paleolithic art

The Upper Paleolithic is rich with evidence of what we recognize as the symbolic artistic expression of modern humans. A persistent question in archeology relates to the dynamic of its origin: were hominins less advanced than *Homo sapiens* also capable of symbolic expression? Archeologists remain divided over the evidence and over its interpretation. (As we saw earlier, this issue is intimately tied to the question of the origin of modern humans.)

*Is there evidence for art before modern humans?*

In the late 1980s, two anthropologists at the University of Pennsylvania, Philip Chase and Harold Dibble,[487] surveyed the evidence for artistic and symbolic expression in the Middle to Upper Paleolithic transition, with the expressed purpose of determining the mode of the transition. Their conclusion was quite firm: "The most striking difference between the Middle and Upper Paleolithic is the contrast between the rich and highly developed art found in the latter period and the almost complete lack of it in the former." Chase and Dibble acknowledge that Middle Paleolithic people undoubtedly cared for one another to a much greater extent than is typical among other primate species, as evidenced by "the survival of aged individuals and of individuals suffering from moderately severe physical handicaps." In addition, "the evidence of burials implies the presence of strong emotional bonds, so that even dead members of one's group were afforded treatment not found among nonhominid primates." No fully expressed symbolism or image making is demonstrable, however.

Not everyone agrees with this position. John Lindly and Geoffrey Clark, of Arizona State University, strongly object: "We are concerned that Chase and Dibble's conclusions might be taken by anthropologists inclined to see marked discontinuity across the Middle/Upper Paleolithic transition as further 'proof' of a major difference between these two periods and, consequently, considerable evolutionary 'distance' between archaic *Homo sapiens* and morphologically modern humans."[488] In their examination of the archeological record, Lindly and Clark see no evidence of the appearance of symbolism coincident with the first appearance of anatomically modern humans (nor does anyone else); they argue that the Middle to Upper Paleolithic transition, as far as artistic expression is concerned, is a gradual, not a punctuational event. According to the two researchers, the complexity of artistic expression in the Upper Paleolithic increases with time, with the Magdalenian being more developed than the Aurignacian.

Randall White disputes Lindly and Chase's contention that the Aurignacian is somehow poorer artistically than later periods in the Upper Paleolithic. "I have been struggling to understand the rich body of Aurignacian and Gravettian evidence, especially body ornamentation, from Western, Central, and Eastern Europe," he says. "The quantity of material is staggering." Paul Mellars, of Cambridge University, is equally unimpressed with this point of Lindly and Clark's. "I have never really understood the argument that the significance of the symbolic and technological 'explosion' at the start of the Upper Paleolithic is in some way diminished by the evidence of further increases in 'cultural complexity' during the later stages of the Upper Paleolithic," he notes.[489] For Mellars, such changes would be expected as population density increases and cultural evolution continues: "To argue that this evidence for later Upper Paleolithic cultural 'intensification' rules out the significance of the far more radical innovations in behavior at the start of the Upper Paleolithic would seem akin to dismissing the significance of the 'Neolithic Revolution' on the grounds that things became even more complicated during the Bronze Age."

Some evidence has been gathered to indicate the existence of image making earlier than the Upper Paleolithic, but it is very limited: a fragment of bone marked with a zigzag motif, from the Bacho Kiro site in Bulgaria, somewhat earlier than 35,000 years ago, for example, and a carved mammoth tooth, worn smooth with use and marked with red ocher, from the 50,000-year-old site of Tata, Hungary. Oldest of all is an ox rib engraved with a series of double arcs, from the French site of Peche de l'Azé, dated as being some 300,000 years old. Ocher has been found at several ancient living sites, including the campsite of Terra Amata, in southern France, which is dated to approximately 250,000 years ago. Nevertheless, argue Chase and Dibble, none of this art betrays modern human symbolism at work, merely weak glimmerings of its eventual development. They deem many of the putative elements of evidence of Neanderthal mythology, such as the supposed use of human skulls in ritual practice, or the Cult of Skulls, to be the products of the overinterpretation of equivocal evidence by eager investigators.

More recently, Robert Bednarik, of the Australian Rock Art Association, has been promulgating the cause of pre-Upper Paleolithic art, arguing that it has not been recognized because archeologists believed it to be non-existent. Marshack has been applying microscopic analysis to incised flint pieces from the 54,000-year-old site of Quenitra, Israel, and a shaped piece of volcanic tuff from the Acheulean site of Berekhat Ram, which is between 233,000 and 800,000 years old. He has concluded that the incisions and the shaping represent the work of human hands. Although his findings may well be correct, many archeologists remain resistant to the notion that non-utilitarian artifacts prior to the Upper Paleolithic in Europe signify substantial symbolic, or abstract, expression.

## BEYOND THE FACTS
# ALL ROADS LEAD TO ROME – THE PROBLEM OF EQUIFINALITY

*The issue: paleoanthropology is to a large extent concerned with inferring life in the past from the few remains that have survived in the present. Is this just a matter of speculation, or are there rules that can guide us? Most people would say that there are rules, but that even with these, more than one interpretation is possible. This is the problem of equifinality, which can apply to simple objects such as stone tools and cutmarks, but is also critical for the appearance of art and language.*

"Once you have excluded the impossible, whatever remains, however improbable, must be the truth." This comes from Sherlock Holmes, lecturing Dr Watson about the finer points of deduction. Although much of what Holmes did was deeply empirical, he was also often at pains to show that facts only made sense in the context of the way they were linked by hypotheses and theories. In his head he was always trying to link the facts he knew in a way that made sense. His point was that there were usually many networks of links that could be made between some of his various observations, but that when you put all the known facts together, one or more would not fit in to a given network. This meant that that particular hypothesis was impossible, and so by a process of elimination, he would be left with a single network that was possible (even if improbable) – and that must be the truth.

In fiction his method works extremely well, and he always solves the case. Archeology has often been likened to detective work, seeking clues to the past. The analogy is a fine one, and Sherlock Holmes' dictum is an excellent guide to how it should be done. But in prehistory – and no doubt in real crime – there is a difficulty that Holmes manages to avoid. This is the problem of equifinality, or different processes or pathways leading to the same end point – all roads leading to Rome. In crime, it would be an example of equifinality if, it having been deduced that the criminal was a left-handed woman with one eye, there were in fact two suspects who fitted the "model." In other words, while it would be possible to exclude many suspects, it would not be possible to narrow it down to one only. The impossible would have been excluded, but there would still be more than one possibility.

Much prehistoric inference is of this sort. It is possible to exclude many interpretations of the past, but seldom to exclude all of them. Usually one is left with a small number of possibilities, each of which may be plausible. For example, a Neanderthal skull may exhibit a trauma on its head, but this could be the product of either an accident or willful violence on the part of another Neanderthal; even if it were possible to show it was violence, it may not be possible to say whether that was the outcome of a fight with another Neanderthal or with a modern human. Each would produce the same outcome.

At one level this might lead to archeological despair – the past is unknowable – but perhaps this should not be the case. Two ideas are important here. The first is that all science is about deduction, and assessing those deductions probabilistically. One must bear in mind that other interpretations may be possible, and it is a question of working out ways of saying which is the more probable, rather than necessarily which is the right one. The second idea is that by following the Sherlock Holmes approach, archeology will focus on those small observations that may be the means of excluding possibilities, rather than on the big pattern.

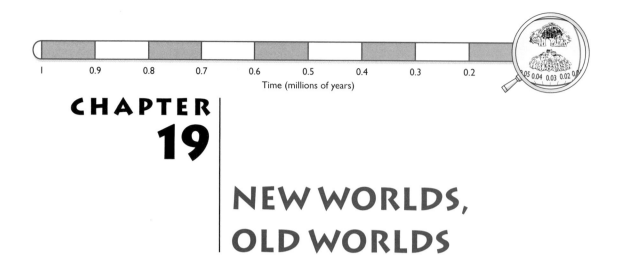

# CHAPTER 19

# NEW WORLDS, OLD WORLDS

The evolution of modern humans constituted a series of events, stretching back in time over nearly a quarter of a million years. Those events include numerous dispersals – that is, events where a population expands and thus part of it moves into new areas and colonizes new habitats, resulting in a larger geographical range. This is part of the general evolutionary process. In modern human origins research, it has become increasingly clear that there have been many such dispersals, occurring over a considerable period of time. Indeed, it is a process that continues to the present day, as the colonization of the New World by European populations from the sixteenth century testifies. Much attention has focused on the earliest dispersals of modern humans, those that occurred with the spread of human populations out of Africa and into Europe and Asia. In both these cases *H. sapiens* was moving into territories that had, at some time in the past, been colonized by earlier species of hominin. However, it is also the case that modern humans reached parts of the globe that had previously not been inhabited, suggesting new capabilities in the human population. Three major zones were reached: the Americas or New World, Australia, and distant oceanic islands such as those of Polynesia (Fig. 19.1). Paleontological, archeological, linguistic, and genetic evidence has been sifted to clarify the issue, the dates at which these dispersals occurred, and the ways in which they did so.

Researchers have often displayed a tendency to contemplate aspects of human history in isolation from that of other groups of animals. Of course, in some respects the path of human history has been determined solely by the rather special behavioral repertoire displayed by the genus *Homo*. Equally, however, the human lineage on occasions must have responded to ecological changes in ways parallel to the responses produced by other animals.

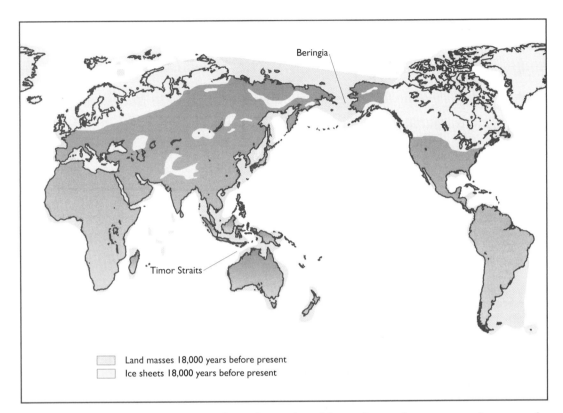

**FIGURE 19.1 Migration routes to Australia and America:** Eighteen thousand years ago was the apogee of the last glaciation (75,000–10,000 years ago). Expanded glacial cover (white areas) lowered sea levels to expose the shallow continental shelf (shaded areas over current coastlines). Although 40,000 years ago and earlier, the glaciation and sea-level lowering were less than at 18,000, the Timor Straits were still considerably narrowed, facilitating the migration into Australia (and Tasmania). The reduced glaciation some 20,000–30,000 years ago might also have left an ice-free corridor linking North America and Siberia.

For example, as Alan Turner, of Liverpool John Moores University, has argued, the initial dispersal from Africa and the later migration to North America can be viewed as territorial expansions in concert with other large predators.[490] Rather than answering some inward spirit's urge for new lands, our ancestors were simply tracking their subsistence potential through new prey populations, as were other predators. One can only speculate, however, about the precise motivations of the first Australian colonists when they struck out in small boats for a land unseen. Whatever their goal, it was not simply taking part in a more general spread of other animals. In some cases, such as the colonization of the distant islands of Polynesia, some researchers would claim that this was a deliberate and highly risky act by particular groups in search of new lands as their own became crowded and exhausted.

## COMPLETING COLONIZATION

# The Americas

**KEY QUESTION** What were the chronology, pattern, and driving force of the colonization of the Americas and Australia?

The origin of the first Americans has been a matter of scientific debate for centuries, with President Thomas Jefferson, for instance, inferring an Asiatic linkage on the basis of linguistic and archeological evidence. Although source (Asia) and the route (across the Bering Strait that separates Alaska and Siberia) are undisputed, no consensus has been reached over the timing of this migration. One school of thought argues for a date close to 12,000 years ago. By 11,500 years ago, the Americas had clearly been peopled, as evidenced by the extensive archeological finds of distinctive "fluted" stone points, characterizing the Clovis and then Folsom cultures (Fig. 19.2), evidence of which was first unearthed in the 1930s. But were the Clovis people the first Americans? Not according to the second school of thought, which argues for a date in the region of 30,000 years ago. Moreover, no agreement has been reached over whether Native Americans are descended from a single migration or from multiple migrations.

Whenever they arrived, the first Americans found a land very different from the one we know today. Between 75,000 and 10,000 years ago, the Earth was held in the pulsating grip of the Ice Age, its frigid grasp being tightest at 65,000 and 21,000 years ago (Fig. 19.3). Throughout this time, at least part of North America was mantled in ice. The Laurentide ice sheet, 3 kilometers thick in places, buried much of Canada and the northern United States from the Atlantic coast to just east of the Rockies. The Cordilleran ice sheet ran ribbon-like up the Pacific coast from Washington State toward Alaska, submerging all but the highest peaks of the Rockies and the mountains of western Canada.

Except during a period between 20,000 and 13,000 years ago, an ice-free corridor appears to have linked southern North America with the ice-free regions of Alaska and Canada's Yukon and Northwest Territories, providing a potential migration route for people coming from Siberia. These individuals would have made the intercontinental crossing dry-shod, because the Beringia landbridge, which linked Siberia with Alaska, was exposed for much of that time as the result of a drop in sea level; this fall in sea level measured as much as 100 meters at the glacial maxima, with the water being locked up in the greatly expanded polar ice caps.

In principle, colonists from Eurasia could have made their way into the Americas for

**FIGURE 19.2 Clovis and after:** Although their skeletal remains are few, the Clovis people left their trademark – the Clovis point (far left) – spread widely over North America. The Clovis point, which usually measured about 7 centimeters in length, was apparently inserted into the split end of a spear shaft and bound in place by hide. Following in close succession after Clovis were the (second-left to right) Folsom, Scottsbluff, and Hell Gap cultures.

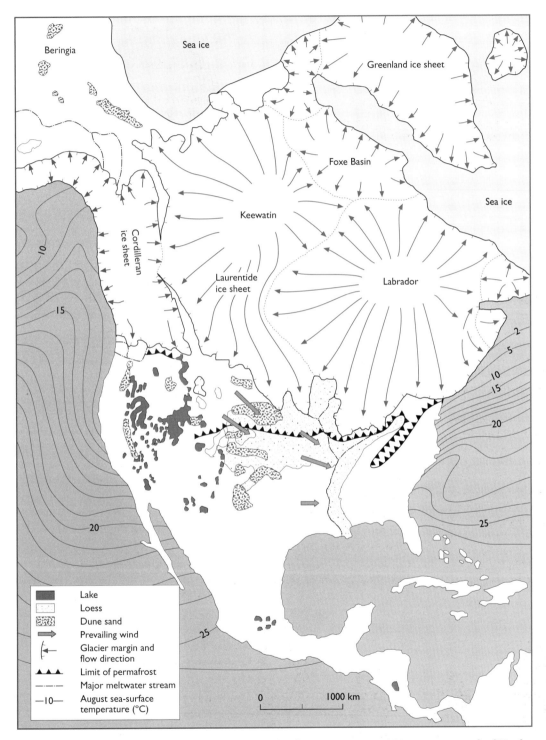

**FIGURE 19.3 In the grip of the ice:** At the peak of the last glaciation, some 18,000 years ago, much of North America was covered by thick ice sheets. To the west was the Cordilleran ice sheet; in the center and east, the Laurentide ice sheet covered the land. There is still dispute as to whether an ice-free corridor existed throughout the period or was temporarily closed. (Courtesy of Stephen C. Porter.)

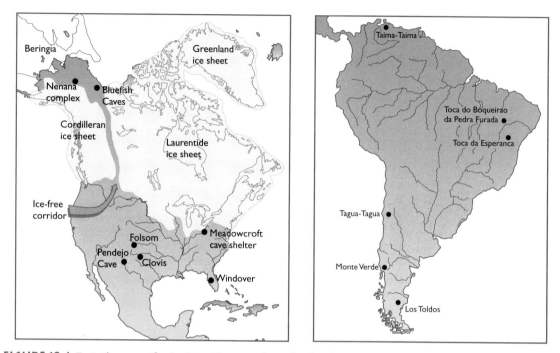

**FIGURE 19.4 Putative pre-Clovis sites:** The maps show the distribution of the sites in North America (left) and South America (right) that have the strongest claims for dating to pre-Clovis times. Pedra Furada, in Brazil, is the least likely candidate of those shown.

most of the time between 75,000 and 10,000 years ago. Some scholars argue that, although the ice-free corridor may not have represented a physical wall in the path of migrants, it might well have been so inhospitable as to be virtually biologically sterile. Thus, the ice-free corridor may have formed an invisible barrier. The time range for possible migration can probably be narrowed somewhat, because archeological evidence gathered to date appears to show that Siberia remained uninhabited until some 40,000 years ago.

*When were the Americas colonized?*

Archeologists are faced with a perplexing question: who – if anybody – preceded the Clovis people into the Americas? Over the past few decades, many claims have been made for archeological evidence earlier than 11,500 years ago south of the area that was submerged under the ice sheets (Fig. 19.4). Most of these claims have been viewed skeptically, with only a few being accepted as valid. Nevertheless, some people preceded the Clovis culture in the Americas, but the paucity of reliable sites indicates that this population was small. The explosion of sites from 11,500 years ago onward presumably reflects an explosion of populations, either from people already present in the continents or from a new migration.

Recently, some of the more famous "old" sites have lost their claims to predate Clovis. Calico Hills, California, which its proponents claim yields stone artifacts dating between 100,000 and 200,000 years, is no longer taken seriously by most authorities. Del Mar Man, a collection of skulls once dated at 70,000 years, has been redated at approximately 8000 years. And the famous bone deflesher from Old Crow in the Yukon Territories, found in 1966 and dated at 27,000 years, was redated in 1987 at just 1400 years. Nevertheless, Richard Morlan, of the University of Toronto, believes that another Yukon site, Bluefish Caves, may prove to be in the vicinity of 25,000 years old.[353] This last site relates to people north of the ice sheets, however.

The serious pre-Clovis contenders south of the ice are mostly in South America:[491]

♦ Los Toldos Cave in the Argentine Patagonia, dated at 12,600 years;
♦ the site of Tagua-Tagua in central Chile, dated at 11,380 years;
♦ also in central Chile, the site of Monte Verde, dated at 12,500 years;
♦ Taima-Taima in northwestern Venezuela, dated at 13,000 years; and
♦ Pedra Furada in northeastern Brazil, dated to over 20,000 years.

The evidence for Monte Verde's early date has recently become particularly strong, and most skeptics became convinced of its authenticity during a site visit in late 1996.[492] The rock shelter site of Pedra Furada, in northeastern Brazil, has been claimed to have been inhabited as early as 50,000 years ago, which would make it by far the oldest pre-Clovis site in the Americas.[493] Many archeologists remain skeptical that the stone artifacts on which the claim is based are truly human-made; they may actually represent the result of natural stone breakage.

The most important site in North America, and among the strongest pre-Clovis contenders in all of the Americas, is the Meadowcroft cave shelter near Pittsburgh, Pennsylvania, a site that is said to have been occupied repeatedly since 19,600 years ago. Skeptics point out the possibility that the site's material has suffered contamination with carbon from nearby coal deposits, which would corrupt the radiocarbon dating used at the site. James Adovasio of the University of Pittsburgh, the site's principal investigator, counters by noting that the dates run from the oldest to the youngest in the deposits from the bottom to the top in the site, just as they should if they were uncontaminated. This dating issue remains to be resolved (Fig. 19.5).

Archeologists now agree that a pre-Clovis people existed in the Americas, perhaps as early as 30,000 years ago. If population growth was small, then the number of archeological sites to be discovered would be correspondingly small. As David Meltzer, of Southern Methodist University, recently observed, "Clovis, in that situation, may reflect the visible portion of a population curve that began much earlier."

When Christopher Columbus arrived in the Americas in the fifteenth century, 1000 different languages were spoken among the native Indian peoples. Stanford University linguist Joseph Greenberg has analyzed the 600 languages that survive, tracing them back to just three source languages: Amerind, the most widespread and diversified; Na-Dene, less

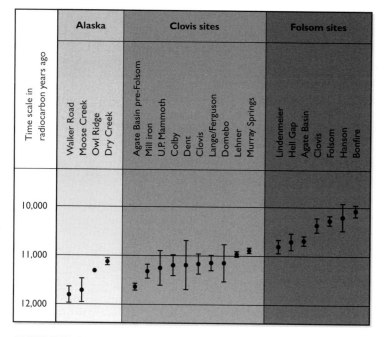

**FIGURE 19.5 The time of Clovis:** Clovis sites are scattered over much of North America (specifically the United States, as most of Canada was under ice at the time). As this diagram shows, dating of the sites lies in a tight range between 11,500 and just less than 11,000 years ago. Folsom sites follow close on behind, but again are confined to North America.

widespread and diversified than Amerind; and Aleut-Eskimo, an even less widespread and diversified language than Na-Dene. It is possible, says Greenberg, that these three linguistic groups signal three separate migrations, with the Amerind group being the first arrivals.

Several molecular biology laboratories are conducting mitochondrial DNA analysis, so far without reaching an agreement as to whether the present population descends from a small founder population or from a large population. Several different mitochondrial DNA lineages have been identified in the modern population, all of Asian origin. The amount of genetic diversity among the lineages has been estimated variously to indicate separation as long as 78,000 years ago. Although humans could have been in the Americas that long, it is more likely that the mitochondrial DNA lineages diverged in Asian populations and were already established in the founding American population at a later date. That later date has yet to be determined, although several estimates close to 30,000 years ago have been made. Douglas Wallace and his colleagues at Emory University have tentatively indicated that the mitochondrial DNA evidence might lend support to Greenberg's three-wave migration hypothesis.[494] Meanwhile, similar work at Oxford University has led to the conclusion that there was a single migration; in Japan, researchers have inferred four migrations from mitochondrial DNA data. Clearly, the matter remains to be resolved.

Work on the Y chromosome has also been recruited to inform on the peopling of the New World. Certainly this shows that that there is a link between the people of the Americas and those in Central and Eastern Asia, but this is more complex than the mitochondrial evidence, for it also shows broader links. Such studies are in fact hampered by the fact that there has been so much movement of human populations in the last ten thousand years or more that sampling populations genetically is not a straightforward job. For example, one could sample a population in North America, and show that it is closely related to a sample taken from Siberia, from which one could infer that Siberia was the source population. However, it may have been the case that at the time of the common ancestral population these "Siberians" were in fact in Central Asia, and only moved into Siberia in the last few hundred years. Studies of recent human evolution have benefited enormously from molecular genetics, but problems remain over precise interpretation.

In this context the skeletal material recovered from archeological sites can be of importance.[495,496] Although there are relatively few of these, they show some interesting results. According to the genetic/linguistic hypothesis outlined earlier, it might be expected that the skulls of the early American populations would be relatively homogeneous, and also similar to those of eastern Asia. However, as American anthropologist Jo Powell has shown, this is not the case for those from North America, and Walter Neves, of the University of Sao Paulo, has argued the same for the early skeletons from the Brazilian site of Lagoa Santa.[497] An example is the recently discovered human remains from Kennewick in Washington.[498] Morphologically this skull has been claimed to show European features, rather than those usually associated with American populations, but more accurately it should be said that it is unlike any living population – least of all the east Asian populations of today, of which it should, according to the genetic theory, be a recent descendant, for it is dated to over 9000 years ago.

American colonization – one, two, or many dispersals?

This, and the other evidence from the New World, perhaps implies something of great interest: that populations living in areas today may not be the descendants of the first colonists, but, as we saw for earlier parts of the record, there may have been multiple dispersals, of which only the final one has left a distinctive genetic and linguistic marker. In the phrase of Marta Mirazón Lahr of Cambridge University, when we look at human populations today and try to infer their Pleistocene ancestry, we should remember that we observe them through a "Holocene filter" – that is, a filter of more recent population movements. Those movements have been considerable, and "regional continuity" even within modern humans cannot be assumed. For example, it is clear that the distinctive morphology of the populations of eastern Asia today does not have a deep ancestry, and, as Peter Brown, of the University of New England in Australia, and others have shown,[499] it is probably related not to the early dispersals of modern humans into Asia, but to the spread of agricultural populations more recently. Lahr's "Holocene filter" is likely to make an understanding of the peopling of the Americas difficult.

## Human impacts of the entry into the Americas

The Americas of the Ice Age differed dramatically from today's world. They teemed with large mammal species, including mammoth, mastodon, giant ground sloth, steppe bison, elk, yak, and lion – 75 species in all, many of which were immigrants from Eurasia. Huge freshwater lakes ponded in the Great Basin. The great equatorial forests of Central and South America survived in sheltered "refuges," having largely been replaced by open grassland and woodland.

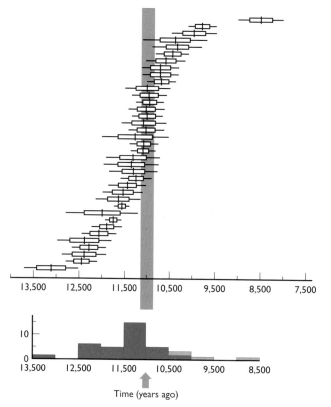

**FIGURE 19.6**

**Extinction profile:** This shows 42 radiocarbon dates on last-appearing Shasta ground sloth dung from various sites in the US southwest. The arrow and shaded column above it indicate the approximate time of activity of Clovis hunters in the region. Last appearance dates cluster at this time.

Did humans cause the extinction of the American megafauna?

Clovis people, who manufactured their characteristic projectile points (an American invention), lived in the narrow archeological window between 11,500 and 10,900 years ago. They were replaced by Folsom people, who produced smaller, more finely crafted projectile points. The Clovis and Folsom worlds were vastly different places, however. Clovis people hunted mammoth and mastodon. By Folsom times, none remained. Gone, too, were the great majority of large mammals, with some 75 species eventually going extinct and a few becoming restricted to South America (Fig. 19.6).

One of the great debates over the peopling of the Americas has centered on this rapid extinction. Some authorities – Paul Martin of the University of Arizona being the most prominent – argue that the animals had been wiped out by a wave of Clovis and then Folsom hunters, advancing north to south for a millennium. Others – with Ernest Lundelius of the University of Texas most prominent – point to the dramatic climatic shift at the end of the Ice Age as the culprit.[500]

Invasion of new lands by humans has been known to cause significant extinctions in relatively recent history. Climate change can certainly drive species to extinction – particularly a change as dramatic as that marked by the Pleistocene/Holocene transition. Thus, while both explanations are plausible, neither has been demonstrated beyond reasonable doubt in the case of the Ice Age mammals of the Americas. (Note, however, that good fossil evidence indicating extinction during the Clovis "window" has been found for only two of these species, mammoth and mastodon. It remains possible that extinction occurred in pre-Clovis times for some of the other species.)

As Donald Grayson of the University of Washington has pointed out, the image of Clovis people as mammoth hunters may have been overemphasized through the bias of the archeological record. "Most Clovis sites have been uncovered following the initial discovery of large bones," he says. "As a result, if Clovis people in the west spent most of their time hunting mice and gathering berries, we would probably not know it." In any case, none of the discoveries of Clovis material east of the Mississippi is associated with unequivocal signs of big-game hunting.

# Australia

Faunal exchange between Asia and the Americas occurred throughout the Cenozoic, and so in that context the spread of humans is perhaps not a surprise – the surprise is maybe that it occurred so late. In contrast, Australia was isolated from the Asian and other continents for many millions of years, and as a result had a unique and distinctive fauna. With the exception of the flying bats, no placental mammal had moved into Australia, and instead the continent had its own distinctive marsupial fauna. The distinction between southeast Asia and greater Australia is one of the best-marked biogeographical barriers. It is known as the Wallace line, after the co-discoverer of evolutionary theory, Alfred Wallace. The dispersal of humans into Australia is thus a major event in evolutionary geography.

The first Australians had to make a water journey to their New World. Even with sea levels at their lowest during glacial maxima, the journey from Sunda Land (the combined landmass of southeast Asia and much of Indonesia) to the Sahul landmass (Australia, Tasmania, and New Guinea) would have required eight sea voyages, the last covering 87 kilometers. So far, no archeological evidence has been recovered from Australian sites of vessels that could have made such a journey. Coastal sites during the Ice Age are mostly now submerged beneath the sea, however. In any case, the ability to construct sea-going craft that could make the required journey to Australia may be taken as proof of modern human behavior.

Although hominins have been present in southeast Asia for almost 2 million years, the first evidence of occupation in the Sahul outside of Australia is just 40,000 years old, taking the form of an archeological site on the northeast coast of New Guinea. Within Australia itself, the principal questions are: when and how was the continent first populated by humans? Fossil and archeological evidence is not extensive, and existing artifacts are often subject to differing interpretations; adding to the uncertainty is the difficulty of dating prehistoric material.

Until recently, the earliest known archeological sites – Malakunanja and nearby Nauwalabila, in Arnhem Land, northern Australia – were approximately 60,000 years old (Fig. 19.7). These dates, obtained with thermoluminescence analysis, are not universally accepted as valid. Claims had also been made for dates in excess of 100,000 years from the site of Jinmium, also in northern Australia; it was claimed that art in the form of peck-marked cups was associated with these dates. These early dates now seem questionable, although they may still indicate the presences of some form of art in Australia as early as 60,000 years ago.[501]

The oldest human fossils come from Lake Mungo, in southern Australia. Several skeletons have been unearthed from this site, one of which appears to have been cremated, making it the oldest example of this form of ritual behavior in the world. These had been dated to between 25,000

*When was Australia colonized?*

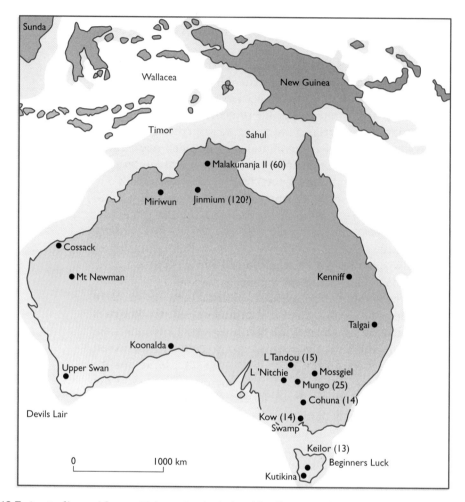

**FIGURE 19.7 Australian evidence:** Major archeological and fossil sites (with dates in thousands of years ago, where known) are shown here. The shaded areas show continental shelf exposed under peak glaciation.

and 30,000 years ago, but recently new dates have been published which suggest that at least one of the burials may be 60,000 years old. Taken together, there does seem to be strong evidence that Australia was colonized by humans between 60,000 and 50,000 years ago, considerably earlier than the first people in Europe. However, some researchers cast doubt on this early date, and prefer to see the colonization of the continent being part of a single worldwide event around 40,000 years ago.

The questions of who the first Australian populations were, and what their relationship to the broader issues of modern human diversity is, have long been debated. Earlier ideas, such as those put forward by the late Joseph Birdsell, saw Australian Aborigines having several original source populations – the so-called tripartite theory.[502] Multiregionalists such as Alan Thorne (an anthropologist at the Australian National University,

Canberra) have argued that the distinctive morphological characteristics of the populations are the result of some level of continuity and/or admixture with the late surviving *Homo erectus* populations of southeast Asia.[503] More recently, the growing support for the "out-of-Africa" theory of modern human origins has led to the view that the Australian populations are part of what Marta Lahr and Robert Foley have called the "southern dispersals," occurring early in the Upper Pleistocene along the southern coasts of Asia.[399] Whatever model is chosen – and there are good grounds for the last of these being the most probable – it is clear that in view of the distance between Australia and the Asian mainland, Australian population history and prehistory have largely been ones of isolation.

The fossil evidence to map out the history of Australian Aborigines has grown in recent years, but has nonetheless been subject to different interpretations. Part of the problem lies in its diversity. On the one hand, fossils such as those from Lake Mungo, which are chronologically the earliest, are gracile in form. Many others, which are usually later in date, are more robust and show signs of major cranial superstructures. The most extreme of these was discovered in the mid-1980s at Willandra Lakes, near Lake Mungo. Although its morphology may have hinted at an early date, it is now considered to belong to the younger group, at around 15,000 years old. This places it at about the same age as some of the fossils that come from Kow Swamp, also in southern Australia, which date to approximately 12,000 years old. These specimens are more robust than the Lake Mungo people, having thick cranial bone, large projecting faces, prominent brow ridges, and large mandibles.

The anatomical differences between these populations have prompted some anthropologists to propose that Australia was colonized at least twice (the multiple-source hypothesis). These researchers suggest that the gracile people came from China, while the robust colonists migrated from Indonesia. Interbreeding would have blurred the distinctions in later generations and produced the great anatomical variability present in modern Aborigines.

In fact, the division of the earliest fossils into gracile and robust is somewhat artificial, argues Phillip Habgood, of the University of Sydney. Both Habgood and a growing number of Australian scholars suggest that the early colonists were more anatomically homogeneous, with the variable morphology of the modern Aborigines being the result of genetic (and cultural) processes acting upon a small founding population. This concept is known as the homogeneity hypothesis. Peter Brown has shown that there is considerable morphological continuity in the Australian prehistoric populations, and that differences reflect local adaptation and the effect of local changes in a small, dispersed population over a long period of time. Certainly it is generally accepted now, as shown by numerous workers, that the affinities of the Australian population lie with the generalized characteristics of modern humans, rather than any local southeast Asian archaic precursors. This position is also supported by the genetic evidence, which shows that Australians represent a part of the narrow human variation, and a subset of the African diasporas of the Pleistocene.[504]

## Human impacts of the entry into Australia

Did humans have an impact on the Australian environment?

Wherever humans have moved into new environments, they have made a significant environmental impact, particularly in terms of destruction of their habitat. Studies in the 1990s that revealed what appears to be a sudden increase in destructive bush fires in southern Australia more than 120,000 years ago have been used to support the hypothesis of a much earlier human presence in Australia, but these results remain controversial. What is clearer is that following the undisputed evidence for humans on the continent after 50,000 years ago, there is an increase in the rate of extinction of mammals. This pattern does not tie in with any particular pattern of climatic change, and this might appear to support the idea of a human contribution to the extinction of the Australian megafauna. It occurs, however, over quite a considerable period of time, so that it cannot be easily linked to the immediate impact of colonization.

## THE FIRST VILLAGERS

**KEY QUESTION** What prompted some populations to develop control of their resources (domestication) while others remained hunter-gatherers?

The date of 12,000 years ago is usually given as the beginning of what has been called the agricultural (or Neolithic) revolution. Prior to this date, human populations subsisted by various forms of hunting and gathering. After 12,000 years ago, however, a shift toward plant and animal domestication occurred independently in several different parts of the world – first in the Fertile Crescent of the Near East, in Meso America, in eastern and southern Asia, and in Africa (Fig. 19.8). The adoption of agriculture was extremely rapid as measured against the established time scale of human prehistory, and was accompanied by an escalation of the population size, rising from approximately 10 million at the outset of the Neolithic to 100 million some 4000 years ago (Fig. 19.9). The tremendous changes wrought during the Neolithic period can be seen as a prelude to the emergence of cities and city states and, of course, to a further rise in population (which now totals 6 billion).

Until relatively recently, the agricultural revolution was viewed as a rather straightforward – if dramatic – transition. Responding to some kind of stimulus, hunters and gatherers, who were assumed to have lived in small nomadic bands of approximately 25 individuals, developed plant and animal domestication as a way of intensifying food production. As a result, these people began living in large, settled communities, whose

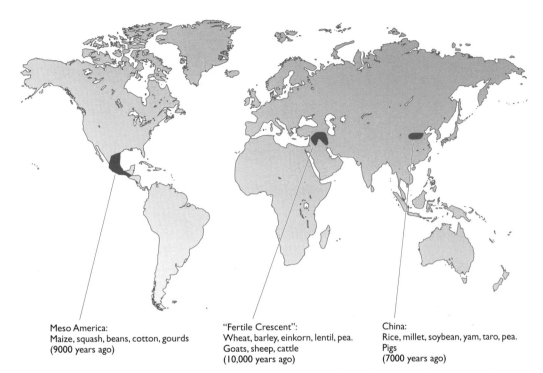

**FIGURE 19.8 Major centers of agricultural innovation:** Plant and animal domestication apparently occurred independently and at different times in many different parts of the world. Three major centers of origin existed, whose influence spread geographically, eventually coming to dominate local innovations. Other areas domesticated particular animals and crops (for example, the grain-crop tef in Ethiopia).

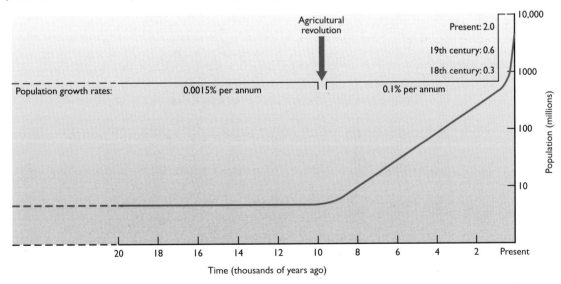

**FIGURE 19.9 Population change since the Neolithic:** The beginnings of substantial population growth coincided with the origin of plant and animal domestication, igniting an explosion that continues today. Controversy continues to swirl over whether population growth itself was a cause or a consequence of domestication.

Traditional view of agricultural revolution

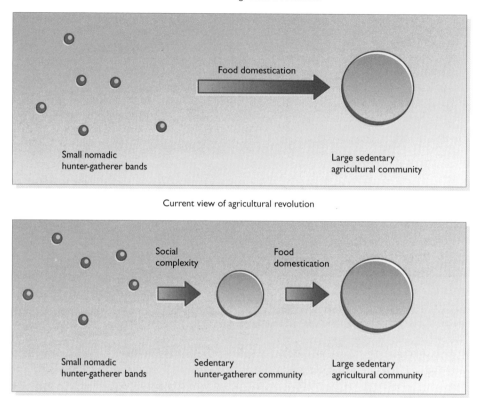

Current view of agricultural revolution

**FIGURE 19.10  Origin seen as more complex:** In the traditional view, sedentism and domestication developed together; small, nomadic, hunter-gatherer bands were viewed as being transformed into large, sedentary, agricultural communities. Recently, scholars have come to realize that the process probably included several steps, in which sedentism and domestication were separated. Intermediate between small nomadic bands and large agricultural communities, therefore, were sedentary communities that subsisted on hunting and gathering.

social and political complexity far exceeded anything achieved earlier in history. In other words, sedentism and social complexity were explained as the consequences of the adoption of agriculture, and the Neolithic transition was characterized as a shift from the simple to the complex (Fig. 19.10).

## New interpretations

Given the discovery of new archeological and ethnographic evidence, and with a reassessment of some existing evidence, the Neolithic transition is now viewed in a different light. Most importantly, it is now clear that many populations established sedentary communities and elaborated complex social systems prior to the advent of agriculture. Hunters and gatherers of the late Pleistocene, it is now realized, were not necessarily living the simple, nomadic life that anthropologists had imagined. Although debate persists about what triggered the Neolithic transition, it is not unreasonable to view some facets of agriculture as a consequence, not the cause, of social complexity (Fig. 19.11).

The traditional characterization of the Neolithic transition as an agricultural revolution rested on two kinds of evidence: archeological and ethnographic. The former was seen as indicating an explosive change in economic organization; the latter was viewed as revealing a shift from simple to complex social organization. The phrase "agricultural revolution" seemed apt for a number of reasons – not least of which was the limited number of archeological data with which to sketch this crucial period in human history. The few major sites, such as the early farming and trading community of Jericho, with its impressive tower and wall, seemed to burst out of an archeological void with dramatic suddenness. Indeed, the remains of Çatal Hüyük, which was occupied by farming people between 8500 and 7800 years ago, have been described as an archeological supernova. Excavated in the 1960s, this Turkish town covering some 12 hectares boasted elaborate architecture and beautiful, symbolic wall paintings and carvings. A British team of archeologists, led by Ian Hodder of Stanford University, began new excavations at this site in 1994.

In the decades since the initial discovery of Çatal Hüyük, further excavations in the Fertile Crescent have uncovered the remains of villages and towns, which collectively make clear that the adoption of agriculture was a much more gradual process than had been envisaged.[505] Such sites include 'Ain Ghazal in Jordan and Abu Hureyra in northern Syria. In particular, a transition is now evident, in which settled communities based entirely on hunting and gathering gave way to a mixed economy of hunting and gathering combined with some domestication, and then to fully committed agriculture. Examination of this more complete archeological record reveals that the Neolithic transition was a step-by-step introduction of domestication, not an overnight revolution.[506]

One of the most informative sites is that of Abu Hureyra, which was occupied from 11,500 to 7000 years ago, with one major break from 10,100 to 9600 years ago. Emergency excavation in 1974 showed that the first period of settlement, Abu Hureyra I, was a hunting and gathering community of 50 to 300 individuals who exploited the rich steppe flora

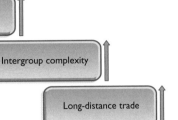

Social complexity

Intergroup complexity

Long-distance trade

Material and ritual culture

Sedentism

Domestication

Day range

Diversity of resources

Population

**FIGURE 19.11 Consequences of sedentism:** A shift from a nomadic to a sedentary way of life necessarily involved a series of potential social and material changes. Although these changes have often been associated exclusively with agricultural societies, it is now evident that sedentism can, by itself, produce at least part of this pattern.

What led to an agricultural way of life?

**FIGURE 19.12**
**Mammoth-bone dwelling:** This dwelling, which measures 5 meters in diameter, is one of five shelters excavated at Mezhirich, in Ukraine. Individually constructed with great technical and esthetic attention, these 15,000-year-old dwellings formed a community that was surely more socially complex than is usually envisaged for pre-agricultural hunter-gatherer peoples. (Courtesy of M. I. Gladkih, N. L. Kornietz, and O. Soffer/*Scientific American*, November 1984. )

(including many wild cereals) and the annually migrating Persian gazelle. A year-round settlement of simple yet substantial single-family houses, Abu Hureyra I confounds the traditional view of hunter-gatherer existence, which posits the existence of small, nomadic bands.

Perhaps because of overexploitation of local resources and an increasingly unfavorable climate, Abu Hureyra I was abandoned 10,100 years ago. It was reoccupied half a millennium later, this time by people who included plant – but not animal – domestication in their economy. For a millennium, the people of Abu Hureyra continued to hunt gazelle as their sole source of meat, after which time they turned to the domestication of sheep and goats. The overall pattern, therefore, is a gradual introduction of domesticated plants and animals.

It should have come as no surprise that late Pleistocene hunters and gatherers led socially complex lives – indications of this way of life have been known from the archeological record for some time. Most notable among this evidence was the art of the European Ice Age. "If one is looking for a single archeological reflection of sociocultural complexity, then presumably attention will continue to focus on the unique and impressive manifestations of Upper Paleolithic cave art from the Franco-Cantabrian region," notes Paul Mellars of Cambridge University. This period of European wall and portable art began approximately 35,000 years ago and ended 10,000 years ago, with the termination of the Ice Age.

More tangible evidence of late Pleistocene social and economic complexity comes from the Central Russian Plain (Fig. 19.12) – specifically, a site near the town of Mezhirich, 1100 kilometers southwest of Moscow. Approximately 15,000 years ago, a settlement of some 50 people lived in a "village" consisting of at least five substantial dwellings, each constructed from mammoth bones. "We are beginning to find evidence of semipermanent dwellings in the Central Russian Plain dating back to nearly 30,000 years ago," notes Olga Soffer, an archeologist at the University of Illinois, Urbana.

Given this and other evidence, it is perhaps surprising that, until relatively recently, late Pleistocene humans were almost universally regarded as simple nomads who wandered endlessly from camp to camp in bands of no more than 25 individuals. This characterization was based on a very important and influential study during the 1960s of the !Kung San (bushmen) of the Kalahari. Organized by Harvard University anthropologists

Richard Lee and Irven DeVore, the !Kung project examined in great detail the socioeconomic life of these people.[507]

The project revealed that, despite living in a marginal environment, the !Kung were able to subsist on simple hunting and gathering, with the expenditure of just a few hours' work each day. In addition, !Kung social life was characterized as an egalitarian, harmonious, sharing environment. The collective results of the Harvard project were presented at a landmark meeting, titled "Man the Hunter," held at the University of Chicago in 1966. For several reasons – including the fact that no other ethnographic project had been so thoroughly and scientifically conducted – the Harvard team's portrayal of the !Kung became the image of the hunting and gathering way of life, both in the modern world and in prehistory, despite existing archeological and ethnographic evidence to the contrary.

For more than a decade, the !Kung model of the hunter-gatherer lifeway dominated anthropological thought. By the early 1980s, however, its shortcomings had been gradually exposed. This shift in perception was driven by new historical, archeological, and behavioral ecology evidence. It indicated that a great deal more variability existed in the hunting and gathering way of life of prehistoric peoples than had been allowed for in the !Kung model; this variability included a degree of social and economic complexity that hitherto had been associated exclusively with agricultural societies. "Many characteristics previously associated solely with farmers – sedentism, elaborate burial and substantial tombs, social inequality, occupational specialization, long-distance exchange, technological innovation, warfare – are to be found among many foraging societies," concluded anthropologists James Brown and T. Douglas Price in 1984, in a classic reassessment of hunters and gatherers.[508]

*How different are hunter-gatherers and agriculturalists?*

In other words, the agricultural revolution was recognized to be neither a revolution nor a movement primarily focused on the adoption of agriculture. Instead, the Neolithic transition involved increasing sedentism and social complexity, usually followed by the gradual adoption of plant and animal domestication. In some cases, however, plant domestication preceded sedentism, particularly in the New World. For instance, Kent Flannery of the University of Michigan has shown that the first plant domesticated in the New World, the bottle gourd, which was grown about 9000 years ago in the southern highlands of Mexico, preceded sedentism by at least 1000 years. Clearly, the Neolithic was a complex period, and must have been influenced substantially by both local and global factors.

One long-standing question of interest in Europe, for instance, has been the mode by which agriculture spread. Was it carried by farmers moving into the region from the Middle East? Or did it develop locally, with the idea, not the farming-oriented people, spreading throughout the continent? This question is amenable to genetic as well as archeological research. Work with classic genetic markers and, more recently, DNA sequences from nuclear genes suggested that population migration was important in the spread of agriculture.[509] This conclusion, known as the demic expansion model, has been challenged by a recent survey of mitochondrial DNA patterns throughout the continent.[510] This work implies that it was principally the idea of agriculture that spread, not a migration of people. At present the genetic evidence is still being developed, but it increasingly looks as if the spread of agriculture in Europe is a combination of population movement and admixture with local populations. The emerging picture in eastern Asia also shows a pattern of demic diffusion, while that in Africa – the Bantu spread – undoubtedly left a major genetic (and linguistic) mark.[511] What these studies imply is that as well as heralding major subsistence shifts, the Neolithic revolution also changed the biological makeup of the human world, and in that sense the development of agriculture has to be considered part of the evolution of humanity.

## Causes of the transition

**FIGURE 19.13**
**Hypotheses of agricultural origins:** Population pressure and climate change have long vied as the most persuasive potential candidates for initiating sedentism and domestication. In recent times, attention has turned to factors concerning internal social complexity.

Because the transition to food production occurred within a few thousand years independently in several different parts of the world, anthropologists have long sought a global cause. Two factors have been candidates for this single, prime mover: population pressure and climate change (Fig. 19.13).

Although a dramatic rise in population numbers undoubtedly accompanied the Neolithic transition, the question of whether this relationship was one of cause or effect remains unanswered. Mark Cohen, of the State University of New York, Plattsburgh, is the main proponent of the population-pressure hypothesis.[512] He argues that the relationship was causal, and adduces signs of nutritional stress in skeletal remains from the late Paleolithic to support his case. In contrast, many anthropologists argue that numerous examples of the adoption of sedentism and agriculture can be found in the apparent absence of high population numbers, as in the southern highlands of Mexico. For these researchers, including Flannery, the population-pressure hypothesis remains unconvincing.

The second major candidate, climatic change, appears more persuasive, as the Neolithic transition coincides with the end of the Pleistocene glaciation. The shift from glacial to interglacial conditions would have driven extensive environmental restructuring, bringing plant and animal communities into areas where they did not previously exist. For instance, warmer, moister climes in the Levant 12,000 years ago likely encouraged the abundant growth of wild cereals on the steppe, allowing foragers to collect them in great numbers and subsequently

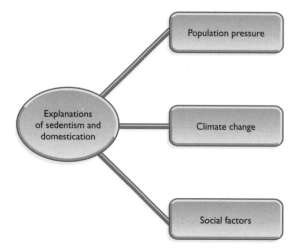

domesticate these plants. Andrew Moore, an archeologist at Yale University, considers this step to have been important in the early establishment of Abu Hureyra and other similar settlements.

Evidence is lacking to prove that climate-driven floral change universally accompanied sedentism. Moreover, some periods earlier than the end of the Pleistocene must have been conducive to intensification of food production. Modern *Homo sapiens* arose more than 100,000 years ago – so why did almost 90,000 years pass before intensification of food production become adopted? Was the delay caused by a combination of population pressure and climate change? Or was it something else entirely?

For some scholars, that "something else" is social complexity. Whereas population pressure and climate change were both "external" factors – the first presenting a problem to be solved, the second an opportunity to be exploited – social complexity would provide an "internal" trigger for change.

Building on the earlier ideas of Robert Braidwood, University of London anthropologist Barbara Bender argues that social complexity is a prerequisite for – not a product of – a sedentary agricultural system.[513] The increasing social complexity, and the stratified social and economic order that accompany it, place demands on food production that cannot be satisfied by the small, nomadic hunter-gatherer society, Bender and her supporters say. In response to this internal pressure, the culture intensifies and formalizes food production; in other words, it creates an agrarian society. Bender does not argue that this internal factor is the sole cause, merely that "technology and demography have been given too much importance in the explanation of agricultural origins, social structure too little."

*Domestication – population pressure or increasing social complexity?*

Although this social focus is gaining popularity among anthropologists, assessing its merits is very difficult. It is analogous to a "black box": you know it is important, but you do not understand how it works. Why, for instance, would social complexity have taken 90,000 years to manifest itself after the origin of anatomically modern humans? One possibility, of course, is that a subtle intellectual evolutionary change occurred relatively recently in human history, but did not manifest itself physically.

In fact, modern humans underwent a biological change between the end of the Pleistocene and the Holocene, but it affected their bodily physique. Not only are post-Pleistocene humans smaller than their immediate ancestors, but the difference in size between males and females – sexual dimorphism – is also significantly reduced. Robert Foley has suggested that this changed body size may have implications for how one views the Neolithic transition.[189]

Inevitably, anthropologists' concepts of hunter-gatherers are influenced by knowledge gleaned from contemporary foragers. These people, whose numbers are rapidly dwindling and who live in the most marginal areas of the globe, generally include a large plant-food component in their diet (notable exceptions exist, of course) and live in egalitarian communities.

Thus, the Neolithic transition is usually seen as a change from this kind of subsistence economy to domestication.

The larger overall body size of late Pleistocene people, and the greater sexual dimorphism in body size, might imply a different socioeconomic context, however. Males may well have engaged in more heated competition for access to females, as well as more big-game hunting and provisioning of their mates and offspring. In this context, what we think of as modern hunter-gathering may be largely a post-Pleistocene phenomenon. Rather than being an adaptation ancestral to food production, it is a parallel development. Both hunter-gatherer and agricultural systems developed as a response to resource depletion at the end of the Pleistocene from the rather different socioecology of Late Pleistocene anatomically modern humans.[189]

Clearly, anthropologists' picture of the Neolithic transition is far from complete. It is fair to say, however, that the search for a single, prime mover is much less popular today. The development of agriculture is likely to have been influenced by many factors, especially demographic pressure and its consequences for resources, but once started it clearly developed an ecological and social momentum of its own.

## Evolutionary consequences of agriculture

Whatever the causes of the shift from a hunter-gatherer to an agricultural way of life, there is little doubt that it represents a very fundamental shift in the nature of the human species. Prior to agriculture, human populations were, for the most part, small, isolated groups, living at low population densities, and subject to considerable fluctuation in numbers and distribution in relation to changes in climate and environment. The demographic flux of the Pleistocene, from the small bottlenecks of the origins of modern humans, to the major dispersals, to the crashes in population size associated with the last glacial maximum, are all evidence of this ecological context of the human species. Socially their groups would have been small, and ethnic units are likely to have consisted of little more than a few communities of large extended families. Agriculture changed the human species, and the world in which it lived, in a way more radical than the actual origin of the species as a whole.

The first major change is in terms of social and ethnic unit size. With sedentism came larger villages, towns and cities, and ultimately state-based political organizations. This meant that human social groups, even if the family remained the fundamental unit, consisted of large numbers of people, in which relationships had to be based on factors other than kinship. Language, religion, and other social and cultural elements became important parts of the way in which populations either were maintained cohesively, or developed more antagonistic relationships. Furthermore, social hierarchies developed in ways that led to greater inequality than was ever possible within a hunter-gatherer mode of organization. Evolutionary psychologists have become particularly interested in the extent to which the social environment in which the human mind evolved may have been different from that in which most people who subsist on agriculture have found themselves.

A second factor is that the spread of agriculture led to major dispersals. These may not have resulted in the complete replacement of populations, but they did produce a major reconfiguration of the linguistic and cultural map of the world. This can be seen in the spread of Indo-European languages, which Colin Renfrew of Cambridge University has suggested is related to the spread of agricultural populations, and also in the congruence between linguistic families – which are mostly agriculturally based – and gene trees.[514] In continents other than Europe the same pattern applies, with the result that the human population is biologically made up of large agglomerations of settled populations, interspersed with increasingly rare hunter-gatherer isolates.

With agriculture also came a whole new epidemiological pattern. Demographically, the transition from low-density hunter-gatherers to settled agriculturalists meant that the scope for pathogens to spread was increased. Furthermore, the close physical proximity of humans and their domestic stock led to a greater exposure to zoonotic diseases. The clearance of forests in the tropics is thought to have led to an expansion of the distribution of malaria. Each and all of these diseases have left their mark on the human population, either in terms of the evolution of genetically based resistance, as is the case with malaria in Africa and other parts of the tropics, or else demographically, by constraining the rate of growth.

*What were the biological consequences of agriculture?*

The development of agriculture also marked a turning point in the impact that humans had on the environment. While there is considerable debate concerning the question of whether Pleistocene hunter-gatherers behaved in ways that resulted in the extinctions of megafauna in all the main landmasses, there is much less controversy concerning the effect of humans once agriculture was established. Forest clearance, salination due to irrigation, and overgrazing are all well-known effects of high-density living. For the most part, hunter-gatherers had little effect on the world in which they found themselves. With agriculture, on the other hand, humans began to shape their own environment on an unprecedented scale.

While the development of agriculture is almost instantaneous in the context of hominin evolution as a whole – the last 12,000 years out of the last 5 million years – the new time scale of human evolution that has emerged since the early 1990s suggests that there is a relationship between the evolution of modern humans and the Neolithic revolution. With the accumulating evidence that humans have a recent origin, and that their behavior is markedly different from that of other species, we can consider the possibility that from the appearance of modern humans 150,000 years ago, to their dispersal across the world in the period from

60,000 to 40,000 years ago, to the population crashes of the last glaciation 18,000 years ago, the events of the last 12,000 years actually occur rapidly relative to the appearance of the species. In this sense, the development of agriculture is in fact an integral part of the human evolutionary story.

# THE HUMAN EVOLUTIONARY HERITAGE

In this book, we have examined briefly the evidence for human evolution and the explanations that have been put forward to account for the observed pattern, and in particular, how evolutionary processes produced a species that is radically different from any other. Knowing about our past, and what makes humans distinctive, is an important part of the growing edifice of scientific knowledge, as well as simply being of intrinsic interest to many people. We can also ask whether a knowledge of human evolution has any broader implications. Does the "old world" of human evolution have any relevance to humans today? Three possible reasons can be suggested.

First, within anthropology more broadly, and indeed across the natural and social sciences, there is a growing interest in looking at humans in terms of Darwinian mechanisms. This is exemplified by the growth of behavioral ecology approaches, by evolutionary psychology, and by Darwinian medicine (that is, the idea that natural responses to illness and disease are themselves the product of selection and evolution, and can act as a guide to medical treatment). Each of these posits the idea that the way humans are today is shaped by evolutionary mechanisms, in the case of behavioral ecology, and by evolutionary history, in the cases of evolutionary psychology and Darwinian medicine. Evolutionary psychologists argue that the human mind is the product of the environment in which humans have evolved – what is referred to as the environment of evolutionary adaptedness. This is not the environment in which most humans live today, but is envisaged as one of small-scale hunter-gatherer bands, or particular habitats such as tropical Africa, where hominins have mostly lived. If this is the case, then it is argued that, when looking for adaptive explanations for the human mind or any element of behavior, it is important that the explanations are rooted in a sound reconstruction of the past. For this reason, a detailed study of human evolution is part of a broader interest in Darwinian evolution and human behavior and affairs. This provides, in turn, an important context for discussions about "nature–nurture" – the debate about whether humans are more influenced by their genes or their environment, or by a more complex interaction between the two.

A second reason for an interest in the details of human evolution lies in evolutionary theory itself. This, and in particular Neo-Darwinian theory, is predicated upon increasingly well-understood principles relating to

*Are humans still the product of their evolutionary history?*

population genetics and natural selection. As we saw at the beginning of this book, selection depends upon an understanding of modes of heritability, which for most species is restricted to genetics. However, as was discussed in chapter 17, during the course of human evolution there emerged aspects of behavior for which the term "culture" acts as a collective shorthand. Evolutionary theorists have recently turned their attention to understanding cultural evolution as a Darwinian process. As the course of human evolution that we have outlined here covers the transition from what was essentially a non-cultural creature to a cultural one, it provides an empirical framework for testing exciting and important new areas of evolutionary theory.

The third reason relates to human diversity. Over the course of the last century and a half, anthropology, including paleoanthropology, has been deeply involved in discussions of how the human species is divided, or not divided, into subgroups – in other words, issues of race. At times, the findings and interpretations of anthropology have been closely linked to broader political movements, both arguing for a racial basis for human diversity (by some anthropologists before World War II) and arguing against it (before but especially after that war). The question of human diversity is ultimately an evolutionary one, for if there are differences between populations, then they must have arisen in the course of human evolution, diversficiation, and dispersals. For this reason, debates about human diversity should be founded on some sort of evolutionary empiricism. As we have seen, information about the origins and evolution of modern humans and their diversity has been a key topic in recent years, and has been greatly enhanced by the emergence of molecular evolutionary genetics. In this context, the increasing acceptance of the fact that all humans share a recent common populational history, and that genetic diversity between populations is very low, is of great significance.

If human evolution has a significance beyond the community of researchers and students who find stones, bones, and genes fascinating, then it is all the more important that it is approached and carried out rigorously and scientifically. Paleoanthropology since the mid-twentieth century has made enormous strides, in terms of both new discoveries and the development of new scientific techniques. Equally important, though, is the theme that we have attempted to carry through this book, that understanding human evolution depends not just upon the fossils – the hard evidence – but also upon applying the general principles of evolutionary biology and theory. Ultimately, human evolutionary studies are comparative, despite the uniqueness of the human species.

## BEYOND THE FACTS
# THE TAPE OF LIFE

*The issue: it is often considered that evolutionary theory is based on randomness, or that chance and historical contingency play a major role. This has led to the idea that the pattern of evolution has no general predictability, and its particular course is unique and would not be repeated. Is this the case, or does evolution proceed in a more predictable way?*

Among scientific ideas, evolutionary theory is perhaps unique in being able to attract criticism from diametrically opposed directions, often simultaneously. One much-repeated criticism is that is impossible for a random process to produce such organized complexity. The analogy is made of a whirlwind rushing through a hanger and constructing a 747 from the pieces lying around. Chance cannot be creative. At the other extreme, Darwinism has been rejected by many social scientists because it is deterministic, and cannot account for the vagaries of human behavior.

Is evolution random or deterministic? According to some, the mass extinctions show that it is random; according to others, the power of selection is such that it is deterministic in its core features. More than any other fundamental issue, this one lies at the heart of much controversy in evolutionary biology. In his book on the Pre-Cambrian fossils of the Burgess Shales, *Wonderful Life* (New York: Norton, 1990), Stephen Jay Gould made a strong case that chance played the major role, and the search for adaptive determinacy was illusory. Gould argues that it was a matter of chance which lineages evolved and diversified, and which ones became extinct, and therefore it was not appropriate to read a pattern in the history of life. In a striking metaphor he asked whether, if the tape of life were rerun, it would play the same story or tune. His answer was "no." Contingency – the effect of small differences in starting conditions – would result in dramatically different outcomes. Slightly different conditions would lead to an entirely different evolutionary history, with no mammals, no vertebrates, no humans. At one extreme, this amounts to what can be called lottery biology.

This view has been challenged. Simon Conway Morris, of Cambridge University, who analyzed the Burgess Shale fossil assemblages, has sharply questioned Gould's interpretation. Conway Morris's case is not that there are no contingent factors, but that once evolutionary trends are in play, the patterns tend to be repetitive. The system is far more constrained by what is biologically possible and optimal, and so the same patterns recur. He points to the very frequent occurrence of convergent evolution – the same features evolving independently – as evidence for the power of selection, and hence the deterministic nature of evolutionary patterns. According to Conway Morris, if the tape of life were rerun, there would be differences in the details, but much of the basic pattern would be the same.

Which of these models is correct – lottery biology or selectionist determinism – is central to understanding the nature of the history of life on this planet – or indeed on any other. For the study of evolution these issues are critical, for they address the question of the probability of intelligent life evolving, and the conditions under which it might.

# REFERENCES

The literature on evolution in general and human evolution in particular is vast, and no bibliography can do justice to the subject without swamping any book. The limited references provided here serve two purposes. First, they indicate the major publications where important discoveries were announced, and where new ideas and theories have been developed. These references provide the historical background necessary to understand the way in which ideas about human evolution have themselves evolved. Second, they indicate where it will be possible to find more detail on a particular subject, and thus gain entry into a broader scientific literature. Many of the references are to major books that themselves have large and more specialized references.

1 Darwin, C. 1859. *The Origin of Species*. London: Murray.

2 Huxley, T. H. 1863. *Evidences as to Man's Place in Nature*. London: Williams and Norgate.

3 Bowler, P. 1984. *Evolution: The History of an Idea*. Berkeley: University of California Press.

4 Paley, W. 1802. *Natural Theology: Or, Evidences of the Existence and Attributes of the Deity, Collected from the Appearances of Nature*.

5 Dawkins, R. 1986. *The Blind Watchmaker*. Harlow: Longman.

6 Gould, S. J. 2002. *The Structure of Evolutionary Theory*. Cambridge, MA: Harvard University Press.

7 Koerner, L. 1999. *Linnaeus: Nature and Nation*. Cambridge, MA: Harvard University Press.

8 Andrews, R. C. 1948. *Meet Your Ancestors*. John Long.

9 Keith, A. 1925. *The Antiquity of Man*. London: Williams and Norgate.

10 Keith, A. 1927. *Concerning Man's Origin*. London: Williams and Norgate.

11 Broom, R. 1933. *The Coming of Man: Was it Accident or Design?* London: Witherby.

12 Foley, R. A. 1995. *Humans Before Humanity: An Evolutionary Perspective*. Oxford: Blackwell.

13 Huxley, J. S. 1940. *The Uniqueness of Man*. London: Allen and Unwin.

14 Diamond, J. M. 1991. *The Rise and Fall of the Third Chimpanzee*. London: Vintage.

15 Lewin, R. 1990. *Bones of Contention*. New York: Simon and Schuster.

16 Reader, J. 1988. *Missing Links*. 2nd edn. London: Pelican.

17 Stocking, G. W. J. 1982. *Race, Culture, and Evolution: Essays in the History of Anthropology*. Chicago: University of Chicago Press.

18 Spencer, F. 1996. *History of Physical Anthropology*. New York: Garland.

19 Darwin, C. 1871. *The Descent of Man and Selection in Relation to Sex*. London: Murray.

20 Gee, H. 1996. Box of bones "clinches" identity of Piltdown palaeontology hoaxer. *Nature* 381: 262.

21  Le Gros Clark, W. E. 1964. *The Fossil Evidence for Human Evolution.* Chicago: University of Chicago Press.

22  Washburn, S. L., Moore, R. 1974. *Ape into Man: A Study of Human Evolution.* Boston, MA: Little, Brown.

23  Pilbeam, D. 1968. The earliest hominids. *Nature* 219: 1335.

24  Goodman, M. 1976. Towards a geneological description of the primates. In *Molecular Anthropology*, eds M. Goodman, R. E. Tashien, pp. 321–53. New York: Plenum Press.

25  Mayr, E. 1950. Taxonomic categories in fossil hominids. *Cold Spring Harbor Symposia on Quantitative Biology* 15: 109–17.

26  Landau, M. 1991. *Narratives of Human Evolution.* New Haven, CT: Yale University Press.

27  Elliot Smith, G. 1927. *The Evolution of Man.* London: Oxford University Press.

28  Dunbar, R. I. M. 1996. *The Trouble with Science.* Cambridge, MA: Harvard University Press.

29  Foley, R. A. 1995. The causes and consequences of human evolution. *Journal of the Royal Anthropological Society* 1: 67–86.

30  Foley, R. A. 1987. *Another Unique Species: Patterns of Human Evolutionary Ecology.* Harlow: Longman.

31  Dawkins, R. 1982. *The Extended Phenotype.* San Francisco: Freeman.

32  Ridley, M. 1996. *Evolution.* Oxford: Blackwell.

33  Futuyama, D. J. 1979. *Evolutionary Biology.* Sunderland, MA: Sinauer Associates.

34  Skelton, P. 1993. *Evolution: A Biological and Palaeontological Approach.* New York: Prentice-Hall.

35  Keller, E. J., ed. 1992. *Keywords in Evolutionary Biology.* Cambridge, MA: Harvard University Press.

36  Cavalli-Sforza, L., Feldman, M. 1981. *Cultural Transmission and Evolution.* Princeton, NJ: Princeton University Press.

37  Boyd, R., Richerson, P. J. 1985. *Culture and the Evolutionary Process.* Chicago: University of Chicago Press.

38  Laland, K. N., Brown, G. R. 2001. *Sense and Nonsense: Evolutionary Perspectives on Human Behaviour.* Oxford: Oxford University Press.

39  Dunbar, R. I. M. 1982. Adaptation, fitness and the evolutionary tautology. In *Current Problems in Sociobiology*, ed. King's College Sociobiology Group, pp. 9–28. Cambridge: Cambridge University Press.

40  Gould, S. J., Lewontin, R. C. 1979. The spandrels of San Marco and the Panglossian paradigm: a critique of the adaptationist programme. *Proceedings of the Royal Society of London, Series B – Biological Sciences* 205: 581–98.

41  Gould, S. J., Vrba, E. S. 1983. Exaptation – a missing term in the science of form. *Paleobiology* 8: 4–15.

42  Dennett, D. 1996. *Darwin's Dangerous Idea.* New York: Touchstone.

43  Tinbergen, N. 1963. On aims and methods in ethology. *Zeitschrift für Tierpsychologie* 20: 410–33.

44  Krebs, J. R., Davies, N. B. 1987. *An Introduction to Behavioural Ecology.* Oxford: Blackwell.

45  Mayr, E. 1985. *The Growth of Biological Thought.* Cambridge, MA: Harvard University Press.

46  Jacob, F. 1977. *The Possible and the Actual.* Seattle: University of Washington Press.

47  Raff, R. A. 1996. *The Shape of Life: Genes, Development, and the Evolution of Animal Form.* Chicago: University of Chicago Press.

48  Kimura, M. 1983. The neutral theory of molecular evolution. In *Evolution of Genes and Proteins*, eds M. Nei, R. Koehn, pp. 208–33. Sunderland, MA: Sinauer Associates.

49  Seyfarth, R. M., Cheney, D. L., Marler, P. 1980. Monkey responses to three different alarm calls: evidence of predator classification and a semantic communication. *Science* 210: 801–3.

50  Wynne Edwards, V. C. 1962. *Animal Dispersal in Relation to Social Behaviour.* Edinburgh: Oliver and Boyd.

51  Hamilton, W. D. 1964. The genetical theory of social behavior. *Journal of Theoretical Biology* 7: 1–52.

52  Trivers, R. L. 1971. The evolution of reciprocal altruism. *Quarterly Review of Biology* 46: 35–57.

53  Wilson, D. S. 1983. The group selection controversy: history and current status. *Annual Review of Ecology and Systematics* 14: 159–87.

54  Maynard Smith, J. 1982. *Evolution Now: A Century after Darwin.* London: Macmillan.

55  Mayr, E. 1963. *Animal Species and Evolution.* Cambridge, MA: Harvard University Press.

56  Cracraft, J. 1989. Species as entities of biological theory. In *What the Philosophy of Biology Is*, ed. M. Ruse, pp. 31–52. Dordrecht: Kluwer.

57 Simpson, G. G. 1953. *The Major Features of Evolution*. New York: Columbia University Press.

58 Paterson, H. E. H. 1985. The species recognition concept. In *Species and Speciation*, ed. E. S. Vrba, pp. 21–9. Pretoria: Transvaal Museum.

59 Eldredge, N., Gould, S. J. 1972. Punctuated equilibria: an alternative to phyletic gradualism. In *Models in Paleobiology*, ed. T. J. M. Schopf, pp. 82–113. San Francisco: Freeman.

60 Gould, S. J. 1980. Is a new and general theory of evolution emerging? *Paleobiology* 6: 119–30.

61 McNamara, K. J., ed. 1990. *Evolutionary Trends*. Tucson, AZ: University of Arizona Press.

62 Vrba, E. 1995. Species as habitat-specific complex systems. In *Speciation and the Recognition Concept: Theory and Application*, eds D. M. Lamber, H. G. Spencer, pp. 3–44. Baltimore: Johns Hopkins University Press.

63 Vrba, E. 1992. Mammals as a key to evolutionary theory. *Journal of Mammalology* 73: 1–18.

64 Foley, R. A. 2001. In the shadow of the modern synthesis: alternative perspectives on the last 50 years of palaeoanthropology. *Evolutionary Anthropology* 10: 5–15.

65 Tattersall, I. 2000. Paleoanthropology: the last half-century. *Evolutionary Anthropology* 9: 2–16.

66 Van Valen, L. 1973. A new evolutionary law. *Evolutionary Theory* 1: 1–30.

67 Foley, R. A. 1984. Early man and the Red Queen: tropical African community evolution and ecology. In *Hominid Evolution and Community Ecology: Prehistoric Human Adaptation in Biological Perspective*, ed. R. A. Foley, pp. 85–110. New York, London: Academic Press.

68 Murphy, W. J., Eizirik, E., O'Brien, S. J., Madsen, O., Scally, M., et al. 2001. Resolution of the early placental mammal radiation using Bayesian phylogenetics. *Science* 294: 2348–51.

69 Vrba, E., ed. 1996. *Palaeoclimate and Neogene Evolution*. New Haven, CT: Yale University Press.

70 Kennett, J. P. 1977. Cenozoic evolution of antarctic glaciation, the Circum-Antarctic ocean, and their implications on global palaeooceanography. *Journal of Geophysical Research* 82: 3843–60.

71 Shackleton, N. J. 1996. New data on the evolution of Pliocene climatic variability. In *Palaeoclimate and Neogene Evolution*, ed. E. Vrba, pp. 282–90. New Haven, CT: Yale University Press.

72 Vrba, E. 1985. Ecological and adaptive changes associated with early hominid evolution. In *Ancestors: The Hard Evidence*, ed. E. Delson, pp. 63–71. New York: Alan Liss.

73 Vrba, E. 1988. Late Pliocene climatic events and human evolution. In *Evolutionary History of the "Robust" Australopithecines*, ed. F. Grine, pp. 405–26. New York: Aldine de Gruyter.

74 Vrba, E. 1974. Chronological and ecological implications of the fossil Bovidae at Sterkfontein archaeological site. *Nature* 250: 19–23.

75 Vrba, E. 1999. Habitat theory in relation to evolution in African Neogene biota and hominids. In *African Biogeography, Climate Change, and Hominid Evolution*, eds T. Bromage, F. Schrenk, pp. 328–48. Oxford: Oxford University Press.

76 Wesselman, H. B. 1996. Of mice and almost men: regional palaeoecology and human evolution in the Turkana Basin. In *Palaeoclimate and Neogene Evolution*, ed. E. Vrba, pp. 356–68. New Haven, CT: Yale University Press.

77 Behrensmeyer, A. K., Todd, N. E., Potts, R., McBrinn, G. E. 1997. Late Pliocene faunal turnover in the Turkana Basin, Kenya and Ethiopia. *Science* 278: 1589–94.

78 Hill, A. 1996. Faunal sequence and environmental change in the Neogene of East Africa. In *Palaeoclimate and Neogene Evolution*, ed. E. Vrba, pp. 178–96. New Haven, CT: Yale University Press.

79 White, T. D. 1996. African omnivores: global climatic change and Plio-Pleistocene hominids and suids. In *Palaeoclimate and Neogene Evolution*, ed. E. Vrba, pp. 369–84. New Haven, CT: Yale University Press.

80 Barry, J. C. 1996. Faunal diversity and turnover in the terrestrial Neogene of Pakistan. In *Palaeoclimate and Neogene Evolution*, ed. E. Vrba, pp. 115–34. New Haven, CT: Yale University Press.

81 Bishop, L. C. 1999. Suid paleoecology and habitat preference at African Pliocene and Pleistocene hominid localities. In *African Biogeography, Climate Change, and Early Hominid Evolution*, eds T. Bromage, F. Schrenk, pp. 216–25. Oxford: Oxford University Press.

82 Foley, R. A. 1999. The evolutionary geography of Pliocene hominids. In *African Biogeography, Climate Change, and Hominid Evolution*, eds

T. Bromage, F. Schrenk, pp. 328–48. Oxford: Oxford University Press.

83 Foley, R. A. 1994. Speciation, extinction and climatic change in hominid evolution. *Journal of Human Evolution* 26: 275–89.

84 Foley, R. A. 1993. African terrestrial primates: the comparative evolutionary biology of *Theropithecus* and hominids. In *Theropithecus – the Rise and Fall of a Primate Genus*, ed. N. Jablonski, pp. 254–72. Cambridge: Cambridge University Press.

85 Tchernov, E. 1992. Eurasian–African biotic exchanges through the Levantine corridor during the Neogene and Quaternary. *Courier Forschungsinstitut Senckenbergensis* 153: 103–23.

86 Lahr, M. M., Foley, R. A. 1998. Towards a theory of modern human origins: geography, demography, and diversity in recent human evolution. In *Yearbook of Physical Anthropology 1998, Vol. 41*, pp. 137–76.

87 Alverez, W., Asaro, F. 1990. An extra-terrestrial impact. *Scientific American* 263: 78–84.

88 Glen, W., ed. 1994. *The Mass Extinction Debates*. Palo Alto, CA: Stanford University Press.

89 Raup, D. M. 1991. *Extinction: Bad Genes or Bad Luck?* New York: Norton.

90 Jablonski, D. 1993. Mass extinctions: new answers, new questions. In *The Last Extinction*, eds L. Kaufman, K. Mallory, pp. 47–68. Cambridge, MA: MIT Press.

91 Aitken, M. J. 1990. *Science-Based Dating in Archaeology*. London: Longman.

92 Cooke, H. B. S. 1984. Horses, elephants and pigs as clues in the African later Cenozoic. In *Late Cainozoic Palaeoclimates of the Southern Hemisphere*, ed. J. C. Vogel, pp. 473–82. Rotterdam: A. A. Balkema.

93 Cande, S. C., Kent, D. V. 1992. A new geomagnetic polarity timescale for the Late Cretaceous and Cenozoic. *Journal of Geophysical Research* 97: 13917–51.

94 Dalrymple, G. B., Lanphere, M. A. 1969. *Potassium–Argon Dating: Principles, Techniques and Applications to Geochronology*. San Francisco: W. H. Freeman.

95 Leakey, L. S. B., Evernden, J. F., Curtis, G. H. 1961. Age of Bed I, Olduvai Gorge, Tanganyika. *Nature* 191: 478.

96 Bard, E., Hamelin, B., Fairbanks, R. G., Zindler, A. 1990. Calibration of the 14C timescale over the past 30,000 years, using mass spectrometric U-Th ages from Barbados corals. *Nature* 345: 405–10.

97 Bartlein, P. J., Edwards, M. E., Shafer, S. L., Barker, E. D. 1995. Calibration of radiocarbon ages and the interpretation of paleoenvironmental records. *Quaternary Research* 44: 417–24.

98 Schwarcz, H. 1992. Uranium series dating in paleoanthropology. *Evolutionary Anthropology* 1: 56–62.

99 Grun, R., Stringer, C. 2000. Tabūn revisited: revised ESR chronology and new ESR and U-series analyses of dental material from Tabūn C1. *Journal of Human Evolution* 39: 601–12.

100 Grun, R., Stringer, C. B. 1991. Electron spin resonance dating and the evolution of modern humans. *Archaeometry* 33: 153–99.

101 Mercier, N., Vallidas, H., Bar-Yosef, O., Vandermeersch, B., Stringer, C. B., Joron, J-L. 1993. Thermoluminescence date for the Mousterian site of Es Skhūl, Mount Carmel. *Journal of Archaeological Science* 20: 169–74.

102 Stringer, C. B. 1989. Documenting the origin of modern humans. In *The Emergence of Modern Humans*, ed. E. Trinkaus, pp. 67–96. Cambridge: Cambridge University Press.

103 Stringer, C. B., Grun, R., Schwarcz, H. P., Goldberg, P. 1989. ESR dates for the hominid burial site of Skhūl in Israel. *Nature* 338: 656–758.

104 Roberts, R. G., Jones, R., Spooner, N. A., Head, M. J., Murray, A. S., Smith, M. A. 1994. The human colonisation of Australia: optical dates of 53,000 and 60,000 years bracket human arrival at Deaf Adder Gorge, Northern Territory. *Quaternary Science Reviews* 13: 575–86.

105 Delson, E. 1988. Chronology of the South African australopith site units. In *Evolutionary History of the "Robust" Australopithecines*, ed. F. Grine, pp. 317–24. New York: Aldine de Gruyter.

106 Behrensmeyer, A. K., Hill, A. P., eds. 1980. *Fossils in the Making*. Chicago: University of Chicago Press.

107 Pope, G. G. 1989. Bamboo and human evolution. *Natural History* October: 48–57.

108 Brown, F. H., Harris, J. M., Leakey, R., Walker, A. 1985. Early *Homo erectus* skeleton from west Lake Turkana. *Nature* 316: 788–92.

109 Behrensmeyer, A. K., Dechant Boaz, D. E. 1980. The recent bones of Amboseli National Park,

Kenya, in relation to east African paleoecology. In *Fossils in the Making*, eds A. K. Behrensmeyer, A. P. Hill, pp. 72–93. Chicago: University of Chicago Press.

110 Brain, C. K. 1981. *The Hunters or the Hunted: An Introduction to African Cave Taphonomy*. Chicago: University of Chicago Press.

111 Leakey, M. D. 1972. *Olduvai Gorge. Vol. 3*. Cambridge: Cambridge University Press.

112 Potts, R. 1982. Lower Pleistocene site formation and hominid activity at Olduvai Gorge, Tanzania. PhD thesis. Harvard University.

113 Behrensmeyer, A. K., Gordon, K. D., Yanagi, G. T. 1986. Trampling as a cause of bone surface damage and pseudo-cutmarks. *Nature* 319: 768–71.

114 Potts, R., Shipman, P. 1981. Cutmarks made by stone tools on bones from Olduvai Gorge, Tanzania. *Nature* 291: 577–80.

115 Sillen, A., Lee-Thorp, J. A. 1993. Diet of *Australopithecus robustus* from Swartkrans. *South African Journal of Science* 89: 174.

116 Kay, R. 1985. Dental evidence for the diet of *Australopithecus*. *Annual Review of Anthropology* 14: 315–43.

117 Keeley, L., Toth, N. 1981. Microwear polishes on early stone tools from Kenya. *Nature* 293: 464–5.

118 Gould, R. 1980. *Living Archaeology*. Cambridge: Cambridge University Press.

119 Binford, L. R. 1978. *Nunamuit Ethnoarchaeology*. New York, London: Academic Press.

120 Binford, L. R. 1981. *Bones: Ancient Men and Modern Myths*. New York, London: Academic Press.

121 Binford, L. R. 1984. *Faunal Remains from Klasies River Mouth*. New York, London: Academic Press.

122 Tattersall, I. 1986. Species recognition in human palaeontology. *Journal of Human Evolution* 15: 165–75.

123 Wolpoff, M. H., Thorne, A. G., Jelinek, J., Zhang, Z. Y. 1994. The case for sinking *Homo erectus*: one hundred years of *Pithecanthropus* is enough. *Courier Forschunginstitut Senckenbergensis* 171: 341–61.

124 Groves, C. P. 1989. *A Theory of Human and Primate Evolution*. Oxford: Oxford University Press.

125 Ridley, M. 1985. *Evolution and Classification*. Harlow: Longman.

126 Fleagle, J. G. 1999. Homology. *Evolutionary Anthropology* 8: 228–9.

127 Foley, R. A. 1993. Striking parallels in human evolution. *Trends in Evolution and Ecology* 8: 196–7.

128 Lockwood, C. A., Fleagle, J. G. 1999. The recognition and evaluation of homoplasy in primate and human evolution. In *Yearbook of Physical Anthropology* 42: 189–232.

129 Fleagle, J. G. 1997. Beyond parsimony. *Evolutionary Anthropology* 6: 1–3.

130 Forey, P. L., Humphries, C. J., Kitching, I. J., Scotland, R. W., Siebert, D. J., Williams, D. M. 1992. *Cladistics: A Practical Course in Systematics*. Oxford: Clarendon Press, Oxford University Press.

131 Martins, E. P., Garland, J. 1991. Phylogenetic analyses of the correlated evolution of continuous characters: a simulation study. *Evolution* 45: 534–57.

132 Atchley, W., Fitch, W. 1991. Gene trees and the origin of inbred strains of mice. *Science* 254: 554–8.

133 Hillis, D. M. 1994. Application and accuracy of molecular phylogenies. *Science* 264: 667–71.

134 Andrews, P., Martin, R. D. 1987. Cladistic relationships of extant and fossil hominoids. *Journal of Human Evolution* 16: 101–18.

135 Gagneux, P., Wills, C., Gerloff, U., Tautz, D., Morin, P. A., et al. 1999. Mitochondrial sequences show diverse evolutionary histories of African hominoids. *Proceedings of the National Academy of Sciences* 96: 5077–82.

136 Goodman, M., Tagle, D. A., Fitch, D. H. A., Bailey, W., Czelusniak, J., et al. 1990. Primate evolution at the DNA level and a classification of hominoids. *Journal of Molecular Evolution* 30: 260–6.

137 Ruvolo, M., Disotell, T. R., Allard, M. W., Brown, W. M., Honeycutt, R. L. 1991. Resolution of the African hominoid trichotomy by use of a mitochondrial gene sequence. *Proceedings of the National Academy of Sciences* 88: 1570–4.

138 Ruvolo, M., Pan, D., Zehr, S., Goldberg, T., Disotell, T. R., Vondornum, M. 1994. Gene trees and hominoid phylogeny. *Proceedings of the National Academy of Sciences* 91: 8900–4.

139 Sibley, C., Ahlquist, J. 1984. The phylogeny of hominoid primates as indicated by DNA–DNA

hybridization. *Journal of Molecular Evolution* 20: 2–15.

140 Avise, J. C. 1994. *Molecular Markers, Natural History and Evolution*. New York: Chapman and Hall.

141 Page, R. D. M., Holmes, E. 1998. *Molecular Evolution: A Phylogenetic Approach*. Oxford: Blackwell.

142 Sarich, V. M., Wilson, A. C. 1967. Immunological time scale for hominoid evolution. *Science* 158: 1200–3.

143 Goodman, M., Baba, M. L., Darga, L. L. 1983. The bearing of molecular data on the cladogenesis and times of divergence of hominoid lineages. In *New Interpretations of Ape and Human Ancestry*, eds R. Ciochon, R. S. Corruccini, pp. 67–86. New York: Plenum Press.

144 Fitch, W. M., Ayala, F. J. 1995. The superoxide dismutatase clock revisited. In *Tempo and Mode in Evolution*, ed. W. M. Fitch, pp. 235–49. Washington, DC: National Academy Press.

145 Rosenberg, N., Feldman, M. W. 2002. The relationship between coalescence times and population divergence times. In *Modern Developments in Theoretical Population Genetics*, eds M. Slatkin, M. Veuille, pp. 130–64. Oxford: Oxford University Press.

146 Fleagle, J. 1999. *Primate Adaptations and Evolution*. London, New York: Academic Press.

147 Martin, R. D. 1990. *Primate Origins and Evolution*. London: Chapman and Hall.

148 Rowe, N. 1996. *The Pictorial Guide to the Living Primates*. New York: Pogonias Press.

149 Cartmill, M. 1992. New views on primate origins. *Evolutionary Anthropology* 1: 105–11.

150 Ross, C. 1994. The craniofacial evidence for anthropoid and tarsier relationships. In *Anthropoid Origins*, eds J. G. Fleagle, R. F. Kay, pp. 469–547. New York: Plenum Press.

151 Miyamoto, M., Goodman, M. 1990. DNA systematics and the evolution of primates. *Annual Review of Ecology and Systematics* 21: 197–220.

152 Shoshani, J., Groves, C. P., Simons, E. L., Gunnell, G. F. 1996. Primate phylogeny: morphological versus molecular results. *Molecular Phylogenetics and Evolution* 5: 102–54.

153 Napier, J. R., Napier, P. H. 1985. *The Natural History of the Primates*. Cambridge, MA: MIT Press.

154 Martin, R. D. 1986. Primates: a definition. In *Major Topics in Primate and Human Evolution*, eds B. Wood, L. Martin, P. Andrews, pp. 1–31. London: Academic Press.

155 Le Gros Clark, W. E. 1949. *History of the Primates*. London: British Museum (Natural History).

156 Napier, J. 1971. *Roots of Mankind*. London: George Allen and Unwin.

157 Cartmill, M. 1974. Rethinking primate origins. *Science* 184: 436–43.

158 Harvey, P., Pagel, M. 1992. *The Comparative Method in Evolutionary Biology*. Oxford: Oxford University Press.

159 Sussman, R. W. 1991. Primate origins and the evolution of angiosperms. *Journal of Primatology* 23: 209–23.

160 Gingerich, P. 1990. African dawn for primates. *Nature* 346: 411.

161 Beard, K. C., MacPhee, R. D. E. 1994. Cranial anatomy of *Shoshonius* and the antiquity of the Anthropoidea. In *Anthropoid Origins*, eds J. Fleagle, R. F. Kay, pp. 55–97. New York: Plenum Press.

162 Kay, R., Fleagle, J. 1994. *Anthropoid Origins*. New York: Plenum Press.

163 Ciochon, R. L., Chiarelli, A. B., eds. 1980. *Evolutionary Biology of the New World Monkeys and Continental Drift*. New York: Plenum Press.

164 Harvey, P. H., Pagel, M. D. 1991. *The Comparative Method in Evolutionary Biology*. Oxford: Oxford University Press.

165 Standen, V., Foley, R. A., eds. 1989. *Comparative Socioecology: The Behavioural Ecology of Humans and other Mammals*. Oxford: Blackwell.

166 Clutton-Brock, T. H., Harvey, P. H. 1977. Primate ecology and social organization. *Journal of Zoology* 183: 1–39.

167 Harvey, P. H., Clutton-Brock, T. H. 1985. Life history variation in primates. *Evolution* 39: 559–81.

168 Charnov, E., Berigan, D. 1993. Why do female primates have such long lifespans and so few babies? Or life in the slow lane. *Evolutionary Anthropology* 1: 191–4.

169 Strasser, E., Fleagle, J. G., McHenry, H., Rosenberger, A. L., eds. 1998. *Primate Locomotion: Recent Advances*. New York: Plenum Press.

170 MacArthur, R. H., Wilson, E. O. 1967. *The Theory of Island Biogeography*. Princeton, NJ: Princeton University Press.

171 Pianka, E. R. 1970. On *r*- and *K*-selection. *American Naturalist* 104: 592–6.

172 Pianka, E. R. 1994. *Evolutionary Ecology.* New York: HarperCollins.

173 Crook, J. H., Gartlan, J. S. 1966. Evolution of primate societies. *Nature* 210: 1200–3.

174 Crook, J. H. 1970. The socio-ecology of primates. In *Social Behaviour in Birds and Mammals,* ed. J. H. Crook, pp. 79–101. London: Academic Press.

175 Lack, D. 1970. *Natural Regulation of Animal Numbers.* Oxford: Oxford University Press.

176 Foley, R. A. 1991. How many hominid species should there be? *Journal of Human Evolution* 20: 413–27.

177 Tattersall, I. 1995. *The Fossil Trail.* Oxford: Oxford University Press.

178 Fleagle, J. 1995. Too many species. *Evolutionary Anthropology* 4: 37–8.

179 Thomson, D. A. 1917. *On Growth and Form.* Cambridge: Cambridge University Press.

180 Alberch, P., Gould, S. J., Oster, G. F., Wake, D. B. 1979. Size and shape in ontogeny and phylogeny. *Paleobiology* 5: 296–317.

181 Ruff, C., Walker, A. C. 1993. Body size and shape. In *The Nariokotome Skeleton,* eds A. C. Walker, R. E. Leakey, pp. 234–65. Cambridge, MA: Harvard University Press.

182 Ruff, C. B. 2000. Body size, body shape, and long bone strength in modern humans. *Journal of Human Evolution* 38: 269–90.

183 Walker, A. C., Leakey, R. E., eds. 1993. *The Nariokotome Skeleton.* Cambridge, MA: Harvard University Press.

184 Trinkaus, E. 1982. Neanderthal limb proportions and cold adaptation. In *Aspects of Human Evolution,* ed. C. Stringer, pp. 187–224. London: Taylor and Francis.

185 Holliday, T. W., Falsetti, A. B. 1995. Lower limb length of European early modern humans in relation to mobility and climate. *Journal of Human Evolution* 9: 141–53.

186 Stock, J., Pfeiffer, S. 2001. Linking structural variability in long bone diaphyses to habitual behaviors: foragers from the southern African Later Stone Age and the Andaman Islands. *American Journal of Physical Anthropology* 115: 337–48.

187 Lahr, M. M., Wright, R. V. S. 1996. The question of robusticity and the relationship between cranial size and shape in *Homo sapiens. Journal of Human Evolution* 31: 157–91.

188 Brown, P. 1987. Pleistocene homogeneity and Holocene size reduction: the Australian human skeletal evidence. *Archaeology in Oceania* 22: 41–67.

189 Foley, R. A. 1988. Hominids, humans and hunter-gatherers: an evolutionary perspective. In *Hunters and Gatherers 1: History, Evolution and Social Change,* eds T. Ingold, D. Riches, J. Woodburn, pp. 207–21. Oxford: Berg.

190 Brace, C. L. 1967. *The Stages of Human Evolution.* Englewood Cliffs, NJ: Prentice-Hall.

191 Lister, A. 1993. Mammoths in miniature. *Nature* 362: 288–9.

192 Horn, H. S. 1978. Optimal tactics of reproduction and life-history. In *Behavioural Ecology,* eds J. Krebs, N. B. Davies, pp. 411–29. Oxford: Blackwell.

193 Harvey, P. H., Martin, R. D., Clutton-Brock, T. H. 1987. Life histories in comparative perspective. In *Primate Societies,* eds B. B. Smuts, D. L. Cheney, R. M. Seyfarth, R. W. Wrangham, T. T. Struhsaker, pp. 181–96. Chicago: University of Chicago Press.

194 Harvey, P. H., Read, A. F., Promislow, D. E. L. 1989. Life history variation in placental mammals: unifying the data with theory. In *Oxford Surveys in Evolutionary Biology,* eds P. H. Harvey, L. Partridge, pp. 13–31. Oxford: Oxford University Press.

195 Ross, C. 1998. Primate life histories. *Evolutionary Anthropology* 6: 54–63.

196 Slater, P. J., Halliday, T. R., eds. 1994. *Behaviour and Evolution.* Cambridge: Cambridge University Press.

197 Jolly, A. J. 1985. *The Evolution of Primate Behavior.* New York, London: Collier-Macmillan.

198 Smuts, B. B., Cheney, D. L., Seyfarth, R. M., Wrangham, R. W., Struhsaker, T. T. 1987. *Primate Societies.* Chicago: University of Chicago Press.

199 McGrew, W. C., Nishida, T., Marchant, L., eds. 1996. *Great Ape Societies.* Cambridge: Cambridge University Press.

200 Wrangham, R. W. 1980. An ecological model of female-bonded primate groups. *Behaviour* 75: 262–300.

201 Hill, R., Dunbar, R. I. M. 1998. An evaluation of the roles of predation rate and predation risk as selective pressures on primate grouping behaviour. *Behaviour* 135: 411–30.

202 Martin, R. D. 1980. Sexual dimorphism and the evolution of higher primates. *Nature* 287: 273–5.

203 Leutenegger, W., Kelley, J. T. 1977. Relationship of sexual dimorphism in canine size and body size to social, behavioral, and ecological correlates in anthropoid primates. *Primates* 18: 117–36.

204 Leutenegger, W. 1978. Scaling of sexual dimorphism in body size and breeding system in primates. *Nature* 272: 610–11.

205 Clutton-Brock, T. H. 1985. *Size, Sexual Dimorphism, and Polygyny in Primates.* New York: Plenum Press.

206 Lee, P. C., ed. 1999. *Comparative Primate Socioecology.* Cambridge: Cambridge University Press.

207 Smuts, B. B., Cheney, D. L., Seyfarth, R. M., Wrangham, R. W., Struhsaker, T. T., eds. 1987. *Primate Societies.* Chicago: University of Chicago Press.

208 Dunbar, R. I. M. 1988. *Primate Social Systems.* London: Croom Helm.

209 Foley, R. A. 1989. The evolution of hominid social behaviour. In *Comparative Socioecology: The Behavioural Ecology of Humans and Other Mammals*, eds V. Standen, R. A. Foley, pp. 474–93. Oxford: Blackwell.

210 Foley, R. A. 1992. Analogue models in palaeoanthropology. In *Cambridge Encyclopaedia of the Human Evolution*, eds S. Jones, R. D. Martin, D. Pilbeam, pp. 335–41. Cambridge: Cambridge University Press.

211 Tooby, J., DeVore, I. 1987. The reconstruction of hominid behavioral evolution through strategic modeling. In *The Evolution of Human Behaviour: Primate Models*, ed. W. G. Kinzey. Albany, NY: SUNY Press.

212 DeVore, I., Washburn, S. L. 1963. Baboon ecology and human evolution. In *African Ecology and Human Evolution*, eds F. C. Howell, F. Bourliere. Chicago: Aldine.

213 Dunbar, R. I. M. 1993. Socioecology of the extinct theropithecines: a modelling approach. In *Theropithecus – the Rise and Fall of a Primate Genus*, ed. N. Jablonski, pp. 465–86. Cambridge: Cambridge University Press.

214 Jolly, C. 1970. The seed-eaters: a new model of hominid differentiation based on a baboon analogy. *Man* 5: 5–26.

215 Zihlman, A. L., Cronin, J. E., Cramer, D. L., Sarich, V. M. 1978. Pygmy chimpanzees as a possible prototype for the common ancestor of humans, chimpanzees and gorillas. *Nature* 275: 744–6.

216 Susman, R. L. 1987. Pygmy chimpanzees and common chimpanzees: models for the behavioral ecology of the earliest hominids. In *The Evolution of Human Behaviour: Primate Models*, ed. W. G. Kinzey, pp. 72–86. Albany, NY: SUNY Press.

217 Schaller, G., Lowther, G. R. 1969. The relevance of carnivore behaviour to the study of early hominids. *Southwestern Journal of Anthropology* 25: 307–41.

218 Morgan, E. 1982. *The Aquatic Ape.* London: Souvenir Press.

219 Wrangham, R. W. 1987. The significance of African apes for reconstructing human evolution. In *The Evolution of Human Behaviour: Primate Models*, ed. W. G. Kinzey, pp. 28–47. Albany, NY: SUNY Press.

220 Foley, R. A., Lee, P. C. 1996. Finite social space and the evolution of human social behaviour. In *The Archaeology of Human Ancestry*, eds J. Steele, S. Shennan, pp. 47–66. London: Routledge.

221 Foley, R. A., Lee, P. C. 1989. Finite social space, evolutionary pathways and reconstructing hominid behaviour. *Science* 243: 901–6.

222 Rendall, D., DiFiore, A. 1996. The road less travelled: phylogenetic perspectives in primatology. *Evolutionary Anthropology* 4(2): 43–52.

223 Lovejoy, C. O. 1981. The origin of man. *Science* 211: 341–50.

224 Foley, R. A. 1996. An evolutionary and chronological framework for human social behaviour. *Proceedings of the British Academy* 88: 95–117.

225 Harcourt, A., Harvey, P. H., Larson, S. G., Short, R. V. 1981. Testis weight, body weight and breeding system in primates. *Nature* 293: 55–7.

226 Dixson, A. 1998. *Primate Sexuality – Comparative Studies of the Prosimians, Monkeys, Apes, and Human Beings.* Oxford: Oxford University Press.

227 Martin, R. D., May, R. 1981. Outward signs of breeding. *Nature* 293: 7–9.

228 Hawkes, K., O'Connell, J. F., Jones, N. G. B., Alvarez, H., Charnov, E. L. 1998. Grandmothering, menopause, and the evolution of

human life histories. *Proceedings of the National Academy of Sciences* 95: 1336–9.

229  Swindler, D. R. 1976. *Dentition of Living Primates.* New York: Academic Press.

230  Smith, B. H. 1989. Dental development as a measure of life history in primates. *Evolution* 43: 683–8.

231  Smith, B. H. 1993. The physiological age of WT15000. In *The Nariokotome Skeleton*, eds A. C. Walker, R. E. Leakey, pp. 195–220. Cambridge, MA: Harvard University Press.

232  Conroy, G. C., Vannier, M. W. 1991. Dental development in South African australopithecines. Part II: Dental stage assessment. *American Journal of Physical Anthropology* 86: 137–56.

233  Bromage, T. G., Dean, M. C. 1985. Re-evaluation of the age at death of Plio-Pleistocene fossil hominids. *Nature* 317: 525–8.

234  Dean, C., Leakey, M. G., Reid, D., Schrenk, F., Schwartz, G. T., et al. 2001. Growth processes in teeth distinguish modern humans from *Homo erectus* and earlier hominins. *Nature* 414: 628–31.

235  Foley, R. A., Lee, P. C. 1991. Ecology and energetics of encephalization in hominid evolution. *Proceedings of the Royal Society of London, Series B – Biological Sciences* 334: 223–32.

236  Maas, M. C., Dumont, E. R. 1999. Built to last: the structure, function, and evolution of primate dental enamel. *Evolutionary Anthropology* 8: 133–52.

237  Martin, L. 1985. Significance of enamel thickness in hominoid evolution. *Nature* 314: 260–3.

238  Walker, A. C. 1981. Dietary hypotheses on human evolution. *Proceedings of the Royal Society of London, Series B – Biological Sciences* 292: 47–64.

239  Goodman, M. 1996. A personal account of the origin of a new paradigm. *Molecular Phylogenetic Evolution* 5: 269–85.

240  Schultz, A. H. 1969. *The Life of Primates.* New York: Universe Books.

241  Simons, E. 1961. The phyletic position of *Ramapithecus. Postilla* 57: 1–9.

242  Simons, E., Pilbeam, D. 1965. Preliminary revision of the Dryopithecinae (Pongidae, Anthropoidea). *Folia Primatologica* 3: 81–152.

243  Leakey, L. S. B. 1967. An early Miocene member of the Hominidae. *Nature* 213: 155.

244  Rogers, J. 1993. The phylogenetic relationships among *Homo, Pan* and *Gorilla. Journal of Human Evolution* 25: 201–15.

245  Ruvolo, M. 1994. Molecular evolutionary processes and conflicting gene trees: the hominoid case. *American Journal of Physical Anthropology* 94: 89–114.

246  Ruvolo, M. 1995. Seeing the forest and the trees. *American Journal of Physical Anthropology* 98: 218–32.

247  Andrews, P., Martin, L. 1987. Cladistic relationships of extant and fossil hominoid primates. *Journal of Human Evolution* 16: 101–27.

248  Groves, C. P., Patterson, J. 1991. Testing hominoid phylogeny with the PHYLIP programs. *Journal of Human Evolution* 20: 167–83.

249  Begun, D. R. 1994. Relations among the great apes and humans. In *Yearbook of Physical Anthropology* 37: 11–64.

250  Shoshani, J., Groves, C. P., Simons, E. L., Gunnell, G. F. 1996. Primate phylogeny: morphological vs molecular results. *Molecular Phylogenetics and Evolution* 5: 102–54.

251  Brunet, M. 2002. A new hominid from the Upper Miocene of Chad, Central Africa. *Nature* 418: 145–51.

252  Fleagle, J. G. 1999. *Primate Adaptation and Evolution.* New York: Academic Press.

253  Andrews, P. J. 1981. Species diversity and diet in monkeys and apes during the Miocene. In *Aspects of Human Evolution*, ed. C. B. Stringer, pp. 25–61. London: Taylor and Francis.

254  Simons, E. L. 1995. Egyptian Oligocene anthropoid primates: a review. In *Yearbook of Physical Anthropology 1995, Vol. 38*, pp. 199–238.

255  Simons, E. L. 1987. New faces of *Aegyptopithecus*, early human forebear of the Oligocene of Egypt. *Journal of Human Evolution* 16: 273–90.

256  Harrison, T. 1987. The phylogenetic relationship of the early catarrhine primates: a review of the current evidence. *Journal of Human Evolution* 16: 41–80.

257  Walker, A. C., Pickford, M. 1983. New postcranial fossils of *Proconsul africanus* and *Proconsul nyanzae*. In *New Interpretations of Ape and Human Ancestry*, eds R. Ciochon, R. S. Corruccini, pp. 325–54. New York: Plenum Press.

258  Pilbeam, D. R., Young, N. M. 2001. *Sivapithecus* and hominoid evolution: some brief comments. In *Hominoid Evolution and Climatic Change in Europe: Phylogeny of the Neogene Hominoid Primates of Eurasia. Vol. 2*, eds L. de Bonis,

G. D. Koufos, P. Andrews. Cambridge: Cambridge University Press.

259 Köhler, M., Moyà-Solà, S. 1997. Fossil muzzles and other puzzles. *Nature* 388: 327–8.

260 Benefit, B. R., McCrossin, M. L. 1997. Earliest known Old World monkey skull. *Nature* 388: 368–71.

261 de Bonis, L., Koufos, G. D. 1994. Our ancestor's ancestor: *Ouranopithecus* is a Greek link in human ancestry. *Evolutionary Anthropology* 3: 75–83.

262 Köhler, M., Moyà-Solà, S. 1997. Ape-like or hominid-like? The positional behavior of *Oreopithecus bambolii* reconsidered. *Proceedings of the National Academy of Sciences* 94: 11747–50.

263 Kingdon, J. 1989. *Island Africa*. London: Academic Press.

264 de Bonis, L., Koufos, G. D., Andrews, P., eds. 2001. *Hominoid Evolution and Climatic Change in Europe*. Cambridge: Cambridge University Press.

265 Stewart, C. B., Disotell, T. R. 1998. Primate evolution – in and out of Africa. *Current Biology* 8: R582–R587.

266 Jablonski, N. G. 1999. Primate evolution – in and out of Africa. *Current Biology* 9: 119–22.

267 Spencer, F. 1990. *Piltdown: A Scientific Forgery*. London, Oxford, New York: Natural History Museum, Oxford University Press.

268 Johanson, D. C., Edey, M. 1981. *Lucy: The Beginnings of Mankind*. New York: Simon and Schuster.

269 White, T. D., Suwa, G., Asfaw, B. 1994. *Australopithecus ramidus*, a new species of early hominid from Aramis, Ethiopia. *Nature* 366: 261–5.

270 Ward, C., Leakey, M., Walker, A. 1999. The new hominid species *Australopithecus anamensis*. *Evolutionary Anthropology* 7: 197–205.

271 Senut, B., Pickford, H., Gommery, D., Mein, P., Cheboi, K., Coppens, Y. 2001. First hominid from the Miocene (Lukeino Formation, Kenya). *Comptes Rendus de l'Academie des Sciences, Series IIA – Earth and Planetary Science* 332: 137–44.

272 Dart, R. A. 1925. *Australopithecus africanus*: the man–ape of South Africa. *Nature* 115: 195–9.

273 Ardrey, R. 1961. *African Genesis*. London: Atheneum.

274 Backwell, L. R., D'Errico, F. 2001. Evidence of termite foraging by Swartkrans early hominids.

*Proceedings of the National Academy of Sciences* 98: 1358–63.

275 Leakey, L. S. B., Tobias, P. V., Napier, J. R. 1961. A new species of the genus *Homo* from Olduvai Gorge, Tanganyika. *Nature* 202: 308–12.

276 Walker, A. C., Leakey, R. E. 1978. The East Turkana hominids. *Scientific American* 239: 54–66.

277 Leakey, M. G., Spoor, F., Brown, F. H., Gathogo, P. N., Kiarie, C., et al. 2001. New hominin genus from eastern Africa shows diverse middle Pliocene lineages. *Nature* 410: 433–40.

278 Coppens, Y., Howell, F. C., Isaac, G., Leakey, R. E. F., eds. 1976. *Earliest Man and Environments in the Lake Rudolf Basin*. Chicago: University of Chicago Press.

279 Asfaw, B., White, T., Lovejoy, O., Latimer, B., Simpson, S., Suwa, G. 1999. *Australopithecus garhi*: a new species of early hominid from Ethiopia. *Science* 284: 629–35.

280 Asfaw, B., Gilbert, W. H., Beyene, Y., Hart, W. K., Renne, P. R., et al. 2002. Remains of *Homo erectus* from Bouri, Middle Awash, Ethiopia. *Nature* 416: 317–20.

281 Abbate, E., Albianelli, A., Azzaroli, A., Benvenuti, M., Tesfamariam, B., et al. 1998. A one-million-year-old *Homo* cranium from the Danakil (Afar) depression of Eritrea. *Nature* 393: 458–60.

282 Brunet, M., Beauvilain, A., Coppens, Y., Heintz, E., Moutaye, A. H. E., Pilbeam, D. 1995. The first Australopithecine 2,500 kilometers west of the Rift Valley (Chad). *Nature* 378: 273–5.

283 Leakey, M. D. 1978. Pliocene footprints at Laetoli, Northern Tanzania. *Antiquity* 52: 133.

284 Dahlberg, F., ed. 1983. *Woman the Gatherer*. New Haven, CT: Yale University Press.

285 Lee, R. B., DeVore, I., eds. 1968. *Man the Hunter*. Chicago: Aldine.

286 Jablonski, N. G., Chaplin, G. 1993. Origin of habitual terrestrial bipedalism in the ancestor of the Hominidae. *Journal of Human Evolution* 24: 259–80.

287 Hunt, K. D. 1994. The evolution of human bipedality: ecology and functional morphology. *Journal of Human Evolution* 23: 183–202.

288 Rodman, P. S., McHenry, H. M. 1980. Bioenergetics and origins of bipedalism. *American Journal of Physical Anthropology* 52: 103–6.

289 Steudel, K. 1994. Locomotor energetics and hominid evolution. *Evolutionary Anthropology* 3: 42–8.

290 Isbell, L. A., Young, T. 1996. The evolution of bipedalism in hominids and reduced group size in chimpanzees: alternative responses to decreasing resource availability. *Journal of Human Evolution* 30: 389–97.

291 Wheeler, P. E. 1991. The influence of bipedalism on the energy and water budgets of early hominids. *Journal of Human Evolution* 21: 117–36.

292 Wheeler, P. E. 1991. The thermoregulatory advantages of hominid bipedalism in open equatorial environments: the contribution of increased convective heat loss and cutaneous evaporative cooling. *Journal of Human Evolution* 21: 107–16.

293 Dunbar, R. I. M. 1992. Time: a hidden constraint on the behavioural ecology of baboons. *Behavioural Ecology and Sociobiology* 31: 35–49.

294 Foley, R., Elton, S. 1998. Time and energy: the ecological context for the evolutoin of bipedalism. In *Primate Locomotion: Recent Advances*, eds E. Strasser, J. Fleagle, A. Rosenberger, H. McHenry, pp. 419–34. New York: Plenum Press.

295 Foley, R. A. 1993. The influence of seasonality on hominid evolution. In *Seasonality and Human Ecology*, eds S. J. Ulijaszek, S. Strickland, pp. 17–37. Cambridge: Cambridge University Press.

296 Leakey, M. G., Feibel, C. S., McDougall, I., Walker, A. C. 1995. New four million year old hominid species from Kanapoi and Allia Bay, Kenya. *Nature* 376: 565–71.

297 White, T. D., Suwa, G., Hart, W. K., Walters, R. C., Woldegabriel, G., et al. 1993. New discoveries of *Australopithecus* at Maka in Ethiopia. *Nature* 366: 261–5.

298 Leakey, M. D., Hay, R. L., Curtis, G. H., Drake, R. E., Jackes, M. K., White, T. D. 1976. Fossil hominids from the Laetolil beds. *Nature* 262: 460–6.

299 Johanson, D. C., Taieb, M., Coppens, Y. 1982. Pliocene hominids from the Hadar Formation, Ethiopia (1973–1977): stratigraphic, chronologic and palaeoenvironmental contexts. *American Journal of Physical Anthropology* 22: 373–402.

300 Johanson, D. C., White, T. D. 1979. A systematic assessment of early African hominids. *Science* 203: 321–30.

301 McHenry, H. 1992. How big were early hominids? *Evolutionary Anthropology* 1: 15–20.

302 Richmond, B. G., Jungers, W. L. 1995. Size variation and sexual dimorphism in *Australopithecus afarensis* and living hominoids. *Journal of Human Evolution* 29: 229–45.

303 Häusler, M. H., Schmid, P. 1995. Comparisons of the pelvises of Sts 14 and AL 288-1. *Journal of Human Evolution* 29: 591–600.

304 Falk, D. 1995. Did more than one hominid species coexist before 3.0 Ma? *Journal of Human Evolution* 29: 591–600.

305 Lovejoy, C. O. 1979. A reconstruction of the pelvis of AL-288 (Hadar Formation). *American Journal of Physical Anthroplogy* 50: 460.

306 Senut, B., Tardieu, C. 1985. Functional aspects of Plio-Pleistocene hominid limb bones: implications for phylogeny and taxonomy. In *Ancestors: The Hard Evidence*, ed. E. Delson, pp. 193–201. New York: Alan Liss.

307 Jungers, W. L. 1982. Lucy's limbs: skeletal allometry and locomotion in *Australopithecus afarensis*. *Nature* 297: 676–8.

308 McHenry, H. 1991. Sexual dimorphism in *Australopithecus afarensis*. *Journal of Human Evolution* 20: 21–32.

309 Susman, R. L., Stern, J. T., Jungers, W. L. 1985. Locomotor adaptations in the Hadar hominids. In *Ancestors: The Hard Evidence*, ed. E. Delson, pp. 184–92. New York: Alan Liss.

310 Aiello, L. 1994. Variable but singular. *Nature* 368: 399–400.

311 Clarke, R. J., Tobias, P. V. 1995. Sterkfontein Member 2 footbones of the oldest South African hominid. *Science* 269: 521–4.

312 Grine, F. E., ed. 1989. *The Evolutionary History of the "Robust" Australopithecines*. Chicago: Aldine de Gruyter.

313 Teaford, M. F., Ungar, P. S. 2000. Diet and the evolution of the earliest human ancestors. *Proceedings of the National Academy of Sciences* 97: 13506–11.

314 McHenry, H., Berger, L. R. 1998. Body proportions in *Australopithecus afarensis* and *A. africanus* and the origin of the genus *Homo*. *Journal of Human Evolution* 35: 1–22.

315 Spoor, F., Wood, B., Zonneveld, F. 1994. Implications of early hominid labyrinthine morphology for the evolution of human bipedal locomotion. *Nature* 369: 645–8.

316 Susman, R. L. 1994. Fossil evidence for early hominid tool use. *Science* 265: 1570–3.

317 Walker, A., Leakey, R. E., Harris, J. M., Brown, F. H. 1986. 2.5 Myr *Australopithecus boisei* from west of Lake Turkana, Kenya. *Nature* 322: 517–22.

318 Reed, K. E. 1997. Early hominid evolution and ecological change through the African Plio-Pleistocene. *Journal of Human Evolution* 32: 289–322.

319 Tobias, P. V. 1991. *Olduvai Gorge. Vol. 4: The Skulls, Teeth, and Endocasts of Homo habilis.* Cambridge: Cambridge University Press.

320 Johanson, D. C., White, T. D. 1987. New partial skeleton of *Homo habilis* from Olduvai Gorge, Tanzania. *Nature* 327: 205–9.

321 Curnoe, D. 2001. Early *Homo* from southern Africa: a cladistic perspective. *South African Journal of Science* 97: 186–90.

322 Hartwig-Scherer, S., Martin, R. D. 1991. Was "Lucy" more human than her "child"? Observations on early hominid postcranial skeletons. *Journal of Human Evolution* 21: 439–49.

323 Bromage, T., Schrenk, F., eds. 1999. *African Biogeography, Climate Change, and Hominid Evolution.* Oxford: Oxford University Press.

324 Stringer, C. 1986. The credibility of *Homo habilis*. In *Major Topics in Primate and Human Evolution*, eds B. Wood, L. Martin, P. Andrews, pp. 266–94. London: Academic Press.

325 Wood, B. A. 1991. *Koobi Fora Research Project. Vol. 4: The Hominid Cranial Remains.* Oxford: Clarendon Press.

326 Skelton, R. R., McHenry, H. M. 1992. Evolutionary relationships among early hominids. *Journal of Human Evolution* 23: 309–49.

327 Strait, D., Grine, F. 1999. Cladistics and early hominid phylogeny. *Science* 285: 210.

328 Foley, R. A. 1992. Problems in the early evolution of *Homo*: the east African fossil record and the evolutionary relationships of the early hominids. *African Archaeological Review* 10: 35–41.

329 McGrew, W. C. 1992. *Chimpanzee Material Culture.* Cambridge: Cambridge University Press.

330 Clark, J. G. D. 1968. *World Prehistory: A New Outline.* Cambridge: Cambridge University Press.

331 Roche, H. 1980. *Premiers outils tailles d'Afrique.* Paris: Societé d'Ethnographie.

332 Toth, N. 1985. The Oldowan reassessed. *Journal of Archaeological Science* 12: 101–20.

333 Wynn, T., McGrew, W. C. 1989. An ape's view of the Oldowan. *Man* 1989: 383–98.

334 Toth, N., Schick, K., Savage-Rumbaugh, E. 1993. *Pan* the toolmaker: investigations into the stone tool-making and tool-using abilities of a bonobo (*Pan paniscus*). *Journal of Archaeological Science* 20: 81–91.

335 Weidenreich, F. 1943. *The Skull of Sinanthropus pekinensis: A Comparative Study of a Primitive Hominid Skull.* Palaeontologica Sinica D10 1–485.

336 Hublin, J. J. 1992. Recent human evolution in northwestern Africa. *Proceedings of the Royal Society of London Series B – Biological Sciences* 337: 185–91.

337 Rightmire, G. P. 1990. *The Evolution of Homo erectus.* Cambridge: Cambridge University Press.

338 Gabunia, L., Vekua, A. 1995. A Plio-Pleistocene hominid from Dmanisi, east Georgia, Caucasus. *Nature* 373: 509–12.

339 Gabunia, L., Vekua, A., Lordkipanidze, D., Swisher, C. C., Ferring, R., et al. 2000. Earliest Pleistocene hominid cranial remains from Dmanisi, Republic of Georgia: taxonomy, geological setting, and age. *Science* 288: 1019–25.

340 Swisher III, C. C., Curtis, G., Lewin, R. 2000. *Java Man: How Two Geologists Changed Our Understanding of Human Evolution.* Chicago: University of Chicago Press.

341 Wanpo, H., Ciochon, R., Yumin, G., Larick, R., Qiren, F., et al. 1995. Early *Homo* and associated artefacts from Asia. *Nature* 378: 275–8.

342 Manzi, G., Mallegni, F., Ascenzi, A. 2001. A cranium for the earliest Europeans: phylogenetic position of the hominid from Ceprano, Italy. *Proceedings of the National Academy of Sciences* 98: 10011–16.

343 Ascenzi, A., Mallegni, F., Manzi, G., Segre, A. G., Naldini, E. S. 2000. A re-appraisal of *Ceprano calvaria* affinities with *Homo erectus*, after the new reconstruction. *Journal of Human Evolution* 39: 443–50.

344 Groves, C. P., Mazak, V. 1975. An approach to the taxonomy of the Hominidae: gracile Villafrancian hominids of Africa. *Casopis pro Minaralogii Geologii* 20: 225–47.

345 Andrews, P. J. 1984. An alternative interpretation of the characters used to define *Homo erectus*. *Courier Forschungsinstitut Senckenbergensis* 69: 167–75.

346 Wood, B. 1984. The origins of *Homo erectus. Courier Forschungsinstitut Senckenbergensis* 69: 99–111.

347 MacLarnon, A. M., Hewitt, G. P. 1999. The evolution of human speech: the role of enhanced breathing control. *American Journal of Physical Anthropology* 109: 341–63.

348 Asfaw, B., Beyene, Y., Suwa, G., Walter, R. C., White, T. D., et al. 1992. The earliest Acheulian from Konso-Gardula. *Nature* 360: 732–5.

349 Foley, R. A. 2001. The evolutionary consequences of increased carnivory in hominids. In *The Early Human Diet: The Role of Meat*, eds C. B. Stanford, H. T. Bunn, pp. 305–31. Oxford: Oxford University Press.

350 Isaac, G. 1984. The archaeology of human origins: studies of the Lower Pleistocene in East Africa 1971–1981. *Advances in World Archaeology* 3: 1–79.

351 Isaac, G. 1976. *Olorgesailie*. Chicago: University of Chicago Press.

352 Foley, R. A. 1987. Hominid species and stone tools assemblages: how are they related? *Antiquity* 61: 380–92.

353 Klein, R. G. 1999. *The Human Career*. Chicago: University of Chicago Press.

354 Goren-Inbar, N., Feibel, C. S., Verosub, K. L., Melamed, Y., Kislev, M. E., et al. 2000. Pleistocene milestones on the out-of-Africa corridor at Gesher Benot Ya'aqov, Israel. *Science* 289: 944–7.

355 Petraglia, M. D. 1998. The Lower Palaeolithic of India and its bearing on the Asian record. In *Early Human Behaviour in Global Context: The Rise and Diversity of the Lower Palaeolithic Record*, eds M. Petraglia, R. Korisettar, pp. 343–90. London: Routledge.

356 Roebroeks, W. 2001. Hominid behaviour and the earliest occupation of Europe: an exploration. *Journal of Human Evolution* 41: 437–61.

357 Movius, H. 1948. The Lower Palaeolithic cultures of southern and eastern Asia. *Transactions of the American Philosophical Society* 38: 329–429.

358 Foley, R. A., Lahr, M. M. 1997. Mode 3 technologies and the evolution of modern humans. *Cambridge Archaeological Journal* 7: 3–36.

359 Schick, K., Toth, N. 1993. *Making Silent Stones Speak*. New York: Simon and Schuster.

360 Keeley, L. H. 1980. *Experimental Determination of Stone Tool-Use: A Microwear Analysis*. Chicago: University of Chicago Press.

361 Isaac, G. 1978. Food sharing and human evolution: archaeological evidence from the Plio-Pleistocene of East Africa. *Journal of Anthropological Research* 34: 311–25.

362 Potts, R. 1984. Hominid hunters? Problems of identifying the earliest hunter/gatherers. In *Hominid Evolution and Community Ecology: Prehistoric Human Adaptation in Biological Perspective*, ed. R. Foley, pp. 129–66. London: Academic Press.

363 Bunn, H. T. 1981. Archaeological evidence for meat-eating by Plio-Pleistocene hominids from Koobi Fora and Olduvai Gorge. *Nature* 291: 575–7.

364 Bunn, H. T., Kroll, E. M. 1986. Systematic butchery by Plio/Pleistocene hominids at Olduvai Gorge, Tanzania. *Current Anthropology* 27: 431–52.

365 Blumenschine, R. 1987. Characteristics of an early hominid scavanging niche. *Current Anthropology* 38: 383–407.

366 Potts, R. 1984. Home bases and early hominids. *American Scientist* 72: 338–47.

367 Isaac, G. 1981. Bones in contention: competing explanations for the juxtaposition of artefacts and faunal remains. In *Animals and Archaeology: 1. Hunters and their Prey*, eds J. Clutton-Borck, C. Grigson, pp. 3–20. Oxford: British Archaeological Reports, International Series, 163.

368 Stanford, C. 1999. *The Hunting Apes: Meat Eating and the Origins of Human Behavior*. Princeton, NJ: Princeton University Press.

369 Aiello, L. C., Wheeler, P. 1995. The expensive tissue hypothesis. *Current Anthropology* 36: 199–222.

370 Milton, K. 1999. A hypothesis to explain the role of meat-eating in human evolution. *Evolutionary Anthropology* 8: 11–21.

371 Thieme, H. 1997. Lower Palaeolithic hunting spears from Germany. *Nature* 385: 807–10.

372 Stiner, M. C., Munro, N. D., Surovell, T. A., Tchernov, E., Bar-Yosef, O. 1999. Paleolithic population growth pulses evidenced by small animal exploitation. *Science* 283: 190–4.

373 O'Connell, J. F., Hawkes, K., Jones, N. G. B. 1999. Grandmothering and the evolution of *Homo erectus. Journal of Human Evolution* 36: 461–85.

374 O'Connell, J., Hawkes, K., Jones, N. B. 2000. A critical look at the role of carnivory in early

human evolution. *Journal of Human Evolution* 38: A23–A24.

375 Rightmire, G. P. 1998. Human evolution in the middle Pleistocene: the role of *Homo heidelbergensis*. *Evolutionary Anthropology* 6: 218–27.

376 deCastro, J. M. B., Arsuaga, J. L., Carbonell, E., Rosas, A., Martinez, I., Mosquera, M. 1997. A hominid from the lower Pleistocene of Atapuerca, Spain: possible ancestor to Neanderthals and modern humans. *Science* 276: 1392–5.

377 Lahr, M. M., Foley, R. A. 2001. Mode 3, *Homo helmei*, and the pattern of human evolution in the Middle Pleistocene. In *Human Roots: Africa and Asia in the Middle Pleistocene*, eds L. Barham, K. Robson Brown, pp. 23–40. Bristol: Westbury Press.

378 Howell, F. C. 1999. Paleo-demes, species clades, and extinctions in the Pleistocene hominin record. *Journal of Anthropological Research* 55: 191–243.

379 Wolpoff, M. H., Hawks, J., Caspari, R. 2000. Multiregional, not multiple origins. *American Journal of Physical Anthropology* 112: 129–36.

380 Stringer, C. 2002. Modern human origins: progress and prospects. *Proceedings of the Royal Society of London, Series B – Biological Sciences* 357: 563–79.

381 Smith, F. H., Simek, J. F., Harrill, M. S. 1989. Geographic variation in supra-orbital torus reduction during the later Pleistocene. In *The Human Revolution*, eds P. Mellars, C. Stringer, pp. 172–93. Edinburgh: Edinburgh University Press.

382 Brauer, G. 1984. The "Afro-European *sapiens* hypothesis" and hominid evolution in East Asia during the late Middle and Upper Pleistocene. *Courier Forschungsinstitut Senckenbergensis* 69: 145–65.

383 Day, M. H., Stringer, C. B. 1991. Les restes craniens d'Omo-Kibish et leur classification à l'interieur du genre *Homo*. *L'anthropologie* 95: 573–94.

384 Klein, R. G. 2000. Archeology and the evolution of human behavior. *Evolutionary Anthropology* 9: 17–36.

385 O'Connell, J. F., Allen, J. 1998. When did humans first arrive in Greater Australia and why is it important to know? *Evolutionary Anthropology* 6: 132–46.

386 Swisher III, C. C., Rink, W. J., Anton, S. C., Schwarcz, H. P., Curtis, G. H., et al. 1996. Latest *Homo erectus* of Java: potential contemporaneity with *Homo sapiens* in Southeast Asia. *Science* 274: 1870–4.

387 Thorne, A. G., Wolpoff, M. H. 1992. The multi-regional evolution of humans. *Scientific American* April: 76–83.

388 Habgood, P. J. 1989. The origin of anatomically modern humans in Australasia. In *The Human Revolution*, eds P. Mellars, C. B. Stringer, pp. 245–73. Edinburgh: Edinburgh University Press.

389 Groves, C. P. 1989. A regional approach to the problem of the origin of modern humans in Australasia. In *The Human Revolution*, eds Mellars, C. B. Stringer, pp. 274–85. Edinburgh: Edinburgh University Press.

390 Lahr, M. M. 1994. The multiregional model of modern human origins: a reassessment of its morphological basis. *Journal of Human Evolution* 26: 23–56.

391 Simpson, J. J., Grün, R. 1998. Non-destructive gamma spectrometric U-series dating. *Quaternary Science Reviews* 17: 1009–22.

392 Wolpoff, M. H. 1989. Multiregional evolution: the fossil alternative to Eden. In *The Human Revolution*, eds P. Mellars, C. B. Stringer, pp. 62–108. Edinburgh: Edinburgh University Press.

393 Etler, T. L. 1994. New archaic human fossil discoveries in China and their bearing on hominid species definition during the Middle Pleistocene. In *Integrative Paths to the Past: Palaeoanthropological Advances in Honor of F. Clark Howell*, eds R. S. Corruccini, R. L. Ciochon, pp. 639–76. Englewood Cliffs, NJ: Prentice-Hall.

394 Kamminga, J., Wright, R. V. S. 1988. The Upper Cave at Zhoukoudian and the origins of the Mongoloids. *Journal of Human Evolution* 17: 739–67.

395 Duarte, C., Mauricio, J., Pettitt, P. B., Souto, P., Trinkaus, E., et al. 1999. The early Upper Paleolithic human skeleton from the Abrigo do Lagar Velho (Portugal) and modern human emergence in Iberia. *Proceedings of the National Academy of Sciences* 96: 7604–9.

396 McBrearty, S., Brooks, A. S. 2000. The revolution that wasn't: a new interpretation of the origin of modern human behavior. *Journal of Human Evolution* 39: 453–563.

397 Aiello, L. C. 1993. The fossil evidence for modern human origins in Africa. *American Anthropologist* 95: 73–96.

398 Lahr, M. M. 1996. *The Evolution of Human Diversity*. Cambridge: Cambridge University Press.

399 Lahr, M. M., Foley, R. A. 1994. Multiple dispersals and modern human origins. *Evolutionary Anthropology* 3: 48–60.

400 Cann, R. L., Stoneking, M., Wilson, A. C. 1987. Mitochondrial DNA and human evolution. *Nature* 325: 31–5.

401 Templeton, A. R. 1993. The "Eve" hypothesis: a genetic critique and reanalysis. *American Anthropologist* 95: 51–72.

402 Maddison, D. R. 1991. African origin of human mitochondrial DNA reexamined. *Systematics Zoology* 40: 355–63.

403 Vigilant, L., Stoneking, M., Harpending, H., Hawkes, K., Wilson, A. 1991. African populations and the evolution of human mitochondrial DNA. *Science* 253: 1503–7.

404 Ingman, M., Kaessmann, H., Pääbo, S., Gyllensten, U. 2000. Mitochondrial genome variation and the origin of modern humans. *Nature* 408: 708–13.

405 Forster, P., Torroni, A., Renfrew, C., Rohl, A. 2001. Phylogenetic star contraction applied to Asian and Papuan mtDNA evolution. *Molecular Biology and Evolution* 18: 1864–81.

406 Maca-Meyer, N., González, A. M., Larruga, J. M., Flores, J. C., Cabrera, V. M. 2001. Major genomic mitochondrial lineages delineate early human expansions. *Biomed Central Genetics* 2: 13.

407 Goldstein, D. B., Linares, A. R., Cavalli-Sforza, L. L., Feldman, M. W. 1995. Genetic absolute dating based on microsatellites and the origin of modern humans. *Proceedings of the National Academy of Sciences* 92: 6723–7.

408 Linares, A. R., Nayar, K., Goldstein, D. B., Hebert, J. M., Seielstad, M. T., et al. 1996. Geographic clustering of human Y-chromosome haplotypes. *Annals of Human Genetics* 60: 401–8.

409 Reich, D. E., Goldstein, D. B. 1998. Genetic evidence for a Paleolithic human population expansion in Africa. *Proceedings of the National Academy of Sciences* 95: 8119–23.

410 Tarazona-Santos, E., Santos, F. R. 2002. The peopling of the Americas: a second major migration? *American Journal of Human Genetics* 70: 1377–80.

411 Tajima, A., Pan, I. H., Fucharoen, G., Fucharoen, S., Matsuo, M., et al. 2002. Three major lineages of Asian Y chromosomes: implications for the peopling of east and southeast Asia. *Human Genetics* 110: 80–8.

412 Underhill, P., Passarinno, G., Lin, A. A., Shen, P., Lahr, M. M., Foley, R. A., Oefner, P. J., Cavalli-Sforza, L. L. 2001. The phylogeography of Y chromosome binary haplotypes and the origins of modern human populations. *Annals of Human Genetics* 65: 43–62.

413 Hammer, M. F., Karafet, T. M., Redd, A. J., Jarjanazi, H., Santachiara-Benerecetti, S., et al. 2001. Hierarchical patterns of global human Y-chromosome diversity. *Molecular Biology and Evolution* 18: 1189–203.

414 Underhill, P. A., Shen, P. D., Lin, A. A., Passarino, G., Yang, W. H. et al. 2000. Y chromosome sequence variation and the history of human populations. *Nature Genetics* 26: 358–61.

415 Takahata, N., Lee, S. H., Satta, Y. 2001. Testing multiregionality of modern human origins. *Molecular Biology and Evolution* 18: 172–83.

416 Quintana-Murci, L., Veitia, R., Santachiara-Benerecetti, S., McElreavey, K., Fellous, M., Bourgeron, T. 1999. Mitochondrial DNA, Y chromosome and human population history. *M S – Medecine Sciences* 15: 974–82.

417 Hammer, M. F. 1995. A recent common ancestry for human Y chromosomes. *Nature* 378: 376–8.

418 Ruvolo, M. E., Zehr, S., Von Dornum, M. 1993. Mitochondrial COII sequences and modern human origins. *Molecular Biology and Evolution* 10: 1115–35.

419 Fullerton, M., Harding, R. M., Griffiths, R. C., Clegg, J. B. 1997. The genetic ancestry of modern humans: inferences from the analysis of DNA sequence diversity at the beta-globin locus. *American Journal of Human Biology* 9: 118.

420 Harpending, H., Batzer, M. A., Gurven, M., Jorde, L. B., Rogers, A. R., Sherry, S. T. 1998. Genetic traces of ancient demography. *Proceedings of the National Academy of Sciences* 95: 1961–8.

421 Harpending, H., Sherry, S. T., Rogers, A. R., Stoneking, M. 1993. The genetic structure of ancient human populations. *Current Anthropology* 34: 483–96.

422 Rogers, A. R., Harpending, H. 1992. Population growth makes waves in the distribution of pairwise genetic differences. *Molecular Biology and Evolution* 9: 552–69.

423 Ambrose, S. H. 1998. Late Pleistocene human population bottlenecks, volcanic winter, and the differentiation of modern humans. *Journal of Human Evolution* 34: 623–51.

424 Oppenheimer, C. 2002. Limited global change due to largest known Quaternary eruption, Toba ~74 kyr BP? *Quaternary Science Reviews* 21: 1593–609.

425 Alonso, S., Armour, J. A. L. 2001. A highly variable segment of human subterminal 16p reveals a history of population growth for modern humans outside Africa. *Proceedings of the National Academy of Sciences* 98: 864–9.

426 Hawks, J., Hunley, K., Lee, S. H., Wolpoff, M. 2000. Population bottlenecks and Pleistocene human evolution. *Molecular Biology and Evolution* 17: 2–22.

427 Krings, M., Stone, A., Schmitz, R. W., Krainitzki, H., Stoneking, M., Paabo, S. 1997. Neanderthal DNA sequences and the origin of modern humans. *Cell* 90: 19–30.

428 Ovchinnikov, I. V., Gotherstrom, A., Romanova, G. P., Kharitonov, V. M., Liden, K., Goodwin, W. 2000. Molecular analysis of Neanderthal DNA from the northern Caucasus. *Nature* 404: 490–3.

429 Adcock, G., Dennis, E., Easteal, S., Huttley, G., Jermin, L., et al. 2001. Mitochondrial DNA sequences in ancient Australians: implications for modern human origins. *Proceedings of the National Academy of Sciences* 98: 537–42.

430 Bordes, F. 1968. *The Old Stone Age*. New York: McGraw Hill.

431 Binford, L. R. 1973. Interassemblage variability – the Mousterian and the functional argument. In *The Exploration of Culture Change*, ed. A. C. Renfrew, pp. 227–54. Pittsburgh: University of Pittsburgh Press.

432 Stringer, C., Gamble, C. 1993. *In Search of the Neanderthals*. London: Thames and Hudson.

433 Klein, R. G. 1995. Anatomy, behaviour and modern human origins. *Journal of World Prehistory* 9: 167–98.

434 Deacon, H. J. 1992. Southern Africa and modern human origins. *Proceedings of the Royal Society of London, Series B – Biological Sciences* 337: 177–83.

435 Klein, R. G. 1992. The archaeology of modern human origins. *Evolutionary Anthropology* 1: 5–14.

436 White, R. 1989. Production, complexity and standardisation in the early Aurignacian bead and pendant manufacture. In *The Human Revolution*, eds P. Mellars, C. B. Stringer, pp. 366–90. Edinburgh: Edinburgh University Press.

437 Clark, G. A., Lindly, J. M. 1989. The case of continuity: observations on the biocultural transition in Europe and Western Asia. In *The Human Revolution*, eds P. Mellars, C. B. Stringer, pp. 626–76. Edinburgh: Edinburgh University Press.

438 Mercier, N., Valladas, H., Joron, J. L., Reyss, J. L., Leveque, F., Vandermeersch, B. 1991. Thermoluminescence dating of the late Neanderthal remains from Saint-Césaire. *Nature* 351: 737–9.

439 Mellars, P. C. 1996. *The Neanderthal Legacy*. Princeton, NJ: Princeton University Press.

440 D'Errico, F., Zilhao, J., Julien, M., Baffier, D., Pelegrin, J. 1998. Neanderthal acculturation in western Europe? A critical review of the evidence and its interpretation. *Current Anthropology* 39: S1–S44.

441 Bar-Yosef, O., Kuhn, S. L. 1999. The big deal about blades: laminar technologies and human evolution. *American Anthropologist* 101: 322–38.

442 Bar-Yosef, O. 1992. The role of western Asia in the origins of modern humans. *Proceedings of the Royal Society of London, Series B – Biological Sciences* 337: 193–200.

443 Lieberman, D. E., Shea, J. J. 1994. Behavioral differences between archaic and modern humans in the Levantine Mousterian. *American Anthropologist* 96: 300–32.

444 Allsworth-Jones, P. 1993. The archaeology of archaic and early modern *Homo sapiens*: an African perspective. *Cambridge Archaeological Journal* 33: 21–39.

445 Barham, L. S. 1998. Possible early pigment use in south-central Africa. *Current Anthropology* 39: 703–10.

446 Deacon, H. J., Shuurman, R. 1992. The origins of modern people: the evidence from Klasies

River. In *Continuity or Replacement? Controverisies in Homo sapiens Evolution*, eds G. Brauer, F. H. Smith, pp. 121–9. Rotterdam: Balkema.

447 Klein, R. G. 1983. The Stone Age prehistory of southern Africa. *Annual Review of Anthropology* 12: 25–48.

448 Henshilwood, C., Sealy, J. 1997. Bone artefacts from the Middle Stone Age at Blombos Cave, southern cape, South Africa. *Current Anthropology* 38: 890–5.

449 Brooks, A. S., Helgren, D. M., Cramer, J. S., Franklin, A., Hornyak, W., et al. 1995. Dating and context of 3 Middle Stone-Age sites with bone points in the Upper Semliki Valley, Zaire. *Science* 268: 548–53.

450 McBurney, C. B. M. 1967. *The Haua Fteah (Cyrenaica) and the Stone Age of the South East Mediterranean*. Cambridge: Cambridge University Press.

451 Mehlman, M. J. 1987. Provenance, age and associations of the archaic *Homo sapiens* cranium from Lake Eyasi. *Journal of Archaeological Science* 14: 133–62.

452 Jerison, H. J. 1991. Brain size and the evolution of mind. 59th James Arthur Lecture, American Museum of Natural History, New York.

453 Martin, R. D. 1983. *Human Brain Evolution in an Ecological Context*. New York: American Museum of Natural History.

454 Pagel, M. D., Harvey, P. H. 1989. Taxonomic differences in the scaling of brain on body weight among mammals. *Science* 244: 1589–93.

455 Aiello, L., Dunbar, R. I. M. 1993. Neocortex size, group size and the origin of language in the hominids. *Current Anthropology* 34: 184–93.

456 McHenry, H. M. 1982. The pattern of human evolution: studies on bipedalism, mastication and encephalization. *Annual Review of Anthropology* 11: 151–73.

457 Holloway, R. 1975. Early hominid endocasts: volumes, morphology and significance for hominid evolution. In *Primate Functional Morphology and Evolution*, ed. R. Tuttle, pp. 393–415. The Hague: Mouton.

458 Falk, D. 1987. Hominid palaeoneurology. *Annual Review of Anthropology* 16: 13–30.

459 Wynn, T. 1991. Tools, grammar, and the archaeology of cognition. *Cambridge Archaeological Journal* 1: 191–206.

460 Foley, R. A. 1996. Measuring the cognition of extinct hominids. In *Modelling the Early Human Mind*, eds P. Mellars, K. Gibson, pp. 57–66. Cambridge: MacDonald Institute.

461 Byrne, R., Whiten, A. 1986. *Machiavellian Intelligence*. Oxford: Clarendon Press.

462 Harcourt, A. H. 1988. *Alliances, Social Contests and Social Intelligence*. Oxford: Clarendon Press.

463 Cheney, D. L. 1987. Interactions and relationships between groups. In *Primate Societies*, eds B. B. Smuts, D. L. Cheney, R. M. Seyfarth, R. W. Wrangham, T. T. Struhsaker, pp. 267–81. Chicago: University of Chicago Press.

464 Dunbar, R. I. M. 1992. Neocortex size as a constraint on group size in primates. *Journal of Human Evolution* 22: 469–93.

465 Humphrey, N. K. 1976. The social function of intellect. In *Growing Points in Ethology*, eds P. P. G. Bateson, R. A. Hinde, pp. 303–17. Cambridge: Cambridge University Press.

466 Whiten, A., Goodall, J., McGrew, W. C., Nishida, T., Reynolds, V., et al. 1999. Cultures in chimpanzees. *Nature* 399: 682–5.

467 Laland, K. N., Brown, G. 2002. *Sense and Nonsense: Evolutionary Perspectives on Human Behaviour*. Oxford: Oxford University Press.

468 Dawkins, R. 1976. *The Selfish Gene*. London: Oxford University Press.

469 Blackmore, S. 2001. Evolution and memes: the human brain as a selective imitation device. *Cybernetics and Systems* 32: 225–55.

470 Aunger, R., ed. 2001. *Darwinising Culture*. Oxford: Oxford University Press.

471 Foley, R. A., Lahr, M. M. 2003. On stony ground: lithic technology, human evolution and the emergence of culture. *Evolutionary Anthropology* 12: 109–22.

472 Shennan, S. J. 2000. Population, culture history and the dynamics of culture change. *Current Anthropology* 41: 811–35.

473 Barkow, L., Cosmides, L., Tooby, J., eds. 1992. *The Adapted Mind*. New York: Oxford University Press.

474 Barrett, L., Dunbar, R. I. M., Lycett, J. 2002. *Human Evolutionary Psychology*. Princeton, NJ: Princeton University Press.

475 Mithen, S. 1996. *The Prehistory of the Mind*. London: Thames and Hudson.

476 Lieberman, P., Laitman, J. T., Reidenberg, J. S., Gannon, P. J. 1992. The anatomy, physiology,

acoustics and perception of speech: essential elements in analysis of the evolution of human speech. *Journal of Human Evolution* 23: 447–67.

477 Lieberman, P. 1996. On Neanderthal speech and human evolution. *Behavioral and Brain Sciences* 19: 156–7.

478 Arensburg, B. 1989. New skeletal evidence concerning the anatomy of Middle Paleolithic populations in the Middle East: the Kebara skeleton. In *The Human Revolution*, eds P. Mellars, C. B. Stringer, pp. 165–71. Edinburgh: Edinburgh University Press.

479 Deacon, T. 1997. *The Symbolic Species: The Co-Evolution of Language and the Brain*. New York: W. W. Norton.

480 Kay, R. F., Cartmill, M., Balow, M. 1998. The hypoglossal canal and the origin of human vocal behavior. *Proceedings of the National Academy of Sciences* 95: 5417–19.

481 Enard, W., Przeworski, M., Fisher, S. E., Lai, C. S. L., Wiebe, V., et al. 2002. Molecular evolution of FOXP2, a gene involved in speech and language. *Nature* 418: 869–72.

482 Davidson, I., Noble, W. 1991. The evolutionary emergence of modern human behaviour: language and its archaeology. *Man* 26: 223–54.

483 Bahn, P., Vertut, J. 1997. *Journey Through the Ice Age*. Berkeley: Orion Group.

484 Dunbar, R. I. M. 1997. *Gossip, Grooming, and Language*. Cambridge, MA: Harvard University Press.

485 Halverson, J. 1987. Art for art's sake in the Paleolithic. *Current Anthropology* 28: 63–89.

486 Lewis-Williams, J. D. 1986. Cognitive and optical illusions in San rock art research. *Current Anthropology* 27: 171–7.

487 Chase, P. G., Dibble, H. L. 1987. Middle Palaeolithic: a review of current evidence and interpretations. *Journal of Anthropological Archaeology* 6: 263–96.

488 Lindly, J. N., Clark, G. A. 1991. Symbolism and modern human origins. *Current Anthropology* 31: 233–62.

489 Mellars, P. 1991. Cognitive changes and the emergence of modern humans in Europe. *Cambridge Archaeological Journal* 1: 63–76.

490 Turner, A. 1984. Hominids and fellow-travellers: human migration into high latitudes as part of a large mammal community. In *Hominid Evolution and Community Ecology: Prehistoric Human Adaptation in Biological Perspective*, ed. R. Foley, pp. 183–218. London: Academic Press.

491 Meltzer, D. J. 1993. Pleistocene peopling of the Americas. *Evolutionary Anthropology* 1: 157–69.

492 Meltzer, D. J. 1997. Anthropology – Monte Verde and the Pleistocene peopling of the Americas. *Science* 276: 754–5.

493 Guidon, N., Arnaud, B. 1991. The chronology of the New World: two faces of one reality. *World Archaeology* 23: 167–78.

494 Torroni, A., Neel, J. V., Barrantes, R., Schurr, T. G., Wallace, D. C. 1994. Mitochondrial DNA "clock" for the Amerinds and its implications for timing their entry into North America. *Proceedings of the National Academy of Sciences* 91: 1158–62.

495 Lahr, M. M. 1995. Patterns of modern human diversification: implications for Amerindian origins. In *Yearbook of Physical Anthropology 1995, Vol. 38*, pp. 163–98.

496 Steele, D. G., Powell, J. F. 1993. Paleobiology of the first Americans. *Evolutionary Anthropology* 2: 138–46.

497 Powell, J. F., Neves, W. A. 1999. Craniofacial morphology of the first Americans: pattern and process in the peopling of the new world. In *Yearbook of Physical Anthropology* 42: 153–88.

498 Lahr, M. M. 1997. History in the bones. *Evolutionary Anthropology* 6: 2–6.

499 Brown, P. 1999. The first east Asians? In *Interdisciplinary Perspectives on the Origins of the Japanese*, ed. K. Omoto, pp. 105–24. Kyoto: International Centre for Japanese Studies.

500 Martin, P. S., Klein, R. G., eds. 1984. *Quaternary Extinctions: A Prehistoric Revolution*. Tucson, AZ: University of Arizona Press.

501 O'Connell, J. F., Allen, J. 1998. When did humans first arrive in Australia and why is it important to know? *Evolutionary Anthropology* 6: 132–46.

502 Birdsell, J. 1977. The recalibration of a paradigm for the first peopling of Greater Australia. In *Sunda and Sahul*, eds J. Allen, J. Golson, R. Jones, pp. 11–17. London: Academic Press.

503 Wolpoff, M. H., Wu, X. Z., Thorne, A. G. 1984. Modern *Homo sapiens* origins: a general theory of hominid evolution involving the fossil evidence from east Asia. In *The Origin of Modern Humans: A World Survey of the Fossil Evidence*, eds

F. H. Smith, F. Spencer, pp. 411–83. New York: Alan Liss.

504 Stoneking, M., Fontius, J. J., Clifford, S. L., Soodyall, H., Arcot, S. S., et al. 1997. Alu insertion polymorphisms and human evolution: evidence for a larger population size in Africa. *Genome Research* 7: 1061–71.

505 Hodder, I., ed. 2000. *Towards Reflexive Method in Archaeology: The Example at Çatalhöyük*. Monograph No. 28. Cambridge: McDonald Institute for Archaeological Research, British Institute of Archaeology at Ankara.

506 Bar-Yosef, O. 1998. The Natufian culture in the Levant, threshold to the origins of agriculture. *Evolutionary Anthropology* 6: 159–77.

507 Lee, R. B. 1968. What hunters do for a living, or, how to make out on scarce resouces. In *Man the Hunter*, eds R. B. Lee, I. DeVore, pp. 30–48. Chicago: Aldine.

508 Price, D., Brown, D., eds. 1985. *Prehistoric Hunter-Gatherers: The Emergence of Cultural Complexity*. London: Academic Press.

509 Cavalli-Sforza, L. L., Piazza, A., Menozzi, P. 1994. *History and Geography of Genes*. Princeton, NJ: Princeton University Press.

510 Richards, M., Corte-Real, H., Forster, P., Macaulay, V., Wilkinson-Herbots, H., Demaine, A., Papiha, S., Hedges, R., Bandelt, H. J., Sykes, B. 1996. Paleolithic and neolithic lineages in the European mitochondrial gene pool. *American Journal of Human Genetics* 59 (1): 185–203.

511 Bellwood, P. 2001. Early agriculturalist population diasporas? Farming, languages and genes. *Annual Review of Anthropology* 30: 181–207.

512 Cohen, M. N. 1977. *The Food Crisis in Prehistory*. New Haven, CT: Yale University Press.

513 Bender, B. 1985. Prehistoric developments in the American continent and in Brittany, northwest France. In *Prehistoric Hunter-Gatherers*, eds D. Price, J. A. Brown, pp. 21–57. London: Academic Press.

514 Renfrew, C. 1992. Archaeology, genetics and linguistic diversity. *Man* 27: 445–78.

# INDEX

The index is arranged in word-by-word sequence; page numbers in *italics* refer to figures.